ALGEBRA
An Intermediate Course

Raymond A. Barnett

Merritt College

**McGRAW-HILL
BOOK COMPANY**

*New York
St. Louis
San Francisco
Auckland
Bogotá
Hamburg
Johannesburg
London
Madrid
Mexico
Montreal
New Delhi
Panama
Paris
São Paulo
Singapore
Sydney
Tokyo
Toronto*

ALGEBRA: An Intermediate Course

Copyright © 1983 by McGraw-Hill, Inc. All rights reserved.
Portions of this book are from *Intermediate Algebra: Structure and Use*, Second Edition by Raymond A. Barnett, copyright © 1980 by McGraw-Hill, Inc. All rights reserved. Printed in the United States of America. Except as permitted under the United States Copyright Act of 1976, no part of this publication may be reproduced or distributed in any form or by any means, or stored in a data base or retrieval system, without prior written permission of the publisher.

1234567890WEBWEB89876543

ISBN 0-07-003744-2

Library of Congress Cataloging in Publication Data

Barnett, Raymond A.
    Algebra, an intermediate course.

    Includes index.
    1. Algebra.   I. Title.
QA154.2.B347   1983   512.9   82-14048
ISBN 0-07-003744-2

This book was set in Aster by Typothetae Book Composition.
The editors were Peter R. Devine and James S. Amar;
the designer was Joan E. O'Connor;
the production supervisor was Phil Galea.
New drawings were done by Danmark & Michaels, Inc.
Webcrafters, Inc., was printer and binder.

CONTENTS

PREFACE

This is a second course in algebra, written for those who have had basic algebra or for those who need a review before proceeding further. The book is designed for use in self-paced classes, math labs, and lecture-discussion classes.

Special Features

1 **Exposition.** Considerable effort has been directed toward making the exposition clear, informal, accurate, and nonthreatening. Vocabulary is controlled. Technical terms and statements are presented only with substantial motivation and clear, concrete illustrations.

2 **Word Problems and Applications.** The translation of verbal statements into symbolic forms receives much attention, and the process is gradually expanded to significant applications. By the end of Chapter 5 students will have had much experience in both areas. Important applications are liberally distributed throughout the text.

3 **Provisions for Substantial Student Involvement and Feedback.** Students are substantially involved in the instructional process through:

- *Examples and Matched Problems.* Following each example is a matched problem with room for its solution right in the text (see page 35). Complete solutions to the matched problems are found near the end of each section in a specially screened box (see page 36).
- *Practice Exercise Sets.* Near the end of each section is a comprehensive, carefully graded, practice exercise set (see page 36). The problems are presented in matched pairs and the student is directed to work through the odd-numbered problems first, check answers, and then to work through even-numbered problems in areas of weakness. Answers to all practice exercises are in the back of the book.
- *Check Exercise Sets.* At the end of each chapter is grouped a series of five- or ten-problem check exercise sets with multiple-choice answers (see page 49). Each individual Check Exercise corresponds to one section in the chapter. Space is provided for

students to show their work. These check sets can be turned in, easily corrected by hand or machine, and returned to the student. Answers are in an instructor's manual. A color stripe which runs down the right-hand edge of each check exercise page is designed to allow students or instructors to quickly turn to a check exercise set for a given section.

- *Chapter Diagnostic (Review) Exercise.* Near the end of each chapter is a Diagnostic (Review) Exercise set (see page 44). A student should work all the problems, check answers in the back of the book, and then review text sections corresponding to problems missed (corresponding section numbers are in italics following each answer).
- *Chapter Practice Test.* At the end of each chapter is a Practice Test for that chapter (see page 46). Students should take this practice test as if it were an in-class test, allowing 50 minutes or less, check answers in the back of the book, and then review text sections where weaknesses still prevail (corresponding section numbers are in italics following each answer).
- *Formal Chapter Test.* A formal test is available through the instructor for each chapter (eight forms of each chapter test are provided in the Instructor's Manual).

Student Aids

1 **Common student errors** are clearly identified at places where they naturally occur (see Sections 3.1, 4.2, and 4.5).

2 **Think boxes** are dotted boxes used to enclose steps that are usually performed mentally (see Sections 2.5, 3.1, and 4.2).

3 **Annotation** of examples and developments is found throughout the text to help students through critical stages (see Sections 2.5, 3.1, and 4.2).

4 **Functional use of second color** guides students through critical steps (see Sections 2.5, 3.1, and 4.2).

5 **Examples and matched problems** are liberally distributed throughout the text to illustrate concepts and to check students' understanding of the concepts as they are being developed. This actively involves each student in the instruction to learning process.

6 **Practice exercise sets** are found near the end of most sections. These are graded from (A) easy mechanics, to (B) moderately difficult mechanics, to (C) most difficult mechanics and concepts. Answers to all practice exercise sets are in the text. Students should work odd-numbered problems first, then even-numbered problems in areas of weakness.

7 **Check exercise sets** are grouped at the end of each chapter. These should be utilized as directed by the instructor. Answers are not in the book.

8 **Chapter review section** includes a concise summary of terms, formulas, and symbols discussed in each section; a Diagnostic (Review) Exercise set keyed to specific sections through the answers in the

book; and a Practice Test for the chapter, also keyed to specific sections through the answers in the book.

9 **End-page summaries** of formulas and symbols (keyed to sections in which they are introduced) and the metric system are inside the front and back covers of the text for convenient reference.

Instructor Aids

1 **Section checks** can be handled in a simple way by making use of the Check Exercises grouped at the end of each chapter. These check exercises contain either five or ten multiple-choice problems with ample space to show work. The whole check exercise is arranged on the front and back of a single page for easy handling. The perforated page can be torn out, turned in, easily graded by hand or machine, and returned to the student in a short time. (These check exercises also provide an effective and easy-to-use **homework** control in lecture-discussion classes.)

2 **A comprehensive testing program** is included in the Instructor's Manual. There are eight forms of each test (four short answer and four multiple-choice). The test packet includes:

   • Chapter tests for each chapter
   • Cumulative tests for the first half of the text
   • Final examinations

   A separate student solution sheet (easily reproduced) has been designed for use for all the tests. The separate test questions can thus be collected and reused—a particular advantage for self-paced labs or classes.

3 **Answers to check exercises and the tests** discussed in item 2 above are all in the Instructor's Manual in an easy-to-use format.

Acknowledgments

The preparation of a book requires the effort and skills of many people in addition to the author. I wish to thank the many reviewers who offered helpful comments and suggestions in the development of both this book and *Algebra: An Elementary Course*. In particular, I wish to thank Sherry Blackman, College of Staten Island; Donald Brook, Mt. San Antonio College; Robert Hoburg, Western Connecticut State College; Louis Hoelzle, Bucks County Community College; Paul Klein, Boston State College; Steve Marsden, Glendale Community College; Peter Nicholls, Northern Illinois University; Albert Otto, Illinois State University; Donald Reichman, Mercer Community College; Fred Safier, City College of San Francisco; and John Sodano, York College.

A special thanks goes to Fred Safier, City College of San Francisco, for his very detailed checking of the whole manuscript, his careful checking of all the answers in the text and his substantial role in the preparation of the Instructor's Manual. His comments, corrections, and suggestions were most helpful.

Error Check

Because of the careful checking and proofing by a number of very competent people (acting independently), the author and publisher believe this book to be substantially error-free.

Raymond A. Barnett

Chapter One

FUNDAMENTALS

CONTENTS

Are you looking down from above or up from below at the cubes in the figure? Turn the figure upside down and answer the same question.

INSTRUCTIONS FOR STUDENTS IN A
SELF-PACED CLASS OR LAB

yes — **HAVE YOU HAD INTERMEDIATE ALGEBRA BEFORE THIS COURSE?** — no

1. Work Diagnostic (Review) Exercise 1.7 on page 44. Check answers in back of book; then work through text sections corresponding to problems missed. (Section numbers are in italics following each answer.)
2. When finished with step 1, take Practice Test Chapter 1 on page 46 as a final check of your understanding of the chapter. Check answers in the back of the book; then review sections where weakness still prevails. (Corresponding section numbers are in italics following each answer.)
3. When you think you are ready, ask your instructor for a graded test for Chapter 1.
4. If your instructor approves, after the test is corrected, go to the next chapter.

1. Work through each section in the chapter as follows:
   (a) Read discussion.
   (b) Read each example and work the corresponding matched problem. Check your solution to the matched problem in Solutions to Matched Problems on the indicated page.
   (c) At the end of a section work the odd-numbered problems in the Practice Exercise and check answers; then work even-numbered problems in areas of weakness. (Answers to *all* Practice Exercise sets are in the back of the book.)
   (d) Work Check Exercise as instructed. Tear out and turn in as directed by your instructor. (Answers are not in the text.)
2. Repeat each step in item 1 for each section in the chapter.
3. After the instructional part of the chapter is completed, proceed with steps 1 to 4 in the box above.

1.1 SETS

• **Set representations**
•• **Subsets and equality**
••• **Set operations**

• Set representations

The use of the word "set" in mathematics is not too much different from its use in everyday language. Words such as "set," "collection," "bunch," and "flock" all suggest the same idea. In this course we will usually be interested in certain sets of numbers. Capital letters, such as A, B, and C, are often used to represent sets. For example,

$$A = \{3, 5, 7\} \qquad B = \{4, 5, 6\}$$

specify sets A and B.

Each object in a set is called a **member** or **element** of the set. Symbolically,

$a \in A$	means	"a is an element of set A"
$a \notin A$	means	"a is not an element of set A"

Referring to sets A and B above, we see that

$5 \in A$ 5 is an element of set A.

$3 \notin B$ 3 is not an element of set B.

A set without any elements is called the **empty** or **null set.** For example, the set of all solutions to the equation $x + 2 = x + 5$ is the empty set. Symbolically,

$\emptyset$ represents "the empty set"

A set is usually described in one of two ways:

1. By **listing** the elements between braces { }:

$\{3, 5, 7\}$

2. By enclosing a **rule** within braces { } that determines the elements in the set:

$\{x \mid x^2 = 81\}$ Read: "The set of all x such that $x^2 = 81$."

        rule
      such that
    all x
   The set of

EXAMPLE 1
Let

$M = $ the set of all numbers x such that $x^2 = 25$.

Set M may be described either by the listing method or the rule method:

Listing method: $\quad M = \{-5, 5\}$

Rule method: $\quad M = \{x \mid x^2 = 25\}$

Work Problem 1 and check solution in Solutions to Matched Problems on page 6.

PROBLEM 1   Let $N = $ The set of all numbers x such that $x^2 = 49$.

   (A)   Write N, using the listing method.

   $N = $

   (B)   Write N, using the rule method.

   $N = $

   (C)   Indicate true (T) or false (F):

   $7 \in N$ _____ $7 \notin N$ _____ $49 \in N$ _____

The letter x introduced above is a variable. In general, a **variable** is a symbol used as a placeholder for elements out of a set with two or more elements. A **constant,** on the other hand, is a symbol that names exactly one object. The symbol "8" is a constant, since it always names the number eight.

••Subsets and equality

If each element of a set A is also an element of a set B, we say that A is a **subset** of B. The set of all women in a class is a subset of the whole class. (Note that the definition of subset allows a set to be a subset of itself.) If two sets have exactly the same elements (the order of listing does not matter), the sets are said to be **equal.** Symbolically,

$$
\begin{array}{lll}
A \subset B & \text{means} & \text{``}A \text{ is a subset of } B\text{''} \\
\{2, 3\} \subset \{1, 2, 3\} & & \\
A = B & \text{means} & \text{``}A \text{ is equal to } B\text{''} \\
\{5, 7\} = \{7, 5\} & &
\end{array}
$$

It can be shown that

$$\varnothing \text{ is a subset of every set}$$

EXAMPLE 2

Let $A = \{-7, 7\}$, $B = \{7, -7\}$, and $C = \{-7\}$. Then each of the following is a true statement:

$$C \subset A \qquad C \subset B \qquad A \subset B \qquad \varnothing \subset A \qquad A = B \qquad B \neq C$$

Work Problem 2 and check solution in Solutions to Matched Problems on page 6.

PROBLEM 2   Let $M = \{5, -5\}$, $N = \{-5, 5\}$, and $Q = \{5\}$. Indicate true (T) or false (F):

(A) $M = N$ ____

(B) $Q \subset N$ ____

(C) $N \neq Q$ ____

(D) $M \subset N$ ____

(E) $Q \subset M$ ____

(F) $\varnothing \subset M$ ____

•••Set operations

The **union** of sets A and B, denoted by $A \cup B$, is the set of all elements formed by combining all the elements of A and all the elements of B into one set. The **intersection** of sets A and B, denoted by $A \cap B$, is the set of elements in A that are also in B. Symbolically,

UNION: $A \cup B = \{x \mid x \in A \text{ or } x \in B\}$
                $\{2, 3\} \cup \{3, 4\} = \{2, 3, 4\}$

INTERSECTION: $A \cap B = \{x \mid x \in A \text{ and } x \in B\}$
                $\{2, 3\} \cap \{3, 4\} = \{3\}$

The word "or" is used in the way it is generally used in mathematics; that is, x may be an element of set A or set B or both. If $A \cap B = \varnothing$ (that is, if the set of all elements common to both A and B is empty), then sets A and B are said to be **disjoint.**

Venn diagrams are useful aids in visualizing set relationships. Union and intersection of sets are illustrated in Figure 1.

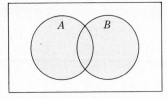

(a) Union of two sets

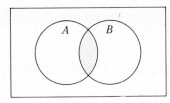

(b) Intersection of two sets

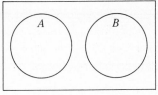

(c) Two disjoint sets

Figure 1   Venn Diagrams

EXAMPLE 3

If $A = \{1, 2, 3, 4\}$, $B = \{1, 3, 5, 7\}$, and $C = \{2, 4, 6\}$, then

$A \cup B = \{1, 2, 3, 4, 5, 7\}$ Elements in A combined with elements in B.

$A \cap B = \{1, 3\}$ Elements in A that are also in B.

$B \cap C = \varnothing$ B and C are disjoint.

Work Problem 3 and check solution in Solutions to Matched Problems on page 6.

PROBLEM 3   If $A = \{3, 6, 9\}$, $B = \{3, 4, 5, 6, 7\}$, and $C = \{4, 5, 7, 8\}$, find

(A) $A \cup B =$ (B) $A \cap B =$ (C) $A \cap C =$

Our approach to sets will remain informal. We will use set concepts and notation only where clarity is improved, but not otherwise.

SOLUTIONS TO MATCHED PROBLEMS _____

1.   (A) $N = \{-7, 7\}$ (B) $N = \{x \mid x^2 = 49\}$ (C) T, F, F
2.   (A) T   (B) T   (C) T   (D) T   (E) T   (F) T
3.   (A) $\{3, 4, 5, 6, 7, 9\}$ (B) $\{3, 6\}$ (C) $\varnothing$ (null set)

PRACTICE EXERCISE 1.1 _____

Work odd-numbered problems first, check answers, and then work even-numbered problems in areas of weakness. Answers to all problems are in the back of the book. Make every effort to work a problem yourself before you look at an answer.

A *Indicate true (T) or false (F).*

1. $3 \in \{2, 3, 4\}$ _____

2. $4 \in \{3, 5, 7\}$ _____

3. $5 \notin \{2, 3, 4\}$ _____

4. $6 \notin \{2, 4, 6, 8\}$ _____

5. $\{2, 3\} \subset \{1, 2, 3\}$ _____

6. $\{2, 4\} \subset \{1, 2, 3\}$ _____

7. $\{7, 3, 5\} = \{3, 5, 7\}$ _____

8. $\{3, 1, 2\} \subset \{1, 2, 3\}$ _____

9. $\varnothing \subset \{2, 5\}$ _____

10. $\varnothing \subset \{1, 3\}$ _____

Write each set in Problems 11 to 28, using the listing method; that is, list the elements between braces. If the set is empty, write $\varnothing$.

11. $\{1, 3, 5\} \cup \{2, 3, 4\}$ _____

12. $\{3, 4, 6, 7\} \cup \{3, 4, 5\}$ _____

13. $\{1, 3, 5\} \cap \{2, 3, 4\}$ _____

14. $\{3, 4, 6, 7\} \cap \{3, 4, 5\}$ _____

15. $\{1, 5, 9\} \cap \{3, 4, 6, 8\}$ _____

16. $\{6, 8, 9\} \cap \{4, 5, 7\}$ _____

B 17. $\{x \mid x - 5 = 0\}$ _____

18. $\{x \mid x + 3 = 0\}$ _____

19. $\{x \mid x + 9 = x + 1\}$ _____

20. $\{x \mid x - 3 = x + 2\}$ _____

21. $\{x \mid x^2 = 4\}$ _____

22. $\{x \mid x^2 = 9\}$ _____

23. Let A be the set of all numbers x such that $x^2 = 36$. Indicate true (T) or false (F) for the following:

(A) $A = \{36\}$ _____

(B) $A = \{-6, 6\}$ _____

(C) $6 \in A$ _____

(D) $A = \{x \mid x^2 = 36\}$ _____

(E) $7 \notin A$ _____

(F) $A = \{x \mid x = 6\}$ _____

24.   Let B be the set of all numbers x such that $x^2 - 16 = 0$. Indicate true (T) or false (F) for the following:

(A) $B = \{-8, 8\}$ _____

(B) $B = \{x \mid x^2 = 16\}$ _____

(C) $4 \in B$ _____

(D) $8 \notin B$ _____   (E) $B = \{x \mid x = 4\}$ _____   (F) $B = \{-4, 4\}$ _____

25. Let B be the set of all numbers x such that $x^2 = 100$.

(A) Denote B by the listing method. _____

(B) Denote B by the rule method (set-builder notation). _____

26. Let M be the set of all numbers x such that $x^2 = 64$.

(A) Denote M by the listing method. _____

(B) Denote M by the rule method (set-builder notation). _____

C 27. If $A = \{1, 2, 3, 4\}$ and $B = \{2, 4, 6\}$, find $\{x \mid x \in A \text{ or } x \in B\}$. _____

28. If $A = \{1, 2, 3, 4\}$ and $B = \{2, 4, 6\}$, find $\{x \mid x \in A \text{ and } x \in B\}$. _____

APPLICATIONS

29. The executive committee of a student council consists of a president, vice president, secretary, and treasurer, and is denoted by the set $\{P, V, S, T\}$. How many two-person subcommittees are possible; that is, how many two-element subsets can be formed?

_____

30. How many three-person subcommittees are possible in Problem 29?

_____

The Check Exercise for this section is on page 49.

1.2 ALGEBRAIC EXPRESSIONS; EQUALITY

• **Variables and constants**
•• **Algebraic expressions**
••• **Evaluation**
•••• **Equality**

•Variables and constants

In Section 1.1 we introduced the concepts of variable and constant. Recall, a **variable** is a symbol used as a placeholder for elements out of a set containing at least two elements. A **constant** is a symbol that names exactly one object.

EXAMPLE 4
In the formula

$$P = 2a + 2b \qquad \text{perimeter of a rectangle}$$

P, a, and b are variables and 2 is a constant. The values of P, a, and b vary from rectangle to rectangle, but 2 never changes.

Work Problem 4 and check solution in Solutions to Matched Problems on page 12.

PROBLEM 4 List the variables and constants in each formula.

(A) $P = 4s$ perimeter of a square

Variables: Constants:

(B) $A = s^2$ area of a square

Variables: Constants:

The introduction of variables into mathematics occurred about A.D. 1600. A French mathematician, François Vieta (1540–1603), is singled out as the one mainly responsible for this new idea. Many mark this point as the beginning of modern mathematics.

••Algebraic expressions

An algebraic expression is a symbolic form involving constants, variables, mathematical operations, and grouping symbols. For example,

$$2 + 8 \qquad 4 \cdot 3 - 7 \qquad 16 - 3(7 - 4)$$
$$5x - 3y \qquad 7(x + 2y) \qquad 4\{u - 3[u - 2(u + 1)]\} \qquad \text{algebraic expressions}$$

are all algebraic expressions.

Two or more algebraic expressions joined by plus or minus signs are called **terms;** two or more algebraic expressions joined by multiplication are called **factors.** For example,

$$\underbrace{3(x - y)}_{\text{term}} - \underbrace{(x + y)(x - y)}_{\text{term}}$$

has two terms, $3(x - y)$ and $(x + y)(x - y)$, and each term has two factors. The first term has factors 3 and $(x - y)$, and the second term has factors $(x + y)$ and $(x - y)$.

•••Evaluation

When evaluating numerical expressions involving various operations and grouping symbols, we follow the convention:

Order of Operations

1. Simplify inside the innermost symbols of grouping first, then the next innermost, and so on.

2. Multiplication and division are performed before addition and subtraction. (In both cases we proceed from left to right.)

EXAMPLE 5

Evaluate each expression.

(A) $10 - 2 \cdot 4$

Solution

$$10 - 2 \cdot 4 = 10 - 8 \qquad \text{Do not subtract the 2 from the 10;}$$
$$ \qquad \text{multiplication is performed first.}$$
$$= 2$$

(B) $8 - 3(6 - 2 \cdot 2)$

Solution

$$8 - 3(6 - 2 \cdot 2) = 8 - 3(6 - 4) \qquad \text{Simplify inside parentheses first.}$$
$$= 8 - 3 \cdot 2 = 8 - 6 = 2$$

(C) $(8 - 3)(6 - 2 \cdot 2)$

Solution

$$(8 - 3)(6 - 2 \cdot 2) = 5(6 - 4) \qquad \text{Note how parts } \mathbf{B} \text{ and } \mathbf{C} \text{ differ.}$$
$$= 5 \cdot 2$$
$$= 10$$

(D) $3\{6 + 2[8 - 2(3 - 1)]\}$

Solution

$3\{6 + 2[8 - 2(3 - 1)]\}$

$= 3\{6 + 2[8 - 2 \cdot 2]\}$

$= 3\{6 + 2 \cdot 4\}$

$= 3\{14\}$

$= 42$

Simplify inside parentheses ( ) first,
then brackets [ ], and then braces { };
that is, work from the inside out.

Work Problem 5 and check solution in Solutions to Matched Problems on page 12.

PROBLEM 5 Evaluate each:

(A) $8 + 3 \cdot 6 =$

(B) $6 + 2(8 - 2 \cdot 3) =$

(C) $(6 + 2)(8 - 2 \cdot 3) =$

(D) $2\{12 - 2[10 - 3(7 - 5)]\} =$

EXAMPLE 6

Evaluate each algebraic expression for $x = 10$ and $y = 3$.

(A) $x - 2y$

Solution

$10 - 2 \cdot 3 = 10 - 6 = 4$

(B) $x - 3(2y - 4)$

Solution

$10 - 3(2 \cdot 3 - 4) = 10 - 3(6 - 4) = 10 - 3 \cdot 2 = 10 - 6 = 4$

(C) $3[29 - x(y - 1)]$

Solution

$3[29 - 10(3 - 1)] = 3[29 - 10 \cdot 2] = 3(29 - 20) = 3 \cdot 9 = 27$

(D) $2\{2x - [3 + 2(3y - 8)]\}$

Solution

$2\{2 \cdot 10 - [3 + 2(3 \cdot 3 - 8)]\} = 2\{20 - [3 + 2(9 - 8)]\}$

$\qquad\qquad = 2\{20 - [3 + 2 \cdot 1]\}$

$\qquad\qquad = 2\{20 - 5\} = 2 \cdot 15 = 30$

Work Problem 6 and check solution in Solutions to Matched Problems on page 12.

PROBLEM 6 Evaluate each algebraic expression for $x = 11$ and $y = 2$.

(A) $x - 4y =$

(B) $x - 2(3y - 3) =$

(C) $2[35 - x(y + 1)] =$

(D) $3\{3y + [1 + 2(x - 10)]\} =$

••••Equality

An **equal sign,** $=$, is used to join two expression if the two expressions are names or descriptions of exactly the same object. Thus,

$$a = b$$

means a and b are names for the same object. Of course,

$$a \neq b$$

means a **is not equal to** b.

    If two algebraic expressions involving at least one variable are joined with an equal sign, the resulting form is called an **algebraic equation.** The following are algebraic equations in one or more variables.

$$2x - 3 = 3(x - 5)$$
$$3x + 5y = 7$$

    Formulating algebraic equations is an important first step in solving many practical problems using algebraic methods.

EXAMPLE 7

Translate each statement into an algebraic equation, using x as the only variable.

(A)   5 times a number is 3 more than twice the number.

Solution

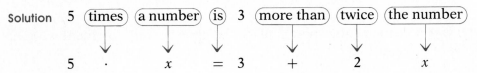

Thus,

$$5x = 3 + 2x$$

(B)   4 times a number is 5 less than twice the number.

Solution   4 (times) (a number) (is) (5) (less than) (twice the number)

4 · x = 2x −   → 5

NOTE: "Less than" means "subtracted from." Thus,

$$4x = 2x - 5 \qquad \text{not } 4x = 5 - 2x$$

    Work Problem 7 and check solution in Solutions to Matched Problems on page 12.

PROBLEM 7   Translate each statement into an algebraic equation, using x as the only variable.

    (A)   7 is 3 more than a certain number.

(B) 12 is 9 less than a certain number.

(C) 3 times a certain number is 6 less than twice that number.

(D) If 6 is subtracted from a certain number, the difference is twice a number that is 4 less than the original number.

To complete this discussion, we state several important properties of the equality symbol, $=$, that must hold at any time it is used.

1. **Symmetric property:** If $a = b$, then $b = a$.

We may reverse the left and right members of an equation any time we wish—a useful process when solving certain types of equations. For example, if

$A = P + Prt$, then $P + Prt = A$

2. **Transitive property:** If $a = b$ and $b = c$, then $a = c$.

This property is used extensively throughout algebra. For example, if

$2x + 3x = (2 + 3)x$ and $(2 + 3)x = 5x$ then

$2x + 3x = 5x$

3. **Substitution property:** If $a = b$, then either may replace the other in any statement without changing the truth or falsity of the statement.

The substitution property is also used a great deal throughout algebra. For example, for a circle with circumference C, diameter D, and radius R, we know that

$C = \pi D$ and $D = 2R$

In the first formula D can be replaced by $2R$ from the second formula to obtain

$C = \pi(2R)$ or $C = 2\pi R$

SOLUTIONS TO MATCHED PROBLEMS

4. **(A)** Variables: P, s; constant: 4   **(B)** Variables: A, s; constant: 2
5. **(A)** $8 + 3 \cdot 6 = 8 + 18 = 26$ **(B)** $6 + 2(8 - 2 \cdot 3) = 6 + 2(8 - 6) = 6 + 2(2) = 6 + 4 = 10$
   (C) $(6 + 2)(8 - 2 \cdot 3) = (8)(8 - 6) = (8)(2) = 16$
   (D) $2\{12 - 2[10 - 3(7 - 5)]\} = 2\{12 - 2[10 - 3(2)]\}$
   $= 2\{12 - 2[10 - 6]\} = 2\{12 - 2(4)\} = 2(12 - 8) = 2(4) = 8$
6. **(A)** $11 - 4(2) = 11 - 8 = 3$ **(B)** $11 - 2[3(2) - 3] = 11 - 2[6 - 3] = 11 - 2(3) = 11 - 6 = 5$
   (C) $2[35 - 11(2 + 1)] = 2[35 - 11(3)] = 2(35 - 33) = 2(2) = 4$
   (D) $3\{3(2) + [1 + 2(11 - 10)]\} = 3\{6 + [1 + 2(1)]\} = 3(6 + 3) = 3(9) = 27$
7. **(A)** $7 = 3 + x$ **(B)** $12 = x - 9$ **(C)** $3x = 2x - 6$ **(D)** $x - 6 = 2(x - 4)$

PRACTICE EXERCISE 1.2 _____

Work odd-numbered problems first, check answers, and then work even-numbered problems in areas of weakness. Answers to all problems are in the back of the book. Make every effort to work a problem yourself before you look at an answer.

A *Evaluate each expression.*

1. $9 - 4 \cdot 2$ _____
2. $10 - 3 \cdot 2$ _____
3. $(9 - 4) \cdot 2$ _____

4. $(10 - 3) \cdot 2$ _____
5. $10 - 3(7 - 4)$ _____
6. $8 + 2(7 + 3)$ _____

7. $(10 - 3)(7 - 5)$ _____
8. $(8 + 2)(7 + 3)$ _____
9. $(10 - 2) - 2(7 - 4)$ _____

10. $(3 + 5) - 4(5 - 3)$ _____
11. $12 - 2(7 - 5)$ _____
12. $15 - 3(9 - 5)$ _____

Evaluate each algebraic expression for x = 9 and y = 2.

13. $x - 2y$ _____
14. $x + 3y$ _____
15. $(x - 2)y$ _____
16. $(x + 3)y$ _____

17. $x - 3(y + 1)$ ___
18. $y + 2(x - 6)$ ___
19. $(x - 3)(y + 1)$ __
20. $(y + 2)(x - 6)$ __

21. $y(x - 2y)$ _____
22. $x(y + 2x)$ _____
23. $3(x - 2) - xy$ __
24. $xy - 2(y + 3)$ __

B *Evaluate each expression.*

25. $6(11 - 8) - 2 \cdot 5$ _____
26. $3 \cdot 8 - 2(8 - 3)$ _____

27. $2[12 - 4(8 - 6)]$ _____
28. $3[18 - 6(12 - 10)]$ _____

29. $2[(3 + 2) + 2(7 - 4) + 6 \cdot 2]$ _____
30. $3[(6 - 4) + 4 \cdot 3 + 3(1 + 6)]$ _____

31. $5\{32 - 5[(10 - 2) - 2 \cdot 3]\}$ _____
32. $2\{26 - 3[12 - 2(8 - 5)]\}$ _____

Evaluate each for u = 2, v = 3, w = 4, and x = 5.

33. $5w - 2(u + v)$ _____
34. $4(x - u) - w$ _____
35. $3[x - 2(w - v)]$ _____

36. $2[x + 3(x - u)]$ _____
37. $2\{w + 2[7 - (u + v)]\}$ __
38. $3\{x - 3[9 - 4(x - v)]\}$ __

39. $2\{[(w - u) + 3(v + 2u)] - w\}$ _____
40. $3\{w + 3[(x - u) + 2(u + v)]\}$ _____

Translate each statement into an algebraic equation, using x as the only variable.

41. 18 is 3 times a certain number. _____

42. 80 is 3 more than twice a certain number. _____

43. 26 is 12 less than a certain number. _____

44. 32 is 5 less than a certain number. _____

45. 43 is 71 less than 4 times a certain number. _____

46. 62 is 9 less than 5 times a certain number. _____

47. 6 times a number is 4 more than 3 times the number. _____

48. 7 times a number is 12 more than 4 times the number. _____

49. 6 less than a certain number is 5 times the number which is 7 more than the certain number.

_____

50. 5 more than a certain number is 3 times the number which is 4 less than the certain number.

_____

C *Evaluate each expression.*

51. $2(3 + 2 \cdot 3) + 2 \cdot 4 + 2[2 \cdot 5 - (7 - 5)]$ _____

52. $2\{12 - 2[7 - 2(12 - 10)]\} - 3(9 - 6)$ _____

Evaluate each expression for u = 12, v = 3, w = 8, and x = 5.

53. $5\{18 - 2[u - v(x - v)]\} - vw$ _____

54. $ux - 2\{2[(w - v) + (u - x)] + 1\}$ _____

55. What is wrong with the following argument? "4 is an even number and 8 is an even number; therefore, we can write "4 = even number" and "8 = even number." By the symmetric property of equality, we can write "even number = 8." Thus, using the transitive property of equality (since 4 = even number and even number = 8), we can conclude that 4 = 8.

_____

APPLICATIONS
56. In a rectangle with area 90 square meters the length is 3 meters less than twice the width. Write an equation relating the area with the length and width, using x as the only variable. $(A = ab)$

_____

57. In a rectangle with perimeter 210 cm the length is 10 cm less than 3 times the width. Write an equation relating the perimeter with the length and width, using x as the only variable. $(P = 2a + 2b)$

_____

The Check Exercise for this section is on page 51.

1.3 REAL NUMBERS AND BASIC PROPERTIES

•The real number system
••Real number properties

In algebra, one is interested in manipulating constants and variables in order to simplify algebraic expressions and to solve algebraic equations. Since the constants and variables usually represent real numbers, it is important that we briefly review the real number system and some of its important properties. These properties are behind many of the operations in algebra.

•The real number system

The real number system is the number system you use must of the time. The following are some examples of real numbers. Most should be familiar to you.

$$5 \qquad -3 \qquad 3.7 \qquad 0 \qquad \frac{3}{4} \qquad -\frac{4}{7}$$

$$-3.14159 \qquad \sqrt{2} \qquad \pi \qquad \sqrt[3]{4} \qquad -\sqrt{7} \qquad 1$$

The set of real numbers has several important subsets:

Natural Numbers (N): Counting numbers (also called positive integers). $\{1, 2, 3, \ldots\}$

Integers (I): Set of all natural numbers, their opposites (negatives), and zero. $\{\ldots, -2, -1, 0, 1, 2, \ldots\}$

Rational Numbers (Q): Any number that can be written in the form m/n, where m and n are integers with $n \neq 0$.

Real Numbers (R): All rational and irrational numbers. All rational numbers have repeating decimal representations, while all irrational numbers have infinite nonrepeating decimal representations.

Rational numbers—repeating decimals. (The overbar indicates the block of numbers that repeats indefinitely.)

$$\frac{7}{8} = 0.875\overline{0} \qquad \frac{4}{3} = 1.\overline{3} \qquad \frac{5}{7} = 0.\overline{714285}$$

Irrational numbers—infinite nonrepeating decimals.

$$\sqrt{2} = 1.41421356\cdots \qquad \pi = 3.14159265\cdots$$

Figure 2 illustrates the real number system, showing how its important subsets are related to each other. Note that N is a subset of I, I is a subset of Q, and Q is a subset of R. Study Figure 2 carefully—it contains a lot of useful information.

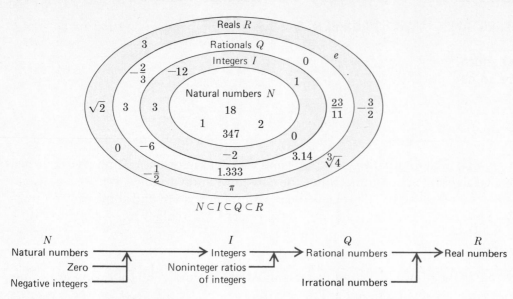

$$N \subset I \subset Q \subset R$$

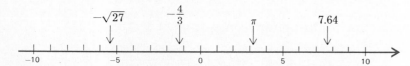

Figure 2   The Set of Real Numbers

A one-to-one correspondence exists between the set of real numbers and the set of points on a line; that is, each real number corresponds to exactly one point, and each point to exactly one real number. A line with a real number associated with each point, as in Figure 3, is called a **real number line,** or, simply, a **real line.**

Figure 3   A Real Number Line

If you have trouble with some of these numbers, don't despair. It took mathematicians over 2,000 years to come up with a precise definition of number, and this didn't happen until the past century. Actually, at this point, how variables and constants that name numbers are manipulated is of greater interest to us than a precise knowledge of numbers as objects. The properties of real numbers provide us with the basic manipulative rules that are used in algebra.

••Real number properties

We now take a closer look at the real number system to uncover a basic list of properties that will enable us to convert algebraic expressions into equivalent forms. (Two algebraic expressions are **equivalent** if they are equal for each replacement of the variables by numbers for which each expression is defined.) These assumed properties are referred to as **axioms,** and they form the basic rules for the "game of algebra" we are about to play.

Does it matter in which order we perform addition, subtraction, multiplication, or division? Which of the following is true?

$$8 + 4 = 4 + 8$$
$$8 - 4 = 4 - 8$$
$$8 \cdot 4 = 4 \cdot 8$$
$$8 \div 4 = 4 \div 8$$

Compute the left and right sides of each equation to obtain:

$$12 = 12$$
$$4 \neq -4$$
$$32 = 32$$
$$2 \neq \tfrac{1}{2}$$

We see that the first and third are true, but the second and fourth are false. In general, it turns out we may add in any order and multiply in any order. This property is referred to as the **commutative property** for real numbers.

> *Commutative Property*
> ───────────────────────
>
> For all real numbers a and b
>
> $a + b = b + a$ $ab = ba$
> 8 + 4 = 4 + 8   8·4 = 4·8

EXAMPLE 8

If x and y are real numbers, then, using the commutative properties for addition or multiplication, we know about

$x + 9 = 9 + x$ $7y = y7$

$2x + 3y = 3y + 2x$ $3 + yx = 3 + xy$

Work Problem 8 and check solution in Solutions to Matched Problems on page 21.

PROBLEM 8 Fill in each square bracket with an appropriate expression, using the commutative property.

(A) $x + 3 = 3 + [\ \ ]$ **(B)** $3x + 5y = 5y + [\ \ ]$

(B) $5(y + x) = 5([\ \ ] + y)$ **(D)** $y + x4 = y + 4[\ \ ]$

When computing

$8 + 4 + 2$

$8 - 4 - 2$

$8 \cdot 4 \cdot 2$

$8 \div 4 \div 2$

does it matter how the numbers are grouped? That is, which of the following are true?

$(8 + 4) + 2 = 8 + (4 + 2)$

$(8 - 4) - 2 = 8 - (4 - 2)$ Compute the left and right sides of each equation to obtain:

$(8 \cdot 4) \cdot 2 = 8 \cdot (4 \cdot 2)$

$(8 \div 4) \div 2 = 8 \div (4 \div 2)$

$$\begin{cases} 14 = 14 \\ 2 \neq 6 \\ 64 = 64 \\ 1 \neq 4 \end{cases}$$

We see that the first and third are true, but the second and fourth are false. In general, it turns out we may group terms in addition and factors in multiplication in any way we please. This property is referred to as the **associative property** for real numbers.

Associative Property

For all real numbers a, b, and c

$$(a + b) + c = a + (b + c) \qquad (a \cdot b) \cdot c = a \cdot (b \cdot c)$$
$$(8 + 4) + 2 = 8 + (4 + 2) \qquad (8 \cdot 4) \cdot 2 = 8 \cdot (4 \cdot 2)$$

EXAMPLE 9

If x, y, and z represent real numbers, then, using the associative property, we know that

$$(x + 7) + 2 = x + (7 + 2) \qquad 3(5y) = (3 \cdot 5)y$$
$$x + (x + 3) = (x + x) + 3 \qquad (2y)y = 2(yy)$$

Work Problem 9 and check solution in Solutions to Matched Problems on page 21.

PROBLEM 9 Fill in each set of parentheses with an appropriate expression; using the associative property.

(A) $(x + 3) + 9 = x + ($ $)$ (B) $6(3y) = ($ $)y$

(C) $(5 + z) + z = 5 + ($ $)$ (D) $(3z)z = 3($ $)$

Conclusion

In addition, commutativity and associativity permit us to change the order at will and insert or remove parentheses as we please. The same is true for multiplication, but not for subtraction and division.

EXAMPLE 10

Remove parentheses and simplify mentally.

(A) $(3x + 5) + (2y + 7) = 3x + 5 + 2y + 7$

$$= 3x + 2y + 5 + 7 \quad †$$

$$= 3x + 2y + 12$$

(B) $(6x)(3y) = 6x3y$

$$= (6 \cdot 3)xy$$

$$= 18xy$$

†Dotted boxes are "think boxes" and are used to indicate steps that are usually done mentally.

It is important to remember that the associative and commutative properties are behind the mental steps in Example 10.

Work Problem 10 and check solution in Solutions to Matched Problems on page 21.

PROBLEM 10   Remove parentheses and simplify mentally.

(A) $(2a + 3) + (3b + 11) =$

(B) $(4m)(6n) =$

We now turn to an important real number property that involves both multiplication and addition. Consider the following computations:

$2(5 + 3)$	$2 \cdot 5 + 2 \cdot 3$
$= 2 \cdot 8$	$= 10 + 6$
$= 16$	$= 16$

Thus, we conclude that

$$2(5 + 3) = 2 \cdot 5 + 2 \cdot 3$$

and say that the factor 2 distributes over the sum $(5 + 3)$. In general, it turns out that multiplication always **distributes** over addition.

Distributive Property

For all real numbers a, b, and c

$a(b + c) = ab + ac$ multiplying
  $2(x + y) = 2x + 2y$

NOTE: Using equality properties discussed in Section 1.2, we can also write:

$ab + ac = a(b + c)$ factoring
  $2x + 2y = 2(x + y)$

EXAMPLE 11
Multiply, using the distributive property.

(A) $3(r + t) = 3r + 3t$
(B) $5(2x + 3) = 10x + 15$
(C) $m(u + v) = mu + mv$

Work Problem 11 and check solution in Solutions to Matched Problems on page 21.

PROBLEM 11   Multiply, using the distributive property.

(A) $5(m + n) =$

20

(B) $4(3y + 5) =$

(C) $a(u + v) =$

EXAMPLE 12

Take out all factors common to both terms.

(A) $6x + 6y = 6x + 6y = 6(x + y)$
(B) $7x^2 + 14 = 7x^2 + 7 \cdot 2 = 7(x^2 + 2)$
(C) $ax + ay = ax + ay = a(x + y)$

Work Problem 12 and check solution in Solutions to Matched Problems on page 21.

PROBLEM 12 Take out all factors common to both terms.

(A) $9m + 9n =$

(B) $12y + 6 =$

(C) $du + dy =$

Several other useful distributive properties follow from the basic one above and real number properties just discussed.

Additional Distributive Properties

All variables represent real numbers.

(1) $(ba + ca) = (b + c)a$ $\qquad$ $(2x + 3x) = (2 + 3)x$

(2) $a(b + c + d) = ab + ac + ad$ $\qquad$ $3(w + x + y) = 3w + 3x + 3y$

(3) $ab + ac + ad = a(b + c + d)$ $\qquad$ $2x + 2y + 2z = 2(x + y + z)$

(4) $ba + ca + da = (b + c + d)a$ $\qquad$ $3x + 4x + 7x = (3 + 4 + 7)x$

and so on.

Thus, we observe that common factors may be taken out either on the left or on the right.

EXAMPLE 13
(A) Multiply:

$$3(u + v + 2) = 3u + 3v + 6$$

(B) Take out all common factors on the left:

$$6m + 3n + 12 = 3(2m + n + 4)$$

(C) Take out all common factors on the right:

$$7w + 3w + w = (7 + 3 + 1)w$$

Work Problem 13 and check solution in Solutions to Matched Problems on page 21.

PROBLEM 13 **(A)** Multiply:

$$6(2w + x + 2) =$$

(B) Take out all common factors on the left:

$$12u + 8v + 4 =$$

(C) Take out all common factors on the right:

$$4z + z + 3z =$$

SOLUTIONS TO MATCHED PROBLEMS

8. **(A)** $x + 3 = 3 + [x]$ **(B)** $3x + 5y = 5y + [3x]$ **(C)** $5(y + x) = 5([x] + y)$
   (D) $y + x4 = y + 4[x]$
9. **(A)** $(x + 3) + 9 = x + (3 + 9)$ **(B)** $6(3y) = (6 \cdot 3)y$ **(C)** $(5 + z) + z = 5 + (z + z)$
   (D) $(3z)z = 3(zz)$
10. **(A)** $(2a + 3) + (3b + 11) = 2a + 3 + 3b + 11 = 2a + 3b + 14$
   (B) $(4m)(6n) = 4m6n = 24mn$
11. **(A)** $5(m + n) = 5m + 5n$ **(B)** $4(3y + 5) = 12y + 20$ **(C)** $a(u + v) = au + av$
12. **(A)** $9m + 9n + 9(m + n)$ **(B)** $12y + 6 = 6(2y + 1)$ **(C)** $du + dy = d(u + y)$
13. **(A)** $6(2w + x + 2) = 12w + 6x + 12$ **(B)** $12u + 8v + 4 = 4(3u + 2v + 1)$
   (C) $4z + z + 3z = (4 + 1 + 3)z$

PRACTICE EXERCISE 1.3

Work odd-numbered problems first, check answers, and then work even-numbered problems in areas of weakness. Answers to all problems are in the back of the book. Make every effort to work a problem yourself before you look at an answer.

All variables represent real numbers.

A *State the justifying real number property for each statement. (For example, $3 + y = y + 3$ is justified by commutative add.)*

1. $12 + w = w + 12$ _____

2. $2x + 3 = 3 + 2x$ _____

3. $m + (n + 3) = (m + n) + 3$ _____

4. $(3x + y) + 5 = 3x + (y + 5)$ _____

5. $x20 = 20x$ _____

6. $MN = NM$ _____

7. $4(8y) = (4 \cdot 8)y$ _____

8. $(12u)v = 12(uv)$ _____

9. $2(a + b) = 2a + 2b$ _____

10. $3(x + 2) = 3x + 6$ _____

11. $4x + 7x = (4 + 7)x$ _____

12. $4m + 4n = 4(m + n)$ _____

Remove parentheses and simplify mentally.

13. $(x + 7) + 2$ _____

14. $3 + (5 + m)$ _____

15. $4(5y)$ _____

16. $6(8n)$ _____

17. $12 + (u + 3)$ _____

18. $(4 + x) + 13$ _____

19. $(3x)7$ _____

20. $4(y3)$ _____

B *State the justifying real number property for each statement.* [*For example, (3m)2 = 2(3m) is justified by commutative mult.*]

21. $2 + (y + 3) = 2 + (3 + y)$ _____

22. $(3m)n = n(3m)$ _____

23. $7(y4) = 7(4y)$ _____

24. $5 + (y + 2) = (y + 2) + 5$ _____

25. $5(w + x + y) = 5w + 5x + 5y$ _____

26. $7x + 14y + 7 = 7(x + 2y + 1)$ _____

27. $3x + 2y = 2y + 3x$ _____

28. $3x + (2x + 5y) = (3x + 2x) + 5y$ _____

29. $4m + m + 3m = (4 + 1 + 3)m$ _____

30. $u + 3u + 2u = (1 + 3 + 2)u$ _____

Remove parentheses and simplify mentally.

31. $(x + 7) + (y + 4) + (z + 1)$ _____

32. $(7 + m) + (8 + n) + (3 + p)$ _____

33. $(3x + 5) + (4y + 6)$ _____

34. $(3a + 7) + (5b + 2)$ _____

35. $(12m)(3n)(1p)$ _____

36. $(8x)(4y)(2z)$ _____

C **37.** *Indicate whether true (T) or false (F), and for each false statement find real number replacements for a, b, and c that will illustrate its falseness: For all real numbers a and b,*

(A) $a + b = b + a$ _____

(B) $a - b = b - a$ _____

(C) $ab = ba$ _____

(D) $a \div b = b \div a$ _____

38. *Indicate whether true (T) or false (F), and for each false statement find real number replacements for a, b, and c that will illustrate its falseness: For all real numbers a, b, and c,*

(A) $(a + b) + c = a + (b + c)$ _____

(B) $(a - b) - c = a - (b - c)$ _____

(C) $a(bc) = (ab)c$ _____

(D) $(a \div b) \div c = a \div (b \div c)$ _____

The Check Exercise for this section is on page 53.

1.4 ADDITION AND SUBTRACTION OF SIGNED NUMBERS

•The negative of a number
••The absolute value of a number
•••Addition of signed numbers
••••Subtraction of signed numbers
•••••Combined addition and subtraction

Before we review addition and subtraction we will say a few words about the operations "the negative of" and "the absolute value of." These operations are useful in describing addition, subtraction, multiplication, and division for signed numbers.

•The negative of a number

For each real number x, we denote its **negative** by

$-x$ the negative of x

"Negative of," "opposite of," and "additive inverse" all describe the same thing and are used interchangeably. You will recall that the negative of a number is obtained by changing its sign. The negative of 0 is 0.

EXAMPLE 14

(A) $-(+5) = -5$
(B) $-(-8) = +8$ or 8
(C) $-(0) = 0$
(D) $-[-(-4)] = -(+4) = -4$

Work Problem 14 and check solution in Solutions to Matched Problems on page 28.

PROBLEM 14   Find each:

(A) $-(+11) =$ (B) $-(-12) =$

(C) $-(0) =$ (D) $-[-(+6)] =$

It is now important to note the three distinct uses of the minus sign $-$.

Multiple Uses of $-$

1. As the operation "subtract": $9 \overset{\downarrow}{-} 3 = 6$
2. As the operation "the negative or opposite of": $\overset{\downarrow}{-}(-8) = 8$
3. As part of a number symbol: $\overset{\downarrow}{-}4$

••The absolute value of a number

The **absolute value** of a number x is an operation on x, denoted by the symbol

$$|x| \qquad \text{the absolute value of } x$$

(not square brackets). The absolute value of a number can be thought of geometrically as the distance of the number from 0 on the real number line, expressed as a positive number or 0. For example, both 5 and -5 are 5 units from 0 (see Figure 4).

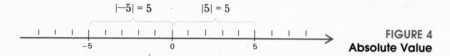

FIGURE 4
Absolute Value

Symbolically, and more formally, we define absolute value as follows:

> **Absolute Value**
>
> $$|x| = \begin{cases} x & \text{if } x \text{ is positive} \\ 0 & \text{if } x \text{ is } 0 \\ -x & \text{if } x \text{ is negative} \end{cases}$$
>
> NOTE: $-x$ is positive if x is negative.

It is important to remember that

The absolute value of a number is never negative.

EXAMPLE 15

(A) $|24| = 24$
(B) $|-7| = 7$
(C) $|0| = 0$
(D) $-(|-8| + |-3|) = -(8 + 3) = -(11) = -11$

Work Problem 15 and check solution in Solutions to Matched Problems on page 28.

PROBLEM 15 Evaluate:

    (A) $|-13| =$ (B) $|43|$

    (C) $-|-14| =$ (D) $-(|-6| - |+2|) =$

•••Addition of signed numbers

We are now ready to review addition of signed numbers. We will consider four cases: addition involving zero, addition of positive numbers, addition of negative numbers, and addition of numbers with unlike signs.

Zero in Addition

For each real number a

$$0 + a = a + 0 = a$$

For example,

$$0 + (-3) = -3 \qquad 5 + 0 = 5$$

Addition of Positive Numbers

Add positive numbers as in arithmetic.
(We assume you can do this.)

For example,

$$18 + 12 = 30 \qquad 6.32 + 1.04 = 7.36$$

Addition of Negative Numbers

If a and b are both negative, their sum is the negative of the sum of their absolute values.

Mechanically, mentally block out the signs of the two numbers (take absolute values), add as in arithmetic, and then attach a minus sign to the result. For example,

$$(-5) + (-3) = -(5 + 3) = -8$$

Addition of Numbers with Unlike Signs

To add two numbers with unlike signs, subtract the smaller absolute value from the larger absolute value; then attach the sign of the number with the largest absolute value to the result.

Mechanically, mentally block out the signs of the two numbers (take their absolute values), subtract the smaller from the larger, and then attach the sign of the number with the largest absolute value. For example,

$$(-3) + (+9) = +(9 - 3) = 6$$

$$(+3) + (-9) = -(9 - 3) = -6$$

EXAMPLE 16
Add:

(A) $0 + 5 = 0$

(B) $4 + 7 = 11$

(C) $(-8) + (-4) = -(8 + 4) = -12$

(D) $(-12) + 7 = -(12 - 7) = -5$

Work Problem 16 and check solution in Solutions to Matched Problems on page 28.

PROBLEM 16   Add:

       (A) $12 + 0 =$ **(B)** $12 + (-7) =$

       (C) $8 + 12 =$ **(D)** $(-8) + (-12) =$

> ### *Adding Three or More Signed Numbers*
>
> To add three or more numbers, add all the positive numbers together, add all the negative numbers together (the commutative and associative justify this procedure), and then add the two resulting sums.

EXAMPLE 17
$$3 + (-6) + 8 + (-4) + (-5) = (3 + 8) + [(-6) + (-4) + (-5)]$$
$$= 11 + (-15) = -(15 - 11) = -4$$

Work Problem 17 and check solution in Solutions to Matched Problems on page 28.

PROBLEM 17   Add:

       $6 + (-8) + (-4) + 10 + (-3) + 1 =$

••••Subtraction of signed numbers

We define subtraction as follows:

> ### *Subtraction*
>
> For a and b any real numbers
>
> $$a - b = a + (-b)$$
>
> To subtract b from a add the negative of b to a (that is, change the sign of b and add).

For example,

$$(-3) - (-9) = (-3) + 9 = 6$$

with "negative of –9" labeling the –9, and "change to addition" labeling the conversion.

You should get to the point where you can perform this type of subtraction mentally, and simply write down the answer.

EXAMPLE 18

(A) $8 - (-5) = 8 + 5 = 13$

(B) $(-8) - 5 = (-8) + (-5) = -13$

(C) $(-8) - (-5) = (-8) + 5 = -3$

(D) $0 - 5 = 0 + (-5) = -5$

Work Problem 18 and check solution in Solutions to Matched Problems on page 28.

PROBLEM 18 Subtract:

(A) $4 - 7 =$ **(B)** $7 - (-4) =$

(C) $(-7) - 4 =$ **(D)** $(-7) - (-4) =$

(E) $(-4) - (-7) =$ **(E)** $0 - (-4) =$

•••••Combined addition and subtraction

When three or more terms are combined by addition and subtraction and symbols of grouping are omitted, we convert (mentally) any subtraction to addition and add. Thus,

$$8 - 5 + 3 = 8 + (-5) + 3 = 11 + (-3) = 6$$
think

EXAMPLE 19

(A) $2 - 3 - 7 + 4 = 2 + (-3) + (-7) + 4 = (-10) + 6 = -4$
think

(B) $-4 - 8 + 2 + 9 = (-4) + (-8) + 2 + 9 = (-12) + 11 = -1$
think

Work Problem 19 and check solution in Solutions to Matched Problems on page 28.

PROBLEM 19 Evaluate:

(A) $5 - 8 + 2 - 6 =$

(B) $-6 + 12 - 2 - 1 =$

SOLUTIONS TO MATCHED PROBLEMS

14. **(A)** $-(+11) = -11$ **(B)** $-(-12) = 12$ **(C)** $-(0) = 0$
(D) $-[-(+6)] = -(-6) = 6$
15. **(A)** $|-13| = 13$ **(B)** $|43| = 43$ **(C)** $-|-14| = -(+14) = -14$
(D) $-(|-6| - |+2|) = -(6 - 2) = -(+4) = -4$
16. **(A)** $12 + 0 = 12$ **(B)** $12 + (-7) = 5$ **(C)** $8 + 12 = 20$ **(D)** $(-8) + (-12) = -20$
17. $6 + (-8) + (-4) + 10 + (-3) + 1 = 17 + (-15) = 2$
18. **(A)** $4 - 7 = -3$ **(B)** $7 - (-4) = 11$ **(C)** $(-7) - 4 = -11$
(D) $(-7) - (-4) = -3$ **(E)** $(-4) - (-7) = 3$ **(F)** $0 - (-4) = 4$
19. **(A)** $5 - 8 + 2 - 6 = 7 - 14 = -7$ **(B)** $-6 + 12 - 2 - 1 = 12 - 9 = 3$

PRACTICE EXERCISE 1.4 _____

Work odd-numbered problems first, check answers, and then work even-numbered problems in areas of weakness. Answers to all problems are in the back of the book. Make every effort to work a problem yourself before you look at an answer.

A *Evaluate:*

1. $-(+7)$ _____ **2.** $-(+12)$ _____ **3.** $-(-6)$ _____ **4.** $-(-8)$ _____

5. $|+2|$ _____ **6.** $|+9|$ _____ **7.** $|-27|$ _____ **8.** $|-32|$ _____

9. $|0|$ _____ **10.** $-(0)$ _____

11. $(-7) + (-3)$ _____ **12.** $(-7) + (+3)$ _____ **13.** $(+7) + (-3)$ _____

14. $(-12) + (+8)$ _____ **15.** $(+3) - (+9)$ _____ **16.** $(+3) - (-9)$ _____

17. $(+9) - (-3)$ _____ **18.** $(-9) - (-3)$ _____

19. The negative of a number is (*always, sometimes, never*) a negative number. _____

20. The absolute value of a number is (*always, sometimes, never*) a positive number. _____

B *Evaluate:*

21. $-[-(-3)]$ _____ **22.** $-[-(+6)]$ _____ **23.** $-|-(+2)|$ _____

24. $-|-(-3)|$ _____ **25.** $-(|-9| - |-3|)$ _____ **26.** $-(|-14| - |-8|)$ _____

27. $(-2) + (-6) + 3$ _____ **28.** $(-2) + (-8) + 5$ _____ **29.** $5 - 7 - 3$ _____

30. $3 - 2 + 4$ _____ **31.** $-7 + 6 - 4$ _____ **32.** $-4 + 7 - 6$ _____

33. $-2 - 3 + 6 - 2$ _____ **34.** $-4 + 7 - 3 - 2$ _____ **35.** $6 - [3 - (-9)]$ _____

36. $(-10) - [(-6) + 3]$ _____ **37.** $[6 - (-8)] - [(-8) - 6]$ _____

38. $[3 - 5] + [(-5) - (-2)]$ _____

Replace each question mark with an appropriate real number.

39. $-(?) = 5$ _____   **40.** $-(?) = -8$ _____   **41.** $|?| = 7$ _____

42. $|?| = -4$ _____   **43.** $(-3) + ? = -8$ _____   **44.** $? + 5 = -6$ _____

45. $(-3) - ? = -8$ _____   **46.** $? - (-2) = -4$ _____

Evaluate for $x = 3$, $y = -8$, and $z = -2$.

47. $x + y$ _____   **48.** $y + z$ _____   **49.** $(x - z) + y$ _____

50. $y - (z - x)$ _____   **51.** $|(-z) - |y||$ _____   **52.** $||-y| - 12|$ _____

C 53. You own a stock that is traded on the New York Stock Exchange. On Monday it closed at $23 per share; it fell $3 on Tuesday and another $6 on Wednesday; it rose $2 on Thursday; and it finished strongly on Friday by rising $7. Use addition of signed numbers to determine the closing price of the stock on Friday.

_____

54. Find, using subtraction of signed numbers, the difference in the height between the highest point in the United States, Mount McKinley (20,270 ft) and the lowest point in the United States, Death Valley (-280 ft).

_____

| The Check Exercise for this section is on page 55. |

1.5 MULTIPLICATION AND DIVISION OF SIGNED NUMBERS

•**Multiplication of signed numbers**
••**Division of signed numbers**
•••**Combined operations**

•Multiplication of signed numbers

We will consider four cases: multiplication involving zero, multiplication of positive numbers, multiplication of negative numbers, and multiplication of numbers with unlike signs.

Multiplication Involving 0

For any real number a,

$0 \cdot a = a \cdot 0 = 0$

For example,

$0 \cdot 3 = 0 \qquad (-7)(0) = 0$

Multiplication of Positive Numbers

Multiply positive numbers as in arithmetic. (We assume you can do this.)

For example,

$5 \cdot 7 = 35 \qquad (3.24)(6.13) = 19.8612$

Multiplication of Negative Numbers

The product of two negative numbers is positive and is found by multiplying the absolute values of the two numbers.

For example,

$(-3)(-7) = 3 \cdot 7 = 21$

Multiplication of Numbers with Unlike Signs

The product of two numbers with unlike signs is negative and is found by taking the negative of the product of the absolute values of the two numbers.

For example,

$8(-4) = -(8 \cdot 4) = -32$

EXAMPLE 20

(A) $2(-7) = -(2 \cdot 7) = -14$

(B) $(-2)(-7) = 2 \cdot 7 = 14$

(C) $0(-7) = 0$

Work Problem 20 and check solution in Solutions to Matched Problems on page 36.

PROBLEM 20 Evaluate:

(A) $4(-3) =$ (B) $(-4)3 =$

(C) $(-4)(-3) =$ (D) $0(-3) =$

Several important sign properties for multiplication are summarized in the following box.

Sign Properties for Multiplication

For a and b any real numbers

(A) $(-1)a = -a$
(B) $(-a)b = a(-b) = -(ab)$
(C) $(-a)(-b) = ab$

EXAMPLE 21

Evaluate $(-a)b$ and $-(ab)$ for $a = -5$ and $b = 4$.

Solution

$(-a)b = [-(-5)]4 = 5 \cdot 4 = 20$

$-(ab) = -[(-5)4] = -(-20) = 20$

Work Problem 21 and check solution in Solutions to Matched Problems on page 36.

PROBLEM 21 Evaluate $(-a)(-b)$ and ab for $a = -5$ and $b = 4$.

Solution $(-a)(-b) =$

$ab =$

Expressions such as

$-ab$

occur frequently and at first glance are confusing to students. If you were asked to evaluate $-ab$ for $a = -3$ and $b = +2$, how would you proceed? Would you take the negative of a and then multiply it by b, or multiply a and b first and then take the negative of the product? Actually it does not matter! Because of the sign properties of multiplication, we get the same result either way since $(-a)b = -(ab)$. Furthermore, if we consider other material in this section, we find that

$$-ab = \begin{cases} (-a)b \\ a(-b) \\ -(ab) \\ (-1)ab \end{cases}$$

and we are at liberty to replace any one of these five forms with another from the same group. NOTE: It is not correct to say that in $-ab$ the minus sign applies to both a and b to yield $(-a)(-b)$. The latter is equal to ab, not $-ab$.

••Division of signed numbers

The two division symbols "÷" and "⟩‾‾" from arithmetic are not used a great deal in algebra and higher mathematics. The horizontal bar "—" and slash mark "/" are the symbols most frequently used. Thus

$$a/b \qquad \frac{a}{b} \qquad a \div b \qquad \text{and} \qquad b\overline{)a} \qquad \text{\small In each case } b \text{ is the divisor.}$$

all name the same number (assuming the quotient is defined), and we can write

$$a/b = \frac{a}{b} = a \div b = b\overline{)a}$$

Now we proceed to the mechanics of division.

Division Involving 0

For a any nonzero real number,

$$\frac{0}{a} = 0 \qquad \frac{a}{0} \text{ is not defined} \qquad \frac{0}{0} \text{ is not defined}$$

For example,

$$\frac{0}{-3} = 0 \qquad \frac{-3}{0} \text{ is not defined} \qquad \frac{0}{0} \text{ is not defined}$$

Zero cannot be used as a divisor—ever!

Division and Positive Numbers

Divide positive numbers as in arithmetic. (We assume you can do this.)

For example,

$$\frac{8}{4} = 2 \qquad \frac{5}{7} = 0.\overline{714285}$$

Division of Negative Numbers

The quotient of two negative numbers is positive and is found by dividing the absolute values of the two numbers.

For example,

$$\frac{-12}{-6} = \frac{12}{6} = 2$$

Division of Numbers with Unlike Signs

The quotient of two numbers with unlike signs is negative and is found by taking the negative of the quotient of the absolute values of the two numbers.

For example,

$$\frac{-18}{3} = -\frac{18}{3} = -6 \qquad \frac{10}{-5} = -\frac{10}{5} = -2$$

EXAMPLE 22

(A) $\dfrac{0}{-7} = 0$

(B) $\dfrac{22}{0}$ not defined

(C) $\dfrac{-36}{-12} = 3$

(D) $\dfrac{30}{-5} = -6$

Work Problem 22 and check solution in Solutions to Matched Problems on page 36.

PROBLEM 22   Divide where possible.

(A) $\dfrac{-12}{-4} =$

(B) $\dfrac{0}{10} =$

(C) $\dfrac{-26}{0} =$

(D) $\dfrac{44}{-11} =$

Several important sign properties for division are summarized in the following box.

Sign Properties for Division

For all real numbers a and b, $b \neq 0$:

(A) $\dfrac{-a}{-b} = \dfrac{a}{b} = -\dfrac{-a}{b} = -\dfrac{a}{-b}$

(B) $\dfrac{-a}{b} = \dfrac{a}{-b} = -\dfrac{a}{b}$

NOTE: It is not correct to say that in $-\dfrac{a}{b}$ the minus sign applies to both a and b to yield $\dfrac{-a}{-b}$. The latter is equal to $\dfrac{a}{b}$, not $-\dfrac{a}{b}$.

EXAMPLE 23

Evaluate $\dfrac{-a}{b}, \dfrac{a}{-b}$, and $-\dfrac{a}{b}$ for $a = -6$ and $b = 2$.

Solution

$$\frac{-a}{b} = \frac{-(-6)}{2} = \frac{6}{2} = 3$$

$$\frac{a}{-b} = \frac{(-6)}{-(2)} = \frac{-6}{-2} = 3$$

$$-\frac{a}{b} = -\frac{-6}{2} = -(-3) = 3$$

Work Problem 23 and check solution in Solutions to Matched Problems on page 36.

PROBLEM 23 Evaluate $\dfrac{-a}{-b}, \dfrac{a}{b}$, and $-\dfrac{-a}{b}$ for $a = -6$ and $b = 2$.

Solution $\dfrac{-a}{-b} =$

$\dfrac{a}{b} =$

$-\dfrac{-a}{b} =$

•••Combined operations

Now let us consider problems where various combinations of $+$, $-$, $\cdot$, and $\div$ are used.

EXAMPLE 24

Evaluate:

(A) $\dfrac{-12}{-3} - \dfrac{24}{-4} = 4 - (-6) = 10$

(B) $(-5)(-3) + 4(-4) + 0(-1) = 15 + (-16) + 0 = -1$

(C) $\dfrac{(-2)(-8) - (4)(-2)}{2 - 14} = \dfrac{16 - (-8)}{-12} = \dfrac{24}{-12} = -2$ Simplify numerator and denominator first.

Work Problem 24 and check solution in Solutions to Matched Problems on page 36.

PROBLEM 24   Evaluate:

(A) $3(-7) - \dfrac{15}{-5} =$

(B) $\dfrac{0}{-3} + \dfrac{-16}{-4} =$

(C) $\dfrac{0(-3) - 5(-2)}{2 - (-3)} =$

EXAMPLE 25

Evaluate for $u = -3$, $v = 2$, and $w = -36$.

(A) $\dfrac{w}{v} - uv + \dfrac{0}{u}$

Solution

$\dfrac{-36}{2} - (-3)(2) + \dfrac{0}{-3} = -18 - (-6) + 0 = -12$

(B) $\dfrac{|w|}{|u|} - \dfrac{u - w}{u - 4v}$

Solution

$\dfrac{|-36|}{|-3|} - \dfrac{-3 - (-36)}{-3 - 4(2)} = \dfrac{36}{3} - \dfrac{33}{-11} = 12 - (-11) = 23$

Work Problem 25 and check solution in Solutions to Matched Problems on page 36.

PROBLEM 25   Evaluate for $x = -24$, $y = -4$, and $z = 2$.

(A) $\dfrac{x}{yz} - 3y + \dfrac{0}{x} =$

(B) $|yz| - \dfrac{y - x}{y + z} =$

SOLUTIONS TO MATCHED PROBLEMS

20. **(A)** $4(-3) = -12$ **(B)** $(-4)3 = -12$ **(C)** $(-4)(-3) = 12$ **(D)** $0(-3) = 0$

21. $(-a)(-b) = [-(-5)][-(4)] = (5)(-4) = -20; ab = (-5)(4) = -20$

22. **(A)** $\dfrac{-12}{-4} = 3$ **(B)** $\dfrac{0}{10} = 0$ **(C)** $\dfrac{-26}{0}$ not defined **(D)** $\dfrac{44}{-11} = -4$

23. $\dfrac{-a}{-b} = \dfrac{-(-6)}{-(2)} = \dfrac{6}{-2} = -3; \dfrac{a}{b} = \dfrac{-6}{2} = -3; -\dfrac{-a}{b} = -\dfrac{-(-6)}{2} = -\dfrac{6}{2} = -3$

24. **(A)** $3(-7) - \dfrac{15}{-5} = (-21) - (-3) = -18$ **(B)** $\dfrac{0}{-3} + \dfrac{-16}{-4} = 0 + 4 = 4$

(C) $\dfrac{0(-3) - 5(-2)}{2 - (-3)} = \dfrac{0 - (-10)}{5} = \dfrac{10}{5} = 2$

25. **(A)** $\dfrac{x}{yz} - 3y + \dfrac{0}{x} = \dfrac{-24}{(-4)(2)} - 3(-4) + \dfrac{0}{-24} = \dfrac{-24}{-8} - (-12) + 0 = 3 - (-12) = 15$

(B) $|yz| - \dfrac{y - x}{y + z} = |(-4)(2)| - \dfrac{(-4) - (-24)}{(-4) + (2)} = |-8| - \dfrac{20}{-2} = 8 - (-10) = 18$

PRACTICE EXERCISE 1.5

Work odd-numbered problems first, check answers, and then work even-numbered problems in areas of weakness. Answers to all problems are in the back of the book. Make every effort to work a problem yourself before you look at an answer.

Perform the indicated operations when possible.

A **1.** $(-3)(-5)$ _____ **2.** $(-7)(-4)$ _____ **3.** $(-18) \div (-6)$ _____

4. $(-20) \div (-4)$ _____ **5.** $(-2)(+9)$ _____ **6.** $(+6)(-3)$ _____

7. $\dfrac{-9}{+3}$ _____ **8.** $\dfrac{+12}{-4}$ _____ **9.** $0(-7)$ _____ **10.** $(-6)0$ _____

11. $0/5$ _____ **12.** $0/(-2)$ _____ **13.** $3/0$ _____ **14.** $-2/0$ _____

15. $0 \div 0$ _____ **16.** $\dfrac{0}{0}$ _____ **17.** $\dfrac{-21}{3}$ _____ **18.** $\dfrac{-36}{-4}$ _____

19. $(-4)(-2) + (-9)$ _____ **20.** $(-7) + (-3)(+2)$ _____ **21.** $(+5) - (-2)(+3)$ _____

22. $(-7) - (-3)(-4)$ _____ **23.** $5 - \dfrac{-8}{2}$ _____ **24.** $7 - \dfrac{-16}{-2}$ _____

25. $(-1)(-8)$ and $-(-8)$ _____ **26.** $(-1)(+3)$ and $-(+3)$ _____

B **27.** $-12 + \dfrac{-14}{-7}$ _____ **28.** $\dfrac{-10}{5} + (-7)$ _____ **29.** $\dfrac{6(-4)}{-8}$ _____

30. $\dfrac{5(-3)}{3}$ _____ **31.** $\dfrac{22}{-11} - (-4)(-3)$ _____ **32.** $3(-2) - \dfrac{-10}{-5}$ _____

33. $\dfrac{-16}{2} - \dfrac{3}{-1}$ _____

34. $\dfrac{27}{-9} - \dfrac{-21}{-7}$ _____

35. $(+5)(-7)(+2)$ _____

36. $(-6)(-3(+4)$ _____

37. $(-22)(+36)(0)$ _____

38. $(+19)(0)(-35)$ _____

39. $[(+2) + (-7)][(+8) - (+10)]$ _____

40. $[(-3) - (+8)][(+4) + (-2)]$ _____

Evaluate for w = +2, x = −3, y = 0, and z = −24.

41. z/w _____

42. z/x _____

43. w/y _____

44. $\dfrac{y}{x}$ _____

45. $\dfrac{z}{x} - wz$ _____

46. $wx - \dfrac{z}{w}$ _____

47. $\dfrac{xy}{w} - xyz$ _____

48. $wxy - \dfrac{y}{z}$ _____

49. $-|w|\,|x|$ _____

50. $(|x|\,|z|)$ _____

51. $\dfrac{|z|}{|x|}$ _____

52. $-\dfrac{|z|}{|w|}$ _____

53. Any integer divided by 0 is (*always, sometimes, never*) 0. _____

54. 0 divided by *any* integer is (*always, sometimes, never*) 0. _____

55. A product made up of an odd number of negative factors is (*sometimes, always, never*) negative.

_____

56. A product made up of an even number of negative factors is (*sometimes, always, never*) negative.

_____

C *Evaluate for w = +2, x = −3, y = 0, and z = −24.*

57. $wx + \dfrac{z}{wx} + wz$ _____

58. $xyz + \dfrac{y}{z} + x$ _____

59. $\dfrac{8x}{z} - \dfrac{z - 6x}{wx}$ _____

60. $\dfrac{w - x}{w + x} - \dfrac{z}{2x}$ _____

The Check Exercise for this section is on page 57.

1.6 NATURAL NUMBER EXPONENTS

•**Definition and five properties**
••**Use of the five properties**

•**Definition and five properties**

Recall that

$$a^5 = a \cdot a \cdot a \cdot a \cdot a \qquad \text{five factors of } a$$

and, in general:

Natural Number Exponent

For n a natural number and a a real number

$$\overset{\text{exponent}}{a}{}^{n} = a \cdot a \cdot \cdots \cdot a \qquad n \text{ factors of } a$$
$$\underset{\text{base}}{}$$

$$2^3 = 2 \cdot 2 \cdot 2 = 8$$

Exponent forms are encountered so frequently in algebra it is essential for you to become completely familiar with the following five properties of exponents and their uses. These properties follow primarily from the definition of natural number exponent in the box and the commutative and associative properties of real numbers. We will give only informal arguments for each property. In all cases a and b represent real numbers and m and n represent natural numbers.

Consider:

$$a^3 a^4 = \overset{3\,\text{factors}}{(a \cdot a \cdot a)}\overset{4\,\text{factors}}{(a \cdot a \cdot a \cdot a)} = \overset{3+4\,\text{factors}}{(a \cdot a \cdot a \cdot a \cdot a \cdot a \cdot a)} = a^{3+4} = a^7$$

which suggests:

Property 1

$$a^m a^n = a^{m+n}$$

$$a^5 a^2 = a^{5+2} = a^7$$

Consider:

$$(a^3)^4 = a^3 \cdot a^3 \cdot a^3 \cdot a^3 = \overset{4\,\text{groups of 3 factors each}}{(a \cdot a \cdot a)(a \cdot a \cdot a)(a \cdot a \cdot a)(a \cdot a \cdot a)}$$

$$= \overset{4 \cdot 3\,\text{factors}}{(a \cdot a \cdot a \cdot a \cdot a \cdot a \cdot a \cdot a \cdot a \cdot a \cdot a \cdot a \cdot)} = a^{4 \cdot 3} = a^{12}$$

which suggests:

Property 2

$$(a^n)^m = a^{mn}$$

$$(a^2)^5 = a^{5 \cdot 2} = a^{10}$$

Consider:

$$\overset{\text{4 factors of } (ab)}{(ab)^4 = (ab)(ab)(ab)(ab)} = \overset{\text{4 factors of } a}{(a \cdot a \cdot a \cdot a)}\overset{\text{4 factors of } b}{(b \cdot b \cdot b \cdot b)} = a^4 b^4$$

which suggests:

Property 3

$$(ab)^m = a^m b^m$$

$$(ab)^7 = a^7 b^7$$

Consider:

$$\left(\frac{a}{b}\right)^5 = \overset{\text{5 factors of } a/b}{\left(\frac{a}{b} \cdot \frac{a}{b} \cdot \frac{a}{b} \cdot \frac{a}{b} \cdot \frac{a}{b}\right)} = \frac{\overset{\text{5 factors of } a}{a \cdot a \cdot a \cdot a \cdot a \cdot}}{\underset{\text{5 factors of } b}{b \cdot b \cdot b \cdot b \cdot b \cdot}} = \frac{a^5}{b^5}$$

which suggests:

Property 4

$$\left(\frac{a}{b}\right)^m = \frac{a^m}{b^m}$$

$$\left(\frac{a}{b}\right)^3 = \frac{a^3}{b^3}$$

Consider:

(A) $\dfrac{a^7}{a^3} = \dfrac{a \cdot a \cdot a \cdot a \cdot a \cdot a \cdot a}{a \cdot a \cdot a}$

$$= \frac{(a \cdot a \cdot a)(a \cdot a \cdot a \cdot a)}{(a \cdot a \cdot a)} = a^{7-3} = a^4$$

(B) $\dfrac{a^3}{a^3} = \dfrac{a \cdot a \cdot a}{a \cdot a \cdot a} = 1$

(C) $\dfrac{a^4}{a^7} = \dfrac{a \cdot a \cdot a \cdot a}{a \cdot a \cdot a \cdot a \cdot a \cdot a \cdot a}$

$$= \frac{(a \cdot a \cdot a \cdot a)}{(a \cdot a \cdot a \cdot a)(a \cdot a \cdot a)} = \frac{1}{a^{7-4}} = \frac{1}{a^3}$$

which suggests:

Property 5

$$\frac{a^m}{a^n} = \begin{cases} a^{m-n} & \text{if } m \text{ is larger than } n \\ 1 & \text{if } m = n \\ \dfrac{1}{a^{n-m}} & \text{if } n \text{ is larger than } m \end{cases}$$

$\dfrac{a^8}{a^3} = a^{8-3} = a^5$ $\dfrac{a^8}{a^8} = 1$ $\dfrac{a^3}{a^8} = \dfrac{1}{a^{8-3}} = \dfrac{1}{a^5}$

   It is very important to observe and remember that the properties of exponents apply to products and quotients, and not to sums and diffferences. Many mistakes are made in algebra by people applying a property of exponents to the wrong algebraic form. For example,

$(ab)^2 = a^2b^2$ but $(a + b)^2 \neq a^2 + b^2$

Order of Operations

1.   Multiplication and division precede addition and subtraction.

$2 \cdot 3 - \dfrac{8}{4} = 6 - 2 = 4$

2.   Powers (and later roots) precede multiplication and division.

$\dfrac{2^4}{8} + 3^2 \cdot 2^3 = \dfrac{16}{8} + 9 \cdot 8 = 2 + 72 = 74$

The exponent properties are summarized below for m and n natural numbers.

Exponent Properties

For m and n positive integers

1. $a^m a^n = a^{m+n}$
2. $(a^n)^m = a^{mn}$
3. $(ab)^m = a^m b^m$
4. $\left(\dfrac{a}{b}\right)^m = \dfrac{a^m}{b^m}$

5. $\dfrac{a^m}{a^n} = \begin{cases} a^{m-n} & \text{if } m \text{ is larger than } n \\ 1 & \text{if } m = n \\ \dfrac{1}{a^{n-m}} & \text{if } n \text{ is larger than } m \end{cases}$

••Use of the five properties

As before, the "dotted boxes" in the examples given below are used to indicate steps that are usually carried out mentally.

EXAMPLE 26

(A) $x^{12}x^{13} = x^{12+13} = x^{25}$

(B) $(t^7)^5 = t^{5 \cdot 7} = t^{35}$

(C) $(xy)^5 = x^5 y^5$

(D) $\left(\dfrac{u}{v}\right)^3 = \dfrac{u^3}{v^3}$

(E) $\dfrac{x^{12}}{x^4} = x^{12-4} = x^8$

(F) $\dfrac{t^4}{t^9} = \dfrac{1}{t^{9-4}} = \dfrac{1}{t^5}$

Work Problem 26 and check solution in Solutions to Matched Problems on page 42.

PROBLEM 26 Simplify, using natural number exponents only.

(A) $x^8 x^6 =$ **(B)** $(u^4)^5 =$ **(C)** $(xy)^9 =$

(D) $\left(\dfrac{x}{y}\right)^4 =$ **(E)** $\dfrac{x^{10}}{x^3} =$ **(F)** $\dfrac{x^3}{x^{10}} =$

EXAMPLE 27

(A) $(x^2 y^3)^4 = (x^2)^4 (y^3)^4 = x^8 y^{12}$

(B) $\left(\dfrac{u^3}{v^4}\right)^3 = \dfrac{(u^3)^3}{(v^4)^3} = \dfrac{u^9}{v^{12}}$

(C) $\dfrac{2x^9 y^{11}}{4x^{12} y^7} = \dfrac{2}{4} \cdot \dfrac{x^9}{x^{12}} \cdot \dfrac{y^{11}}{y^7} = \dfrac{1}{2} \cdot \dfrac{1}{x^3} \cdot \dfrac{y^4}{1} = \dfrac{y^4}{2x^3}$

Work Problem 27 and check solution in Solutions to Matched Problems on page 42.

PROBLEM 27 Simplify, using natural number exponents only.

(A) $(u^3 v^4)^2 =$ **(B)** $\left(\dfrac{x^4}{y^3}\right)^2 =$

(C) $\dfrac{9x^7 y^2}{3x^5 y^3} =$ **(D)** $\dfrac{(2x^2 y)^3}{(4xy^3)^2} =$

COMMON ERROR

$-4^2 \neq 16$ $\begin{cases} -4^2 = -(4)(4) = -16 \\ (-4)^2 = (-4)(-4) = 16 \end{cases}$

Knowing the rules of the game of chess doesn't make one good at playing chess; similarly, memorizing the properties of exponents doesn't necessarily make one good at using these properties. To acquire skill in their use, one must use these properties in a fairly large variety of problems. Practice Exercise 1.6 should help you acquire this skill.

SOLUTIONS TO MATCHED PROBLEMS

26. **(A)** $x^8 x^6 = x^{14}$ **(B)** $(u^4)^5 = u^{20}$ **(C)** $(xy)^9 = x^9 y^9$ **(D)** $\left(\dfrac{x}{y}\right)^4 = \dfrac{x^4}{y^4}$ **(E)** $\dfrac{x^{10}}{x^3} = x^7$

(F) $\dfrac{x^3}{x^{10}} = \dfrac{1}{x^7}$

27. **(A)** $(u^3 v^4)^2 = u^6 v^8$ **(B)** $\left(\dfrac{x^4}{y^3}\right)^2 = \dfrac{x^8}{y^6}$ **(C)** $\dfrac{9x^7 y^2}{3x^5 y^3} = \dfrac{3x^2}{y}$ **(D)** $\dfrac{(2x^2 y)^3}{(4xy^3)^2} = \dfrac{2^3 x^6 y^3}{4^2 x^2 y^6} = \dfrac{x^4}{2y^3}$

PRACTICE EXERCISE 1.6

Work odd-numbered problems first, check answers, and then work even-numbered problems in areas of weakness. Answers to all problems are in the back of the book. Make every effort to work a problem yourself before you look at an answer.

A *Replace the question marks with appropriate symbols.*

1. $y^2 y^7 = y^?$ _____

2. $x^7 x^5 = x^?$ _____

3. $y^8 = y^3 y^?$ _____

4. $x^{10} = x^? x^6$ _____

5. $(u^4)^3 = u^?$ _____

6. $(v^2)^3 = ?$ _____

7. $x^{10} = (x^?)^5$ _____

8. $y^{12} = (y^6)^?$ _____

9. $(uv)^7 = ?$ _____

10. $(xy)^5 = x^5 y^?$ _____

11. $p^4 q^4 = (pq)^?$ _____

12. $m^3 n^3 = (mn)^?$ _____

13. $\left(\dfrac{a}{b}\right)^8 = ?$ _____

14. $\left(\dfrac{x}{y}\right)^4 = \dfrac{x^?}{y^4}$ _____

15. $\dfrac{m^3}{n^3} = \left(\dfrac{m}{n}\right)^?$ _____

16. $\dfrac{x^7}{y^7} = \left(\dfrac{x}{y}\right)^?$ _____

17. $\dfrac{n^{14}}{n^8} = n^?$ _____

18. $\dfrac{x^7}{x^3} = x^?$ _____

19. $m^6 = \dfrac{m^8}{m^?}$ _____

20. $x^3 = \dfrac{x^?}{x^4}$ _____

21. $\dfrac{x^4}{x^{11}} = \dfrac{1}{x^?}$ _____

22. $\dfrac{a^5}{a^9} = \dfrac{1}{a^?}$ _____

23. $\dfrac{1}{x^8} = \dfrac{x^4}{x^?}$ _____

24. $\dfrac{1}{u^2} = \dfrac{u^?}{u^9}$ _____

Simplify, using appropriate properties of exponents.

25. $(5x^2)(2x^9)$ _____

26. $(2x^3)(3x^7)$ _____

27. $\dfrac{9x^6}{3x^4}$ _____

28. $\dfrac{4x^8}{2x^6}$ _____

29. $\dfrac{6m^5}{8m^7}$ _____

30. $\dfrac{4u^3}{2u^7}$ _____

31. $(xy)^{10}$ _____

32. $(cd)^{12}$ _____

33. $\left(\dfrac{m}{n}\right)^5$ _____

34. $\left(\dfrac{x}{y}\right)^6$ _____

B 35. $(4y^3)(3y)(y^6)$ _____

36. $(2x^2)(3x^3)(x^4)$ _____

37. $(5 \times 10^8)(7 \times 10^9)$ _____

38. $(2 \times 10^3)(3 \times 10^{12})$ _____

39. $(10^7)^2$ _____

40. $(10^4)^5$ _____

41. $(x^3)^2$ _____ **42.** $(y^4)^5$ _____ **43.** $(m^2n^5)^3$ _____ **44.** $(x^2y^3)^4$ _____

45. $\left(\dfrac{c^2}{d^5}\right)^3$ _____ **46.** $\left(\dfrac{a^3}{b^2}\right)^4$ _____ **47.** $\dfrac{9u^8v^6}{3u^4v^8}$ _____ **48.** $\dfrac{2x^3y^8}{6x^7y^2}$ _____

49. $(2s^2t^4)^4$ _____ **50.** $(3a^3b^2)^3$ _____ **51.** $6(xy^3)^5$ _____ **52.** $2(x^2y)^4$ _____

53. $\left(\dfrac{mn^3}{p^2q}\right)^4$ _____ **54.** $\left(\dfrac{x^2y}{2w^2}\right)^3$ _____ **55.** $\dfrac{(4u^3v)^3}{(2uv^2)^6}$ _____ **56.** $\dfrac{(2xy^3)^2}{(4x^2y)^3}$ _____

57. $\dfrac{(9x^3)^2}{(3x)^2}$ _____ **58.** $\dfrac{-(2x^2)^3}{(2^2x)^4}$ _____ **59.** $\dfrac{-x^2}{(-x)^2}$ _____ **60.** $\dfrac{-2^2}{(-2)^2}$ _____

C 61. $\dfrac{3(x+y)^3(x-y)^4}{(x-y)^26(x+y)^5}$ _____ **62.** $\dfrac{10(u-v+w)^8}{5(u-v+w)^{11}}$ _____

The Check Exercise for this section is on page 59.

1.7 Chapter Review

Important terms and symbols

1.1 SETS   set, element, member, empty set, null set, listing representation, rule representation, subset, equal sets, union, intersection, disjoint, Venn diagram, $a \in A$, $a \notin A$, $\varnothing$, $A \subset B$, $A = B$, $A \cup B$, $A \cap B$

1.2 ALGEBRAIC EXPRESSIONS; EQUALITY   variable, constant, algebraic expression, term, factor, order of operations, equal sign, algebraic equation, symmetric property, transitive property, substitution principle, $=$, $\neq$

1.3 REAL NUMBERS AND BASIC PROPERTIES   natural numbers, integers, rational numbers, irrational numbers, real numbers, real (number) line, equivalent algebraic expressions, commutative property, associative property, distributive property, $a + b = b + a$, $ab = ba$, $(a + b) + c = a + (b + c)$, $(ab)c = a(bc)$, $a(b + c) = ab + ac$

1.4 ADDITION AND SUBTRACTION OF SIGNED NUMBERS   negative of, absolute value of, addition of signed numbers, subtraction of signed numbers, $-x$, $|x|$

1.5 MULTIPLICATION AND DIVISION OF SIGNED NUMBERS multiplication of signed numbers, division of signed numbers

Zero cannot be used as a divisor—ever!

1.6 NATURAL NUMBER EXPONENTS natural number exponent, properties of exponents, $a^n = a \cdot a \cdots a$ (n factors of a), $a^m a^n = a^{m+n}$, $(a^n)^m = a^{mn}$, $(ab)^m = a^m b^m$, $(a/b)^m = a^m/b^m$,

$$\frac{a^m}{a^n} = \begin{cases} a^{m-n} & \text{if } m \text{ is larger than } n \\ 1 & \text{if } m = n \\ \dfrac{1}{a^{n-m}} & \text{if } n \text{ is larger than } m \end{cases}$$

DIAGNOSTIC (REVIEW) EXERCISE 1.7 _____

Work through all the problems in this chapter review and check answers in the back of the book. (Answers to all problems are there, and following each answer is a number in italics indicating the section in which that type of problem is discussed.) Where weaknesses show up, review appropriate sections in the text. When you are satisfied that you know the material, take the practice test following this review.

A *Problems 1 and 2 refer to the following sets:*

$A = \{1, 2, 3, 4, 5, 6\}$
$B = \{1, 3, 5\}$
$C = \{3, 4, 5\}$
$D = \{2, 4, 6\}$

1. Indicate true (T) or false (F):

    (A) $3 \notin A$ _____

    (B) $3 \notin D$ _____

    (C) $B \subset A$ _____

    (D) $C \subset B$ _____

2. Find each of the following:

    (A) $B \cup C$ _____

    (B) $B \cap C$ _____

    (C) $B \cap D$ _____

    (D) $A \cap D$ _____

B *Evaluate.*

3. $3 \cdot 7 - 4$ _____

4. $7 + 2 \cdot 3$ _____

5. $(-8) + 3$ _____

6. $(-9) + (-4)$ _____

7. $(-3) - (-9)$ _____

8. $4 - 7$ _____

9. $0 - (-3)$ _____

10. $(-12) - 0$ _____

11. $(-7)(-4)$ _____

12. $3(-6)$ _____

13. $(-16)/4$ _____

14. $(-12)/(-2)$ _____

15. $(-6)/0$ _____

16. $0/(-3)$ _____

17. $10 - 3(6 - 4)$ _____

18. $(-8) - (-2)(-3)$ _____

19. $(-9) - [(-12)/3]$ _____

20. $(4 - 8) + 4(-2)$ _____

21. $|-8|$ _____

22. $-(-5)$ _____

23. $-[-(-3)]$ _____

24. $-|-(-2)|$ _____

25. $-(|-3| + |-2|)$ _____

26. $-(|8| - |3|)$ _____

Remove parentheses and simplify, using commutative and associative properties mentally.

27. $7 + (x + 3)$ _____

28. $(3x)5$ _____

29. $(2x)(4y)$ _____

30. $(y + 7) + (x + 2) + (z + 3)$ _____

Simplify, using natural number exponents only.

31. $5^6 \cdot 5^8$ _____

32. $x^4 x^8$ _____

33. $(x^4)^3$ _____

34. $(2xy)^3$ _____

35. $\left(\dfrac{c}{d}\right)^4$ _____

36. $\left(\dfrac{x^2}{y^3}\right)^2$ _____

37. $\dfrac{u^8}{u^3}$ _____

38. $\dfrac{y}{y^5}$ _____

Evaluate.

39. $2[9 - 3(3 - 1)]$ _____

40. $6 - 2 - 3 - 4 + 5$ _____

41. $[(-3) - (-3)] - (-4)$ _____

42. $[-(-4)] + (-|-3|)$ _____

43. $\dfrac{-16}{2} - (-3)(4)$ _____

44. $(-2)(-4)(-3) - \dfrac{-36}{(-2)(9)}$ _____

45. $2\{9 - 2[(3 + 1) - (1 + 1)]\}$ _____

46. $(-3) - 2\{5 - 3[2 - 2(3 - 6)]\}$ _____

47. $3[14 - x(x + 1]$ for $x = 3$ _____

48. $-(-x)$ for $x = -2$ _____

49. $-(|x| - |w|)$ for $x = -2$ and $w = -10$ _____

50. $(x + y) - z$ for $x = 6, y = -8, z = 4$ _____

51. $\left(2x - \dfrac{z}{x}\right) - \dfrac{w}{x}$ for $w = -10, x = -2, z = 0$ _____

52. $\dfrac{(xyz + xz) - z}{z}$ for $x = -6, y = 0, z = -3$ _____

State the real number property (commutative +, commutative ·, associative +, associative ·, distributive) that justifies each statement.

53. $5 + (x + 3) = 5 + (3 + x)$ _____

54. $5 + (3 + x) = (5 + 3) + x$ _____

55. $5(x3) = 5(3x)$ _____

56. $5(3x) = (5 \cdot 3)x$ _____

57. $5(x + y) = 5x + 5y$ _____

58. $(6x + 7x) = (6 + 7)x$ _____

Simplify, using natural number exponents only.

59. $(3x^3 y^4)^2$ _____

60. $(4u^4 v)(3u^2 v^3)$ _____

61. $\dfrac{6x^8 y}{3x^7 y^5}$ _____

62. $\dfrac{(2x^2y)^3}{(4xy^4)^2}$ _____

63. $\left(\dfrac{2x^2}{y^5}\right)^3$ _____

64. $(3x^3)(2x)(x^7)$ _____

65. $\dfrac{-4^2}{(-2)^2}$ _____

66. $(3 \times 10^4)(4 \times 10^3)$ _____

In Problems 67 to 69 translate into an equation, using only x as a variable.

67. 3 times a certain number is 8 less than twice the number. _____

68. 4 more than a certain number is twice the number that is 3 less than the certain number. _____

69. If the length of a rectangle is 5 meters longer than its width x and the perimeter is 43 meters, write an algebraic equation relating the sides and the perimeter.

_____

In Problems 70 to 72 write the elements in each set, using the listing method. If the set is empty, write $\varnothing$.

70. $\{x \mid x^2 = 100\}$ _____

71. $\{x \mid x + 8 = x - 3\}$ _____

72. The set of all integers from -2 to 2, inclusive. _____

73. Evaluate $uv - 3\{x - 2[(x + y) - (x - y)] + u\}$ for $u = -2, v = 3, x = 2,$ and $y = -3$. _____

C 74. Evaluate $\dfrac{5w}{x - 7} - \dfrac{wx - 4}{x - w}$ for $w = -4$ and $x = 2$. _____

Simplify Problems 75 and 76, using natural number exponents only.

75. $\dfrac{12(x + y - z)^5}{3(x + y - z)^2}$ _____

76. $\dfrac{(x + y)^4(x - y)}{(x + y)^3(x - y)^5}$ _____

77. Translate into an algebraic equation, using only one variable x: Subtracting 3 from a given number and multiplying the difference by 2 yields a number 5 less than 3 times the given number.

_____

PRACTICE TEST CHAPTER 1 _____

Take this as if it were a graded test. Allow yourself up to 50 minutes. Work the problems without looking back in the chapter. Correct your work, using the answers (keyed to appropriate sections) in the back of the book.

Evaluate Problems 1 to 7.

1. $-12 - 3 + 7 - 4 + 2$ _____

2. $9 - 3(6 - 2 \cdot 2)$ _____

3. $3\{4 + 2[8 - (5 - 1)]\}$ ___

4. $(-4)(-8) - \dfrac{-14}{2}$ _____

5. $-(|-3| - |8|)$ _____

6. $\left(xy - \dfrac{z}{x}\right) - \dfrac{x}{y}$ for $x = -10, y = 2,$ and $z = 0$. _____

7. $\dfrac{x-y}{x+3y} - \dfrac{x}{y}$ for $x = -8$ and $y = 2$. _____

8. One of the following is false. Indicate which one. _____
   (A) $\dfrac{0}{-9} = 0$ **(B)** $\dfrac{-9}{0} = 0$ **(C)** $\dfrac{5 \cdot 0}{5 + 0} = 0$ **(D)** $(-8)(0)(3) = 0$

9. One of the following is false. Indicate which one. _____
   (A) $5 \notin \{2, 4, 6\}$ **(B)** $\{2, 3\} \cup \{3, 4\} = \{2, 3, 4\}$
   (C) $\{3, 4, 5\} \subset \{3, 4\}$ **(D)** $\{2, 3\} \cap \{1, 4\} = \varnothing$

10. One of the following is false. Indicate which one. _____
    (A) $3 + (x + 5) = 3 + (5 + x)$ illustrates the use of the associative property for addition.
    (B) $8(u + v) = 8u + 8v$ illustrates the use of the distributive property.
    (C) $3 + (x + 5) = 3 + (5 + x)$ illustrates the use of the commutative property for addition.
    (D) $(yz)(wx) = (wx)(yz)$ illustrates the use of the commutative property for multiplication.

11. Take out all factors common to all terms on the right: $(5x + x + 4x)$ _____

12. Write "The set of all integers between -3 and 1, inclusive," using the listing method. _____

13. Write "The set of all x such that $x^2 = 9$," using the rule method. _____

In Problems 14 and 15 translate into an equation using x as the only variable.

14. 5 more than a certain number is 6 less than twice the number. _____

15. The perimeter of a rectangle is 74 centimeters (cm). If the length is 2 cm less than twice the width, write an equation, using x as the width that relates the perimeter with the length and width.

    _____

Simplify Problems 16 to 20, using natural number exponents only.

16. $(4m^2 n^4)(3m^5 n)$ _____

17. $\left(\dfrac{4x^4}{2y^2}\right)^2$ _____

18. $\dfrac{(4u^2 v)^2}{(2u^3 v^2)^3}$ _____

19. $\dfrac{-6^2}{(-3)^2}$ _____

20. $\dfrac{(u + 2v)^3 (u - v)^2}{(u + 2v)(u - v)^5}$ _____

CHECK EXERCISE 1.1

NAME _____

CLASS _____

SCORE _____

Work the following problems without looking at any text examples. Show your work in the space provided. Write the letter that best indicates your answer in the answer column.

1. One of the statements on the right is false. Indicate which one.

   (A) $\{-2, 7, 3\} = \{3, 7, -2\}$
   (B) $-2 \in \{-2, 7, 3\}$
   (C) $3 \notin \{-2, 7, 3\}$
   (D) $5 \notin \{-2, 7, 3\}$

2. One of the statements on the right is false. Indicate which one.

   (A) $\{2, 5\} \subset \{1, 2, 3, 5\}$
   (B) $\{2, 3, 4\} \subset \{1, 2, 3, 5\}$
   (C) $\varnothing \subset \{1, 2, 3, 5\}$
   (D) $\{1, 2, 3, 5,\} \subset \{1, 2, 3, 5\}$

3. Let M be the set of all numbers x such that $x^2 = 1$. Write M, using the listing method.

   (A) $M = \{x \mid x^2 = 1\}$
   (B) $M = \{1\}$
   (C) $M = \{-1\}$
   (D) $M = \{-1, 1\}$

4. Let M be the set of all numbers x such that $x^2 = 1$. Write M, using the rule method.

   (A) $M = \{x \mid x^2 = 1\}$
   (B) $M = \{1\}$
   (C) $M = \{-1\}$
   (D) $M = \{-1, 1\}$

5. $\{3, 5, 7\} \cup \{5, 7, 9\} =$

   (A) $\{5, 7\}$
   (B) $\{3, 5, 7, 9\}$
   (C) $\{5\}$
   (D) $\varnothing$

6. $\{3, 5, 7\} \cap \{5, 7, 9\} =$

(A) $\{5, 7\}$
(B) $\{3, 5, 7, 9\}$
(C) $\{5\}$
(D) $\varnothing$

7. $\{6, 8, 10\} \cap \{7, 9\} =$

(A) $\{6, 7, 8, 9, 10\}$
(B) $\{7, 8, 9\}$
(C) $\{7, 9\}$
(D) $\varnothing$

8. If $A = \{3, 5, 7\}$ and $B = \{5, 7, 9\}$, then $\{x \mid x \in A \text{ or } x \in B\} =$

(A) $\{5, 7\}$
(B) $\{3, 5, 7, 9\}$
(C) $\{5\}$
(D) $\varnothing$

9. If $A = \{3, 5, 7\}$ and $B = \{5, 7, 9\}$, then $\{x \mid x \in A \text{ and } x \in B\} =$

(A) $\{5, 7\}$
(B) $\{3, 5, 7, 9\}$
(C) $\{5\}$
(D) $\varnothing$

10. How many subsets does the set $\{a, b, c\}$ have? List and count them.

(A) 6
(B) 7
(C) 3
(D) 8

CHECK EXERCISE 1.2

Work the following problems without looking at any text examples. Show your work in the space provided. Write the letter that best indicates your answer in the answer column.

ANSWER COLUMN

1. _____
2. _____
3. _____
4. _____
5. _____
6. _____
7. _____
8. _____
9. _____
10. _____

Evaluate Problems 1 to 4.

1. $8 + 2(5 - 2) =$

    (A) 30
    (B) 14
    (C) 16
    (D) None of these

2. $10 - 3(2 + 1) =$

    (A) 21
    (B) 3
    (C) 15
    (D) None of these

3. $12 - 2[9 - 2(3 - 1)] =$

    (A) 140
    (B) 2
    (C) 50
    (D) None of these

4. $2\{1 + 2[7 - 2(3 - 1)] + 2\} =$

    (A) 18
    (B) 64
    (C) 9
    (D) None of these

Evaluate Problems 5 to 7 for $x = 2$ and $y = 8$.

5. $y - 2(x + 1) =$

    (A) 18
    (B) 5
    (C) 13
    (D) None of these

6. $x + 2(y - x) =$

(A) 14
(B) 24
(C) 16
(D) None of these

7. $2\{x + 3[(y - x) - 2x]\} =$

(A) 80
(B) 20
(C) 16
(D) None of these

Translate Problems 8 and 9 into algebraic equations, using x as the only variable.

8. 7 times a number is 5 less than 3 times the number.

(A) $7 + x = 3 + x - 5$
(B) $7x = 3x - 5$
(C) $7x = 5 - 3x$
(D) None of these

9. 3 subtracted from a certain number is twice the number, which is 2 less than the certain number.

(A) $x - 3 = 2(x - 2)$
(B) $3 - x = 2(2 - x)$
(C) $3 - x = 2(x - 2)$
(D) None of these

10. In a rectangle with perimeter 100 centimeters (cm) the length is 5 cm less than twice the width. Write an equation relating the perimeter with the length and width, using x as the only variable. ($P = 2a + 2b$)

(A) $100 = 2(2x - 5) + 2x$
(B) $100 = 2(5 - 2x) + 2x$
(C) $100 = (2x - 5) + x$
(D) None of these

CHECK EXERCISE 1.3

Work the following problems without looking at any text examples. Show your work in the space provided. Write the letter that best indicates your answer in the answer column.

In Problems 1 to 6 state the justifying real number property for each statement.

1. $5 + 7y = 7y + 5$

    (A) Distributive
    (B) Commutative mult.
    (C) Associative add.
    (D) None of these

2. $m(x + y) = mx + my$

    (A) Associative mult.
    (B) Associative add.
    (C) Distributive
    (D) None of these

3. $5 + (A + 3) = 5 + (3 + A)$

    (A) Associative add.
    (B) Commutative add.
    (C) Distributive
    (D) None of these

4. $(x + 2)(x + 3) = (x + 3)(x + 2)$

    (A) Commutative mult.
    (B) Commutative add.
    (C) Associative add.
    (D) None of these

5. $7m + m + 3m = (7 + 1 + 3)m$

    (A) Associative add.
    (B) Associative mult.
    (C) Distributive
    (D) None of these

53

6. $5x + 10y + 15z = 5(x + 2y + 3z)$

(A) Distributive
(B) Associative add.
(C) Associative mult.
(D) None of these

In Problems 7 to 9 remove parentheses and simplify mentally.

7. $(5u + 3) + (2v + 7) =$

(A) $7uv + 10$
(B) $10uv + 10$
(C) $5u + 2v + 10$
(D) None of these

8. $(a + 3) + (b + 2) + (c + 1) =$

(A) $a + b + c + 6$
(B) $abc + 6$
(C) $3abc + 6$
(D) None of these

9. $(4u)(3v)(2w) =$

(A) $14uvw$
(B) $20uvw$
(C) $18uvw$
(D) None of these

10. Each of the statements on the right is true for all real numbers a, b, and c, except for one. Which one?

(A) $a + b = b + a$
(B) $(ab)c = a(bc)$
(C) $a - b = b - a$
(D) $a(b + c) = ab + ac$

CHECK EXERCISE 1.4

NAME

CLASS

SCORE

*Work the following problems without looking at any text examples. Show
your work in the space provided. Write the letter that best indicates your
answer in the answer column.*

Evaluate Problems 1 to 8.

ANSWER
COLUMN

1. _____

2. _____

3. _____

4. _____

5. _____

6. _____

7. _____

8. _____

9. _____

10. _____

1. $-[-(-4)] =$

   (A) -4
   (B) 4
   (C) 0
   (D) None of these

2. $8 + (-7) =$

   (A) -56
   (B) -1
   (C) 2
   (D) None of these

3. $-|-8| =$

   (A) -8
   (B) 8
   (C) 0
   (D) None of these

4. $(-12) - (-9) =$

   (A) -21
   (B) 3
   (C) 21
   (D) None of these

5. $6 - 10 =$

   (A) -60
   (B) 4
   (C) -4
   (D) None of these

6. $-7 - 4 + 6 - 8 + 9 =$

    (A)   4
    (B) -5
    (C) -4
    (D)   None of these

7. $-(|-9| - |-12|) =$

    (A)   3
    (B) -21
    (C) -3
    (D)   None of these

8. $[(-8) - 3] - [3 - (-9)] =$

    (A) -1
    (B) -23
    (C)   1
    (D)   None of these

9. Replace the question mark with an appropriate number:
$(-7) - (?) = 5$

    (A)   2
    (B) -2
    (C)   12
    (D)   None of these

10. Evaluate for $x = -9$ and $y = 2$: $|x - y| - |x| =$

    (A)   2
    (B) -20
    (C)   20
    (D)   None of these

CHECK EXERCISE 1.5

Work the following problems without looking at any text examples. Show your work in the space provided. Write the letter that best indicates your answer in the answer column.

ANSWER
COLUMN

1. _____
2. _____
3. _____
4. _____
5. _____
6. _____
7. _____
8. _____
9. _____
10. _____

1. $(-3)(-2) - 8 =$

        (A) -14
        (B) $+2$
        (C) -2
        (D) None of these

2. $(-5)(0)(-3)(2) =$

        (A) 30
        (B) -30
        (C) -6
        (D) None of these

3. $\dfrac{-45}{9} - (-3)(3) =$

        (A) 4
        (B) -4
        (C) -9
        (D) None of these

4. $\dfrac{8 - 5}{0} =$

        (A) 0
        (B) 3
        (C) Not defined
        (D) None of these

5. $\dfrac{0}{(-2)(-5)} =$

        (A) Not defined
        (B) 0
        (C) $\frac{1}{10}$
        (D) None of these

6. $\dfrac{27}{-3} - \dfrac{-8}{-4} =$

(A) -11
(B) -7
(C) 11
(D) None of these

7. $[(-8) - (-2)][9 - (-1)] =$

(A) -80
(B) -60
(C) -100
(D) None of these

8. $\dfrac{(-4) - (+8)}{(-2)(3)} - \dfrac{0}{-12} =$

(A) Not defined
(B) 14
(C) 10
(D) None of these

9. Evaluate for $u = -12$, $v = 0$, and $w = 4$:

$$uvw - \dfrac{v}{u} - |u|\,|w| =$$

(A) Not defined
(B) -48
(C) 0
(D) None of these

10. Evaluate for $x = -30$, $y = 4$, and $z = -3$:

$$\dfrac{2x}{yz} - \dfrac{x - 6}{z - 3} =$$

(A) -15
(B) 25
(C) -1
(D) None of these

CHECK EXERCISE 1.6

NAME _____

CLASS _____

SCORE _____

Work the following problems without looking at any text examples. Show your work in the space provided. Write the letter that best indicates your answer in the answer column.

ANSWER
COLUMN

1. _____
2. _____
3. _____
4. _____
5. _____
6. _____
7. _____
8. _____
9. _____
10. _____

Simplify, using appropriate properties of exponents.

1. $2^3 \cdot 2^2 =$

    (A) 4^6

    (B) 2^6

    (C) 4^5

    (D)   None of these

2. $(3x^4)(2x^5) =$

    (A) $5x^9$

    (B) $6x^{20}$

    (C) $6x^9$

    (D)   None of these

3. $(2u^3)(4u)(u^5) =$

    (A) $8u^9$

    (B) $8u^8$

    (C) $8u^{15}$

    (D)   None of these

4. $\dfrac{12x^6}{8x^4} =$

    (A) $\dfrac{3x^{10}}{2}$

    (B) $\dfrac{3x^2}{2}$

    (C) $\dfrac{3}{2x^2}$

5. $\left(\dfrac{m^4}{n^3}\right)^3 =$

    (A) $\dfrac{m^{12}}{n^{18}}$

    (B) $\dfrac{m^7}{n^9}$

    (C) $\dfrac{3m^4}{3n^6}$

    (D)   None of these

6. $(2 \times 10^6)(4 \times 10^7) =$

    (A) 8×10^{42}
    (B) 8×10^{13}
    (C) 6×10^{42}
    (D) None of these

7. $\left(\dfrac{x^2y}{u^3v^2}\right)^3 =$

    (A) $\dfrac{x^6y^3}{u^9v^6}$

    (B) $\dfrac{x^6y}{u^9v^6}$

    (C) $\dfrac{x^5y^4}{u^6v^5}$

    (D) None of these

8. $\dfrac{(2x^3y)^2}{(4xy^2)^2} =$

    (A) $\dfrac{x^4}{2y^2}$

    (B) $\dfrac{x^4}{4y^2}$

    (C) $\dfrac{x^7}{4y^2}$

    (D) None of these

9. $\dfrac{-3^2}{(-3)^2}$

    (A) 1
    (B) 3
    (C) -1
    (D) None of these

10. $\dfrac{(a+b)^7(a-b)^4}{(a+b)^9(a-b)} =$

    (A) $\dfrac{(a-b)^4}{(a+b)^2}$

    (B) $\dfrac{(a+b)^2}{(a-b)^3}$

    (C) $\dfrac{(a-b)^3}{(a+b)^2}$

    (D) None of these

Chapter Two   POLYNOMIALS

CONTENTS

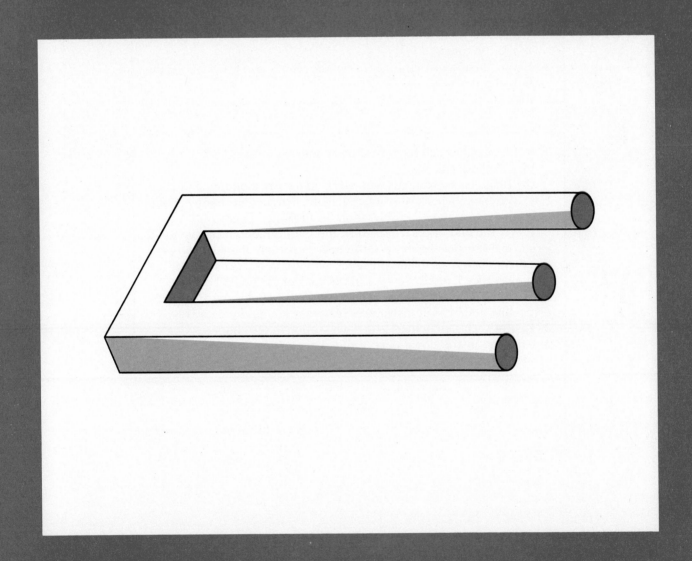

Could you build this three-pronged "tuning fork"?

INSTRUCTIONS FOR STUDENTS IN A
SELF-PACED CLASS OR LAB

yes ← **HAVE YOU HAD INTERMEDIATE ALGEBRA BEFORE THIS COURSE?** → no

1. Work Diagnostic (Review) Exercise 2.9 on page 103. Check answers in back of book; then work through text sections corresponding to problems missed. (Section numbers are in italics following each answer.)
2. When finished with step 1, take Practice Test Chapter 2 on page 105 as a final check of your understanding of the chapter. Check answers in the back of the book; then review sections where weakness still prevails. (Corresponding section numbers are in italics following each answer.)
3. When you think you are ready, ask your instructor for a graded test for Chapter 2.
4. If your instructor approves, after the test is corrected, go to the next chapter.

1. Work through each section in the chapter as follows:
   (a) Read discussion.
   (b) Read each example and work the corresponding matched problem. Check your solution to the matched problem in Solutions to Matched Problems on the indicated page.
   (c) At the end of a section work the odd-numbered problems in the Practice Exercise and check answers; then work even-numbered problems in areas of weakness. (Answers to *all* Practice Exercise sets are in the back of the book.)
   (d) Work Check Exercise as instructed. Tear out and turn in as directed by your instructor. (Answers are not in the text.)
2. Repeat each step in item 1 for each section in the chapter.
3. After the instructional part of the chapter is completed, proceed with steps 1 to 4 in the box above.

2.1 DEFINITIONS

•Polynomials
••Degree

In this chapter we consider basic operations (addition, subtraction, multiplication, division, and factoring) on polynomials. Polynomial forms are encountered with great frequency throughout mathematics; hence, these operations must be mastered early.

•Polynomials

Let us list some polynomial and nonpolynomial expressions and see if you can determine what the polynomial forms have in common that are not shared by the nonpolynomial forms.

POLYNOMIALS

$$3x - 1 \qquad x \qquad 2x^2 - 3x + 2 \qquad 5$$

$$x^3 - 3x^2y - 4y^2 \qquad 0 \qquad x^2 - \tfrac{2}{3}xy + 2y^2$$

NONPOLYNOMIALS

$$\frac{2x + 1}{3x^2 - 5x + 7} \qquad 3^x \qquad x^3 - 2\sqrt{x} + \frac{1}{x^3} \qquad |2x^3 - 5|$$

We see that **polynomials** (in one and two variables) can be constructed using terms of the form ax^n, by^m, cx^py^q, and d (where $a, b, c,$ and d are real numbers and $p, q, m,$ and n are positive integers) and the operations of addition and subtraction. It follows that in a polynomial a variable cannot appear in a denominator, as an exponent, within a radical sign, or within absolute-value bars.

••Degree

It is convenient to identify certain types of polynomials for more efficient study. The concept of degree is often used for this purpose. If a term in a polynomial has only one variable as a factor, then the **degree of that term** is the power of the variable. If two or more variables are present in a term as factors, then the **degree of the term** is the sum of the powers of the variables. The **degree of a polynomial** is the degree of the nonzero term with the highest degree in the polynomial. (Recall that we add or subtract *terms* and multiply *factors*.) Any nonzero real constant is defined to be a **polynomial of degree 0.** Thus, 6 is a polynomial of degree 0. The number 0 is also a polynomial, but it is not assigned a degree.

EXAMPLE 1
(A) $4x^3$ is of degree 3.
(B) $3x^3y^2$ is of degree 5.
(C) In $3x^5 - 2x^4 + x^2 - 3$, the highest-degree term is the first, with degree 5; thus, the degree of the polynomial is 5.
(D) The degree of each of the first three terms in $x^2 - 2xy + y^2 + 2x - 3y + 2$ is 2; the fourth and fifth terms are each of degree 1; thus, this is a second-degree polynomial.

Work Problem 1 and check solution in Solutions to Matched Problems on page 64.

PROBLEM 1   Fill in the blanks:

(A) $7x^5$ is of degree _____.

(B) $3x^3y^4$ is of degree _____.

(C) $5x^3 - 2x^2 + x - 9$ is of degree _____.

(D) $x^2 - 3xy + y^2$ is of degree _____.

(E) $2x^2 - 5xy - y^2 - 3x + 2y + 6$ is of degree _____.

We also call a one-term polynomial a **monomial,** a two-termed polynomial a **binomial,** and a three-termed polynomial a **trinomial.**

$4x^3 - 3x + 7$	$5x - 2y$	$6x^4y^3$	7
trinomial	binomial	monomial	monomial
degree 3	degree 1	degree 7	degree 0

SOLUTIONS TO MATCHED PROBLEMS _____

1.   (A) $7x^5$ is of degree 5.   (B) $3x^3y^4$ is of degree 7.   (C) $5x^3 - 2x^2 + x - 9$ is of degree 3.
(D) $x^2 - 3xy + y^2$ is of degree 2.   (E) $2x^2 - 5xy - y^2 - 3x + 2y + 6$ is of degree 2.

2.2 ADDITION AND SUBTRACTION

•Combining like terms
••Removing symbols of grouping
•••Addition and subtraction of polynomials

•Combining like terms

A constant present as a factor in a term is called the **numerical coefficient, or, simply, the coefficient** of the term. If no constant appears in the term, then the coefficient is understood to be 1. The coefficient of a term in a polynomial includes the sign that precedes it. Be careful not to confuse "coefficient" with "exponent."

Given the polynomial.

$3x^4 - 2x^3 + x^2 - x + 3$

the coefficient of the first term is 3; of the second term, -2; of the third term, 1; and of the fourth term, -1. The exponent of the variable in the first term is 4; in the second term, 3; in the third term, 2; and in the fourth term, 1.

Two terms are called **like terms** if they are exactly alike, except the numerical coefficients may or may not be the same. Like terms must have exactly the same variable factors to the same powers. If an algebraic expression contains two or more like terms, these terms can always be combined into a single term. The distributive property is the principal tool behind this process.

EXAMPLE 2

(A) $3x + 7x = (3 + 7)x = 10x$

(B) $6m - 9m = (6 - 9)m = -3m$

(C) $5x^3y - 2xy - x^3y - 2x^3y = 5x^3y - x^3y - 2x^3y - 2xy$

$$= (5 - 1 - 2)x^3y - 2xy$$

$$= 2x^3y - 2xy$$

Work Problem 2 and check solution in Solutions to Matched Problems on page 69.

PROBLEM 2 Combine like terms, as in Example 2:

(A) $5y + 4y =$

(B) $2u - 6u =$

(C) $6mn^2 - m^2n - 3mn^2 - mn^2 =$

In Example 2 free use was made of the commutative, associative, and distributive properties of real numbers discussed in Section 1.3. In practice, most of the steps shown in the "dotted boxes" are done mentally. The process is quickly mechanized as follows:

> *Mechanics of Combining Like Terms*
>
> Like terms are combined by adding their numerical coefficients.

EXAMPLE 3

Combine like terms mentally.

(A) $3x - 5y + 6x + 2y = 9x - 3y$

(B) $x^3y^2 - 2x^2y^3 + 5x^2y^2 - 4x^2y^3 - x^3y^2 - 5x^2y^2 = -6x^2y^3$

Work Problem 3 and check solution in Solutions to Matched Problems on page 69.

PROBLEM 3 Combine like terms mentally.

(A) $7m + 8n - 5m - 10n =$

(B) $2u^4v^2 - 3uv^3 - u^4v^2 + 6u^4v^2 + 2uv^3 - 6u^4v^2 =$

••Removing symbols of grouping

How can we simplify expressions such as

$2(3x - 5y) - 2(x + 3y)$

We use distributive properties discussed in Section 1.3 to remove the parentheses:

$6x - 10y - 2x - 6y$

We then combine like terms to obtain

$4x - 16y$

EXAMPLE 4

Remove parentheses and combine like terms.

(A) $2(x^2 - 3x + 1) - 3(2x^2 - x + 2) = 2x^2 - 6x + 2 - 6x^2 + 3x - 6$

$$= -4x^2 - 3x - 4$$

(B) $2(3x^2 - 2x + 5) + (x^2 + 3x - 7)$

$= 2(3x^2 - 2x + 5) + 1(x^2 + 3x - 7)$
        think

$= 6x^2 - 4x + 10 + x^2 + 3x - 7$

$= 7x^2 - x + 3$

(C) $(x^3 - 2x - 6) - (2x^3 - x^2 + 2x - 3)$

$= 1(x^3 - 2x - 6) - 1(2x^3 - x^2 + 2x - 3)$
        think

$= x^3 - 2x - 6 - 2x^3 + x^2 - 2x + 3$

$= -x^3 + x^2 - 4x - 3$

Work Problem 4 and check solution in Solutions to Matched Problems on page 69.

PROBLEM 4   Remove parentheses and combine like terms.

(A) $3(2y^2 - 4) - 2(y^2 - 5y + 1) =$

(B) $3(u^2 - 2v^2) + (u^2 + 5v^2) =$

(C) $(m^3 - 3m^2 + m - 1) - (2m^3 - m + 3) =$

In Example **4B** and **C** we observe that if parentheses are preceded by a + sign, we can remove the parentheses without changing any sign within the parentheses. However, if parentheses are preceded by a − sign, then every sign within the parentheses must be changed when the parentheses are removed. The reverse of these statements also holds, and this is useful when inserting parentheses in certain algebraic expressions. We will be interested in grouping certain terms, using parentheses, when we consider factoring later in this chapter.

EXAMPLE 5

Insert parentheses around the last two terms in the following algebraic expressions.

(A) $5x + 2y - 3 = 5x + (2y - 3)$
                                       Parentheses preceded by a + sign require no sign change within the parentheses.

(B) $7u - 5v + w = 7u - (5v - w)$
                                        Parentheses preceded by a − sign requires a sign change for all terms placed within the parentheses.

Work Problem 5 and check solution in Solutions to Matched Problems on page 69.

PROBLEM 5   Replace each question mark with an appropriate algebraic expression.

      (A) $7u + 2v - w = 7u + (\ ?\ )$

      (B) $4r - s - 2t = 4r - (\ ?\ )$

EXAMPLE 6

Remove symbols of grouping and combine like terms.

(A) $3x - 2[x - 4(x + 3)] = 3x - 2[x - 4x - 12]$
                                 $= 3x - 2[-3x - 12]$
                                 $= 3x + 6x + 24$
                                 $= 9x + 24$

Work from the inside out, combining like terms where possible.

(B) $2\{t - [2t - (t + 3)] + 4\} = 2\{t - [2t - t - 3] + 4\}$
                                          $= 2\{t - [t - 3] + 4\}$
                                          $= 2\{t - t + 3 + 4\}$
                                          $= 2\{7\} = 14$

Work from the inside out, combining like terms where possible.

Work Problem 6 and check solution in Solutions to Matched Problems on page 69.

PROBLEM 6   Remove symbols of grouping and combine like terms.

      (A) $7u - [2u - 3(u + 2)] =$

(B) $3\{5 - 2[x - (3x - 1)] - 1\} =$

•••Addition and subtraction of polynomials

Addition and subtraction of polynomials can be thought of in terms of removing parentheses and combining like terms, as illustrated in Example 4 above. Horizontal and vertical arrangements are illustrated in the next two examples. You should be able to work either way, letting the situation dictate the choice.

EXAMPLE 7
Add:

$$x^4 - 3x^3 + x^2 \qquad -x^3 - 2x^2 + 3x \qquad \text{and} \qquad 3x^2 - 4x - 5$$

Solution
Add horizontally:

$$(x^4 - 3x^3 + x^2) + (-x^3 - 2x^2 + 3x) + (3x^2 - 4x - 5)$$
$$= x^4 - 3x^3 + x^2 - x^3 - 2x^2 + 3x + 3x^2 - 4x - 5$$
$$= x^4 - 4x^3 + 2x^2 - x - 5$$

or add vertically by lining up like terms and adding their coefficients:

$$
\begin{array}{r}
x^4 - 3x^3 + x^2 \\
- x^3 - 2x^2 + 3x \\
3x^2 - 4x - 5 \\
\hline
x^4 - 4x^3 + 2x^2 - x - 5
\end{array}
$$

Work Problem 7 and check solution in Solutions to Matched Problems on page 69.

PROBLEM 7   Add horizontally and vertically.

$$3x^4 - 2x^3 - 4x^2 \qquad x^3 - 2x^2 - 5x \qquad \text{and} \qquad x^2 + 7x - 2$$

Solution

EXAMPLE 8

Subtract

$$4x^2 - 3x + 5 \qquad \text{from} \qquad x^2 - 8$$

Solution

$$(x^2 - 8) - (4x^2 - 3x + 5) \qquad \text{or}$$

$$= x^2 - 8 - 4x^2 + 3x - 5$$

$$= -3x^2 + 3x - 13$$

$$\begin{array}{r} x^2 - 8 \\ \underline{4x^2 - 3x + 5} \quad \leftarrow \text{Change signs and} \\ -3x^2 + 3x - 13 \text{add.} \end{array}$$

Work Problem 8 and check solution in Solutions to Matched Problems on page 69.

PROBLEM 8 Subtract

$$2x^2 - 5x + 4 \qquad \text{from} \qquad 5x^2 - 6$$

Solution

SOLUTIONS TO MATCHED PROBLEMS

2. **(A)** $5y + 4y = (5 + 4)y = 9y$ **(B)** $2u - 6u = (2 - 6)u = -4u$

(C) $6mn^2 - m^2n - 3mn^2 - mn^2 = 6mn^2 - 3mn^2 - mn^2 - m^2n$
$= (6 - 3 - 1)mn^2 - m^2n$
$= 2mn^2 - m^2n$

3. **(A)** $7m + 8n - 5m - 10n = 2m - 2n$
(B) $2u^4v^2 - 3uv^3 - u^4v^2 + 6u^4v^2 + 2uv^3 - 6u^4v^2 = u^4v^2 - uv^3$

4. **(A)** $3(2y^2 - 4) - 2(y^2 - 5y + 1) = 6y^2 - 12 - 2y^2 + 10y - 2 = 4y^2 + 10y - 14$
(B) $3(u^2 - 2v^2) + (u^2 + 5v^2) = 3u^2 - 6v^2 + u^2 + 5v^2 = 4u^2 - v^2$
(C) $(m^3 - 3m^2 + m - 1) - (2m^3 - m + 3) = m^3 - 3m^2 + m - 1$
$- 2m^3 + m - 3 = -m^3 - 3m^2 + 2m - 4$

5. **(A)** $7u + 2v - w = 7u + (2v - w)$ **(B)** $4r - s - 2t = 4r - (s + 2t)$

6. **(A)** $7u - [2u - 3(u + 2)] = 7u - (2u - 3u - 6) = 7u - (-u - 6) = 7u + u + 6 = 8u + 6$
(B) $3\{5 - 2[x - (3x - 1)] - 1\} = 3[5 - 2(x - 3x + 1) - 1]$
$= 3[5 - 2(-2x + 1) - 1] = 3(5 + 4x - 2 - 1) = 3(4x + 2) = 12x + 6$

7. Horizontal: $(3x^4 - 2x^3 - 4x^2) + (x^3 - 2x^2 - 5x) + (x^2 + 7x - 2)$
$= 3x^4 - 2x^3 - 4x^2 + x^3 - 2x^2 - 5x + x^2 + 7x - 2 = 3x^4 - x^3 - 5x^2 + 2x - 2$
Vertical: $\begin{array}{r} 3x^4 - 2x^3 - 4x^2 \\ x^3 - 2x^2 - 5x \\ x^2 + 7x - 2 \\ \hline 3x^4 - x^3 - 5x^2 + 2x - 2 \end{array}$

8. $(5x^2 - 6) - (2x^2 - 5x + 4) = 5x^2 - 6 - 2x^2 + 5x - 4 = 3x^2 + 5x - 10$

PRACTICE EXERCISE 2.2 ───

Work odd-numbered problems first, check answers, and then work even-numbered problems in areas of weakness. Answers to all problems are in the back of the book. Make every effort to work a problem yourself before you look at an answer.

A *Given the polynomial* $7x^4 - 3x^3 - x^2 + x - 3$, *indicate:*

1. The coefficient of the second term _____

2. The coefficient of the third term _____

3. The exponent of the variable in the second term _____

4. The exponent of the variable in the fourth term _____

5. The coefficient of the fourth term _____

6. The coefficient of the first term _____

Simplify by removing parentheses, if any, and combining like terms.

7. $9x + 8x$ _____   8. $7x + 3x$ _____   9. $9x - 8x$ _____

10. $7x - 3x$ _____   11. $5x + x + 2x$ _____   12. $3x + 4x + x$ _____

13. $4t - 8t - 9t$ _____   14. $2x - 5x + x$ _____   15. $4y + 3x + y$ _____

16. $2x + 3y + 5x$ _____

17. $5m + 3n - m - 9n$ _____   18. $2x + 8y - 7x - 5y$ _____

19. $3(u - 2v) + 2(3u + v)$ _____   20. $2(m + 3n) + 4(m - 2n)$ _____

21. $4(m - 3n) - 3(2m + 4n)$ _____   22. $2(x - y) - 3(3x - 2y)$ _____

23. $(2u - v) + (3u - 5v)$ _____   24. $(x + 3y) + (2x - 5y)$ _____

25. $(2u - v) - (3u - 5v)$ _____   26. $(x + 3y) - (2x - 5y)$ _____

Add.

27. $6x + 5$ and $3x - 8$ _____   28. $3x - 5$ and $2x + 3$ _____

29. $7x - 5, -x + 3$, and $-8x - 2$ _____   30. $2x + 3, -4x - 2$, and $7x - 4$ _____

31. $5x^2 + 2x - 7, 2x^2 + 3$, and $-3x - 8$ _____   32. $2x^2 - 3x + 1, 2x - 3$, and $4x^2 + 5$ _____

Subtract.

33. $3x - 8$ from $2x - 7$ _____   34. $4x - 9$ from $2x + 3$ _____

35. $2y^2 - 6y + 1$ from $y^2 - 6y - 1$ _____   36. $x^2 - 3x - 5$ from $2x^2 - 6x - 5$ _____

B *Simplify by removing symbols of grouping, if any, and combining like terms.*

37. $-x^2y + 3x^2y - 5x^2y$ _____

38. $-4r^3t^3 - 7r^3t^3 - 7r^3t^3 + 9r^3t^3$ _____

39. $y^3 + 4y^2 - 10 + 2y^3 - y + 7$ _____

40. $3x^2 - 2x + 5 - x^2 + 4x - 8$ _____

41. $a^2 - 3ab + b^2 + 2a^2 + 3ab - 2b^2$ _____

42. $2x^2y + 2xy^2 - 5xy + 2xy^2 - xy - 4x^2y$ _____

43. $x - 3y - 4(2x - 3y)$ _____

44. $a + b - 2(a - b)$ _____

45. $y - 2(-y) - 3x$ _____

46. $x - 3(x + 2y) + 5y$ _____

47. $-2(-3x + 1) - (2x + 4)$ _____

48. $-3(-t + 7) - (t - 1)$ _____

49. $2(x - 1) - 3(2x - 3) - (4x - 5)$ _____

50. $-2(y - 7) - 3(2y + 1) - (-5y + 7)$ _____

51. $4t - 3[4 - 2(t - 1)]$ _____

52. $3x - 2[2x - (x - 7)]$ _____

Replace each question mark with an appropriate algebraic expression.

53. $5 + m - 2n = 5 + (\ ?\ )$ _____

54. $2 + 3x - y = 2 + (\ ?\ )$ _____

55. $5 + m - 2n = 5 - (\ ?\ )$ _____

56. $2 + 3x - y = 2 - (\ ?\ )$ _____

57. $w^2 - x + y - z = w^2 + (\ ?\ )$ _____

58. $w^2 - x + y - z = w^2 - (\ ?\ )$ _____

Add.

59. $2x^4 - x^2 - 7, 3x^3 + 7x^2 + 2x,$ and $x^2 - 3x - 1$ _____

60. $3x^3 - 2x^2 + 5, 3x^2 - x - 3,$ and $2x + 4$ _____

Subtract.

61. $5x^3 - 3x + 1$ from $2x^3 + x^2 - 1$ _____

62. $3x^3 - 2x^2 - 5$ from $2x^3 - 3x + 2$ _____

63. Subtract the sum of the first two polynomials from the sum of the last two: $3m^3 - 2m + 5, 4m^2 - m,$ $3m^2 - 3m - 2,$ and $m^3 + m^2 + 2$

_____

64. Subtract the sum of the last two polynomials from the sum of the first two: $2x^2 - 4xy + y^2, 3xy - y^2,$ $x^2 - 2xy - y^2,$ and $-x^2 + 3xy - 2y^2$

_____

C *Remove symbols of grouping and combine like terms.*

65. $2t - 3\{t + 2[t - (t + 5)] + 1\}$ _____

66. $x - \{x - [x - (x - 1)]\}$ _____

67. $w - \{x - [z - (w - x) - z] - (x - w)\} + x$ _____

68. $3x^2 - 2\{x - x[x + 4(x - 3)] - 5\}$ _____

APPLICATIONS

69. The width of a rectangle is 5 meters less than its length. If x is the length of the rectangle, write an algebraic expression that represents the perimeter P of the rectangle and simplify the expression.

_____

70. Repeat Problem 69 if the length of the rectangle is 3 meters more than twice its width. _____

The Check Exercise for this section is on page 107.

2.3 MULTIPLICATION

•**Multiplication of monomials**
••**Multiplication of polynomials**
•••**Mental multiplication of first-degree polynomials**
••••**Squaring binomials**

•Multiplication of monomials

You have already had experience in multiplying monomials in Section 1.6 on exponents. We will review the process in the following example.

EXAMPLE 9

(A) $x^3x^5 = x^{3+5} = x^8 \qquad x^3x^5 \neq x^{3\cdot5}$

(B) $(3m^{12})(5m^{23}) = 3 \cdot 5m^{12+23} = 15m^{35}$

(C) $(-3x^3y^4)(2x^2y^3) = (-3)(2)x^{3+2}y^{4+3} = -6x^5y^7$

Work Problem 9 and check solution in Solutions to Matched Problems on page 76.

PROBLEM 9   Multiply:

(A) $y^4y^7 =$

(B) $(9x^4)(3x^2) =$

(C) $(4u^3v^2)(-3uv^3) =$

••Multiplication of polynomials

How do we multiply polynomials with more than one term? The distribtive property plays a central role in the process and leads directly to the following mechanical rule:

Mechanics of Multiplying Two Polynomials

To multiply two polynomials, multiply each term of one by each term of the other, then add like terms.

EXAMPLE 10

(A) $3x^2(2x^2 - 3x + 4) = 6x^4 - 9x^3 + 12x^2$

(B) $(2x - 3)(3x^2 - 2x + 3)$

$= 2x(3x^2 - 2x + 3) - 3(3x^2 - 2x + 3)$

$= 6x^3 - 4x^2 + 6x - 9x^2 + 6x - 9$

$= 6x^3 - 13x^2 + 12x - 9$

or

$$\begin{array}{r} 3x^2 - 2x + 3 \\ 2x - 3 \\ \hline 6x^3 - 4x^2 + 6x \\ -9x^2 + 6x - 9 \\ \hline 6x^3 - 13x^2 + 12x - 9 \end{array}$$

Note that either way, each term in $3x^2 - 2x + 3$ is multiplied by each term in $2x - 3$. In the vertical arrangement, by multiplying by $2x$ first the like terms line up more conveniently. Students usually prefer a vertical arrangement for this type of problem.

Work Problem 10 and check solution in Solutions to Matched Problems on page 76.

PROBLEM 10   Multiply:

(A) $2m^3(3m^2 - 4m - 3) =$

(B) $\begin{array}{l} 2x^2 + 3x - 1 \\ 3x - 4 \end{array}$ (C) $\begin{array}{l} 2x^2 + 3x - 2 \\ 3x^2 - 2x + 1 \end{array}$

•••Mental multiplication of first-degree polynomials

For reasons that will become clear shortly, it is essential that you learn to multiply first-degree polynomials of the type $(3x + 2)(2x - 1)$ and $(3x - y)(x + 2y)$ mentally. To discover relationships that will make this possible, let us first multiply $(3x + 2)$ and $(2x - 1)$ using a vertical arrangement.

$$\begin{array}{r} 3x + 2 \\ 2x - 1 \\ \hline 6x^2 + 4x \\ -3x - 2 \\ \hline 6x^2 + x - 2 \end{array}$$

Now let us use a horizontal arrangement and try to discover a method that will enable us to carry out the multiplication mentally. We start by multiplying each term in the first binomial times each term in the second binomial:

F
First product
$(2x - 1)(3x + 2)$

O
Outer product
$(2x - 1)(3x + 2)$

I
Inner product
$(2x - 1)(3x + 2)$

L
Last product
$(2x - 1)(3x + 2)$

Performing these four operations on one line, we obtain:

F — First product
O — Outer product
I — Inner product
L — Last product

$$(2x - 1)(3x + 2) = 6x^2 + 4x - 3x - 2$$

The inner and outer products are like terms and hence combine into one term. Thus,

$$(2x - 1)(3x + 2) = 6x^2 + x - 2$$

To speed up the process we combine the inner and outer products above mentally. The method just described is called the **FOIL method.** We note again that the product of two first-degree polynomials is a second-degree polynomial.

A simple three-step process for carrying out the FOIL method is illustrated in Example 11.

EXAMPLE 11

(A) $(2x - 1)(3x + 2) = 6x^2 + x - 2$

The like terms are obtained in step 2 by multiplying the inner and outer products, and are combined mentally.

(B) $(2a - b)(a + 3b) = 2a^2 + 5ab - 3b^2$

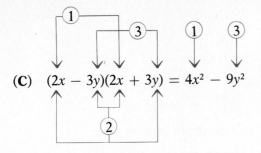

(C) $(2x - 3y)(2x + 3y) = 4x^2 - 9y^2$

Notice that a middle term does not appear since its coefficient is 0.

Work Problem 11 and check solution in Solutions to Matched Problems on page 76.

PROBLEM 11   Mentally multiply:

   (A) $(3x - 2)(2x - 1) =$ **(B)** $(a - 3b)(2a + 3b) =$

   (C) $(5x - y)(5x + y) =$ **(D)** $(4x - 3y)(2x + 5y) =$

In Section 2.6 we will consider the reverse problem: Given a second-degree polynomial, such as $2x^2 - 5x - 3$, find first-degree factors with integer coefficients that will produce this second-degree polynomial as a product. To be able to factor second-degree polynomial forms with any degree of efficiency, it is important that you know how to mentally multiply quickly and accurately first-degree factors of the types illustrated in this section.

••••Squaring binomials

Since

$$(a + b)^2 = (a + b)(a + b) = a^2 + 2ab + b^2$$
$$(a - b)^2 = (a - b)(a - b) = a^2 - 2ab + b^2$$

we can formulate a simple mechanical rule for squaring any binomial directly.

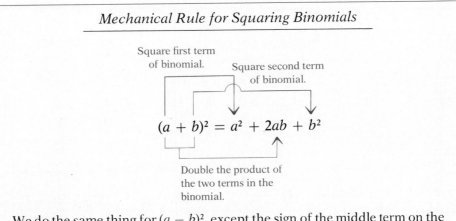

Mechanical Rule for Squaring Binomials

Square first term of binomial.   Square second term of binomial.

$$(a + b)^2 = a^2 + 2ab + b^2$$

Double the product of the two terms in the binomial.

We do the same thing for $(a - b)^2$, except the sign of the middle term on the right becomes negative.

EXAMPLE 12

Square each binomial, using the mechanical rule.

(A) $(2x + 3)^2 = (2x)^2 + 2[(2x)(3)] + (3)^2 = 4x^2 + 12x + 9$

(B) $(3x - 2y)^2 = (3x)^2 - 2[(3x)(2y)] + (2y)^2 = 9x^2 - 12xy + 4y^2$

Work Problem 12 and check solution in Solutions to Matched Problems on page 76.

> **PROBLEM 12** Square each binomial, using the mechanical rule.
>
> **(A)** $(3y + 4)^2 =$
>
> **(B)** $(2u - 3v)^2 =$

SOLUTIONS TO MATCHED PROBLEMS

9. **(A)** $y^4y^7 = y^{11}$ **(B)** $(9x^4)(3x^2) = 27x^6$ **(C)** $(4u^3v^2)(-3uv^3) = -12u^4v^5$

10. **(A)** $2m^3(3m^2 - 4m - 3) = 6m^5 - 8m^4 - 6m^3$

(B)
$$
\begin{array}{r}
2x^2 + 3x - 1 \\
3x - 4 \\
\hline
6x^3 + 9x^2 - 3x \\
-8x^2 - 12x + 4 \\
\hline
6x^3 + x^2 - 15x + 4
\end{array}
$$

(C)
$$
\begin{array}{r}
2x^2 + 3x - 2 \\
3x^2 - 2x + 1 \\
\hline
6x^4 + 9x^3 - 6x^2 \\
-4x^3 - 6x^2 + 4x \\
2x^2 + 3x - 2 \\
\hline
6x^4 + 5x^3 - 10x^2 + 7x - 2
\end{array}
$$

11. **(A)** $(3x - 2)(2x - 1) = 6x^2 - 7x + 2$ **(B)** $(a - 3b)(2a + 3b) = 2a^2 - 3ab - 9b^2$

(C) $(5x - y)(5x + y) = 25x^2 - y^2$ **(D)** $(4x - 3y)(2x + 5y) = 8x^2 + 14xy - 15y^2$

12. **(A)** $(3y + 4)^2 = 9y^2 + 24y + 16$ **(B)** $(2u - 3v)^2 = 4u^2 - 12uv + 9v^2$

PRACTICE EXERCISE 2.3

Work odd-numbered problems first, check answers, and then work even-numbered problems in areas of weakness. Answers to all problems are in the back of the book. Make every effort to work a problem yourself before you look at an answer.

A *Multiply.*

1. y^2y^3 _____

2. x^3x^2 _____

3. $(5y^4)(2y)$ _____

4. $(2x)(3x^4)$ _____

5. $(8x^{11})(-3x^9)$ _____

6. $(-7u^9)(5u^7)$ _____

7. $(-3u^4)(2u^5)(-u^7)$ _____

8. $(2x^3)(-3x)(-4x^5)$ _____

9. $(cd^2)(c^2d^2)$ _____

10. $(a^2b)(ab^2)$ _____

11. $(-3xy^2z^3)(-5xyz^2)$ _____

12. $(-2xy^3z)(3x^3yz)$ _____

13. $y(y + 7)$ _____

14. $x(1 + x)$ _____

15. $5y(2y - 7)$ _____

16. $3x(2x - 5)$ _____

17. $3a^2(a^3 + 2a^2)$ _____

18. $2m^2(m^2 + 3m)$ _____

19. $2y(y^2 + 2y - 3)$ _____

20. $2x(2x^2 - 3x + 1)$ _____

21. $7m^3(m^3 - 2m^2 - m + 4)$ _____

22. $3x^2(2x^3 + 3x^2 - x - 2)$ _____

23. $5uv^2(2u^3v - 3uv^2)$ _____

24. $4m^2n^3(2m^3n - mn^2)$ _____

25. $2cd^3(c^2d - 2cd + 4c^3d^2)$ _____

26. $3x^2y(2xy^3 + 4x - y^2)$ _____

B 27. $(3y + 2)(2y^2 + 5y - 3)$ _____

28. $(2x - 1)(x^2 - 3x + 5)$ _____

29. $(m + 2n)(m^2 - 4mn - n^2)$ _____

30. $(x - 3y)(x^2 - 3xy + y^2)$ _____

31. $(2m^2 + 2m - 1)(3m^2 - 2m + 1)$ _____

32. $(x^2 - 3x + 5)(2x^2 + x - 2)$ _____

33. $(a + b)(a^2 - ab + b^2)$ _____

34. $(a - b)(a^2 + ab + b^2)$ _____

35. $(2x^2 - 3xy + y^2)(x^2 + 2xy - y^2)$ _____

36. $(a^2 - 2ab + b^2)(a^2 + 2ab + b^2)$ _____

Multiply mentally.

37. $(x + 3)(x + 2)$ _____

38. $(m - 2)(m - 3)$ _____

39. $(a + 8)(a - 4)$ _____

40. $(m - 12)(m + 5)$ _____

41. $(t + 4)(t - 4)$ _____

42. $(u - 3)(u + 3)$ _____

43. $(m - n)(m + n)$ _____

44. $(a + b)(a - b)$ _____

45. $(4t - 3)(t - 2)$ _____

46. $(3x - 5)(2x + 1)$ _____

47. $(3x + 2y)(x - 3y)$ _____

48. $(2x - 3y)(x + 2y)$ _____

49. $(2m - 7)(2m + 7)$ _____

50. $(3y + 2)(3y - 2)$ _____

51. $(6x - 4y)(5x + 3y)$ _____

52. $(3m + 7n)(2m - 5n)$ _____

53. $(2s - 3t)(3s - t)$ _____

54. $(2x - 3y)(3x - 2y)$ _____

Square each binomial, using the mechanical rule.

55. $(3x + 2)^2$ _____

56. $(4x + 3y)^2$ _____

57. $(2x - 5y)^2$ _____

58. $(2x - 7)^2$ _____

59. $(6u + 5v)^2$ _____

60. $(7p + 2q)^2$ _____

61. $(2m - 5n)^2$ _____

62. $(4x - 1)^2$ _____

C *Simplify.*

63. $(3x - 1)(x + 2) - (2x - 3)^2$ _____

64. $(2x + 3)(x - 5) - (3x - 1)^2$ _____

65. $(2x - 1)^2 - (x + 3)^2$ _____

66. $(x - y)^2 - (x + y)^2$ _____

67. $(x + 2y)^3$ _____

68. $(2m - n)^3$ _____

APPLICATIONS

69. The length of a rectangle is 8 meters more than its width. If y is the length of the rectangle, write an algebraic expression that represents its area. Change the expression to a form without parentheses.

_____

70. Repeat Problem 69 if the length of the rectangle is 3 meters less than twice the width.

_____

The Check Exercise for this section is on page 109.

2.4 DIVISION

There are times when it is useful to find quotients of polynomials by a long-division process similar to that used in arithmetic. Several examples will illustrate the process.

EXAMPLE 13
Divide: $2x^2 + 5x - 12$ by $x + 4$

Solution

$x + 4 \overline{\smash{)}2x^2 + 5x - 12}$ — Both polynomials are arranged in descending powers of the variable if this is not already done.

$\begin{array}{r} 2x \\ x + 4 \overline{\smash{)}2x^2 + 5x - 12} \end{array}$ — Divide the first term of the divisor into the first term of the dividend; i.e., what must x be multiplied by so that the product is exactly $2x^2$? Answer: $2x$.

$\begin{array}{r} 2x \\ x + 4 \overline{\smash{)}2x^2 + 5x - 12} \\ \underline{2x^2 + 8x} \\ -3x - 12 \end{array}$ — Multiply the divisor by $2x$, line up like terms as indicated, subtract, and bring down -12 from above.

$\begin{array}{r} 2x - 3 \\ x + 4 \overline{\smash{)}2x^2 + 5x - 12} \\ \underline{2x^2 + 8x} \\ -3x - 12 \\ \underline{-3x - 12} \\ 0 \end{array}$ — Repeat the process above until the degree of the remainder is less than that of the divisor, or the remainder is zero.

CHECK
$(x + 4)(2x - 3) = 2x^2 + 5x - 12$

Work Problem 13 and check solution in Solutions to Matched Problems on page 80.

PROBLEM 13   Divide $2x^2 + 7x + 3$ by $x + 3$ and check.

Solution   **CHECK**

$\overline{)\qquad\qquad}$

EXAMPLE 14

Divide $(x^3 + 8)/(x + 2)$.

Solution

$$
\begin{array}{r}
x^2 - 2x\ + 4 \\
x + 2 \overline{)\ x^3 + 0x^2 + 0x + 8} \\
\underline{x^3 + 2x^2} \\
-2x^2 + 0x \\
\underline{-2x^2 - 4x} \\
4x + 8 \\
\underline{4x + 8} \\
0
\end{array}
$$

Insert, with 0 coefficients, any missing terms of lower degree than 3, and proceed as in Example 13.

Can you check this problem?

Work Problem 14 and check solution in Solutions to Matched Problems on page 80.

PROBLEM 14   Divide $(x^3 - 8)/(x - 2)$ and check.

Solution   **CHECK**

$\overline{)\qquad\qquad}$

EXAMPLE 15

Divide $(3 + 6x^2 - 7x)/(3x + 1)$ and check.

Solution

$$
\begin{array}{r}
2x\ - 3 \\
3x + 1 \overline{)\ 6x^2 - 7x + 3} \\
\underline{6x^2 + 2x} \\
-9x + 3 \\
\underline{-9x - 3} \\
6 = R \qquad \text{(remainder)}
\end{array}
$$

Arrange $3 - 7x + 6x^2$ in descending powers of x; then proceed as above until the degree of the remainder is less than the degree of the divisor.

CHECK
Just as in arithmetic, when there is a remainder we check by adding the remainder to the product of the divisor and quotient. Thus

$$(3x + 1)(2x - 3) + 6 \overset{?}{=} 6x^2 - 7x + 3$$

$$6x^2 - 7x - 3 + 6 \overset{?}{=} 6x^2 - 7x + 3$$

$$6x^2 - 7x + 3 \overset{\checkmark}{=} 6x^2 - 7x + 3$$

Work Problem 15 and check solution in Solutions to Matched Problems on page 80.

PROBLEM 15 Divide $(2 - x + 6x^2)/(3x - 2)$ and check.

Solution

$\overline{)}$

CHECK

SOLUTIONS TO MATCHED PROBLEMS

13.

$$\begin{array}{r} 2x + 1 \\ x + 3 \overline{)2x^2 + 7x + 3} \\ \underline{2x^2 + 6x} \\ x + 3 \\ \underline{x + 3} \\ 0 = R \end{array}$$

CHECK

$$(x + 3)(2x + 1) = 2x^2 + 7x + 3$$

14.

$$\begin{array}{r} x^2 + 2x + 4 \\ x - 2 \overline{)x^3 + 0x^2 + 0x - 8} \\ \underline{x^3 - 2x^2} \\ 2x^2 + 0x \\ \underline{2x^2 - 4x} \\ 4x - 8 \\ \underline{4x - 8} \\ 0 = R \end{array}$$

CHECK

$$\begin{array}{r} x^2 + 2x + 4 \\ \underline{x - 2} \\ x^3 + 2x^2 + 4x \\ \underline{-2x^2 - 4x - 8} \\ x^3 \qquad\qquad - 8 \end{array}$$

15.

$$\begin{array}{r} 2x + 1 \\ 3x - 2 \overline{)6x^2 - x + 2} \\ \underline{6x^2 - 4x} \\ 3x + 2 \\ \underline{3x - 2} \\ 4 = R \end{array}$$

CHECK

$$(3x - 2)(2x + 1) + 4$$
$$= 6x^2 - x - 2 + 4$$
$$= 6x^2 - x + 2$$

PRACTICE EXERCISE 2.4 _____

Work odd-numbered problems first, check answers, and then work even-numbered problems in areas of weakness. Answers to all problems are in the back of the book. Make every effort to work a problem yourself before you look at an answer.

Divide, using the long-division process. Check the answers.

A **1.** $(3x^2 - 5x - 2)/(x - 2)$ _____

    3. $(2y^3 + 5y^2 - y - 6)/(y + 2)$ _____

    5. $(3x^2 - 11x - 1)/(x - 4)$ _____

    7. $(8x^2 - 14x + 3)/(2x - 3)$ _____

    9. $(6x^2 + x - 13)/(2x + 3)$ _____

    11. $(x^2 - 4)/(x - 2)$ _____

B **13.** $(12x^2 + 11x - 2)/(3x + 2)$ _____

    15. $(8x^2 + 7)/(2x - 3)$ _____

    17. $(-7x + 2x^2 - 1)/(2x + 1)$ _____

    19. $(x^3 - 1)/(x - 1)$ _____

    21. $(x^4 - 81)/(x - 3)$ _____

    23. $(4a^2 - 22 - 7a)/(a - 3)$ _____

    25. $(x + 5x^2 - 10 + x^3)/(x + 2)$ _____

    27. $(3 + x^3 - x)/(x - 3)$ _____

C **29.** $(9x^4 - 2 - 6x - x^2)/(3x - 1)$ _____

    31. $(8x^2 - 7 - 13x + 24x^4)/(3x + 5 + 6x^2)$ _____

    32. $(16x - 5x^3 - 8 + 6x^4 - 8x^2)/(2x - 4 + 3x^2)$ _____

    33. $(9x^3 - 2x + 2x^5 + 9x^3 - 2)/(2 + x^2 - 3x)$ _____

    34. $(12x^2 - 19x^3 - 4x - 3 + 12x^5)/(4x^2 - 1)$ _____

    2. $(2x^2 + x - 6)/(x + 2)$ _____

    4. $(x^3 - 5x^2 + x + 10)/(x - 2)$ _____

    6. $(2x^2 - 3x - 4)/(x - 3)$ _____

    8. $(6x^2 + 5x - 6)/(3x - 2)$ _____

    10. $(6x^2 + 11x - 12)/(3x - 2)$ _____

    12. $(y^2 - 9)/(y + 3)$ _____

    14. $(8x^2 - 6x + 6)/(2x - 1)$ _____

    16. $(9x^2 - 8)/(3x - 2)$ _____

    18. $(13x - 12 + 3x^2)/(3x - 2)$ _____

    20. $(a^3 + 27)/(a + 3)$ _____

    22. $(x^4 - 16)/(x + 2)$ _____

    24. $(8c + 4 + 5c^2)/(c + 2)$ _____

    26. $(5y^2 - y + 2y^3 - 6)/(y + 2)$ _____

    28. $(3y - y^2 + 2y^3 - 1)/(y + 2)$ _____

    30. $(4x^2 - 10x - 9x^2 - 10)/(2x + 3)$ _____

The Check Exercise for this section is on page 111.

2.5 FACTORING OUT COMMON FACTORS; FACTORING BY GROUPING

•Removing common factors
••Factoring by grouping

•Removing common factors

You have already had experience in factoring out common factors from polynomials. The distributive property of real numbers in the form

$$ab + ac = a(b + c)$$

is the important property behind the process. Note that factoring is just the opposite of multiplication.

EXAMPLE 16
Factor out all factors common to all terms:

(A) $6x^2 + 15x = 3x \cdot 2x + 3x \cdot 5 = 3x(2x + 5)$

(B) $2u^3v - 6u^2v^2 + 8uv^3 = 2uv \cdot u^2 - 2uv \cdot 3uv + 2uv \cdot 4v^2$
$$= 2uv(u^2 - 3uv + 4v^2)$$

Work Problem 16 and check solution in Solutions to Matched Problems on page 84.

PROBLEM 16   Factor out all factors common to all terms.

   (A) $12y^2 - 28y =$

   (B) $6m^4 - 15m^3n + 9m^2n^2 =$

Now look closely at the following four examples and try to determine what they all have in common!

$$2x + 3y = y(2x + 3)$$
$$2x + 3A = A(2x + 3)$$
$$2x(x - 4) + 3(x - 4) = (x - 4)(2x + 3)$$
$$2x(3x + 1) + 3(3x + 1) = (3x + 1)(2x + 3)$$

Because of the commutative property, common factors may be taken out either on the left or on the right.

The factoring involved in each example is essentially the same. The only difference is the nature of the common factors being taken out. In the last two examples think of $(x - 4)$ and $(3x + 1)$ as single numbers, just as A represents a single number in the second example.

EXAMPLE 17
Remove all factors common to all terms.

(A) $5x(x - 1) - (x - 1) = 5x(x - 1) - 1(x - 1) = (x - 1)(5x - 1)$

(B) $3x(2x - y) - 2y(2x - y) = (2x - y)(3x - 2y)$

Work Problem 17 and check solution in Solutions to Matched Problems on page 84.

PROBLEM 17   Remove all factors common to all terms.

(A) $3m(m + 2) - (m + 2) =$

(B) $2u(u + 3v) - 3v(u + 3v) =$

••Factoring by grouping

Some polynomials can be factored by grouping terms in such a way that we obtain results that look like Example 17. We can then complete the factoring following the procedures used there. This process will prove useful in Section 2.7, where an efficient method is developed for factoring a second-degree polynomial as the product of two first-degree polynomials.

EXAMPLE 18
Factor by grouping.

(A) $2x^2 - 8x + 3x - 12$

Solution

$2x^2 - 8x + 3x - 12$	Group the first two and last two terms.
$= (2x^2 - 8x) + (3x - 12)$	Remove common factors from each group.
$= 2x(x - 4) + 3(x - 4)$	The common factor $(x - 4)$ can be taken out.
$= (x - 4)(2x + 3)$	The factoring is complete.

(B) $5x^2 - 5x - x + 1$

Solution

$5x^2 - 5x - x + 1$	Group first two and last two terms. Signs change inside second set of parentheses.
$= (5x^2 - 5x) - (x - 1)$	Remove common factors from each group.
$= 5x(x - 1) - 1(x - 1)$	The common factor $(x - 1)$ can be taken out.
$= (x - 1)(5x - 1)$	The factoring is complete.

(C) $6x^2 - 3xy - 4xy + 2y^2$

Solution

$6x^2 - 3xy - 4xy + 2y^2$	
$= (6x^2 - 3xy) - (4xy - 2y^2)$	Signs change inside second set of parentheses.
$= 3x(2x - y) - 2y(2x - y)$	
$= (2x - y)(3x - 2y)$	

Work Problem 18 and check solution in Solutions to Matched Problems on page 84.

PROBLEM 18   Factor by grouping.

(A) $6x^2 + 2x + 9x + 3 =$

(B) $3m^2 + 6m - m - 2 =$

(C) $2u^2 + 6uv - 3uv - 9v^2 =$

SOLUTIONS TO MATCHED PROBLEMS

16. (A) $12y^2 - 28y = 4y(3y - 7)$ (B) $6m^4 - 15m^3n + 9m^2n^2 = 3m^2(2m^2 - 5mn + 3n^2)$
17. (A) $3m(m + 2) - (m + 2) = (m + 2)(3m - 1)$
(B) $2u(u + 3v) - 3v(u + 3v) = (u + 3v)(2u - 3v)$
18. (A) $6x^2 + 2x + 9x + 3 = (6x^2 + 2x) + (9x + 3) = 2x(3x + 1) + 3(3x + 1) = (3x + 1)(2x + 3)$
(B) $3m^2 + 6m - m - 2 = (3m^2 + 6m) - (m + 2) = 3m(m + 2) - 1(m + 2) = (m + 2)(3m - 1)$
(C) $2u^2 + 6uv - 3uv - 9v^2 = (2u^2 + 6uv) - (3uv + 9v^2) = 2u(u + 3v) - 3v(u + 3v)$
$= (u + 3v)(2u - 3v)$

PRACTICE EXERCISE 2.5

Work odd-numbered problems first, check answers, and then work even-numbered problems in areas of weakness. Answers to all problems are in the back of the book. Make every effort to work a problem your-self before you look at an answer.

Write in factored form by removing all factors common to all terms.

A 1. $2xA + 3A$ _____ 2. $xM - 4M$ _____ 3. $10x^2 + 15x$ _____

4. $9y^2 - 6y$ _____ 5. $14u^2 - 6u$ _____ 6. $20m^2 + 12m$ _____

7. $6u^2 - 10uv$ _____ 8. $14x^2 - 21xy$ _____ 9. $10m^2n - 15mn^2$ _____

10. $9u^2v + 6uv^2$ _____ 11. $2x^3y - 6x^2y^2$ _____ 12. $6x^2y^2 - 6xy^3$ _____

13. $3x(x + 2) + 5(x + 2)$ _____ 14. $4y(y + 3) + 7(y + 3)$ _____

15. $3m(m - 4) - 2(m - 4)$ _____ 16. $x(x - 1) - 4(x - 1)$ _____

17. $x(x + y) - y(x + y)$ _____ 18. $m(m - n) + n(m - n)$ _____

B 19. $6x^4 - 9x^3 + 3x^2$ _____ 20. $6m^4 - 8m^3 - 2m^2$ _____

21. $8x^3y - 6x^2y^2 + 4xy^3$ _____ 22. $10u^3v + 20u^2v^2 - 15uv^3$ _____

23. $8x^4 - 12x^3y + 4x^2y^2$ _____

24. $9m^4 - 6m^3n - 6m^2n^2$ _____

25. $3x(2x + 3) - 5(2x + 3)$ _____

26. $2u(3u - 8) - 3(3u - 8)$ _____

27. $x(x + 1) - (x + 1)$ _____

28. $3u(u - 1) - (u - 1)$ _____

29. $4x(2x - 3) - (2x - 3)$ _____

30. $3y(4y - 5) - (4y - 5)$ _____

Replace question marks with algebraic expressions that will make both sides equal.

31. $3x^2 - 3x + 2x - 2 = (3x^2 - 3x) + (\ ? \ )$ _____

32. $2x^2 + 4x + 3x + 6 = (2x^2 + 4x) + (\ ? \ )$ _____

33. $3x^2 - 12x - 2x + 8 = (3x^2 - 12x) - (\ ? \ )$ _____

34. $2y^2 - 10y - 3y + 15 = (2y^2 - 10y) - (\ ? \ )$ _____

35. $8u^2 + 4u - 2u - 1 = (8u^2 + 4u) - (\ ? \ )$ _____

36. $6x^2 + 10x - 3x - 5 = (6x^2 + 10x) - (\ ? \ )$ _____

Factor out all common factors from each group; then complete the factoring if possible.

37. $(3x^2 - 3x) + (2x - 2)$ _____

38. $(2x^2 + 4x) + (3x + 6)$ _____

39. $(3x^2 - 12x) - (2x - 8)$ _____

40. $(2y^2 - 10y) - (3y - 15)$ _____

41. $(8u^2 + 4u) - (2u + 1)$ _____

42. $(6x^2 + 10x) - (3x + 5)$ _____

Factor as the product of two first-degree polynomials, using grouping. (These problems are related to Problems 31 to 36 and 37 to 42.)

43. $3x^2 - 3x + 2x - 2$ _____

44. $2x^2 + 4x + 3x + 6$ _____

45. $3x^2 - 12x - 2x + 8$ _____

46. $2y^2 - 10y - 3y + 15$ _____

47. $8u^2 + 4u - 2u - 1$ _____

48. $6x^2 + 10x - 3x - 5$ _____

Factor as the product of two first-degree factors, using grouping.

49. $2m^2 - 8m + 5m - 20$ _____

50. $5x^2 - 10x + 2x - 4$ _____

51. $6x^2 - 9x - 4x + 6$ _____

52. $12x^2 + 8x - 9x - 6$ _____

C 53. $3u^2 - 12u - u + 4$ _____

54. $6m^2 + 4m - 3m - 2$ _____

55. $6u^2 + 3uv - 4uv - 2v^2$ _____

56. $2x^2 - 4xy - xy + 2y^2$ _____

57. $6x^2 + 3xy - 10xy - 5y^2$ _____

58. $4u^2 - 16uv - 3uv + 12v^2$ _____

The Check Exercise for this section is on page 113.

2.6 FACTORING SECOND-DEGREE POLYNOMIALS

We now turn our attention to factoring second-degree polynomials of the form

$$2x^2 - 5x - 3$$
$$2x^2 + 3xy - 2y^2$$

into the product of two first-degree polynomials with integer coefficients. By now it should be very easy for you to obtain (by mental multiplication) the products

$$(x - 3)(x + 2) = x^2 - x - 6$$
$$(x - 3y)(x + 2y) = x^2 - xy - 6y^2$$

but can you reverse the process? Can you, for example, find integers a, b, c, and d so that

$$2x^2 - 5x - 3 = (ax + b)(cx + d)$$

Representing a second-degree polynomial with integers as coefficients as the product of two first-degree polynomials with integer coefficients is not as easy as multiplying first-degree polynomials. In this section we will develop a method of attack that is relatively easy to understand, but not always easy to apply. In the next section we will develop an approach to the problem that builds on factoring by grouping discussed in the last section. This approach is a little harder to understand, but once the method is understood, it is fairly easy to apply.

Let us start with a very simply polynomial whose factors you will likely guess at once:

$$x^2 + 6x + 8$$

Our problem is to find two first-degree factors with integer coefficients, if they exist. To start we write

$$x^2 + 6x + 8 = (x + \quad)(x + \quad) \qquad \text{Why must both signs be positive?}$$

Now, what are the constant terms? Since they are positive-integer factors of 8, we write

$$\begin{array}{c} 8 \\ \hline 1 \cdot 8 \\ 2 \cdot 4 \end{array}$$

If we try the first pair (mentally), we obtain the first and last terms, but not the middle term. We next try 2 and 4, which gives us the middle term as well as the first and last terms. Thus,

$$x^2 + 6x + 8 = (x + 2)(x + 4)$$

Because of the commutative property, we could also write

$$x^2 + 6x + 8 = (x + 4)(x + 2)$$

Let us try another polynomial. Find first-degree factors with integer coefficients for

$$2x^2 - 7x + 6$$

Again, we write

$$2x^2 - 7x + 6 = (2x - \quad)(x - \quad) \qquad \text{Why must both signs be negative?}$$

Both constant terms must be factors of 6. The possibilities are

$$\underline{6}$$

$$1 \cdot 6 \qquad \text{All pairs give the first and last terms}$$
$$6 \cdot 1 \qquad \text{in } 2x^2 - 7x + 6, \text{ but will any give the}$$
$$2 \cdot 3 \qquad \text{middle term, } -7x?$$
$$3 \cdot 2$$

Testing each pair (this is why you need to do binomial multiplication mentally), we find that the last pair gives the middle term. Thus,

$$2x^2 - 7x + 6 = (2x - 3)(x - 2)$$

Before you conclude that all second-degree polynomials with integer coefficients have first-degree factors with integer coefficients, consider the following simple polynomial:

$$x^2 + x + 2$$

Proceeding as above, we write

$$x^2 + x + 2 = (x + \quad)(x + \quad)$$

$$\underline{2}$$
$$1 \cdot 2$$
$$2 \cdot 1$$

and find that neither pair produces the middle term, x. Hence, we conclude that

$$x^2 + x + 2$$

has no first-degree factors with integer coefficients, and we say the polynomial is not factorable (using integers).

EXAMPLE 19
Factor each polynomial, if possible, using integer coefficients.

(A) $2x^2 + 3xy - 2y^2$

Solution
$$2x^2 + 3xy - 2y^2$$

$$= (2x + \quad y)(x - \quad y) \qquad \begin{array}{l} \text{Put in what we know. Signs must be opposite. (We can} \\ \text{reverse this choice if we get } -3xy \text{ instead of } +3xy \text{ for the} \\ \text{middle term.)} \end{array}$$

$$\qquad \uparrow \qquad \uparrow$$
$$\qquad ? \qquad ?$$

Now what are the factors of 2 (the coefficient of y^2)?

$$\frac{2}{\begin{array}{l}1 \cdot 2 \\ 2 \cdot 1\end{array}}$$

The first choice gives us $-3xy$ for the middle term—close, but not there—so we reverse our choice of signs to obtain

$$2x^2 + 3xy - 2y^2 = (2x - y)(x + 2y)$$

(B) $x^2 - 3x + 4$

Solution
$x^2 - 3x + 4 = (x - \quad)(x - \quad)$

$$\frac{4}{\begin{array}{l}2 \cdot 2 \\ 1 \cdot 4 \\ 4 \cdot 1\end{array}}$$

No choice produces the middle term; hence,

$$x^2 - 3x + 4$$

is not factorable using integer coefficients.

(C) $6x^2 + 5xy - 4y^2$

Solution
$6x^2 + 5xy - 4y^2 = (?x + ?y)(?x - ?y)$

The signs must be opposite in the factors, since the third term is negative. (We can reverse our choice of signs later, if necessary.)

We now write all factors of 6 and of 4:

6	4
$2 \cdot 3$	$2 \cdot 2$
$3 \cdot 2$	$1 \cdot 4$
$1 \cdot 6$	$4 \cdot 1$
$6 \cdot 1$	

and try each choice on the left with each on the right—a total of 12 combinations that give us the first and last terms in $6x^2 + 5xy - 4y^2$. The question is, does any combination also give us the middle term, $5xy$? After trial and error, and, perhaps, some educated guessing among the choices, we find that $3 \cdot 2$ matched with $4 \cdot 1$ gives us the correct middle term. Thus,

$$6x^2 + 5xy - 4y^2 = (3x + 4y)(2x - y)$$

If none of the 24 combinations (including reversing our sign choice) had produced the middle term, then we would conclude that the polynomial is not factorable.

Work Problem 19 and check solution in Solutions to Matched Problems on page 89.

PROBLEM 19   Factor each polynomial, if possible, using integer coefficients.

(A) $x^2 - 8x + 12 =$

(B) $x^2 + 2x + 5 =$

(C) $2x^2 + 7xy - 4y^2 =$

(D) $4x^2 - 15xy - 4y^2 =$

Concluding remarks

It is about here that many students begin to lose interest in factoring, particularly with problems like Example 19**C**. They quickly observe that if the coefficients of the first and last terms get larger and larger with more and more factors, the number of combinations that needs to be checked increases very rapidly. And it is quite possible in most practical situations that none of the combinations will work. It is important, however, that you understand the approach presented above, since it will work for most of the simpler factoring problems you will encounter. If you have the time and inclination, read the next (optional) section for a systematic approach to the problem of factoring that will reduce the amount of trial and error substantially and even tell you whether a polynomial can be factored before you proceed too far.

In conclusion, we point out that if a, b, and c are selected at random out of the integers, the probability that

$$ax^2 + bx + c$$

is not factorable in the integers is much greater than the probability that it is. But even being able to factor some second-degree polynomials leads to marked simplification of some algebraic expressions and an easy way to solve some second-degree equations, as will be seen later.

SOLUTIONS TO MATCHED PROBLEMS

19.   **(A)** $x^2 - 8x + 12 = (x - 2)(x - 6)$ **(B)**   Not factorable
     (C) $2x^2 + 7xy - 4y^2 = (2x - y)(x + 4y)$ **(D)** $4x^2 - 15xy - 4y^2 = (4x + y)(x - 4y)$

PRACTICE EXERCISE 2.6 _____

Work odd-numbered problems first, check answers, and then work even-numbered problems in areas of weakness. Answers to all problems are in the back of the book. Make every effort to work a problem yourself before you look at an answer.

Factor in the integers, if possible. If not factorable, say so.

A 1. $x^2 + 5x + 4$ _____   2. $x^2 + 4x + 3$ _____

3. $x^2 + 5x + 6$ _____   4. $x^2 + 7x + 10$ _____

5. $x^2 - 4x + 3$ _____   6. $x^2 - 5x + 4$ _____

7. $x^2 - 7x + 10$ _____   8. $x^2 - 5x + 6$ _____

9. $y^2 + 3y + 3$ _____   10. $y^2 + 2y + 2$ _____

11. $y^2 - 2y + 6$ _____   12. $x^2 - 3x + 5$ _____

13. $x^2 + 8xy + 15y^2$ _____   14. $x^2 + 9xy + 20y^2$ _____

15. $x^2 - 10xy + 21y^2$ _____   16. $x^2 - 10xy + 16y^2$ _____

17. $u^2 + 4uv + v^2$ _____   18. $u^2 + 5uv + 3v^2$ _____

19. $3x^2 + 7x + 2$ _____   20. $2x^2 + 7x + 3$ _____

21. $3x^2 - 7x + 4$ _____   22. $2x^2 - 7x + 6$ _____

B 23. $3x^2 - 14x + 8$ _____   24. $2y^2 - 13y + 15$ _____

25. $3x^2 - 11xy + 6y^2$ _____   26. $2x^2 - 7xy + 6y^2$ _____

27. $n^2 - 2n - 8$ _____   28. $n^2 + 2n - 8$ _____

29. $x^2 - 4x - 6$ _____   30. $x^2 - 3x - 8$ _____

31. $3x^2 - x - 2$ _____   32. $6m^2 + m - 2$ _____

33. $x^2 + 4xy - 12y^2$ _____   34. $2x^2 - 3xy - 2y^2$ _____

35. $3u^2 - 11u - 4$ _____   36. $8u^2 + 2u - 1$ _____

37. $6x^2 + 7x - 5$ _____   38. $2m^2 - 3m - 20$ _____

39. $3s^2 - 5s - 2$ _____   40. $2s^2 + 5s - 3$ _____

41. $3x^2 + 2xy - 3y^2$ _____   42. $2x^2 - 3xy - 4y^2$ _____

43. $5x^2 - 8x - 4$ _____   44. $12x^2 + 16x - 3$ _____

45. $6u^2 - uv - 2v^2$ _____   46. $6x^2 - 7xy - 5y^2$ _____

47. $8x^2 + 6x - 9$ _____   **48.** $6x^2 - 13x + 6$ _____

49. $3u^2 + 7uv - 6v^2$ _____   **50.** $4m^2 + 11mn - 3n^2$ _____

51. $4u^2 - 19uv + 12v^2$ _____   **52.** $12x^2 - xy - 6y^2$ _____

C 53. $12x^2 - 40xy - 7y^2$ _____   **54.** $15x^2 + 17xy - 4y^2$ _____

55. $12x^2 + 19xy - 10y^2$ _____   **56.** $24x^2 - 31xy - 15y^2$ _____

The Check Exercise for this section is on page 115.

2.7 ac TEST AND FACTORING

We continue our discussion of factoring second-degree polynomials of the type

$$
\begin{aligned}
ax^2 + bx + c \\
ax^2 + bxy + cy^2
\end{aligned}
\qquad (1)
$$

with integer coefficients into the product of two first-degree factors with integer coefficients. We now provide a test, called the *ac* **test for factorability,** that not only tells us if the polynomials in (1) can be factored using integer coefficients, but, in addition, leads to a direct way of factoring those that are factorable.

---

ac Test for Factorability

If in equations (1) the product *ac* has two integer factors *p* and *q* whose sum is the coefficient of the middle term *b*; that is, if integers *p* and *q* exist so that

$$pq = ac \quad \text{and} \quad p + q = b \qquad (2)$$

then equations (1) have first-degree factors with integer coefficients. If no integers *p* and *q* exist that satisfy (2), then equations (1) will not have first-degree factors with integer coefficients.

Once we find integers p and q in the *ac* test, if they exist, then our work is almost finished, since we can then write equations (1), splitting the middle term, into the forms:

$$
\boxed{
\begin{aligned}
&ax^2 + px + qx + c \\
&ax^2 + pxy + qxy + cy^2
\end{aligned}
\qquad (3)
}
$$

and the factoring can be completed in a couple of steps, using factoring by grouping discussed at the end of Section 2.5.

Let us make the discussion concrete through several examples.

EXAMPLE 20

Factor $2x^2 + 11x - 6$, if possible, using integer coefficients.

Solution

Step 1. Test for factorability, using the *ac* test. Since $ax^2 + bx + c = 2x^2 + 11x - 6$, $a = 2$, $b = 11$, and $c = -6$. Thus,

$$ac = (2)(-6) = -12$$

Try to find two integer factors of -12 whose sum is $b = 11$. We write (or think) of all two-integer factors of -12:

$$
\begin{array}{l}
\underline{\quad pq \quad} \\
(-3)(4) \\
(3)-(4) \\
(2)(-6) \\
(-2)(6) \\
(-1)(12) \\
(1)(-12)
\end{array}
$$

and see if any of these pairs adds up to 11, the coefficient of the middle term. We see that the next to the last pair works; that is,

$$
\overset{p}{(-1)}\overset{q}{(12)} = \overset{ac}{-12} \qquad \text{and} \qquad \overset{p}{(-1)} + \overset{q}{(12)} = \overset{b}{11}
$$

We can conclude, because of the *ac* test, that $2x^2 + 11x - 6$ can be factored using integer coefficients.

Step 2. Now split the middle term in the original equation, using $p = -1$ and $q = 12$ as coefficients for x. This is possible, since $p + q = b$.

$$
2x^2 + \overset{b}{11x} - 6 = 2x^2 - \overset{p}{1x} + \overset{q}{12x} - 6
$$

Step 3. Factor the result obtained in step 2 by grouping (this will always work if we can get to step 2, and it doesn't matter whether the values for p and q are reversed).

$2x^2 - x + 12x - 6$ Group first two and last two terms.

$= (2x^2 - x) + (12x - 6)$ Factor out common factors in each term.

$= x(2x - 1) + 6(2x - 1)$ Factor out the common factor $(2x - 1)$.

$= (2x - 1)(x + 6)$ The factoring is complete.

Thus,

$2x^2 + 11x - 6 = (2x - 1)(x + 6)$

The process above can be reduced to a few key operational steps when all of the commentary is eliminated and some of the process is done mentally. The only trial and error occurs in step 1, and with a little practice that step will go fairly fast.

Work Problem 20 and check solution in Solutions to Matched Problems on page 95.

PROBLEM 20   Factor $4x^2 + 4x - 3$, if possible, using integer coefficients. Proceed as in Example 20, using the *ac* test.

Solution

EXAMPLE 21

Factor $4x^2 - 7x + 4$, if possible, using integer coefficients.

Solution

Compute ac: $ac = (4)(4) = 16$.

Write (or think) all two-integer factors of 16, and try to find a pair whose sum is -7, the coefficient of the middle term.

pq
(4)(4)
(−4)(−4)
(2)(8)
(−2)(−8)
(1)(16)
(−1)(−16)

None of these adds up to $-7 = b$; thus, according to the *ac* test,

$$4x^2 - 7x + 4$$

is not factorable using integer coefficients.

Work Problem 21 and check solution in Solutions to Matched Problems on page 95.

PROBLEM 21   Factor $6x^2 - 3x - 4$, if possible, using integer coefficients.

Solution

EXAMPLE 22

Factor $6x^2 + 5xy - 4y^2$, if possible, using integer coefficients.

Solution

Compute *ac*: $ac = (6)(-4) = -24$.

Does -24 have two integer factors whose sum is $5 = b$? A little trial and error (either mentally or by listing) will turn up

$$\overset{p}{(8)}\overset{q}{(-3)} = \overset{ac}{-24} \qquad \text{and} \qquad \overset{p}{8} + \overset{q}{(-3)} = \overset{b}{5}$$

We now split the middle term, $5xy$, into two terms and write

$$6x^2 + \overset{b}{5xy} - 4y^2 = 6x^2 + \overset{p}{8xy} - \overset{q}{3xy} - 4y^2$$

Then complete the factoring by grouping:

$$6x^2 + 8xy - 3xy - 4y^2 = (6x^2 + 8xy) - (3xy + 4y^2)$$
$$= 2x(3x + 4y) - y(3x + 4y)$$
$$= (3x + 4y)(2x - y)$$

Thus,

$$6x^2 + 5xy - 4y^2 = (3x + 4y)(2x - y)$$

Work Problem 22 and check solution in Solutions to Matched Problems below.

PROBLEM 22   Factor $6x^2 - 25xy + 4y^2$, if possible, using integer coefficients.

Solution

SOLUTIONS TO MATCHED PROBLEMS _____

20. $ax^2 + bx + c = 4x^2 + 4x - 3$; thus $a = 4$, $b = 4$, and $c = -3$.

$ac = (4)(-3) = -12$

$$
\begin{array}{l}
\underline{pq} \\
(-2)(6) \quad \leftarrow \quad \text{This pair works; that is, } (-2)(6) = -12 \text{ and} \\
(2)(-6) \qquad\quad (-2) + 6 = 4 = b. \\
(3)(-4) \\
(-3)(4) \\
(1)(-12) \\
(-1)(12)
\end{array}
$$

Thus, we write

$$
\begin{aligned}
4x^2 + 4x - 3 &= 4x^2 - 2x + 6x - 3 \\
&= (4x^2 - 2x) + (6x - 3) \\
&= 2x(2x - 1) + 3(2x - 1) \\
&= (2x - 1)(2x + 3)
\end{aligned}
$$

21. $ax^2 + bx + c = 6x^2 - 3x - 4$; thus $a = 6$, $b = -3$, and $c = -4$.

$ac = (6)(-4) = -24$

No two-integer factors of -24 add up to $-3 = b$. We conclude that the polynomial cannot be factored using integer coefficients.

22. $ax^2 + bx + c = 6x^2 - 25xy + 4y^2$; thus $a = 6$, $b = -25$, and $c = 4$.

$ac = (6)(4) = 24$

Testing two-integer factors of 24, we find that $(-1)(-24) = 24$ and $(-1) + (-24) = -25 = b$; thus the polynomial is factorable.

$$
\begin{aligned}
6x^2 - 25xy + 4y^2 &= 6x^2 - xy - 24xy + 4y^2 \\
&= (6x^2 - xy) - (24xy - 4y^2) \\
&= x(6x - y) - 4y(6x - y) = (6x - y)(x - 4y)
\end{aligned}
$$

PRACTICE EXERCISE 2.7

Work odd-numbered problems first, check answers, and then work even-numbered problems in areas of weakness. Answers to all problems are in the back of the book. Make every effort to work a problem yourself before you look at an answer.

Factor, if possible, using integer coefficients. Use the ac test and proceed as in Examples 20 to 22.

A 1. $3x^2 - 7x + 4$ _____

2. $2x^2 - 7x + 6$ _____

3. $x^2 + 4x - 6$ _____

4. $x^2 - 3x - 8$ _____

5. $2x^2 + 5x - 3$ _____

6. $3x^2 - 5x - 2$ _____

7. $3x^2 - 5x + 4$ _____

8. $2x^2 - 11x + 6$ _____

B 9. $3x^2 - 14x + 8$ _____

10. $2y^2 - 13y + 15$ _____

11. $6x^2 + 7x - 5$ _____

12. $5x^2 - 8x - 4$ _____

13. $6x^2 - 4x - 5$ _____

14. $5x^2 - 7x - 4$ _____

15. $2m^2 - 3m - 20$ _____

16. $12x^2 + 16x - 3$ _____

17. $3u^2 - 11u - 4$ _____

18. $8u^2 + 2u - 1$ _____

19. $6u^2 - uv - 2v^2$ _____

20. $6x^2 - 7xy - 5y^2$ _____

21. $3x^2 + 2xy - 3y^2$ _____

22. $2x^2 - 3xy - 4y^2$ _____

23. $8x^2 + 6x - 9$ _____

24. $6x^2 - 5x - 6$ _____

25. $2m^2 + 11mn - 6n^2$ _____

26. $3u^2 + 7uv - 6v^2$ _____

27. $3u^2 - 8uv - 6v^2$ _____

28. $4m^2 - 9mn - 6n^2$ _____

C 29. $4u^2 - 19uv + 12v^2$ _____

30. $12x^2 - xy - 6y^2$ _____

31. $12x^2 - 40xy - 7y^2$ _____

32. $15x^2 + 17xy - 4y^2$ _____

The Check Exercise for this section is on page 117.

2.8 MORE FACTORING

•Sum and difference of two squares
••Sum and difference of two cubes
•••Combined forms
••••More factoring by grouping

In this final section on factoring, we will consider a couple of additional factoring forms as well as combinations of those already learned.

•Sum and difference of two squares

If we multiply $(A - B)$ and $(A + B)$, we obtain

$$(A - B)(A + B) = A^2 - B^2 \qquad \text{The middle term dropped out.}$$

a difference of two squares. Writing this result in reverse order, we obtain a very useful factoring formula. If we try to factor the sum of two squares, $A^2 + B^2$, we find that it cannot be factored using integer coefficients unless A and B have common factors. (Try it to see why.)

Sum and Difference of Two Squares

$A^2 + B^2$ Cannot be factored using integer coefficients, unless   (4)
            A and B have common factors.

$A^2 - B^2 = (A - B)(A + B)$ (5)

EXAMPLE 23

(A) $x^2 - y^2 = (x - y)(x + y)$

(B) $9x^2 - 4 = (3x)^2 - (2)^2 = (3x - 2)(3x + 2)$

(C) $m^2 + n^2$ is not factorable using integer coefficients.

(D) $9u^2 - 25v^2 = (3u)^2 - (5v)^2 = (3u - 5v)(3u + 5v)$

(E) $9x^2 + 36y^2 = 9(x^2 + 4y^2)$ Cannot be factored further.

Work Problem 23 and check solution in Solutions to Matched Problems on page 101.

PROBLEM 23   Factor, if possible, using integer coefficients.

     (A) $x^2 - 25 =$ **(B)** $16u^2 - v^2 =$

     (C) $9x^2 + y^2 =$ **(D)** $7x^2 - 3 =$

     (E) $36u^2 + 4v^2 =$

••Sum and difference of two cubes

It is easy to verify, by direct multiplication of the right sides, the following factoring formulas for the sum and difference of two cubes:

$$
\begin{array}{|c|}
\hline
\textit{Sum and Difference of Two Cubes} \\
\\
A^3 + B^3 = (A + B)(A^2 - AB + B^2) \qquad (6) \\
\\
A^3 - B^3 = (A - B)(A^2 + AB + B^2) \qquad (7) \\
\hline
\end{array}
$$

These formulas are used in the same way as the factoring formula for the difference of two squares. (Notice that neither $A^2 - AB + B^2$ nor $A^2 + AB + B^2$ factors further using integer coefficients.) Both formulas should be memorized.

To factor

$$y^3 - 27$$

we first note that it can be written in the form

$$y^3 - 3^3$$

Thus we are dealing with the difference of two cubes. If in the factoring formula (7) we let $A = y$ and $B = 3$, we obtain

$$A^3 - B^3 = (A - B)(A^2 + AB + B^2)$$

$$y^3 - 27 = y^3 - 3^3 = (y - 3)(y^2 + y3 + 3^2)$$
$$\qquad\qquad\quad = (y - 3)(y^2 + 3y + 9)$$

EXAMPLE 24

$$A^3 + B^3 = (A + B)(A^2 - AB + B^2)$$

(A) $\quad 8x^3 + 27 = (2x)^3 + 3^3 = (2x + 3)[(2x)^2 - (2x)3 + 3^2]$
$$\qquad\qquad\qquad\qquad\qquad = (2x + 3)(4x^2 - 6x + 9)$$

(B) $\quad t^3 - 1 = t^3 - 1^3 = (t - 1)(t^2 + t + 1)$

Work Problem 24 and check solution in Solutions to Matched Problems on page 101.

PROBLEM 24   Factor as far as possible, using integer coefficients.

(A) $\quad u^3 + 8 =$

(B) $\quad m^3 + 1 =$

(C) $\quad u^3 - v^3 =$

(D) $\quad x^3 - 8y^3 =$

•••Combined forms

We now consider some examples that involve removing common factors as well as factoring second- and third-degree forms.

General Factoring Principle

Remove common factors first before proceeding further.

EXAMPLE 25

Factor as far as possible, using integer coefficients.

(A) $18x^3 - 8x$

Solution

$18x^3 - 8x$ Remove common factors first.

$= 2x(9x^2 - 4)$ Factor the difference of two squares.

$= 2x(3x - 2)(3x + 2)$ Factoring is complete.

(B) $3y^3 + 6y^2 + 6y$

Solution

$3y^3 + 6y^2 + 6y$ Remove common factors.

$= 3y(y^2 + 2y + 2)$ Cannot be factored further using integer coefficients.

(C) $3x^4 - 24xy^3$

Solution

$3x^4 - 24xy^3$ Remove common factors.

$= 3x(x^3 - 8y^3)$ Factor difference of two cubes.

$= 3x(x - 2y)(x^2 + 2xy + 4y^2)$

(D) $8x^3y + 20x^2y^2 - 12xy^3$

Solution

$8x^3y + 20x^2y^2 - 12xy^3$

$= 4xy(2x^2 + 5xy - 3y^2)$

$= 4xy(2x - y)(x + 3y)$

Work Problem 25 and check solution in Solutions to Matched Problems on page 101.

PROBLEM 25   Factor as far as possible, using integer coefficients.

(A) $3x^3 - 48x =$

(B) $3x^3 - 15x^2y + 18xy^2 =$

(C) $4x^3 + 12x^2 + 12x =$

(D) $2u^4 - 16u =$

••••More factoring by grouping

Occasionally, polynomial forms of a more general nature than we considered in Section 2.5 can be factored by appropriate grouping of terms. The following example illustrates the process.

EXAMPLE 26

(A) $x^2 + xy + 2x + 2y$ Group the first two and last two terms.

$= (x^2 + xy) + (2x + 2y)$ Remove common factors from each term.

$= x(x + y) + 2(x + y)$ Remove the common factor $(x + y)$.

$= (x + y)(x + 2)$ Notice the factors are not first-degree factors of the same type.

(B) $x^2 - 2x - xy + 2y$

$= (x^2 - 2x) - (xy - 2y)$ Be careful of signs here.

$= x(x - 2) - y(x - 2)$

$= (x - 2)(x - y)$

(C) $y^3 - y^2 - y + 1$

$= (y^3 - y^2) - (y - 1)$

$= y^2(y - 1) - (y - 1)$

$= (y - 1)(y^2 - 1)$

$= (y - 1)(y - 1)(y + 1) = (y - 1)^2(y + 1)$

(D) $4v^2 - u^2 + 6u - 9$ Group the last three terms.

$= 4v^2 - (u^2 - 6u + 9)$

$= 4v^2 - (u - 3)^2$

$= [2v - (u - 3)][2v + (u + 3)]$

$= (2v - u + 3)(2v + u + 3)$

Work Problem 26 and check solution in Solutions to Matched Problems on page 101.

PROBLEM 26   Factor by grouping terms.

(A) $x^2 - xy + 5x - 5y =$

(B) $x^2 + 4x - xy - 4y =$

(C) $u^3 + 2u^2 - 9u - 18 =$

(D) $x^2 - 8x + 16 - 9y^2 =$

SOLUTIONS TO MATCHED PROBLEMS

23. (A) $x^2 - 25 = (x - 5)(x + 5)$ (B) $16u^2 - v^2 = (4u - v)(4u + v)$
(C) $9x^2 + y^2$ not factorable   (D) $7x^2 - 3$ not factorable   (E) $36u^2 + 4v^2 = 4(9u^2 + v^2)$
24. (A) $u^3 + 8 = (u + 2)(u^2 - 2u + 4)$ (B) $m^3 + 1 = (m + 1)(m^2 - m + 1)$
(C) $u^3 - v^3 = (u - v)(u^2 + uv + v^2)$ (D) $x^3 - 8y^3 = (x - 2y)(x^2 + 2xy + 4y^2)$
25. (A) $3x^3 - 48x = 3x(x^2 - 16) = 3x(x - 4)(x + 4)$
(B) $3x^3 - 15x^2y + 18xy^2 = 3x(x^2 - 5xy + 6y^2) = 3x(x - 2y)(x - 3y)$
(C) $4x^3 + 12x^2 + 12x = 4x(x^2 + 3x + 3)$
(D) $2u^4 - 16u = 2u(u^3 - 8) = 2u(u - 2)(u^2 + 2u + 4)$
26. (A) $x^2 - xy + 5x - 5y = (x^2 - xy) + (5x - 5y) = x(x - y) + 5(x - y) = (x - y)(x + 5)$
(B) $x^2 + 4x - xy - 4y = (x^2 + 4x) - (xy + 4y) = x(x + 4) - y(x + 4) = (x + 4)(x - y)$
(C) $u^3 + 2u^2 - 9u - 18 = (u^3 + 2u^2) - (9u + 18) = u^2(u + 2) - 9(u + 2) = (u + 2)(u^2 - 9)$
$= (u + 2)(u - 3)(u + 3)$
(D) $x^2 - 8x + 16 - 9y^2 = (x^2 - 8x + 16) - 9y^2 = (x - 4)^2 - 9y^2 = [(x - 4) - 3y][(x - 4) + 3y]$
$= (x - 4 - 3y)(x - 4 + 3y)$

PRACTICE EXERCISE 2.8

Work odd-numbered problems first, check answers, and then work even-numbered problems in areas of weakness. Answers to all problems are in the back of the book. Make every effort to work a problem yourself before you look at an answer.

Factor as far as possible, using integer coefficients.

A **1.** $v^2 - 25$ _____   **2.** $x^2 - 81$ _____   **3.** $9x^2 - 4$ _____

4. $4m^2 - 1$ _____   **5.** $x^2 + 49$ _____   **6.** $y^2 + 64$ _____

7. $9x^2 - 16y^2$ _____   **8.** $25u^2 - 4v^2$ _____   **9.** $x^3 + 1$ _____

10. $y^3 - 1$ _____   **11.** $m^3 - n^3$ _____   **12.** $p^3 + q^3$ _____

13. $8x^3 + 27$ _____   **14.** $u^3 - 8v^3$ _____   **15.** $6u^2v^2 - 3uv^3$ _____

16. $2x^3y - 6x^2y^3$ _____   **17.** $2x^2 - 8$ _____   **18.** $3y^2 - 27$ _____

19. $2x^3 + 8x$ _____   **20.** $3x^4 + 27x^2$ _____   **21.** $12x^3 - 3xy^2$ _____

22. $2u^3v - 2uv^3$ _____   **23.** $2x^4 + 2x$ _____   **24.** $xy^3 + x^4$ _____

B **25.** $6x^2 + 36x + 48$ _____   **26.** $4x^2 - 4x - 24$ _____

27. $3x^3 - 6x^2 + 15x$ _____

28. $2x^3 - 2x^2 + 8x$ _____

29. $x^2y^2 - 16$ _____

30. $m^2n^2 - 36$ _____

31. $a^3b^3 + 8$ _____

32. $27 - x^3y^3$ _____

33. $4x^3y + 14x^2y^2 + 6xy^3$ _____

34. $3x^3y - 15x^2y^2 + 18xy^3$ _____

35. $4u^3 + 32v^3$ _____

36. $54x^3 - 2y^3$ _____

37. $60x^2y^2 - 200xy^3 - 35y^4$ _____

38. $60x^4 + 68x^3y - 16x^2y^2$ _____

39. $xy + 2x + y^2 + 2y$ _____

40. $x^2 + 3x + xy + 3y$ _____

41. $x^2 - 5x + xy - 5y$ _____

42. $x^2 - 3x - xy + 3y$ _____

43. $ax - 2bx - ay + 2by$ _____

44. $mx + my - 2nx - 2ny$ _____

45. $15ac - 20ad + 3bc - 4bd$ _____

46. $2am - 3an + 2bm - 3bn$ _____

47. $x^3 - 2x^2 - x + 2$ _____

48. $x^3 - 2x^2 + x - 2$ _____

49. $(y - x)^2 - y + x$ _____

50. $x^2(x - 1) - x + 1$ _____

C 51. $r^4 - s^4$ _____

52. $16a^4 - b^4$ _____

53. $x^4 - 3x^2 - 4$ _____

54. $x^4 - 7x^2 - 18$ _____

55. $(x - 3)^2 - 16y^2$ _____

56. $(x + 2)^2 - 9y^2$ _____

57. $(a - b)^2 - 4(c - d)^2$ _____

58. $(x^2 - x)^2 - 9(y^2 - y)^2$ _____

59. $25(4x^2 - 12xy + 9y^2) - 9a^2b^2$ _____

60. $18a^3 - 8a(x^2 + 8x + 16)$ _____

61. $x^6 - 1$ _____

62. $a^6 - 64b^6$ _____

63. $2x^3 - x^2 - 8x + 4$ _____

64. $4y^3 - 12y^2 - 9y + 27$ _____

65. $25 - a^2 - 2ab - b^2$ _____

66. $x^2 - 2xy + y^2 - 9$ _____

67. $16x^4 - x^2 + 6xy - 9y^2$ _____

68. $x^4 - x^2 + 4x - 4$ _____

The Check Exercise for this section is on page 119.

2.9 CHAPTER REVIEW

Important terms and symbols

2.1 POLYNOMIALS   polynomial, degree of a term, degree of a polynomial, monomial, binomial, trinomial

2.2 ADDITION AND SUBTRACTION   coefficient, like terms, combining like terms, removing symbols of grouping, addition of polynomials, subtraction of polynomials

2.3 MULTIPLICATION   multiplication of monomials, multiplication of polynomials, mental multiplication of first-degree polynomials, squaring binomials

2.4 DIVISION   algebraic long division

2.5 FACTORING OUT COMMON FACTORS; FACTORING BY GROUPING   removing common factors, factoring by grouping

2.6 FACTORING SECOND-DEGREE POLYNOMIALS   factoring second-degree polynomials of the form $ax^2 + bx + c$ and $ax^2 + bxy + cy^2$

2.7 ac TEST AND FACTORING ac test for factorability

2.8 MORE FACTORING   sum and difference of squares, sum and difference of cubes, combined forms

$$A^2 - B^2 = (A - B)(A + B)$$

$$a^3 + B^3 = (A + B)(A^2 - AB + B^2)$$

$$A^3 - B^3 = (A - B)(A^2 + AB + B^2)$$

DIAGNOSTIC (REVIEW) EXERCISE 2.9

Work through all the problems in this chapter review and check answers in the back of the book. (Answers to all problems are there, and following each answer is a number in italics indicating the section in which that type of problem is discussed.) Where weaknesses show up, review appropriate sections in the text. When you are satisfied that you know the material, take the practice test following this review.

A **1.** Add: $3x^2 - 2x + 1$, $3x - 2$, and $2x^2 - 3$ _____

2. Subtract: $5x^2 - 2x + 5$ from $3x^2 - x - 2$ _____

3. Multiply: $3x^2y(2x^3 - 3x^2y + y^2)$ _____

4. Multiply: $(3x - 2)(2x + 5)$ _____

5. Multiply: $(2x - 1)(x^2 - 3x + 5)$ _____

6. Divide, using long division: $(6x^2 + 5x - 2)/(2x - 1)$ _____

Factor as far as possible in the integers.

7. $4x^2y - 6xy^2$ _____

8. $x^2 - 9x + 14$ _____

9. $9x^2 - 12x + 4$ _____

10. $t^2 + 4t - 6$ _____

11. $u^2 - 64$ _____

12. $3x^2 - 10x + 8$ _____

13. $x^3 - 5x^2 + 6x$ _____

14. $x(x + y) + y(x + y)$ _____

B 15. Multiply: $(9x^2 - 4)(3x^2 + 7x - 6)$ _____

16. Subtract $2x^2 - 5x - 6$ from the product $(2x - 1)(2x + 1)$. _____

17. Divide, using long division: $(2 - 10x + 9x^3)/(3x - 2)$ _____

18. Given the polynomial $3x^5 - 2x^3 + 7x^2 - x + 2$,

   (A) What is its degree? _____

   (B) What is the degree of the second term? _____

Factor as far as possible in the integers.

19. $m^2 - 3mn - 4n^2$ _____

20. $2m^2 - 8n^2$ _____

21. $12x^3y + 27xy^3$ _____

22. $2x^2 - xy - 3y^2$ _____

23. $6n^3 - 9n^2 - 15n$ _____

24. $3x^2 + 2xy - 7y^2$ _____

25. $xp + xq + yp + yq$ _____

26. $x^2 - xy - 4x + 4y$ _____

27. $(y - b)^2 - y + b$ _____

28. $3x^3 - 24y^3$ _____

29. Multiply: $(4x^2 - 1)(3x^3 - 4x + 3)$ _____

30. Divide, using long division: $(5x - 5 + 8x^4)/(x + 2x^2 - 1)$ _____

31. Simplify: $[(3x^2 - x + 1) - (x^2 - 4)] - [(2x - 5)(x + 3)]$ _____

32. Simplify: $-2x\{(x^2 + 2)(x - 3) - x[x - x(3 - x)]\}$ _____

C *Factor as far as possible in the integers.*

33. $36x^3y + 24x^2y^2 - 45xy^3$ _____

34. $12u^4 - 12u^3v - 20u^2v^2$ _____

35. $(x - y)^2 - x^2$ _____

36. $a^4 - 2a^2b^2 + b^4$ _____

37. $m^6 - n^6$ _____

38. $4x^2 - 9m^2 + 6m - 1$ _____

PRACTICE TEST CHAPTER 2

Take this as if it were a graded test. Allow yourself up to 50 minutes. Work the problems without looking back in the chapter. Correct your work, using the answers (keyed to appropriate sections) in the back of the book.

1. Given the polynomial $7x^5 - x^3 + 2x - 5$, one of the following is false. Which one? _____
   (A) The degree of the second term is 3.
   (B) The degree of the third term is 1.
   (C) The degree of the polynomial is 7.
   (D) The coefficient of the second term is -1.

Problems 2 to 9 refer to the following polynomials:
(1) $x - 2$ (2) $2x - 1$ (3) $2x^2 - 5x + 2$
(4) $3x^2 - 5$ (5) $x^2 - 2x - 3$ (6) $3x^3 - 2x^2 + 5$

2. Add (2), (3), (4), and (5). _____

3. Subtract (3) from (5). _____

4. Multiply (1) and (2). _____

5. Multiply (2) and (3). _____

6. Divide (3) by (2). _____

7. Subtract (5) from the product of (1) and (2). _____

8. Divide (6) by (1). _____

9. Subtract (6) from the product of (2) and (4). _____

10. Divide: $(5 - 5x + 4x^3)/(2x - 1)$. _____

In Problems 11 to 20 factor as far as possible, using integer coefficients.

11. $6x^2 + 5x - 6$ _____

12. $2x^2 - 4x - 3$ _____

13. $12x^3y - 14x^2y - 10xy$ _____

14. $8u^2 - 18$ _____

15. $4x^2 + 16$ _____

16. $m^3 + 8$ _____

17. $2x^4 - 16x$ _____

18. $3x^2 - xy - 6x + 2y$ _____

19. $(x - 3)^2 - 4y^2$ _____

20. $x^4 - 16$ _____

CHECK EXERCISE 2.2

Work the following problems without looking at any text examples. Show your work in the space provided. Write the letter that best indicates your answer in the answer column.

ANSWER COLUMN

1. _____
2. _____
3. _____
4. _____
5. _____
6. _____
7. _____
8. _____
9. _____
10. _____

Simplify Problems 1 to 5 by removing symbols of grouping, if any, and combine like terms.

1. $(x + 3y) - (2x - y) =$

    (A) $x + 2y$
    (B) $-x - 2y$
    (C) $x - 2y$
    (D)   None of these

2. $2m^3n - 4mn^3 + 3mn^3 - m^3n =$

    (A) $m^3n - mn^3$
    (B) $-2m^3n + 2mn^3$
    (C) $2m^3n - 2mn^3$
    (D)   None of these

3. $4(y + 3) - (y - 2) + 3(2y - 1) =$

    (A) $9y$
    (B) $9y - 7$
    (C) $9y + 11$
    (D)   None of these

4. $4[3 - 2(x - 3) - 3x] =$

    (A) $36 - 20x$
    (B) $-12 - 20x$
    (C) $-8x - 12$
    (D)   None of these

5. $2\{3x - [x - (2x - 1)] - 5\} =$

    (A) $8x - 8$
    (B) $8x - 12$
    (C) $8x + 8$
    (D)   None of these

107

Fill in the blank parentheses in Problems 6 and 7 with appropriate algebraic expressions.

6. $7x + 6y^2 + 3y + 2 = 7x + ($? $)$

    (A) $3y + 2$
    (B) $-6y^2 - 3y - 2$
    (C) $6y^2 + 3y + 2$
    (D) None of these

7. $3(x + y) - x - y = 3(x + y) - ($? $)$

    (A) $x - y$
    (B) $x + y$
    (C) $-x + y$
    (D) None of these

Problems 8 to 10 refer to the following three polynomials:

$$2x^2 - 3x + 1 \qquad 3x - 5 \qquad x^2 + 4x - 2$$

8. Add all three polynomials.

    (A) $3x^2 + 4x + 6$
    (B) $3x^2 + 4x - 1$
    (C) $3x^2 + 4x - 6$
    (D) None of these

9. Subtract the sum of the last two polynomials from the first.

    (A) $3x^2 + 4x - 6$
    (B) $x^2 - 10x + 8$
    (C) $-x^2 + 10x - 8$
    (D) None of these

10. Subtract the first from the third.

    (A) $x^2 - 7x + 3$
    (B) $-x^2 + 7x - 3$
    (C) $4x^2 - 6x + 2$
    (D) None of these

CHECK EXERCISE 2.3

NAME _____

CLASS _____

SCORE _____

Work the following problems without looking at any text examples. Show your work in the space provided. Write the letter that best indicates your answer in the answer column.

ANSWER COLUMN

1. _____
2. _____
3. _____
4. _____
5. _____
6. _____
7. _____
8. _____
9. _____
10. _____

Multiply in Problems 1 to 9.

1. $(-2x^2y)(3xy^2 - 2x^2y) =$

   (A) $-5x^3y^3 + 4x^4y^2$
   (B) $-6x^3y^3 + 4x^4y^2$
   (C) $-6x^3y^3 - 4x^4y^2$
   (D) None of these

2. $(3x - 2)(x + 4) =$

   (A) $3x^2 + 16x - 8$
   (B) $3x^2 + 10x - 6$
   (C) $3x^2 - 3x - 8$
   (D) None of these

3. $(5x - y)(2x - 3y) =$

   (A) $10x^2 - 17xy - 3y^2$
   (B) $10x^2 - 17xy + 3y^2$
   (C) $10x^2 - 13xy + 3y^2$
   (D) None of these

4. $(2m - 5n)(3m + 2n) =$

   (A) $6m^2 - 11m - 10n^2$
   (B) $6m^2 - 19m - 10n^2$
   (C) $6m^2 - 11m + 10n^2$
   (D) None of these

5. $2x^2 - 3x + 2$
       $\underline{\quad\quad 3x - 2}$

   (A) $6x^3 + x^2 - 4$
   (B) $6x^3 - 13x - 6x - 4$
   (C) $6x^3 - 13x^2 + 12x - 4$
   (D) None of these

6. $2x^2 - xy + 3y^2$
    $\underline{x^2 + 2xy - y^2}$

    (A) $2x^4 + x^3y + 3x^2y^2 + 7xy^3 - 3y^4$
    (B) $2x^4 + 3x^3y + 7xy^3 - 3y^4$
    (C) $2x^4 + 3x^3y - x^2y^2 + 7xy^3 - 3y^4$
    (D)   None of these

7. $(5a - 2)^2 =$

    (A) $25a^2 - 4$
    (B) $25a^2 + 4$
    (C) $25a^2 - 20a + 4$
    (D)   None of these

8. $(6m - n)(6m + n)$

    (A) $36m^2 - n^2$
    (B) $36m^2 - 12mn - n^2$
    (C) $36m^2 + n^2$
    (D)   None of these

9. $(4x + 5y)^2 =$

    (A) $16x^2 + 25y^2$
    (B) $16x^2 + 40xy + 25y^2$
    (C) $8x^2 + 40xy + 10y^2$
    (D)   None of these

10. Simplify:

    $(4x - 1)(2x + 3) - (2x - 3)^2 =$

    (A) $4x^2 + 22x - 12$
    (B) $4x^2 - 2x + 6$
    (C) $4x^2 + 10x + 6$
    (D)   None of these

CHECK EXERCISE 2.4

Work the following problems without looking at any text examples. Show your work in the space provided. Write the letter that best indicates your answer in the answer column.

Divide, using the long-division process, and check answers.

1. $(6x^2 - 5x - 2)/(2x - 3)$

(**A**) $3x + 2$, R $= -8$
(**B**) $3x - 7$, R $= 19$
(**C**) $3x + 2$, R $= 4$
(**D**) None of these

2. $(x^2 + 4)/(x - 2)$

(**A**) $x + 2$
(**B**) $x - 2$
(**C**) $x + 2$, R $= -4$
(**D**) None of these

3. $(8x^3 - 8x + 3)/(2x - 1)$

(**A**) $4x^2 - 2$, R $= 1$
(**B**) $4x^2 + 2x - 3$
(**C**) $4x^2 - 2x - 3$
(**D**) None of these

111

4. $(1 - x^2 - 7x + x^3)/(x - 3)$

    (A) $x^2 + 2x - 1$, $R = -2$
    (B) $x^2 + 2x - 1$, $R = 2$
    (C) $x^2 - 4x + 5$, $R = -14$
    (D) None of these

5. $(2x^4 - 3 + 5x + 3x^2 - 3x^3)/(3 - 2x + x^2)$

    (A) $2x^2 + x - 1$
    (B) $2x^2 + x - 7$, $R = -12x - 24$
    (C) $2x^2 - 7x - 5$, $R = 14x + 12$
    (D) None of these

CHECK EXERCISE 2.5

Work the following problems without looking at any text examples. Show your work in the space provided. Write the letter that best indicates your answer in the answer column.

Write Problems 1 to 5 in factored form by removing all factors common to all terms.

ANSWER COLUMN

1. _____
2. _____
3. _____
4. _____
5. _____
6. _____
7. _____
8. _____
9. _____
10. _____

1. $8u^3v - 12uv^2 =$

    (A) $2u(4u^2v - 6v^2)$
    (B) $4(2u^3v - 3uv^2)$
    (C) $4u(2u^2v - 3v^2)$
    (D) None of these

2. $6m^2n - 6mn - 8mn^2 =$

    (A) $2mn(3m - 3 - 4n)$
    (B) $6mn(m - 1) - 8mn^2$
    (C) Not factorable
    (D) None of these

3. $7x(x - 3) + 2(x - 3) =$

    (A) $7x(x - 3) + 2$
    (B) $(x - 3)(7x + 2)$
    (C) Not factorable
    (D) None of these

4. $5y(2y + 1) - (2y + 1) =$

    (A) $5y(2y + 1)$
    (B) $(2y + 1)(5y - 1)$
    (C) Not factorable
    (D) None of these

5. $3m(2n - 1) - 2(2n + 1) =$

    (A) $(2n - 1)(3m - 2)$
    (B) $(2n + 1)(3m - 2)$
    (C) Not factorable
    (D) None of these

113

6. Fill in the open parentheses with an appropriate algebraic expression:

   $6x^2 + 9x - 2x - 3 = (6x^2 + 9x) - ($ $)$

   (A) $2x - 3$
   (B) $2x + 3$
   (C) $-2x + 3$
   (D) None of these

7. Remove all common factors from each group; then complete the factoring if possible.

   $(3x^2 - 3x) - (2x - 2) =$

   (A) $(x - 1)(3x - 2)$
   (B) $(x - 1)(3x - 1)$
   (C) Not factorable
   (D) None of these

Factor Problems 8 to 10 as the product of two first-degree factors, using grouping.

8. $6x^2 - 9x + 4x - 6 =$

   (A) $(2x + 3)(3x - 2)$
   (B) $(2x - 3)(3x + 2)$
   (C) Not factorable
   (D) None of these

9. $6x^2 + 15x - 2x - 5 =$

   (A) $(2x - 5)(3x - 1)$
   (B) $3x(2x + 5)$
   (C) Not factorable
   (D) None of these

10. $10m^2 + 15mn - 4mn - 6n^2$

    (A) $(2m + 3n)(5m - 2n)$
    (B) $(2m - 3n)(5m - 2n)$
    (C) Not factorable
    (D) None of these

CHECK EXERCISE 2.6

Work the following problems without looking at any text examples. Show your work in the space provided. Write the letter that best indicates your answer in the answer column.

Factor, using integer coefficients. If not factorable, say so.

1. $2x^2 - 7x + 3 =$

    (A) $(2x - 3)(x - 1)$
    (B) $(2x - 1)(x - 3)$
    (C) Not factorable
    (D) None of these

2. $x^2 - 4x + 6 =$

    (A) $(x - 2)(x - 3)$
    (B) $(x - 1)(x - 6)$
    (C) Not factorable
    (D) None of these

3. $x^2 + x - 6 =$

    (A) $(x + 2)(x - 3)$
    (B) $(x - 2)(x - 3)$
    (C) Not factorable
    (D) None of these

4. $x^2 - 5xy + 6y^2 =$

    (A) $(x - 3y)(x - 2y)$
    (B) $(x - 6y)(x - y)$
    (C) Not factorable
    (D) None of these

5. $2u^2 + uv + 3v^2 =$

    (A) $(2u + v)(u + 3v)$
    (B) $(2u + 3v)(u + v)$
    (C) Not factorable
    (D) None of these

6. $8x^2 + 27x + 12 =$

   (A) $(8x + 3)(x + 4)$
   (B) $(4x + 3)(2x + 4)$
   (C) Not factorable
   (D) None of these

7. $4x^2 + 4x - 3 =$

   (A) $(2x - 1)(2x + 3)$
   (B) $(4x - 1)(x + 3)$
   (C) Not factorable
   (D) None of these

8. $4m^2 + 15mn - 4n^2 =$

   (A) $(4m - n)(m - 4n)$
   (B) $(m - 4n)(4m + n)$
   (C) Not factorable
   (D) None of these

9. $12x^2 + xy - 6y^2 =$

   (A) $(6x + 2y)(2x - 3y)$
   (B) $(3x - 2y)(4x + 3y)$
   (C) Not factorable
   (D) None of these

10. $16w^2 + 14w - 15 =$

   (A) $(4w - 5)(4w + 3)$
   (B) $(8w - 5)(2w + 3)$
   (C) Not factorable
   (D) None of these

CHECK EXERCISE 2.7

Work the following problems without looking at any text examples. Show your work in the space provided. Write the letter that best indicates your answer in the answer column.

Factor, if possible, using integer coefficients. Use the ac test.

1. $2x^2 + 7x - 4$

    (A) $(2x + 1)(x - 4)$
    (B) $(2x - 1)(x + 4)$
    (C) Not factorable
    (D) None of these

2. $4x^2 + 5x - 6$

    (A) $(4x - 3)(x + 2)$
    (B) $(2x - 1)(2x + 6)$
    (C) Not factorable
    (D) None of these

3. $6u^2 - 3u - 4$

    (A) $(3u - 2)(2u + 2)$
    (B) $(6u - 1)(u + 4)$
    (C) Not factorable
    (D) None of these

4. $3x^2 - 11xy + 6y^2$

    (A) $(3x - y)(x - 6y)$
    (B) $(x - 3y)(3x - 2y)$
    (C)   Not factorable
    (D)   None of these

5. $4u^2 + 9uv - 4v^2$

    (A) $(4u - v)(u + 4v)$
    (B) $(2u - 2v)(2u + 2v)$
    (C)   Not factorable
    (D)   None of these

CHECK EXERCISE 2.8

NAME _____

CLASS _____

SCORE _____

Work the following problems without looking at any text examples. Show your work in the space provided. Write the letter that best indicates your answer in the answer column.

Factor as far as possible, using integer coefficients.

ANSWER COLUMN

1. _____
2. _____
3. _____
4. _____
5. _____
6. _____
7. _____
8. _____
9. _____
10. _____

1. $4u^2 - 1 =$

    (A) $(2u - 1)^2$
    (B) $(2u - 1)(2u + 1)$
    (C) Not factorable
    (D) None of these

2. $m^3 + n^3 =$

    (A) $(m + n)^3$
    (B) $(m + n)(m^2 - n^2)$
    (C) Not factorable
    (D) None of these

3. $2u^2 + 8v^2 =$

    (A) $2(u + 2v)^2$
    (B) $2(u^2 + 4v^2)$
    (C) Not factorable
    (D) None of these

4. $8y^3 - 28y^2 - 16y =$

    (A) $4y(2y^2 - 7y - 4)$
    (B) $4y(y - 4)(2y + 1)$
    (C) Not factorable
    (D) None of these

5. $2x^3y - 6x^2y + 12xy =$

    (A) $2xy(x^2 - 3x + 6)$
    (B) $2xy(x - 2)(x - 3)$
    (C) Not factorable
    (D) None of these

6. $2m^4 - 16m =$

    (A) $2m(m - 2)(m^2 + 2m + 4)$
    (B) $2m(m^3 - 8)$
    (C) Not factorable
    (D) None of these

7. $ac + ad - bc - bd =$

    (A) $(a - b)(c + d)$
    (B) $(a + b)(c - d)$
    (C) Not factorable
    (D) None of these

8. $x^4 - y^4 =$

    (A) $(x - y)^4$
    (B) $(x - y)(x + y)(x + y)^2$
    (C) $(x - y)(x + y)(x^2 + y^2)$
    (D) None of these

9. $(u - 2)^2 - 4v^2 =$

    (A) $[(u - 2) - 2v][(u - 2) + 2v]$
    (B) $[(u - 2) - 2v]^2$
    (C) Not factorable
    (D) None of these

10. $x^3 - 3x^2 - 4x + 12 =$

    (A) $(x - 3)(x^2 - 4)$
    (B) $(x - 3)(x - 2)(x + 2)$
    (C) $(x + 2)(x^2 - 5x + 6)$
    (D) $(x - 2)(x^2 - x + 6)$

Chapter Three

ALGEBRAIC FRACTIONS

CONTENTS

Is the "box" on the left or right side of the figure?

INSTRUCTIONS FOR STUDENTS IN A
SELF-PACED CLASS OR LAB

yes — **HAVE YOU HAD INTERMEDIATE ALGEBRA BEFORE THIS COURSE?** — no

1. Work Diagnostic (Review) Exercise 3.5 on page 144. Check answers in back of book; then work through text sections corresponding to problems missed. (Section numbers are in italics following each answer.)
2. When finished with step 1, take Practice Test Chapter 3 on page 145 as a final check of your understanding of the chapter. Check answers in the back of the book; then review sections where weakness still prevails. (Corresponding section numbers are in italics following each answer.)
3. When you think you are ready, ask your instructor for a graded test for Chapter 3.
4. If your instructor approves, after the test is corrected, go to the next chapter.

1. Work through each section in the chapter as follows:
   (a) Read discussion.
   (b) Read each example and work the corresponding matched problem. Check your solution to the matched problem in Solutions to Matched Problems on the indicated page.
   (c) At the end of a section work the odd-numbered problems in the Practice Exercise and check answers; then work even-numbered problems in areas of weakness. (Answers to *all* Practice Exercise sets are in the back of the book.)
   (d) Work Check Exercise as instructed. Tear out and turn in as directed by your instructor. (Answers are not in the text.)
2. Repeat each step in item 1 for each section in the chapter.
3. After the instructional part of the chapter is completed, proceed with steps 1 to 4 in the box above.

3.1 RATIONAL EXPRESSIONS

- •Rational expressions
- ••Fundamental principle of fractions
- •••Reducing to lowest terms
- ••••Raising to higher terms

•Rational expressions

Fractional forms in which the numerator and denominator are polynomials are called **rational expressions.** For example,

$$\frac{1}{x} \qquad \frac{-6}{y-3} \qquad \frac{x-2}{x^2-2x+5} \qquad \frac{x^2-3xy+y^2}{3x^3y^4} \qquad \frac{3x^2-8x-4}{5}$$

are all rational expressions. (Recall that a nonzero constant is a polynomial of degree 1.)

Each rational expression names a real number for real number replacements of the variables, division by zero excluded. Hence, all properties of the real numbers apply to those expressions.

••Fundamental principle of fractions

You will recall from arithmetic that

$$\frac{8}{12} = \frac{8 \div 4}{12 \div 4} = \frac{2}{3} \qquad \text{and} \qquad \frac{3}{5} = \frac{2 \cdot 3}{2 \cdot 5} = \frac{6}{10}$$

The first example illustrates reducing a fraction to lowest terms, while the second example illustrates raising a fraction to higher terms. Both illustrate the use of the **fundamental principle of fractions.**

The Fundamental Principle of Fractions

For all polynomials P, Q, and $K(Q, K \neq 0)$,

$$\frac{PK}{QK} = \frac{P}{Q}$$ We may cancel a common factor out of the numerator and denominator (which is the same as dividing the top and bottom by the same nonzero quantity).

$$\frac{P}{Q} = \frac{KP}{KQ}$$ We may multiply the numerator and denominator (top and bottom) by the same nonzero quantity.

•••Reducing to lowest terms

To reduce a fraction to **lowest terms** is to cancel *all* common factors from the numerator and denominator.

COMMON ERROR

Common terms do not cancel: $\quad \dfrac{2+y}{3+y} \neq \dfrac{2+\overset{1}{\cancel{y}}}{3+\underset{1}{\cancel{y}}} \neq \dfrac{2}{3}$

Common factors do cancel:
$$\frac{2y}{3y} = \frac{2\overset{1}{\cancel{y}}}{3\underset{1}{\cancel{y}}} = \frac{2}{3}$$

EXAMPLE 1
Reduce to lowest terms.

(A) $\dfrac{8x^2y}{12xy^2} = \dfrac{\overset{1}{\cancel{(4xy)}}(2x)}{\underset{1}{\cancel{(4xy)}}(3y)}$

Cancel common factors (and/or use exponent properties from Section 1.6).

$$= \frac{2x}{3y}$$

(B) $\dfrac{6x^2 - 3x}{3x} = \dfrac{\overset{1}{\cancel{3x}}(2x - 1)}{\underset{1}{\cancel{3x}}}$

Factor the top; then cancel common factors.

$$= 2x - 1$$

NOTE: $\dfrac{6x^2 - \overset{1}{\cancel{3x}}}{\underset{1}{\cancel{3x}}}$ is wrong (why?)

(C) $\dfrac{x^2y - xy^2}{x^2 - xy} = \dfrac{\overset{1}{\cancel{xy}}\overset{1}{\cancel{(x-y)}}}{\underset{1}{\cancel{x}}\underset{1}{\cancel{(x-y)}}}$

Factor the top and bottom; then cancel common factors

$$= y$$

(D) $\dfrac{3x^3 - 2x^2 - 8x}{x^4 - 8x} = \dfrac{\overset{1}{\cancel{x}}(3x^2 - 2x - 8)}{\underset{1}{\cancel{x}}(x^3 - 8)}$

$$= \frac{(3x + 4)\overset{1}{\cancel{(x-2)}}}{\underset{1}{\cancel{(x-2)}}(x^2 + 2x + 4)}$$

$$= \frac{3x + 4}{x^2 + 2x + 4}$$

Work Problem 1 and check solution in Solutions to Matched Problems on page 126.

PROBLEM 1   Reduce to lowest terms.

(A) $\dfrac{16u^5v^3}{24u^3v^6} =$

(B) $\dfrac{4x}{8x^2 - 4x} =$

(C) $\dfrac{x^2 - 3x}{x^2y - 3xy} =$

(D) $\dfrac{2x^3 - 8x}{2x^4 + 16x} =$

••••Raising to higher terms

The following example shows how to use the fundamental principle of fractions to raise fractions to higher terms.

EXAMPLE 2

(A) $\dfrac{3}{2x} = \dfrac{(4xy)(3)}{(4xy)(2x)} = \dfrac{12xy}{8x^2y}$

(B) $\dfrac{2x}{3y} = \dfrac{(x - y)2x}{(x - y)3y} = \dfrac{2x^2 - 2xy}{3xy - 3y^2}$

(C) $\dfrac{x - 2y}{2x + y} = \dfrac{(2x - y)(x - 2y)}{(2x - y)(2x + y)} = \dfrac{2x^2 - 5xy + 2y^2}{4x^2 - y^2}$

Work Problem 2 and check solution in Solutions to Matched Problems on page 126.

PROBLEM 2   Complete the raising-to-higher-terms process by filling in the unshaded areas with appropriate expressions.

(A) $\dfrac{3u^2}{4v} = \dfrac{\boxed{}}{(5u^2v)(4v)} = \dfrac{\boxed{}}{20u^2v^2}$

(B) $\dfrac{3m}{2n} = \dfrac{3m(m - 1)}{\boxed{}} = \dfrac{3m^2 - 3m}{\boxed{}}$

(C) $\dfrac{2x - 3}{x - 4} = \dfrac{\boxed{}}{(3x - 2)(x - 4)} = \dfrac{\boxed{}}{3x^2 - 14x + 8}$

SOLUTIONS TO MATCHED PROBLEMS

1. **(A)** $\dfrac{16u^5v^3}{24u^3v^6} = \dfrac{2u^2}{3v^3}$ **(B)** $\dfrac{4x}{8x^2 - 4x} = \dfrac{\overset{1}{\cancel{4x}}}{\underset{1}{\cancel{4x}}(2x - 1)} = \dfrac{1}{2x - 1}$ **(C)** $\dfrac{x^2 - 3x}{x^2y - 3xy} = \dfrac{\overset{1}{\cancel{x}}\overset{1}{\cancel{(x - 3)}}}{\underset{1}{\cancel{xy}}\underset{1}{\cancel{(x - 3)}}} = \dfrac{1}{y}$

   (D) $\dfrac{2x^3 - 8x}{2x^4 + 16x} = \dfrac{2x(x^2 - 4)}{2x(x^3 + 8)} = \dfrac{\overset{1}{\cancel{2x}}(x - 2)\overset{1}{\cancel{(x + 2)}}}{\underset{1}{\cancel{2x}}\underset{1}{\cancel{(x + 2)}}(x^2 + 2x + 4)} = \dfrac{x - 2}{x^2 + 2x + 4}$

2. **(A)** $\dfrac{3u^2}{4v} = \dfrac{(5u^2v)(3u^2)}{(5u^2v)(4v)} = \dfrac{15u^4v}{20u^2v^2}$ **(B)** $\dfrac{3m}{2n} = \dfrac{3m(m - 1)}{2n(m - 1)} = \dfrac{3m^2 - 3m}{2mn - 2n}$

   (C) $\dfrac{2x - 3}{x - 4} = \dfrac{(3x - 2)(2x - 3)}{(3x - 2)(x - 4)} = \dfrac{6x^2 - 13x + 6}{3x^2 - 14x + 8}$

PRACTICE EXERCISE 3.1

Work odd-numbered problems first, check answers, and then work even-numbered problems in areas of weakness. Answers to all problems are in the back of the book. Make every effort to work a problem yourself before you look at an answer.

A *Reduce to lowest terms.*

1. $\dfrac{3x^3}{6x^5}$ _____

2. $\dfrac{9u^5}{3u^6}$ _____

3. $\dfrac{14x^3y}{21xy^2}$ _____

4. $\dfrac{20m^4n^6}{15m^5n^2}$ _____

5. $\dfrac{15y^3(x - 9)^3}{5y^4(x - 9)^2}$ _____

6. $\dfrac{2x^2(x + 7)}{6x(x + 7)}$ _____

7. $\dfrac{(2x - 1)(2x + 1)}{3x(2x + 1)}$ _____

8. $\dfrac{(x + 3)(2x + 5)}{2x^2(2x + 5)}$ _____

9. $\dfrac{x^2 - 2x}{2x - 4}$ _____

10. $\dfrac{2x^2 - 10x}{4x - 20}$ _____

11. $\dfrac{m^2 - mn}{m^2n - mn^2}$ _____

12. $\dfrac{a^2b + ab^2}{ab + b^2}$ _____

Complete the raising-to-higher-terms process by replacing the question marks with appropriate expressions.

13. $\dfrac{3}{2x} = \dfrac{?}{8x^2y}$ _____

14. $\dfrac{5x}{3} = \dfrac{10x^3y^2}{?}$ _____

15. $\dfrac{7}{3y} = \dfrac{?}{6x^3y^2}$ _____

16. $\dfrac{5u}{4v^2} = \dfrac{20u^3v}{?}$ _____

B *Reduce to lowest terms.*

17. $\dfrac{x^2 + 6x + 8}{3x^2 + 12x}$ _____

18. $\dfrac{x^2 + 5x + 6}{2x^2 + 6x}$ _____

19. $\dfrac{x^2 - 9}{x^2 + 6x + 9}$ _____

20. $\dfrac{x^2 - 4}{x^2 + 4x + 4}$ _____

21. $\dfrac{4x^2 - 9y^2}{4x^2y + 6xy^2}$ _____

22. $\dfrac{a^2 - 16b^2}{4ab - 16b^2}$ _____

23. $\dfrac{x^2 - xy + 2x - 2y}{x^2 - y^2}$ _____   **24.** $\dfrac{u^2 + uv - 2u - 2v}{u^2 + 2vu + v^2}$ _____   **25.** $\dfrac{6x^3 + 28x^2 - 10x}{12x^3 - 4x^2}$ _____

26. $\dfrac{12x^3 - 78x^2 - 42x}{16x^4 + 8x^3}$ _____   **27.** $\dfrac{x^3 - 8}{x^2 - 4}$ _____   **28.** $\dfrac{y^3 + 27}{2y^3 - 6y^2 + 18y}$ _____

Complete the raising-to-higher-terms process by replacing the question marks with appropriate expressions. (Factor any expressions on the right as a first step.)

29. $\dfrac{3x}{4y} = \dfrac{?}{4xy + 4y^2}$ _____   **30.** $\dfrac{4m}{5n} = \dfrac{4m^2 - 4mn}{?}$ _____

31. $\dfrac{x - 2y}{x + y} = \dfrac{x^2 - 3xy + 2y^2}{?}$ _____   **32.** $\dfrac{2x + 5}{x - 3} = \dfrac{?}{3x^2 - 8x - 3}$ _____

C *Reduce to lowest terms.*

33. $\dfrac{x^3 - y^3}{3x^3 + 3x^2y + 3xy^2}$ _____   **34.** $\dfrac{2u^3v - 2u^2v^2 + 2uv^3}{u^3 + v^3}$ _____

35. $\dfrac{ux + vx - uy - vy}{2ux + 2vx + uy + vy}$ _____   **36.** $\dfrac{mx - 2my + nx - 2ny}{mx - 2my - nx + 2ny}$ _____

37. $\dfrac{x^4 - y^4}{(x^2 - y^2)(x + y)^2}$ _____   **38.** $\dfrac{x^4 - 2x^2y^2 + y^4}{x^4 - y^4}$ _____

The Check Exercise for this section is on page 147.

3.2 MULTIPLICATION AND DIVISION

•Multiplication
••Division

•Multiplication

We start with multiplication of rational forms.

> *Multiplication*
>
> If P, Q, R, and S are polynomials (Q, $S \neq 0$), then
>
> $$\frac{P}{Q} \cdot \frac{R}{S} = \frac{P \cdot R}{Q \cdot S}$$

EXAMPLE 3

(A) $\dfrac{3a^2b}{4c^2d} \cdot \dfrac{8c^2d^3}{9ab^2} = \dfrac{(3a^2b) \cdot (8c^2d^3)}{(4c^2d) \cdot (9ab^2)} = \dfrac{24a^2bc^2d^3}{36ab^2c^2d}$

$$= \dfrac{(2ad^2)\overset{1}{\cancel{(12abc^2d)}}}{(3b)\underset{1}{\cancel{(12abc^2d)}}} = \dfrac{2ad^2}{3b}$$

This process is easily shortened to the following when it is realized that, in effect, any factor in either numerator may "cancel" any like factor in either denominator. Thus,

$$\dfrac{\overset{1 \cdot a \cdot 1}{\cancel{3a^2b}}}{\underset{1 \cdot 1 \cdot 1}{\cancel{4c^2d}}} \cdot \dfrac{\overset{2 \cdot 1 \cdot d^2}{\cancel{8c^2d^3}}}{\underset{3 \cdot 1 \cdot b}{\cancel{9ab^2}}} = \dfrac{2ad^2}{3b}$$

The process is not as confusing as it looks, when you proceed one step at a time.

(B) $(x^2 - 4) \cdot \dfrac{2x - 3}{x + 2} = \dfrac{(x + 2)(x - 2)}{1} \cdot \dfrac{(2x - 3)}{\cancel{(x + 2)}}$

Factor where possible; cancel common factors; then write the answer.

$$= (x - 2)(2x - 3)$$

(C) $\dfrac{4a^2 - 9b^2}{4a^2 + 12ab + 9b^2} \cdot \dfrac{6a^2b}{8a^2b^2 - 12ab^3}$

$$= \dfrac{\cancel{(2a - 3b)}\cancel{(2a + 3b)}}{\underset{(2a + 3b)}{\cancel{(2a + 3b)^2}}} \cdot \dfrac{\overset{3a}{\cancel{6a^2b}}}{\underset{2b \quad 1}{\cancel{4ab^2}\cancel{(2a - 3b)}}}$$

$$= \dfrac{3a}{2b(2a + 3b)}$$

Work Problem 3 and check answer in Solutions to Matched Problems on page 130.

PROBLEM 3 Multiply and reduce to lowest terms:

(A) $\dfrac{4x^2y^3}{9w^2z} \cdot \dfrac{3wz^2}{2xy^4} =$

(B) $\dfrac{x + 5}{x^2 - 9} \cdot (x + 3) =$

(C) $\dfrac{x^2 - 9y^2}{x^2 - 6xy + 9y^2} \cdot \dfrac{6x^2y}{2x^2 + 6xy} =$

••Division

We now turn to division of rational forms.

Division

If $P, Q, R,$ and S are polynomials ($Q, R, S \neq 0$), then

Divisor$\quad$reciprocal of divisor

$$\frac{P}{Q} \div \frac{R}{S} = \frac{P}{Q} \cdot \frac{S}{R}$$

That is, to divide one fraction by another, multiply by the reciprocal of the divisor.

To show that the indicated division process is a valid procedure, one has only to show that the product of the divisor and quotient is the dividend:

(Divisor) · (quotient) = $\quad$ (dividend)

$$\frac{R}{S} \cdot \left(\frac{P}{Q} \cdot \frac{S}{R} \right) = \frac{\overset{1}{\cancel{R}}}{\cancel{S}} \cdot \frac{P}{Q} \cdot \frac{\overset{1}{\cancel{S}}}{\cancel{R}} = \frac{P}{Q}$$

EXAMPLE 4

(A) $\quad \dfrac{6a^2b^3}{5cd} \div \dfrac{3a^2c}{10bd} = \dfrac{6a^2b^3}{5cd} \cdot \dfrac{10bd}{3a^2c} = \dfrac{4b^4}{c^2}$

(B) $\quad (x + 4) \div \dfrac{2x^2 - 32}{6xy} = \dfrac{x + 4}{1} \cdot \dfrac{6xy}{2(x - 4)(x + 4)} = \dfrac{3xy}{x - 4}$

(C) $\quad \dfrac{10x^3y}{3xy + 9y} \div \dfrac{4x^2 - 12x}{x^2 - 9} = \dfrac{10x^3y}{3y(x + 3)} \cdot \dfrac{(x + 3)(x - 3)}{4x(x - 3)} = \dfrac{5x^2}{6}$

Work Problem 4 and check solution in Solutions to Matched Problems on page 130.

PROBLEM 4 Divide and reduce to lowest terms:

(A) $\quad \dfrac{8w^2z^2}{9x^2y} \div \dfrac{4wz}{6xy^2} =$

(B) $\quad \dfrac{2x^2 - 8}{4x} \div (x + 2) =$

(C) $\quad \dfrac{x^2 - 4x + 4}{4x^2y - 8xy} \div \dfrac{x^2 + x - 6}{6x^2 + 18x} =$

SOLUTIONS TO MATCHED PROBLEMS

3. **(A)** $\dfrac{4x^2y^3}{9w^2z}\cdot\dfrac{3wz^2}{2xy^4}=\dfrac{2xz}{3wy}$

(B) $\dfrac{x+5}{x^2-9}\cdot\dfrac{x+3}{1}=\dfrac{x+5}{(x-3)(x+3)}\cdot\dfrac{x+3}{1}=\dfrac{x+5}{x-3}$

(C) $\dfrac{x^2-9y^2}{x^2-6xy+9y^2}\cdot\dfrac{6x^2y}{2x^2+6xy}=\dfrac{(x-3y)(x+3y)}{(x-3y)^2}\cdot\dfrac{6x^2y}{2x(x+3y)}=\dfrac{3xy}{x-3y}$

4. **(A)** $\dfrac{8w^2z^2}{9x^2y}\div\dfrac{4wz}{6xy^2}=\dfrac{8w^2z^2}{9x^2y}\cdot\dfrac{6xy^2}{4wz}=\dfrac{4wzy}{3x}$

(B) $\dfrac{2x^2-8}{4x}\div\dfrac{x+2}{1}=\dfrac{2(x-2)(x+2)}{4x}\cdot\dfrac{1}{x+2}=\dfrac{x-2}{2x}$

(C) $\dfrac{x^2-4x+4}{4x^2y-8xy}\div\dfrac{x^2+x-6}{6x^2+18x}=\dfrac{(x-2)^2}{4xy(x-2)}\cdot\dfrac{6x(x+3)}{(x+3)(x-2)}=\dfrac{3}{2y}$

PRACTICE EXERCISE 3.2

Work odd-numbered problems first, check answers, and then work even-numbered problems in areas of weakness. Answers to all problems are in the back of the book. Make every effort to work a problem yourself before you look at an answer.

In answers do not change improper fractions to mixed fractions; that is, write $\frac{7}{2}$, not $3\frac{1}{2}$.

A *Multiply and reduce to lowest terms.*

1. $\dfrac{10}{9}\cdot\dfrac{12}{15}$ _____

2. $\dfrac{3}{7}\cdot\dfrac{14}{9}$ _____

3. $\dfrac{2a}{3bc}\cdot\dfrac{9c}{a}$ _____

4. $\dfrac{2x}{3yz}\cdot\dfrac{6y}{4x}$ _____

5. $\dfrac{3x^2}{4}\cdot\dfrac{16y}{12x^3}$ _____

6. $\dfrac{2x^2}{3y^2}\cdot\dfrac{9y}{4x}$ _____

Divide and reduce to lowest terms.

7. $\dfrac{9m}{8n}\div\dfrac{3m}{4n}$ _____

8. $\dfrac{6x}{5y}\div\dfrac{3x}{10y}$ _____

9. $\dfrac{a}{4c}\div\dfrac{a^2}{12c^2}$ _____

10. $\dfrac{2x}{3y}\div\dfrac{4x}{6y^2}$ _____

11. $\dfrac{x}{3y}\div 3y$ _____

12. $2xy\div\dfrac{x}{y}$ _____

Perform the indicated operations and reduce to lowest terms.

13. $\dfrac{8x^2}{3xy}\cdot\dfrac{12y^3}{6y}$ _____

14. $\dfrac{6a^2}{7c}\cdot\dfrac{21cd}{12ac}$ _____

15. $\dfrac{21x^2y^2}{12cd}\div\dfrac{14xy}{9d}$ _____

16. $\dfrac{3uv^2}{5w}\div\dfrac{6u^2v}{15w}$ _____

17. $\dfrac{9u^4}{4v^3}\div\dfrac{-12u^2}{15v}$ _____

18. $\dfrac{-6x^3}{5y^2}\div\dfrac{18x}{10y}$ _____

19. $\dfrac{3c^2d}{a^3b^3}\div\dfrac{3a^3b^3}{cd}$ _____

20. $\dfrac{uvw}{5xyz}\div\dfrac{5vy}{uwxz}$ _____

B 21. $\dfrac{3x^2y}{x-y}\cdot\dfrac{x-y}{6xy}$ _____

22. $\dfrac{x+3}{2x^2}\cdot\dfrac{4x}{x+3}$ _____

23. $\dfrac{x+3}{x^3+3x^2} \cdot \dfrac{x^3}{x-3}$ _____

24. $\dfrac{a^2-a}{a-1} \cdot \dfrac{a+1}{a}$ _____

25. $\dfrac{x-2}{4y} \div \dfrac{x^2+x-6}{12y^2}$ _____

26. $\dfrac{4x}{x-4} \div \dfrac{8x^2}{x^2-6x+8}$ _____

27. $\dfrac{6x^2}{4x^2y-12xy} \cdot \dfrac{x^2+x-12}{3x^2+12x}$ _____

28. $\dfrac{2x^2+4x}{12x^2y} \cdot \dfrac{6x}{x^2+6x+8}$ _____

29. $(t^2-t-12) \div \dfrac{t^2-9}{t^2-3t}$ _____

30. $\dfrac{2y^2+7y+3}{4y^2-1} \div (y+3)$ _____

31. $\dfrac{m+n}{m^2-n^2} \div \dfrac{m^2-mn}{m^2-2mn+n^2}$ _____

32. $\dfrac{x^2-6x+9}{x^2-x-6} \div \dfrac{x^2+2x-15}{x^2+2x}$ _____

33. $-(x^2-3x) \cdot \dfrac{x-2}{x-3}$ _____

34. $-(x^2-4) \cdot \dfrac{3}{x+2}$ _____

35. $\left(\dfrac{d^5}{3a} \div \dfrac{d^2}{6a^2}\right) \cdot \dfrac{a}{4d^3}$ _____

36. $\dfrac{d^5}{3a} \div \left(\dfrac{d^2}{6a^2} \cdot \dfrac{a}{4d^3}\right)$ _____

37. $\dfrac{2x^2}{3y^2} \cdot \dfrac{-6yz}{2x} \cdot \dfrac{y}{-xz}$ _____

38. $\dfrac{-a}{-b} \cdot \dfrac{12b^2c}{15ac} \cdot \dfrac{-10}{4b}$ _____

C 39. $\dfrac{9-x^2}{x^2+5x+6} \cdot \dfrac{x+2^*}{x-3}$ _____

40. $\dfrac{2-m}{2m+m^2} \cdot \dfrac{m^2+4m+4}{m^2-4}$ _____

41. $\dfrac{x^2-xy}{xy+y^2} \div \left(\dfrac{x^2-y^2}{x^2+2xy+y^2} \div \dfrac{x^2-2xy+y^2}{x^2y+xy^2}\right)$ _____

42. $\left(\dfrac{x^2-xy}{xy+y^2} \div \dfrac{x^2-y^2}{x^2+2xy+y^2}\right) \div \dfrac{x^2-2xy+y^2}{x^2y+xy^2}$ _____

*Note that $9-x^2 = -(x^2-9)$, or, in general, $b-a = -(a-b)$.

The Check Exercise for this section is on page 149.

3.3 ADDITION AND SUBTRACTION

Addition and subtraction of rational expressions are based on the corresponding properties of real number fractions. Thus,

---

Addition and Subtraction

If P, D, Q, and K are polynomials (K, $D \neq 0$), then

$$\frac{P}{D} + \frac{Q}{D} = \frac{P+Q}{D} \tag{1}$$

$$\frac{P}{D} - \frac{Q}{D} = \frac{P-Q}{D} \tag{2}$$

$$\frac{P}{D} = \frac{KP}{KD} \tag{3}$$

Verbally, if the denominators of two rational expressions are the same, we may either add or subtract the expressions by adding or subtracting the numerators and placing the result over the common denominator. If the denominators are not the same, we use property (3) to change the form of each fraction so they have a common denominator, and then use either (1) or (2).

Even though any common denominator will do, the problem will generally become less involved if the least common denominator is used. Recall, the **least common denominator (lcd) is the least common multiple (lcm) of all the denominators; that is, it is the "smallest" quantity exactly divisible by each denominator.**

If the lcd is not obvious, then it is found as follows:

Finding the Least Common Denominator (lcd)

Step 1. Factor each denominator completely, using integer coefficients.

Step 2. The lcm of all the denominators (the lcd) must contain each *different* factor that occurs in all of the denominators to the highest power it occurs in any one denominator.

EXAMPLE 5

(A) Find the least common multiple for $18x^3$, $15x$, $10x^2$.

Solution

Write each expression in completely factored form, including coefficients:

$$18x^3 = 2 \cdot 3^2 x^3$$

$$15x = 3 \cdot 5x$$

$$10x^2 = 2 \cdot 5x^2$$

The lcm must contain each different factor (2, 3, 5, and x) to the highest power it occurs in any one denominator. Thus,

$$\text{lcm} = 2 \cdot 3^2 \cdot 5x^3 = 90x^3$$

(B) Find the lcm for $6(x - 3)$, $x^2 - 9$, $4x^2 + 24x + 36$.

Solution

Factor each expression completely, including coefficients:

$$6(x - 3) = 2 \cdot 3(x - 3)$$

$$x^2 - 9 = (x - 3)(x + 3)$$

$$4x^2 + 24x + 36 = 4(x^2 + 6x + 9) = 2^2(x + 3)^2$$

Thus,

$$\text{lcm} = 2^2 \cdot 3(x - 3)(x + 3)^2 = 12(x - 3)(x + 3)^2$$

Work Problem 5 and check solution in Solutions to Matched Problems on page 135.

PROBLEM 5   **(A)**   Find the lcm for $15y^2$, $12y$, $9y^4$.

Solution

(B)   Find the lcm for $3x^2 - 12$, $x^2 - 4x + 4$, $12(x + 2)$.

Solution

EXAMPLE 6

Combine into a single fraction and simplify:

$$\frac{1}{x - 3} - \frac{x - 2}{x - 3}$$

Solution

$\dfrac{1}{x - 3} - \dfrac{(x - 2)}{x - 3}$ When a numerator has more than one term, place terms in parentheses before proceeding. Since denominators are the same, use property (2) to subtract.

$= \dfrac{1 - (x - 2)}{x - 3}$ Simplify numerator. Watch signs.

$= \dfrac{1 - x + 2}{x - 3}$ Sign errors are frequently made where the arrow points.

$= \dfrac{3 - x}{x - 3}$ $3 - x \neq x - 3$; $3 - x = -(x - 3)$

$= \dfrac{-(x - 3)}{x - 3}$ Reduce to lowest terms.

$= -1$

Work Problem 6 and check solution in Solutions to Matched Problems on page 135.

PROBLEM 6   Combine into a single fraction and simplify:

$$\frac{x+3}{2x-5}-\frac{3x-2}{2x-5}=$$

EXAMPLE 7

$$\frac{3}{2y}-\frac{1}{3y^2}+1 \qquad \text{lcd}=6y^2$$

$$=\frac{3y\cdot3}{3y\cdot2y}-\frac{2\cdot1}{2\cdot3y^2}+\frac{6y^2}{6y^2} \qquad \text{Use property (3) to make each denominator } 6y^2.$$

$$=\frac{9y}{6y^2}-\frac{2}{6y^2}+\frac{6y^2}{6y^2}$$

$$=\frac{9y-2+6y^2}{6y^2} \qquad \text{Arrange numerator in descending powers of } y.$$

$$=\frac{6y^2+9y-2}{6y^2}$$

Work Problem 7 and check solution in Solutions to Matched Problems on page 135.

PROBLEM 7   Combine into a single fraction and simplify:

$$\frac{5}{4x^3}-\frac{1}{3x}+2=$$

EXAMPLE 8

$$\frac{4}{3x^2-27}-\frac{x-1}{4x^2+24x+36} \qquad \text{Factor each denominator completely.}$$

$$=\frac{4}{3(x-3)(x+3)}-\frac{(x-1)}{2^2(x+3)^2} \qquad \text{lcd}=12(x-3)(x+3)^2$$

$$=\frac{4(x+3)\cdot4}{4(x+3)\cdot3(x-3)(x+3)}-\frac{3(x-3)\cdot(x-1)}{3(x-3)\cdot2^2(x+3)^2} \qquad \begin{array}{l}\text{Use property (3)}\\\text{to make each}\\\text{denominator}\\12(x-3)(x+3)^2.\end{array}$$

$$=\frac{16(x+3)}{12(x-3)(x+3)^2}-\frac{3(x-3)(x-1)}{12(x-3)(x+3)^2}$$

$$=\frac{16(x+3)-3(x-3)(x-1)}{12(x-3)(x+3)^2}$$

$$= \frac{16x + 48 - 3(x^2 - 4x + 3)}{12(x - 3)(x + 3)^2}$$

$$= \frac{16x + 48 - 3x^2 + 12x - 9}{12(x - 3)(x + 3)^2}$$

$$= \frac{-3x^2 + 28x + 39}{12(x - 3)(x + 3)^2}$$

Work Problem 8 and check solution in Solutions to Matched Problems on page 135.

PROBLEM 8 Combine into a single fraction and simplify:

$$\frac{3}{2x^2 - 8x + 8} - \frac{x + 1}{3x^2 - 12} =$$

SOLUTIONS TO MATCHED PROBLEMS

5. **(A)** $15y^2 = 3 \cdot 5y^2$
   $12y = 2^2 \cdot 3y$
   $9y^4 = 3^2y^4$
   $\text{lcm} = 2^2 \cdot 3^2 \cdot 5y^4 = 180y^4$

   (B) $3x^2 - 12 = 3(x - 2)(x + 2)$
   $x^2 - 4x + 4 = (x - 2)^2$
   $12(x + 2) = 2^2 \cdot 3(x + 2)$
   $\text{lcm} = 2^2 \cdot 3(x - 2)^2(x + 2) = 12(x - 2)^2(x + 2)$

6. $\frac{x + 3}{2x - 5} - \frac{3x - 2}{2x - 5} = \frac{(x + 3) - (3x - 2)}{2x - 5} = \frac{x + 3 - 3x + 2}{2x - 5} = \frac{5 - 2x}{2x - 5} = \frac{-(2x - 5)}{2x - 5} = -1$

7. $\frac{5}{4x^3} - \frac{1}{3x} + \frac{2}{1} = \frac{3(5)}{3(4x^3)} - \frac{4x^2(1)}{4x^2(3x)} + \frac{12x^3(2)}{12x^3(1)} = \frac{15}{12x^3} - \frac{4x^2}{12x^3} + \frac{24x^3}{12x^3} = \frac{24x^3 - 4x^2 + 15}{12x^3}$
   $\text{lcd} = 12x^3$

8. $\frac{3}{2x^2 - 8x + 8} - \frac{x + 1}{3x^2 - 12} = \frac{3}{2(x - 2)^2} - \frac{(x + 1)}{3(x - 2)(x + 2)}$ $\text{lcd} = 6(x - 2)^2(x + 2)$

   $= \frac{3(x + 2)(3)}{3(x + 2)[2(x - 2)^2]} - \frac{2(x - 2)(x + 1)}{2(x - 2)[3(x - 2)(x + 2)]}$

   $= \frac{9(x + 2)}{6(x - 2)^2(x + 2)} - \frac{2(x - 2)(x + 1)}{6(x - 2)^2(x + 2)} = \frac{9x + 18 - (2x^2 - 2x - 4)}{6(x - 2)^2(x + 2)}$

   $= \frac{9x + 18 - 2x^2 + 2x + 4}{6(x - 2)^2(x + 2)} = \frac{-2x^2 + 11x + 22}{6(x - 2)^2(x + 2)}$

PRACTICE EXERCISE 3.3

Work odd-numbered problems first, check answers, and then work even-numbered problems in areas of weakness. Answers to all problems are in the back of the book. Make every effort to work a problem yourself before you look at an answer.

A *Find the least common multiple (lcm) for each group of expressions.*

1. $3, x$ _____

2. $4, y$ _____

3. $x, 1$ _____

4. $y, 1$ _____

5. v^2, v, v^3 _____

6. x, x, x^2 _____

7. $3x, 6x^2, 4$ _____

8. $8u^3, 6u, 4u^2$ _____

9. $x + 1, x - 2$ _____

10. $x - 2, x + 3$ _____

11. $y + 3, 3y$ _____

12. $x - 2, 2x$ _____

Combine into single fractions and simplify.

13. $\dfrac{7x}{5x^2} - \dfrac{2}{5x^2}$ _____

14. $\dfrac{3m}{2m^2} - \dfrac{1}{2m^2}$ _____

15. $\dfrac{4x}{2x - 1} - \dfrac{2}{2x - 1}$ _____

16. $\dfrac{5a}{a - 1} - \dfrac{5}{a - 1}$ _____

17. $\dfrac{y}{y^2 - 9} - \dfrac{3}{y^2 - 9}$ _____

18. $\dfrac{2x}{4x^2 - 9} + \dfrac{3}{4x^2 - 9}$ _____

19. $\dfrac{5}{3k} - \dfrac{6x - 4}{3k}$ _____

20. $\dfrac{1}{2a^2} - \dfrac{2b - 1}{2a^2}$ _____

21. $\dfrac{3x}{y} + \dfrac{1}{4}$ _____

22. $\dfrac{2}{x} - \dfrac{1}{3}$ _____

23. $\dfrac{2}{y} + 1$ _____

24. $x + \dfrac{1}{x}$ _____

25. $\dfrac{u}{v^2} - \dfrac{1}{v} + \dfrac{u^3}{v^3}$ _____

26. $\dfrac{1}{x} - \dfrac{y}{x^2} + \dfrac{y^2}{x^3}$ _____

27. $\dfrac{2}{3x} - \dfrac{1}{6x^2} + \dfrac{3}{4}$ _____

28. $\dfrac{1}{8u^3} + \dfrac{5}{6u} - \dfrac{3}{4u^2}$ _____

29. $\dfrac{2}{x + 1} + \dfrac{3}{x - 2}$ _____

30. $\dfrac{1}{x - 2} + \dfrac{1}{x + 3}$ _____

31. $\dfrac{3}{y + 3} - \dfrac{2}{3y}$ _____

32. $\dfrac{2}{x - 2} - \dfrac{3}{2x}$ _____

B *Find the lcm for each group of expressions.*

33. $12x^3, 8x^2y^2, 3xy^2$ _____

34. $9u^3v^2, 6uv, 12v^3$ _____

35. $15x^2y, 25xy, 5y^2$ _____

36. $18m^4n^2, 12m^2n^4, 9mn$ _____

37. $6(x - 1), 9(x - 1)^2$ _____

38. $8(y - 3)^2, 6(y - 3)$ _____

39. $6(x - 7)(x + 7), 8(x + 7)^2$ _____

40. $3(x - 5)^2, 4(x + 5)(x - 5)$ _____

41. $x^2 - 4, x^2 + 4x + 4$ _____

42. $x^2 - 6x + 9, x^2 - 9$ _____

43. $3x^2 + 3x, 4x^2, 3x^2 + 6x + 3$ _____

44. $3m^2 - 3m, m^2 - 2m + 1, 5m^2$ _____

Combine into a single fraction and simplify.

45. $\dfrac{2}{9u^3v^2} - \dfrac{1}{6uv} + \dfrac{1}{12v^3}$ _____

46. $\dfrac{1}{12x^3} + \dfrac{3}{8x^2y^2} - \dfrac{2}{3xy^2}$ _____

47. $\dfrac{4t - 3}{18t^3} + \dfrac{3}{4t} - \dfrac{2t - 1}{6t^2}$ _____

48. $\dfrac{3y + 8}{4y^2} - \dfrac{2y - 1}{y^3} - \dfrac{5}{8y}$ _____

49. $\dfrac{t + 1}{t - 1} - 1$ _____

50. $2 + \dfrac{x + 1}{x - 3}$ _____

51. $5 + \dfrac{a}{a+1} - \dfrac{a}{a-1}$ _____

52. $\dfrac{1}{y+2} + 3 - \dfrac{2}{y-2}$ _____

53. $\dfrac{2}{3(x-5)^2} - \dfrac{1}{4(x+5)(x-5)}$ _____

54. $\dfrac{1}{6(x-7)(x+7)} + \dfrac{3}{8(x+7)^2}$ _____

55. $\dfrac{5}{6(x-1)} + \dfrac{2}{9(x-1)^2}$ _____

56. $\dfrac{3}{8(y-3)^2} - \dfrac{1}{6(y-3)}$ _____

57. $\dfrac{3}{x+3} - \dfrac{3x+1}{(x-1)(x+3)}$ _____

58. $\dfrac{4}{2x-3} - \dfrac{2x+1}{(2x-3)(x+2)}$ _____

59. $\dfrac{3s}{3s^2-12} + \dfrac{1}{2s^2+4s}$ _____

60. $\dfrac{2t}{3t^2-48} + \dfrac{t}{4t+t^2}$ _____

61. $\dfrac{3}{x^2-4} - \dfrac{1}{x^2+4x+4}$ _____

62. $\dfrac{2}{x^2-6x+9} - \dfrac{1}{x^2-9}$ _____

63. $\dfrac{2}{x+3} - \dfrac{1}{x-3} + \dfrac{2x}{x^2-9}$ _____

64. $\dfrac{2x}{x^2-y^2} + \dfrac{1}{x+y} - \dfrac{1}{x-y}$ _____

C 65. $\dfrac{x}{x^2-x-2} - \dfrac{1}{x^2+5x-14} - \dfrac{2}{x^2+8x+7}$ ___

66. $\dfrac{m^2}{m^2+2m+1} + \dfrac{1}{3m+3} - \dfrac{1}{6}$ _____

67. $\dfrac{1}{3x^2+3x} + \dfrac{1}{4x^2} - \dfrac{1}{3x^2+6x+3}$ _____

68. $\dfrac{1}{3m(m-1)} + \dfrac{1}{m^2-2m+1} - \dfrac{1}{5m^2}$ ___

69. $\dfrac{xy^2}{x^3-y^3} - \dfrac{y}{x^2+xy+y^2}$ _____

70. $\dfrac{x}{x^2-xy+y^2} - \dfrac{xy}{x^3+y^3}$ _____

For the next four problems note that $b - a = -(a - b)$; thus, $3 - y = -(y - 3)$, $1 - x = -(x - 1)$, and so on.

71. $\dfrac{5}{y-3} - \dfrac{2}{3-y}$ _____

72. $\dfrac{3}{x-1} + \dfrac{2}{1-x}$ _____

73. $\dfrac{3}{x-3} + \dfrac{x}{3-x}$ _____

74. $\dfrac{-2}{2-y} - \dfrac{y}{y-2}$ _____

The Check Exercise for this section is on page 151.

3.4 COMPLEX FRACTIONS

A fractional form with fractions in its numerator or denominator is called a **complex fraction.** It is often necessary to represent a complex fraction as a **simple fraction,** that is (in all cases we will consider), as the quotient of two polynomials. The process does not involve any new concepts. It is a matter of applying old concepts in the right way. In particular, we will find the fundamental principle of real fractions

$$\frac{a}{b} = \frac{ka}{kb} \qquad b, k \neq 0 \tag{4}$$

of considerable use. Several examples should clarify the process.

EXAMPLE 9
Express as simple fractions.

(A) $\dfrac{\frac{2}{3}}{\frac{3}{4}}$

Solution
Use property (4) and multiply numerator and denominator by a number divisible by both 3 and 4, that is, 12, the lcd of the two internal fractions.

$$\frac{\frac{2}{3}}{\frac{3}{4}} = \frac{12 \cdot \frac{2}{3}}{12 \cdot \frac{3}{4}} \qquad \text{Multiply top and bottom by 12; then cancel internal denominators.}$$

$$= \frac{4 \cdot 2}{3 \cdot 3}$$

$$= \frac{8}{9}$$

NOTE: We can also treat this as a division problem:

$$\frac{\frac{2}{3}}{\frac{3}{4}} = \frac{2}{3} \div \frac{3}{4} = \frac{2}{3} \cdot \frac{4}{3} = \frac{8}{9}$$

(B) $\dfrac{1\frac{1}{2}}{3\frac{2}{3}}$

Solution
Recall $1\frac{1}{2}$ and $3\frac{2}{3}$ represent sums and not products; that is, $1\frac{1}{2} = 1 + \frac{1}{2}$ and $3\frac{2}{3} = 3 + \frac{2}{3}$. Thus,

$$\frac{1\frac{1}{2}}{3\frac{2}{3}} = \frac{1 + \frac{1}{2}}{3 + \frac{2}{3}} \qquad \text{Write mixed fractions as sums.}$$

$$= \frac{6(1 + \frac{1}{2})}{6(3 + \frac{2}{3})} \qquad \begin{array}{l}\text{Multiply top and bottom by 6,}\\ \text{the lcd of all fractions within the main fraction.}\end{array}$$

$$= \frac{6 \cdot 1 + 6 \cdot \frac{1}{2}}{6 \cdot 3 + 6 \cdot \frac{2}{3}} \qquad \text{The denominators 2 and 3 cancel.}$$

$$= \frac{6 + 3}{18 + 4} = \frac{9}{22} \qquad \text{A simple fraction.}$$

Work Problem 9 and check solution in Solutions to Matched Problems on page 141.

PROBLEM 9   Express as simple fractions:

(A) $\dfrac{\frac{3}{5}}{\frac{1}{4}} =$

(B) $\dfrac{2\frac{3}{4}}{4\frac{1}{3}} =$

EXAMPLE 10

Express as simple fractions.

(A) $\dfrac{1 - \dfrac{1}{x^2}}{1 + \dfrac{1}{x}}$ Multiply top and bottom by x^2, the lcd of all internal fractions.

$= \dfrac{x^2\left(1 - \dfrac{1}{x^2}\right)}{x^2\left(1 + \dfrac{1}{x}\right)}$ These steps can be done mentally after a little practice.

$= \dfrac{x^2 \cdot 1 - x^2 \cdot \dfrac{1}{x^2}}{x^2 \cdot 1 + x^2 \cdot \dfrac{1}{x}}$

$= \dfrac{x^2 - 1}{x^2 + x}$ Factor top and bottom to reduce to lowest terms.

$= \dfrac{\overset{1}{(x - 1)\cancel{(x + 1)}}}{\underset{1}{x\cancel{(x + 1)}}}$

$= \dfrac{x - 1}{x}$

(B) $\dfrac{\dfrac{a}{b} - \dfrac{b}{a}}{\dfrac{a}{b} + 2 + \dfrac{b}{a}}$ lcd of all internal fractions is ab

$= \dfrac{ab\left(\dfrac{a}{b} - \dfrac{b}{a}\right)}{ab\left(\dfrac{a}{b} + 2 + \dfrac{b}{a}\right)}$ Use fundamental principle of fractions; that is, multiply top and bottom by ab to clear internal fractions.

$$= \frac{ab \cdot \dfrac{a}{b} - ab \cdot \dfrac{b}{a}}{ab \cdot \dfrac{a}{b} + ab \cdot 2 + ab \cdot \dfrac{b}{a}}$$

$$= \frac{a^2 - b^2}{a^2 + 2ab + b^2} \qquad \text{Reduce to lowest terms.}$$

$$= \frac{(a - b)(a + b)}{(a + b)^2}$$

$$= \frac{a - b}{a + b} \qquad \text{A simple fraction.}$$

Work Problem 10 and check solution in Solutions to Matched Problems on page 142.

PROBLEM 10 Express as simple fractions.

(A) $\dfrac{1 - \dfrac{1}{3x}}{1 - \dfrac{1}{9x^2}} =$

(B) $\dfrac{\dfrac{x}{y} + 1 - \dfrac{2y}{x}}{\dfrac{x}{y} - \dfrac{y}{x}} =$

EXAMPLE 11

Express as a simple fraction.

$$x - \frac{x}{1 - \dfrac{1}{x}} = x - \frac{x(x)}{x\left(1 - \dfrac{1}{x}\right)} \qquad \begin{array}{l}\text{Express complex fraction} \\ \text{on right as a simple fraction.}\end{array}$$

$$= x - \frac{x^2}{x - 1} \qquad \begin{array}{l}\text{Combine into a single fraction} \\ \text{as in Section 3.3 (lcd} = x - 1).\end{array}$$

$$= \frac{x(x - 1)}{x - 1} - \frac{x^2}{x - 1}$$

$$= \frac{x^2 - x - x^2}{x - 1}$$

$$= \frac{-x}{x - 1} = \frac{(-1)(-x)}{(-1)(x - 1)} \qquad \begin{array}{l}\text{Represent final answer, using} \\ \text{minimal number of negative signs.}\end{array}$$

$$= \frac{x}{-x + 1} = \frac{x}{1 - x}$$

Work Problem 11 and check solution in Solutions to Matched Problems on page 142.

PROBLEM 11   Express as a simple fraction.

$$x - \cfrac{x}{1 + \cfrac{1}{1 - \cfrac{1}{x}}} =$$

REMARKS

Complex fractions can also be thought of in terms of division. Example 10**A**, for instance, could have been worked as follows:

$$\frac{1 - \dfrac{1}{x^2}}{1 + \dfrac{1}{x}} = \left(1 - \frac{1}{x^2}\right) \div \left(1 + \frac{1}{x}\right)$$

$$= \frac{x^2 - 1}{x^2} \div \frac{x + 1}{x}$$

$$= \frac{\overset{1}{(x - 1)\cancel{(x + 1)}}}{\underset{x}{\cancel{x^2}}} \cdot \frac{\overset{1}{\cancel{x}}}{\underset{1}{\cancel{x + 1}}} = \frac{x - 1}{x}$$

The first method involved fewer written steps (compare the two).

     In certain types of problems the division approach may be easier. For example,

$$\frac{\dfrac{x^2 - 5x + 4}{x^2 - 4x}}{\dfrac{x^2 + x - 2}{x^3 + 2x^2}} = \frac{x^2 - 5x + 4}{x^2 - 4x} \div \frac{x^2 + x - 2}{x^3 + 2x^2}$$

$$= \frac{\overset{1}{\cancel{(x - 4)}}\overset{1}{\cancel{(x - 1)}}}{\underset{1}{\cancel{x}}\underset{1}{\cancel{(x - 4)}}} \cdot \frac{\overset{x}{\cancel{x^2}}\overset{1}{\cancel{(x + 2)}}}{\underset{1}{\cancel{(x + 2)}}\underset{1}{\cancel{(x - 1)}}} = x$$

Use the method that is most appropriate for the problem. Most of the problems in Practice Exercise 3.4 are readily worked by the first method.

SOLUTIONS TO MATCHED PROBLEMS

9.   (**A**) $\dfrac{\frac{3}{5}}{\frac{1}{4}} = \dfrac{20(\frac{3}{5})}{20(\frac{1}{4})} = \dfrac{4 \cdot 3}{5 \cdot 1} = \dfrac{12}{5}$ (**B**) $\dfrac{2 + \frac{3}{4}}{4 + \frac{1}{3}} = \dfrac{12(2 + \frac{3}{4})}{12(4 + \frac{1}{3})} = \dfrac{24 + 9}{48 + 4} = \dfrac{33}{52}$

10. (A)
$$\frac{1 - \dfrac{1}{3x}}{1 - \dfrac{1}{9x^2}} = \frac{9x^2\left(1 - \dfrac{1}{3x}\right)}{9x^2\left(1 - \dfrac{1}{9x^2}\right)} = \frac{9x^2 - 3x}{9x^2 - 1} = \frac{\overset{1}{\cancel{3x(3x-1)}}}{\underset{1}{\cancel{(3x-1)}}(3x+1)} = \frac{3x}{3x+1}$$

(B)
$$\frac{\dfrac{x}{y} + 1 - \dfrac{2y}{x}}{\dfrac{x}{y} - \dfrac{y}{x}} = \frac{xy\left(\dfrac{x}{y} + 1 - \dfrac{2y}{x}\right)}{xy\left(\dfrac{x}{y} - \dfrac{y}{x}\right)} = \frac{x^2 + xy - 2y^2}{x^2 - y^2} = \frac{(x+2y)\overset{1}{\cancel{(x-y)}}}{\underset{1}{\cancel{(x-y)}}(x+y)} = \frac{x+2y}{x+y}$$

11.
$$x - \frac{x}{1 + \dfrac{1}{1 - \dfrac{1}{x}}} = x - \frac{x}{1 + \dfrac{x(1)}{x\left(1 - \dfrac{1}{x}\right)}} = x - \frac{x}{1 + \dfrac{x}{x-1}} = x - \frac{(x-1)(x)}{(x-1)\left(1 + \dfrac{x}{x-1}\right)} = x - \frac{x^2 - x}{x - 1 + x}$$

$$= \frac{x(2x-1)}{2x-1} - \frac{x^2 - x}{2x-1} = \frac{2x^2 - x - x^2 + x}{2x-1} = \frac{x^2}{2x-1}$$

PRACTICE EXERCISE 3.4

Work odd-numbered problems first, check answers, and then work even-numbered problems in areas of weakness. Answers to all problems are in the back of the book. Make every effort to work a problem yourself before you look at an answer.

Express as simple fractions reduced to lowest terms.

A 1. $\dfrac{\frac{1}{2}}{\frac{2}{3}}$ _____

2. $\dfrac{\frac{1}{4}}{\frac{2}{3}}$ _____

3. $\dfrac{\frac{3}{8}}{\frac{5}{12}}$ _____

4. $\dfrac{\frac{4}{15}}{\frac{5}{6}}$ _____

5. $\dfrac{1\frac{1}{3}}{2\frac{1}{6}}$ _____

6. $\dfrac{3\frac{1}{10}}{2\frac{1}{5}}$ _____

7. $\dfrac{1\frac{2}{9}}{2\frac{5}{6}}$ _____

8. $\dfrac{2\frac{4}{15}}{1\frac{7}{10}}$ _____

9. $\dfrac{\dfrac{x}{y}}{\dfrac{1}{y^2}}$ _____

10. $\dfrac{\dfrac{1}{b^2}}{\dfrac{a}{b}}$ _____

11. $\dfrac{\dfrac{y}{2x}}{\dfrac{1}{3x^2}}$ _____

12. $\dfrac{\dfrac{2x}{5y}}{\dfrac{1}{3x}}$ _____

B 13. $\dfrac{1 + \dfrac{3}{x}}{x - \dfrac{9}{x}}$ _____

14. $\dfrac{1 - \dfrac{2}{x}}{x - \dfrac{4}{x}}$ _____

15. $\dfrac{1 - \dfrac{y^2}{x^2}}{1 - \dfrac{y}{x}}$ _____

16. $\dfrac{\dfrac{a^2}{b^2} - 1}{\dfrac{a}{b} - 1}$ _____

17. $\dfrac{\dfrac{1}{x} + \dfrac{1}{y}}{\dfrac{y}{x} - \dfrac{x}{y}}$ _____

18. $\dfrac{b - \dfrac{a^2}{b}}{\dfrac{1}{a} - \dfrac{1}{b}}$ _____

19. $\dfrac{\dfrac{x}{y} - 2 + \dfrac{y}{x}}{\dfrac{x}{y} - \dfrac{y}{x}}$ _____

20. $\dfrac{1 + \dfrac{2}{x} - \dfrac{15}{x^2}}{1 + \dfrac{4}{x} - \dfrac{5}{x^2}}$ _____

21. $\dfrac{\dfrac{a^2}{a - b} - a}{\dfrac{b^2}{a - b} + b}$ _____

143

22. $\dfrac{n - \dfrac{n^2}{n - m}}{1 + \dfrac{m^2}{n^2 - m^2}}$ _____

23. $\dfrac{\dfrac{m}{m + 2} - \dfrac{m}{m - 2}}{\dfrac{m + 2}{m - 2} - \dfrac{m - 2}{m + 2}}$ _____

24. $\dfrac{\dfrac{y}{x + y} - \dfrac{x}{x - y}}{\dfrac{x}{x + y} + \dfrac{y}{x - y}}$ _____

C 25. $1 - \dfrac{1}{1 - \dfrac{1}{x}}$ _____

26. $2 - \dfrac{1}{1 - \dfrac{2}{x + 2}}$ _____

27. $1 - \dfrac{x - \dfrac{1}{x}}{1 - \dfrac{1}{x}}$ _____

28. $\dfrac{t - \dfrac{1}{1 + \dfrac{1}{t}}}{t + \dfrac{1}{t - \dfrac{1}{t}}}$ _____

The Check Exercise for this section is on page 153.

3.5 CHAPTER REVIEW

Important terms and symbols

3.1 RATIONAL EXPRESSIONS   rational expressions, fundamental principle of fractions, reducing to lowest terms, raising to higher terms

$$\frac{PK}{QK} = \frac{P}{Q} \qquad \frac{P}{Q} = \frac{KP}{KQ}$$

3.2 MULTIPLICATION AND DIVISION   multiplication, division

$$\frac{P}{Q} \cdot \frac{R}{S} = \frac{P \cdot R}{Q \cdot S} \qquad \frac{P}{Q} \div \frac{R}{S} = \frac{P}{Q} \cdot \frac{S}{R}$$

3.3 ADDITION AND SUBTRACTION   addition, subtraction, least common multiple (lcm), least common denominator (lcd)

$$\frac{P}{D} + \frac{Q}{D} = \frac{P + Q}{D} \qquad \frac{P}{D} - \frac{Q}{D} = \frac{P - Q}{D} \qquad \frac{P}{D} = \frac{KP}{KD}$$

3.4 COMPLEX FRACTIONS   complex fraction, simple fraction

DIAGNOSTIC (REVIEW) EXERCISE 3.5 _____

Work through all the problems in this chapter review and check answers in the back of the book. (Answers to all problems are there, and following each answer is a number in italics indicating the section in which that type of problem is discussed.) Where weaknesses show up, review appropriate sections in the text. When you are satisfied that you know the material, take the practice test following this review.

Perform the indicated operations and simplify.

A 1. $1 + \dfrac{2}{3x}$ _____

2. $\dfrac{2}{x} - \dfrac{1}{6x} + \dfrac{1}{3}$ _____

3. $\dfrac{1}{6x^3} - \dfrac{3}{4x} - \dfrac{2}{3}$ _____

4. $\dfrac{4x^2y^3}{3a^2b^2} \div \dfrac{2xy^2}{3ab}$ _____

5. $\dfrac{6x^2}{3(x-1)} - \dfrac{6}{3(x-1)}$ _____

6. $1 - \dfrac{m-1}{m+1}$ _____

7. $\dfrac{3}{x-2} - \dfrac{2}{x+1}$ _____

8. $(d-2)^2 \div \dfrac{d^2-4}{d-2}$ _____

9. $\dfrac{x+1}{x+2} - \dfrac{x+2}{x+3}$ _____

10. $\dfrac{\frac{2}{3}}{\frac{3}{4}}$ _____

11. $\dfrac{2\frac{3}{4}}{1\frac{1}{2}}$ _____

12. $\dfrac{1 - \frac{2}{y}}{1 + \frac{1}{y}}$ _____

B 13. $\dfrac{2}{5b} - \dfrac{4}{3b^3} - \dfrac{1}{6a^2b^2}$ _____

14. $\dfrac{2}{2x-3} - 1$ _____

15. $\dfrac{4x^2y}{3ab^2} \div \left(\dfrac{2a^2x^2}{b^2y} \cdot \dfrac{6a}{2y^2}\right)$ _____

16. $\dfrac{x}{x^2+4x} + \dfrac{2x}{3x^2-48}$ _____

17. $\dfrac{x^3-x}{x^2-x} \div \dfrac{x^2+2x+1}{x}$ _____

18. $\dfrac{\frac{x}{y} - \frac{y}{x}}{\frac{x}{y} + 1}$ _____

C 19. $\dfrac{x}{x^3-y^3} - \dfrac{1}{x^2+xy+y^2}$ _____

20. $\dfrac{\frac{y^2}{x^2-y^2} + 1}{\frac{x^2}{x-y} - x}$ _____

21. $\dfrac{x^3-1}{x^2+x+1} \div \dfrac{x^2-1}{x^2+2x+1}$ _____

22. $\dfrac{1}{3x^2-27} - \dfrac{x-1}{4x^3+24x^2+36x}$ _____

23. $\dfrac{4}{s^2-4} + \dfrac{1}{2-s}$ _____

24. $\dfrac{y^2-y-6}{(y+2)^2} \cdot \dfrac{2+y}{3-y}$ _____

25. $\dfrac{y}{x^2} \div \left(\dfrac{x^2+3x}{2x^2+5x-3} \div \dfrac{x^3y-x^2y}{2x^2-3x+1}\right)$ _____

PRACTICE TEST CHAPTER 3

Take this as if it were a graded test. Allow yourself up to 50 minutes. Work the problems without looking back in the chapter. Correct your work, using the answers (keyed to appropriate sections) in the back of the book.

Perform the indicated operations and simplify.

1. $\dfrac{1}{6} - \dfrac{1}{2x} + \dfrac{2}{3x^2}$ _____

2. $\dfrac{4a^2b}{3xy^2} \div \dfrac{12ax}{6by}$ _____

3. $\dfrac{x+2}{x-3} - 1$ _____

4. $\dfrac{1 - \dfrac{3}{m}}{1 + \dfrac{3}{m}}$ _____

5. $\dfrac{3}{x-5} - \dfrac{1}{x+3}$ _____

6. $\dfrac{3x^2 - 3xy}{12x^2y} \cdot \dfrac{4xy + 4y^2}{x^2 - y^2}$ _____

7. $\dfrac{1}{4m^3 - 4m^2n} + \dfrac{2}{3m^3 - 6m^2n + 3mn^2}$ _____

8. $\dfrac{x^2 + xy + y^2}{x^2 - y^2} \div \dfrac{x^3 - y^3}{x^2 - 2xy + y^2}$ _____

9. $\dfrac{5}{x-5} + \dfrac{x}{5-x}$ _____

10. $\dfrac{\dfrac{u}{v} - \dfrac{v}{u}}{\dfrac{u}{v} - 1}$ _____

CHECK EXERCISE 3.1

Work the following problems without looking at any text examples. Show your work in the space provided. Write the letter that best indicates your answer in the answer column.

ANSWER
COLUMN

1. _____
2. _____
3. _____
4. _____
5. _____
6. _____
7. _____
8. _____
9. _____
10. _____

In Problems 1 to 5 reduce to lowest terms.

1. $\dfrac{6x^5y^2}{8xy^4} =$

     (A) $\dfrac{6x^4}{8y^2}$

     (B) $\dfrac{2x^4}{3y^2}$

     (C) $\dfrac{3x^4}{4y^2}$

     (D) None of these

2. $\dfrac{(x-y)^2}{(x+y)(x-y)} =$

     (A) -1

     (B) $\dfrac{x-y}{x+y}$

     (C) $\dfrac{(x-y)^2}{x^2-y^2}$

     (D) None of these

3. $\dfrac{3x^2(x-1)}{6x(x-1)^2} =$

     (A) $\dfrac{x}{2(x-1)}$

     (B) $\dfrac{x(x-1)}{2(x-1)^2}$

     (C) $\dfrac{x^2}{2x(x-1)}$

     (D) None of these

4. $\dfrac{u^2v - uv}{u^2 - uv} =$

     (A) v

     (B) $\dfrac{v}{u-v}$

     (C) $v - 1$

     (D) None of these

5. $\dfrac{a^2 - 4b^2}{a^2 - 4ab + 4b^2} =$

     (A) $\dfrac{a - 2b}{a + 2b}$

     (B) $\dfrac{1}{4ab}$

     (C) $\dfrac{a + 2b}{a - 2b}$

     (D) None of these

147

Complete the raising-to-higher-terms process in Problems 6 to 8 by replacing the question marks with appropriate expressions.

6. $\dfrac{3x}{5y} = \dfrac{?}{15xy^2}$

    (A) $3xy$
    (B) y
    (C) $3y$
    (D)   None of these

7. $\dfrac{2}{x} = \dfrac{2x - 2}{?}$

    (A) $x - 1$
    (B) $x^2 - x$
    (C) $\dfrac{1}{x - 1}$
    (D)   None of these

8. $\dfrac{x + y}{x - 2y} = \dfrac{?}{x^2 - 4y^2}$

    (A) $x^2 + 3xy + 2y^2$
    (B) $x + 2y$
    (C) $x^2 - xy - 2y^2$
    (D)   None of these

In Problems 9 and 10 reduce to lowest terms.

9. $\dfrac{ac + bc - ad - bd}{c^2 + dc - 2d^2} =$

    (A) $\dfrac{a + b - a - d}{c - 2d}$
    (B) $\dfrac{a + b - a}{c}$
    (C) $\dfrac{a + b}{c + 2d}$
    (D)   None of these

10. $\dfrac{x^3 - y^3}{3x^3y + 3x^2y^2 + 3xy^3}$

    (A) $\dfrac{x - y}{3xy}$
    (B) $\dfrac{-1}{3}$
    (C) $\dfrac{-1}{3y + 3x^2y^2 + 3x}$
    (D)   None of these

CHECK EXERCISE 3.2

Work the following problems without looking at any text examples. Show your work in the space provided. Write the letter that best indicates your answer in the answer column.

Perform the indicated operations and reduce to lowest terms.

1. $\dfrac{8x^2y}{6z^3} \div \dfrac{4y^3}{3x^2z^2} =$

    (A) $\dfrac{16y^4}{9z^4}$

    (B) $\dfrac{x^4}{zy^2}$

    (C) $\dfrac{16x^4}{9zy^2}$

    (D) None of these

2. $\dfrac{3x^2 + 9x}{x^2 - 2x - 15} \cdot (x - 5) =$

    (A) $\dfrac{3x}{(x - 5)^2}$

    (B) $\dfrac{3x^2 + 9x}{x + 3}$

    (C) $3x$

    (D) None of these

3. $\dfrac{2x^2 + 5x - 3}{16x^3y - 8x^2y} \div \dfrac{x^2 + 6x + 9}{6x^2y^2 + 18xy^2}$

    (A) $\dfrac{3y}{4x}$

    (B) $\dfrac{(x + 3)^2}{48x^3y^3}$

    (C) $\dfrac{4x}{3y}$

    (D) None of these

149

4. $\dfrac{m^3}{ab} \div \left(\dfrac{4m^2}{b^2} \div \dfrac{a^2}{m}\right) =$

(A) $\dfrac{4m^4}{b^3}$

(B) $\dfrac{4m^6}{a^3b^3}$

(C) $\dfrac{ab}{4}$

(D) None of these

5. $\dfrac{3-x}{3x+x^2} \cdot \dfrac{x^2+6x+9}{x^2-9} =$

(A) $\dfrac{-1}{x}$

(B) $\dfrac{3-x}{x(x-3)}$

(C) $\dfrac{(3-x)(x+3)}{x(3+x)(x-3)}$

(D) None of these

CHECK EXERCISE 3.3

Work the following problems without looking at any text examples. Show your work in the space provided. Write the letter that best indicates your answer in the answer column.

In Problems 1 to 4 find the least common multiple (lcm) for each set of algebraic expressions.

1. $8, 3x, 2y$

   (A) $48xy$
   (B) $16xy$
   (C) $24xy$
   (D) None of these

2. $9xy^2, 4x^2y, 6xy$

   (A) $216x^3y^3$
   (B) $72x^3y^3$
   (C) $36x^2y^2$
   (D) None of these

3. $4(x - 1), 12(x - 1)^2, 6$

   (A) $24(x - 1)$
   (B) $24(x - 1)^2$
   (C) $288(x - 1)^3$
   (D) None of these

4. $6x^2 - 6, 4x^2 - 8x + 4$

   (A) $24(x - 1)^3(x + 1)$
   (B) $12(x - 1)^2(x + 1)$
   (C) $12(x - 1)^3(x + 1)$
   (D) None of these

In Problems 5 to 10 combine into a single fraction and simplify.

5. $\dfrac{1}{8} + \dfrac{1}{3x} - \dfrac{1}{2y}$

   (A) $\dfrac{3xy + 8y - 12x}{24xy}$
   (B) $\dfrac{1 + 8y - 12x}{8}$
   (C) $\dfrac{6xy + 16y - 24x}{48xy}$
   (D) None of these

151

6. $\dfrac{2}{9xy^2} - \dfrac{3}{4x^2y} + \dfrac{1}{6xy}$

(A) $\dfrac{8x - 9y + 6xy}{36x^2y^2}$

(B) $\dfrac{8x - 9y + 1}{6xy}$

(C) 0

(D) None of these

7. $\dfrac{x + 2}{x - 3} - 1 =$

(A) $\dfrac{-1}{x - 3}$

(B) $\dfrac{x + 1}{x - 3}$

(C) $-\frac{1}{3}$

(D) None of these

8. $\dfrac{1}{x^2 - 10x + 25} - \dfrac{1}{x^2 - 25} =$

(A) 0

(B) $\dfrac{2}{(x - 5)^2}$

(C) $\dfrac{10}{(x - 5)^2(x + 5)}$

(D) None of these

9. $\dfrac{1}{2u^2 - 4u} - \dfrac{1}{6u^2} - \dfrac{2u - 4}{12u^3 - 48u^2 + 48u} =$

(A) $\dfrac{u^2 - 12u + 4}{6u^2(u - 2)^2}$

(B) $\dfrac{u + 2}{6u^2(u - 2)}$

(C) $\dfrac{2(u - 2)(u + 2)}{12u(u - 2)^2}$

(D) None of these

10. $\dfrac{x}{x - 4} + \dfrac{x - 1}{4 - x} =$

(A) $\dfrac{1}{x - 4}$

(B) $\dfrac{-x + 4}{(x - 4)(4 - x)}$

(C) $\dfrac{-1}{x - 4}$

(D) None of these

CHECK EXERCISE 3.4

Work the following problems without looking at any text examples. Show
your work in the space provided. Write the letter that best indicates your
answer in the answer column.

Express as simple fractions reduced to lowest terms.

1. $\dfrac{1\frac{2}{3}}{3\frac{1}{2}} =$

   (A) $\frac{5}{9}$

   (B) $\frac{4}{9}$

   (C) $\frac{10}{21}$

   (D) None of these

2. $\dfrac{\dfrac{m}{3n}}{\dfrac{1}{2n^2}} =$

   (A) $\dfrac{2m}{3n}$

   (B) $\dfrac{3m}{2n}$

   (C) $\dfrac{3mn}{2}$

   (D) None of these

3. $\dfrac{\dfrac{b}{a} - \dfrac{a}{b}}{\dfrac{1}{a} - \dfrac{1}{b}} =$

   (A) $b - a$

   (B) $b + a$

   (C) $\dfrac{a - b}{ab}$

   (D) None of these

153

4. $\dfrac{1 - \dfrac{b}{a}}{1 - \dfrac{b^2}{a^2}} =$

(A) $\dfrac{a}{a + b}$

(B) $\dfrac{a + b}{a}$

(C) $\dfrac{a^2 - ab}{a^2 - b^2}$

(D) None of these

5. $1 - \dfrac{1}{1 + \dfrac{1}{x}} =$

(A) $\dfrac{1}{x + 1}$

(B) $\dfrac{x}{x + 1}$

(C) $\dfrac{1 - x}{x + 1}$

(D) None of these

Chapter Four

FIRST-DEGREE EQUATIONS AND INEQUALITIES

CONTENTS

Could you build the "box" in the figure?

INSTRUCTIONS FOR STUDENTS IN A
SELF-PACED CLASS OR LAB

yes — **HAVE YOU HAD INTERMEDIATE ALGEBRA BEFORE THIS COURSE?** — no

1. Work Diagnostic (Review) Exercise 4.6 on page 190. Check answers in back of book; then work through text sections corresponding to problems missed. (Section numbers are in italics following each answer.)
2. When finished with step 1, take Practice Test: Chapter 4 on page 192 as a final check of your understanding of the chapter. Check answers in the back of the book; then review sections where weakness still prevails. (Corresponding section numbers are in italics following each answer.)
3. When you think you are ready, ask your instructor for a graded test for Chapter 4.
4. If your instructor approves, after the test is corrected, go to the next chapter.

1. Work through each section in the chapter as follows:
   (a) Read discussion.
   (b) Read each example and work the corresponding matched problem. Check your solution to the matched problem in Solutions to Matched Problems on the indicated page.
   (c) At the end of a section work the odd-numbered problems in the Practice Exercise and check answers; then work even-numbered problems in areas of weakness. (Answers to *all* Practice Exercise sets are in the back of the book.)
   (d) Work Check Exercise as instructed. Tear out and turn in as directed by your instructor. (Answers are not in the text.)
2. Repeat each step in item 1 for each section in the chapter.
3. After the instructional part of the chapter is completed, proceed with steps 1 to 4 in the box above.

4.1 Equations with Integer Coefficients

•Introduction
••Solving equations
•••Summary

•Introduction

In this section we will review methods of solving equations with integer coefficients, such as

$$5x - (7x - 4) - 2 = 5 - (3x + 2)$$

A **solution** or **root** of an equation in one variable is a replacement of the variable by a constant that makes the left side of the equation equal to the right side. For example,

$x = 4$ is a solution of

$2x - 1 = x + 3$

since

$2(4) - 1 = 4 + 3$

$\qquad 7 = 7$

To **solve an equation** is to find all of its solutions.

Knowing what we mean by a solution of an equation is one thing; finding it is another. Our objective now is to develop a systematic approach to solving equations that is free from guess-work. We start by introducing the idea of equivalent equations. We say that two equations are **equivalent** if they both have exactly the same solutions.

The basic idea in solving equations is to perform operations on equations that produce simpler equivalent equations and to continue the process until we reach an equation whose solution is obvious—generally an equation such as

$x = -5$

The following properties of equality produce equivalent equations when applied.

Properties of Equality

(A) *Addition property.* The same quantity may be added to each side of an equation.

$a = b$
$a + c = b + c$

(B) *Subtraction property.* The same quantity may be subtracted from each side of an equation.

$a = b$
$a - c = b - c$

(C) *Multiplication property.* Each side of an equation may be multiplied by the same nonzero quantity.

$a = b$
$ca = cb$

(D) *Division property.* Each side of an equation may be divided by the same nonzero quantity.

$a = b$
$\dfrac{a}{c} = \dfrac{b}{c}$

We can think of the process of solving an equation as a game. The objective of the game is to isolate the variable (with a coefficient of 1) on one side of the equation (usually the left), leaving a constant on the other side.

••Solving equations

We are now ready to solve equations.

Equation-Solving Strategy

1. Simplify the left- and right-hand sides of the equation by removing grouping symbols and combining like terms.
2. Use equality properties above to get all variable terms on one side (usually the left) and all constant terms on the other side (usually the right). Combine like terms in the process.
3. Isolate the variable (with a coefficient of 1), using the division or multiplication property of equality.

EXAMPLE 1

Solve $3x - 2(2x - 5) = 2(x + 3) - 8$ and check.

Solution

$3x - 2(2x - 5) = 2(x + 3) - 8$	Clear parentheses.
$3x - 4x + 10 = 2x + 6 - 8$	Combine like terms.
$-x + 10 = 2x - 2$	Isolate x on left side.
$-x + 10 - 10 = 2x - 2 - 10$	Subtraction property
$-x = 2x - 12$	
$-x - 2x = 2x - 12 - 2x$	Subtraction property
$-3x = -12$	
$\dfrac{-3x}{-3} = \dfrac{-12}{-3}$	Division property
$x = 4$	

You should soon be performing these three steps mentally.

We have produced a string of simpler equivalent equations. Since 4 is a solution to the last equation, it must be a solution to the original equation.

CHECK

$$3x - 2(2x - 5) = 2(x + 3) - 8$$
$$3 \cdot 4 - 2(2 \cdot 4 - 5) \stackrel{?}{=} 2(4 + 3) - 8$$
$$12 - 2 \cdot 3 \stackrel{?}{=} 2 \cdot 7 - 8$$
$$6 \stackrel{\checkmark}{=} 6$$

Work Problem 1 and check solution in Solutions to Matched Problems on page 160.

PROBLEM 1 Solve and check:

$$8x - 3(x - 4) = 3(x - 4) + 6$$

EXAMPLE 2

Solve $2x - (4x - 3) = 2(4 - x) + 1$ and check.

Solution

$$2x - (4x - 3) = 2(4 - x) + 1$$
$$2x - 4x + 3 = 8 - 2x + 1$$
$$-2x + 3 = 9 - 2x$$
$$0 = 6$$

There is no solution (otherwise we have proved that $0 = 6$).

Work Problem 2 and check solution in Solutions to Matched Problems on page 160.

PROBLEM 2 Solve and check:

$$2(5 - 3x) + (2x - 11) = 4(3 - x)$$

•••Summary

If all terms in Example 1 had been transferred to the left side (leaving 0 on the right) and like terms combined, we would have obtained

$$-3x + 12 = 0$$

a special case of

$$ax + b = 0 \qquad\qquad (1)$$

Any equation that can be put in form (1) is called a **first-degree equation in one variable.** Any equation of this form with $a \neq 0$ always has exactly one solution.

$$ax + b = 0$$

$$ax = -b$$

$$x = -\frac{b}{a}$$

In general, it is important to know under what conditions a particular type of equation has a solution and how many solutions are possible. We have now answered both questions for equations of the type

$$ax + b = 0 \qquad a \neq 0$$

Other types of equations will be studied later which have more than one solution. For example,

$$x^2 - 4 = 0 \qquad \text{second-degree equation in one variable}$$

has two solutions, -2 and $+2$.

SOLUTIONS TO MATCHED PROBLEMS

1. $8x - 3(x - 4) = 3(x - 4) + 6$ CHECK: $8(-9) - 3[(-9) - 4] \overset{?}{=} 3[(-9) - 4] + 6$
    $8x - 3x + 12 = 3x - 12 + 6$
    $5x + 12 = 3x - 6$ $-72 - 3(-13) \overset{?}{=} 3(-13) + 6$
    $2x = -18$ $-72 + 39 \overset{?}{=} -39 + 6$
    $x = -9$ $-33 \overset{\checkmark}{=} -33$

2. $2(5 - 3x) + (2x - 11) = 4(3 - x)$
    $10 - 6x + 2x - 11 = 12 - 4x$
    $-4x - 1 = 12 - 4x$
    $0 = 13$
    No solution.

PRACTICE EXERCISE 4.1

Work odd-numbered problems first, check answers, and then work even-numbered problems in areas of weakness. Answers to all problems are in the back of the book. Make every effort to work a problem yourself before you look at an answer.

Solve and check.

A 1. $x - 7 = -9$ _____

2. $x + 4 = -6$ _____

3. $5x = -13$ _____

4. $-8x = 6$ _____

5. $-5m = 0$ _____

6. $3y = 0$ _____

7. $-4m + 5 = -9$ _____

8. $6w + 18 = -2$ _____

9. $3x - 4 = 6x - 19$ _____

10. $2t + 9 = 5t - 6$ _____

11. $4y + 7 = 2y - 6$ _____

12. $3x - 5 = x + 6$ _____

13. $2y + 8 = 2y - 6$ _____

14. $x - 3 = x + 7$ _____

15. $3(x + 2) = 5(x - 6)$ _____

16. $5x + 10(x - 2) = 40$ _____

17. $4(x - 2) = 4x - 8$ _____

B 18. $3y + 6 = 3(y + 2)$ _____

19. $5 + 4(t - 2) = 2(t + 7) + 1$ _____

20. $7x - (8x - 4) - 2 = 5 - (4x + 2)$ _____

21. $10x + 25(x - 3) = 275$ _____

22. $x + (x + 2) + (x + 4) = 54$ _____

23. $5x - (7x - 4) - 2 = 5 - (3x + 2)$ _____

24. $-3(4 - t) = 5 - (t + 1)$ _____

25. $x(x - 1) + 5 = x^2 + x - 3$ _____

26. $x(x + 2) = x(x + 4) - 12$ _____

27. $x(x - 4) - 2 = x^2 - 4(x + 3)$ _____

28. $t(t - 6) + 8 = t^2 - 6t - 3$ _____

29. $-2\{3 + [2x - (x - 4)]\} = 2[(x + 2) - 3]$ _____

30. $-2\{2 - [1 - 2(x + 1)]\} = 2(x + 5) - 4$ _____

C 31. Which of the following are equivalent to $2x + 5 = x - 3$: $2x = x - 8$, $2x = x + 2$, $3x = -8$, $x = -8$?

_____

32. Which of the following are equivalent to $3x - 6 = 6$: $3x = 12$, $3x = 0$, $x = 4$, $x = 0$? _____

> The Check Exercise for this section is on page 193.

4.2 EQUATIONS INVOLVING FRACTIONAL FORMS

•Equations with constants in denominators
••A common error
•••Equations with variables in denominators

We will divide the discussion into two parts. The first part will deal with equations with constants in denominators and the second part will deal with equations with variables in denominators.

•Equations with constants in denominators

To solve equations involving fractions we can start by using the multiplication property of equality to clear the fractions; then we proceed as in the last section. What do we multiply both sides by to clear the fractions? We use the lcm of all denominators present in the equation.

EXAMPLE 3

Solve $\dfrac{x}{3} - \dfrac{1}{2} = \dfrac{5}{6}$.

Solution

$$\frac{x}{3} - \frac{1}{2} = \frac{5}{6}$$
Clear fractions by multiplying both sides by 6, the lcm of all the denominators.

$$6 \cdot \left(\frac{x}{3} - \frac{1}{2}\right) = 6 \cdot \frac{5}{6}$$
Clear ( ) before canceling.

$$6 \cdot \frac{x}{3} - 6 \cdot \frac{1}{2} = 6 \cdot \frac{5}{6}$$
Wrong: $\overset{2}{\cancel{6}} \cdot \left(\frac{x}{\underset{1}{\cancel{3}}} - \frac{1}{2}\right)$

$$2x - 3 = 5$$
Equation is now free of fractions.

$$2x = 8$$
Solve as in Section 4.1.

$$x = 4$$

Work Problem 3 and check solution in Solutions to Matched Problems on page 166.

PROBLEM 3   Solve:

$$\frac{1}{4}x - \frac{2}{3} = \frac{5}{12}x \qquad \text{NOTE:} \quad \frac{1}{4}x = \frac{x}{4} \text{ and } \frac{5}{12}x = \frac{5x}{12}.$$

EXAMPLE 4

Solve $0.2x + 0.3(x - 5) = 13$.

Solution

Some equations involving decimal-fraction coefficients are readily solved by first clearing decimals.

$$0.2x + 0.3(x - 5) = 13$$
Multiply by 10 to clear decimals.

$$10(0.2x) + 10[0.3(x - 5)] = 10 \cdot 13$$

$$2x + 3(x - 5) = 130$$
Solve as in Section 4.1.

$$2x + 3x - 15 = 130$$

$$5x = 145$$

$$x = 29$$

Work Problem 4 and check solution in Solutions to Matched Problems on page 166.

PROBLEM 4   Solve:

$$0.3(x + 2) + 0.5x = 3$$

EXAMPLE 5

Solve $5 - \dfrac{2x - 1}{4} = \dfrac{x + 2}{3}$.

Solution

Before multiplying both sides by 12, the lcm of the denominators, enclose any numerator with more than one term in parentheses.

$5 - \dfrac{(2x - 1)}{4} = \dfrac{(x + 2)}{3}$ Multiply both sides by 12.

$12 \cdot 5 - \overset{3}{\cancel{12}} \cdot \dfrac{(2x - 1)}{\underset{1}{\cancel{4}}} = \overset{4}{\cancel{12}} \cdot \dfrac{(x + 2)}{\underset{1}{\cancel{3}}}$ Cancel denominators.

$60 - 3(2x - 1) = 4(x + 2)$ Solve as in Section 4.1.

$60 - 6x + 3 = 4x + 8$

$-6x + 63 = 4x + 8$

$-10x = -55$

$x = \tfrac{11}{2}$ or 5.5

Work Problem 5 and check solution in Solutions to Matched Problems on page 166.

PROBLEM 5 Solve:

$$\dfrac{x + 3}{4} - \dfrac{x - 4}{2} = \dfrac{3}{8}$$

••A common error

A very common error occurs about now—students tend to confuse *algebraic expressions* involving fractions with *algebraic equations* involving fractions.

Consider the two problems:

(A) Solve: $\dfrac{x}{2} + \dfrac{x}{3} = 10$

(B) Add: $\dfrac{x}{2} + \dfrac{x}{3} + 10$

The problems look very much alike, but are actually very different. To solve the equation in **(A)** we multiply both sides by 6 (the lcm of 2 and 3) to clear the fractions. This works so well for equations that students want to do the same thing for problems like (**B**). The only catch is that (**B**) is not an equation and the multiplication property of equality does not apply. If we multiply (**B**) by 6, we obtain an expression 6 times as large as the original. To add in (**B**) we find the lcd and proceed as in Section 3.3.

Compare the following:

(**A**) $\dfrac{x}{2} + \dfrac{x}{3} = 10$

$$6 \cdot \dfrac{x}{2} + 6 \cdot \dfrac{x}{3} = 6 \cdot 10$$

$$3x + 2x = 60$$

$$5x = 60$$

$$x = 12$$

(**B**) $\dfrac{x}{2} + \dfrac{x}{3} + 10$

$$= \dfrac{3 \cdot x}{3 \cdot 2} + \dfrac{2 \cdot x}{2 \cdot 3} + \dfrac{6 \cdot 10}{6 \cdot 1}$$

$$= \dfrac{3x}{6} + \dfrac{2x}{6} + \dfrac{60}{6}$$

$$= \dfrac{5x + 60}{6}$$

•••Equations with variables in denominators

If an equation involves a variable in one or more denominators, such as

$$\dfrac{2}{3} - \dfrac{2}{x} = \dfrac{4}{x}$$

we may proceed in essentially the same way as above as long as we

avoid any value of x that makes a denominator zero.

EXAMPLE 6

Solve $\dfrac{2}{3} - \dfrac{2}{x} = \dfrac{4}{x}$.

Solution

$$\dfrac{2}{3} - \dfrac{2}{x} = \dfrac{4}{x} \qquad x \neq 0$$

We note that $x \neq 0$; then multiply both sides by $3x$, the lcm of the denominators. If 0 turns up later as a "solution," it must be discarded.

$$3x \cdot \dfrac{2}{3} - 3x \cdot \dfrac{2}{x} = 3x \cdot \dfrac{4}{x}$$

All denominators cancel.

$$2x - 6 = 12$$

$$2x = 18$$

$$x = 9$$

Work Problem 6 and check solution in Solutions to Matched Problems on page 166.

PROBLEM 6   Solve:

$$\dfrac{3}{x} - \dfrac{1}{2} = \dfrac{4}{x}$$

EXAMPLE 7

Solve $\dfrac{3x}{x-2} - 4 = \dfrac{14 - 4x}{x-2}$.

Solution

$$\dfrac{3x}{x-2} - 4 = \dfrac{14 - 4x}{x-2} \qquad x \neq 2$$

If 2 turns up as a "solution," it must be discarded.

$$(x-2)\dfrac{3x}{(x-2)} - 4(x-2) = (x-2)\dfrac{(14 - 4x)}{(x-2)}$$

Multiply by $(x - 2)$, the lcm of the denominators. Also place all binomial numerators and denominators in parentheses.

$$3x - 4(x-2) = 14 - 4x$$

$$3x - 4x + 8 = 14 - 4x$$

$$-x + 8 = 14 - 4x$$

$$3x = 6$$

$$x = 2$$

$x = 2$ cannot be a solution to the original equation (see comments above).

Equation has no solution. (Hence, solution set is empty.)

Work Problem 7 and check solution in Solutions to Matched Problems on page 166.

PROBLEM 7 Solve:

$$\dfrac{2x}{x-1} - 3 = \dfrac{7 - 3x}{x-1}$$

EXAMPLE 8

Solve $2 - \dfrac{3x}{1-x} = \dfrac{8}{x-1}$.

Solution

$$2 - \dfrac{3x}{1-x} = \dfrac{8}{x-1}$$

Recall: $1 - x = -(x - 1)$.

$$2 - \dfrac{3x}{-(x-1)} = \dfrac{8}{x-1}$$

Recall: $-\dfrac{a}{-b} = \dfrac{a}{b}$

$$2 + \dfrac{3x}{x-1} = \dfrac{8}{x-1}$$

Multiply both sides by $(x - 1)$, keeping in mind that $x \neq 1$.

$$(x - 1)(2) + (x - 1)\left(\frac{3x}{x - 1}\right) = (x - 1)\left(\frac{8}{x - 1}\right)$$

$$2x - 2 + 3x = 8 \qquad \text{Solve as in Section 4.1.}$$

$$5x = 10$$

$$x = 2$$

Work Problem 8 and check solution in Solutions to Matched Problems on page 167.

PROBLEM 8   Solve:

$$\frac{5x}{x - 2} - \frac{6}{2 - x} = 3$$

SOLUTIONS TO MATCHED PROBLEMS

3.
$$\frac{x}{4} - \frac{2}{3} = \frac{5x}{12}$$

$$12\left(\frac{x}{4} - \frac{2}{3}\right) = 12\left(\frac{5x}{12}\right)$$

$$3x - 8 = 5x$$
$$-2x = 8$$
$$x = -4$$

4.
$$0.3(x + 2) + 0.5x = 3$$

$$10[0.3(x + 2)] + 10(0.5x) = 10(3)$$

$$3(x + 2) + 5x = 30$$
$$3x + 6 + 5x = 30$$
$$8x = 24$$
$$x = 3$$

5.
$$\frac{x + 3}{4} - \frac{x - 4}{2} = \frac{3}{8}$$

$$8 \cdot \frac{(x + 3)}{4} - 8 \cdot \frac{(x - 4)}{2} = 8 \cdot \frac{3}{8}$$

$$2(x + 3) - 4(x - 4) = 3$$
$$2x + 6 - 4x + 16 = 3$$
$$-2x + 22 = 3$$
$$-2x = -19$$
$$x = \frac{-19}{-2} = \frac{19}{2} \text{ or } 9.5$$

6.
$$\frac{3}{x} - \frac{1}{2} = \frac{4}{x} \qquad x \neq 0$$

$$2x\left(\frac{3}{x} - \frac{1}{2}\right) = 2x\left(\frac{4}{x}\right)$$

$$6 - x = 8$$
$$-x = 2$$
$$x = -2$$

7.
$$\frac{2x}{x - 1} - 3 = \frac{7 - 3x}{x - 1} \qquad x \neq 1$$

$$(x - 1)\left(\frac{2x}{x - 1}\right) - 3(x - 1) = (x - 1)\left(\frac{7 - 3x}{x - 1}\right)$$

$$2x - 3x + 3 = 7 - 3x$$
$$-x + 3 = 7 - 3x$$
$$2x = 4$$
$$x = 2$$

8.

$$\frac{5x}{x-2} - \frac{6}{2-x} = 3 \qquad x \neq 2$$

$$\frac{5x}{x-2} - \frac{6}{-(x-2)} = 3$$

$$\frac{5x}{x-2} + \frac{6}{x-2} = 3$$

$$(x-2)\left(\frac{5x}{x-2}\right) + (x-2)\left(\frac{6}{x-2}\right) = (x-2)(3)$$

$$5x + 6 = 3(x-2)$$
$$5x + 6 = 3x - 6$$
$$2x = -12$$
$$x = -6$$

PRACTICE EXERCISE 4.2

Work odd-numbered problems first, check answers, and then work even-numbered problems in areas of weakness. Answers to all problems are in the back of the book. Make every effort to work a problem before you look at an answer.

Solve each equation.

A 1. $\dfrac{x}{5} - 2 = \dfrac{3}{5}$ _____

2. $\dfrac{x}{7} - 1 = \dfrac{1}{7}$ _____

3. $\dfrac{x}{3} + \dfrac{x}{6} = 4$ _____

4. $\dfrac{y}{4} + \dfrac{y}{2} = 9$ _____

5. $\dfrac{m}{4} - \dfrac{m}{3} = \dfrac{1}{2}$ _____

6. $\dfrac{n}{5} - \dfrac{n}{6} = \dfrac{6}{5}$ _____

7. $\dfrac{5}{12} - \dfrac{m}{3} = \dfrac{4}{9}$ _____

8. $\dfrac{2}{3} - \dfrac{x}{8} = \dfrac{5}{6}$ _____

9. $0.7x = 21$ _____

10. $0.9x = 540$ _____

11. $0.7x + 0.9x = 32$ _____

12. $0.3x + 0.5x = 24$ _____

13. $\dfrac{1}{2} - \dfrac{2}{x} = \dfrac{3}{x}$ _____

14. $\dfrac{2}{x} - \dfrac{1}{3} = \dfrac{5}{x}$ _____

15. $\dfrac{1}{m} - \dfrac{1}{9} = \dfrac{4}{9} - \dfrac{2}{3m}$ _____

16. $\dfrac{1}{2t} + \dfrac{1}{8} = \dfrac{2}{t} - \dfrac{1}{4}$ _____

B 17. $\dfrac{x-2}{3} + 1 = \dfrac{x}{7}$ _____

18. $\dfrac{x+3}{2} - \dfrac{x}{3} = 4$ _____

19. $\dfrac{2x-3}{9} - \dfrac{x+5}{6} = \dfrac{3-x}{2} - 1$ _____

20. $\dfrac{3x+4}{3} - \dfrac{x-2}{5} = \dfrac{2-x}{15} - 1$ _____

21. $0.1(x-7) + 0.05x = 0.8$ _____

22. $0.4(x+5) - 0.3x = 17$ _____

23. $0.02x - 0.5(x-2) = 5.32$ _____

24. $0.3x - 0.04(x+1) = 2.04$ _____

25. $\dfrac{7}{y-2} - \dfrac{1}{2} = 3$ _____

26. $\dfrac{9}{A+1} - 1 = \dfrac{12}{A+1}$ _____

27. $\dfrac{3}{2x-1} + 4 = \dfrac{6x}{2x-1}$ _____

28. $\dfrac{5x}{x+5} = 2 - \dfrac{25}{x+5}$ _____

29. $\dfrac{2E}{E-1} = 2 + \dfrac{5}{2E}$ _____

30. $\dfrac{3N}{N-2} - \dfrac{9}{4N} = 3$ _____

31. $\dfrac{n-5}{6n-6} = \dfrac{1}{9} - \dfrac{n-3}{4n-4}$ _____

32. $\dfrac{1}{3} - \dfrac{s-2}{2s+4} = \dfrac{s+2}{3s+6}$ _____

33. $5 + \dfrac{2x}{x-3} = \dfrac{6}{x-3}$ _____

34. $\dfrac{6}{x-2} = 3 + \dfrac{3x}{x-2}$ _____

35. $\dfrac{x^2+2}{x^2-4} = \dfrac{x}{x-2}$ _____

36. $\dfrac{5}{x-3} = \dfrac{33-x}{x^2-6x+9}$ _____

C 37. $\dfrac{3x}{24} - \dfrac{2-x}{10} = \dfrac{5+x}{40} - \dfrac{1}{15}$ _____

38. $\dfrac{2x}{10} - \dfrac{3-x}{14} = \dfrac{2+x}{5} - \dfrac{1}{2}$ _____

39. $\dfrac{5t-22}{t^2-6t+9} - \dfrac{11}{t^2-3t} - \dfrac{5}{t} = 0$ _____

40. $\dfrac{x^2}{x^2-6x+9} - \dfrac{6x+1}{x^2-3x} = 1$ _____

41. $5 - \dfrac{2x}{3-x} = \dfrac{6}{x-3}$ _____

42. $\dfrac{3x}{2-x} + \dfrac{6}{x-2} = 3$ _____

The Check Exercise for this section is on page 195.

4.3 SOLVING FOR A PARTICULAR VARIABLE

One of the immediate applications you will have for algebra is the changing of formulas or equations to alternate equivalent forms. In the process we will make frequent use of the symmetric property of equality introduced in Section 1.2. Recall

---

Symmetric Property of Equality

If $a = b$, then $b = a$.

An equation can be reversed without changing any signs.

Thus, if we are given the formula

$$c = \frac{wrt}{1,000}$$

then we may reverse it if we wish to obtain

$$\frac{wrt}{1,000} = c$$

We will do exactly this in the next example, where we solve the formula for t in terms of the other variables.

EXAMPLE 9
Solve the formula $c = wrt/1,000$ for t. (The formula gives the cost of using an electrical appliance; w = power in watts, r = rate per kilowatt-hour, t = time in hours.)

Solution

$$c = \frac{wrt}{1,000} \qquad \text{Start with the given formula.}$$

$$\frac{wrt}{1,000} = c \qquad \text{Reverse the equation to get } t \text{ on the left side.}$$

$$\frac{1,000}{wr} \cdot \frac{wrt}{1,000} = \frac{1,000}{wrt} \cdot c \qquad \text{Multiply both sides by } 1,000/wr \text{ to isolate } t \text{ on the left.}$$

$$t = \frac{1,000c}{wr} \qquad \text{We have solved for } t.$$

Work Problem 9 and check solution in Solutions to Matched Problems on page 172.

PROBLEM 9   Solve the formula in Example 9 for w.

Solution

EXAMPLE 10
Solve the formula $A = P + Prt$ for r (simple interest formula).

Solution

$$A = P + Prt$$

Reverse the equation; then perform operations to isolate r on the left side.

$$P + Prt = A$$

$$Prt = A - P$$

$$\frac{Prt}{Pt} = \frac{A - P}{Pt}$$

$$r = \frac{A - P}{Pt}$$

Work Problem 10 and check solution in Solutions to Matched Problems on page 172.

PROBLEM 10   Solve the formula $A = P + Prt$ for t.

Solution

EXAMPLE 11

Solve the formula $A = P + Prt$ for P.

Solution

$$A = P + Prt$$

COMMON ERROR: If we write $P = A - Prt$ for the solution, we will not have solved for P. To solve for P is to isolate P on the left side with a coefficient of 1. **In general, if the variable for which we are solving appears on both sides of the equation, we have not solved for it!**

$$P + Prt = A$$

We start by reversing the equation. Since P is a common factor to both terms on the left, we factor P out and complete the problem.

$$P(1 + rt) = A$$

$$\frac{P(1 + rt)}{1 + rt} = \frac{A}{1 + rt}$$

$$P = \frac{A}{1 + rt}$$

Note that P appears only on the left side.

Work Problem 11 and check solution in Solutions to Matched Problems on page 172.

PROBLEM 11   Solve $A = xy + xz$ for x.

Solution

EXAMPLE 12

Solve $m = \dfrac{2t^2}{n}$ for n.

Solution

$m = \dfrac{2t^2}{n}$ Do not reverse equation. Multiply both sides by n to get n on the left side.

$mn = 2t^2$ Now divide both sides by m to isolate n.

$n = \dfrac{2t^2}{m}$

Work Problem 12 and check solution in Solutions to Matched Problems on page 172.

PROBLEM 12   Solve $L = \dfrac{a + b}{c}$ for c.

Solution

EXAMPLE 13

Solve $\dfrac{1}{r} = \dfrac{1}{s} + \dfrac{1}{t}$ for r.

Solution

$\dfrac{1}{r} = \dfrac{1}{s} + \dfrac{1}{t}$ Multiply both sides by rst to clear fractions.

$rst \cdot \dfrac{1}{r} = rst \cdot \dfrac{1}{s} + rst \cdot \dfrac{1}{t}$

$st = rt + rs$ Reverse equation.

$rt + rs = st$ Factor out r.

$r(t + s) = st$ Divide both sides by $(t + s)$.

$r = \dfrac{st}{t + s}$

Work Problem 13 and check solution in Solutions to Matched Problems on page 172.

PROBLEM 13   Solve $\dfrac{1}{r} = \dfrac{1}{s} + \dfrac{1}{t}$ for s.

Solution

SOLUTIONS TO MATCHED PROBLEMS

9.
$$c = \frac{wrt}{1{,}000}$$
$$\frac{wrt}{1{,}000} = c$$
$$\frac{1{,}000}{rt}\left(\frac{wrt}{1{,}000}\right) = \frac{1{,}000}{rt}c$$
$$w = \frac{1{,}000c}{rt}$$

10.
$$A = P + Prt$$
$$P + Prt = A$$
$$Prt = A - P$$
$$t = \frac{A - P}{Pr}$$

11.
$$A = xy + xz$$
$$xy + xz = A$$
$$x(y + z) = A$$
$$x = \frac{A}{y + z}$$

12.
$$L = \frac{a + b}{c}$$
$$cL = a + b$$
$$c = \frac{a + b}{L}$$

13.
$$\frac{1}{r} = \frac{1}{s} + \frac{1}{t}$$
$$rst \cdot \frac{1}{r} = rst \cdot \frac{1}{s} + rst \cdot \frac{1}{t}$$
$$st = rt + rs$$
$$st - rs = rt$$
$$s(t - r) = rt$$
$$s = \frac{rt}{t - r}$$

PRACTICE EXERCISE 4.3

Work odd-numbered problems first, check answers, and then work even-numbered problems in areas of weakness. Answers to all problems are in the back of the book. Make every effort to work a problem your-self before you look at an answer.

A **1.** Solve $d = rt$ for r _____ (*Distance-rate-time*)

2. Solve $d = 1{,}100t$ for t _____ (*Sound distance in air*)

3. Solve $C = 2\pi r$ for r _____ (*Circumference of a circle*)

4. Solve $I = Prt$ for t _____ (*Simple interest*)

5. Solve $C = \pi D$ for π _____ (*Circumference of a circle*)

6. Solve $e = mc^2$ for m _____ (*Mass-energy equation*)

7. Solve $ax + b = 0$ for x _____ (*First-degree polynomial equation*)

8. Solve $p = 2a + 2b$ for a _____ (*Perimeter of a rectangle*)

9. Solve $y = 2x - 5$ for x _____ (*Slope-intercept equation for a line*)

10. Solve $y = mx + b$ for m _____ (*Slope-intercept equation for a line*)

B 11. Solve $3x - 4y - 12 = 0$ for y _____ (*Linear equation in two variables*)

12. Solve $Ax + By + C = 0$ for y _____ (*Linear equation in two variables*)

13. Solve $I = \dfrac{E}{R}$ for R _____ (*Electric circuits—Ohm's law*)

14. Solve $m = \dfrac{b}{a}$ for a _____ (*Optics—magnification*)

15. Solve $C = \dfrac{100B}{L}$ for B _____ (*Anthropology—cephalic index*)

16. Solve $(IQ) = \dfrac{100(MA)}{(CA)}$ for (CA) _____ (*Psychology—intelligence quotient*)

17. Solve $F = G\dfrac{m_1 m_2}{d^2}$ for G^* _____ (*Gravitational force between two masses*)

18. Solve $F = G\dfrac{m_1 m_2}{d^2}$ for m_1 _____ (*Gravitational force between two masses*)

19. Solve $F = \tfrac{9}{5}C + 32$ for C _____ (*Celsius-Fahrenheit*)

20. Solve $C = \tfrac{5}{9}(F - 32)$ for F _____ (*Celsius-Fahrenheit*)

21. Solve $P = M - Mdt$ for d _____ (*Simple discount*)

22. Solve $P = M - Mdt$ for t _____

23. Solve $P = M - Mdt$ for M _____

24. Solve $A = \dfrac{ah}{2} + \dfrac{bh}{2}$ for h _____ (*Area of a trapezoid*)

C 25. Solve $\dfrac{1}{f} = \dfrac{1}{a} + \dfrac{1}{b}$ for f _____ (*Optics—focal length*)

26. Solve $\dfrac{1}{R} = \dfrac{1}{R_1} + \dfrac{1}{R_2}$ for R _____ (*Electric circuits*)

*The 1 and 2 in m_1 and m_2, respectively, are called **subscripts** and have no operational meaning. Subscripts allow us to use the same letter for more than one quantity. In this case m_1 and m_2 represent two different masses.

27. Solve $a_n = a_1 + (n - 1)d$ for n _____   (*Arithmetic progression*)

28. Solve $a_n = a_1 + (n - 1)d$ for d _____   (*Arithmetic progression*)

29. Solve $\dfrac{P_1 V_1}{T_1} = \dfrac{P_2 V_2}{T_2}$ for T_2 _____   (*Gas law*)

30. Solve $\dfrac{P_1 V_1}{T_1} = \dfrac{P_2 V_2}{T_2}$ for V_1 _____   (*Gas law*)

31. Solve $y = \dfrac{2x - 3}{3x - 5}$ for x _____   (*Rational equation*)

32. Solve $y = \dfrac{3x + 2}{2x - 4}$ for x _____   (*Rational equation*)

The Check Exercise for this section is on page 197.

4.4 INEQUALITIES

 •Inequality symbols
 ••Inequalities and line graphs
 •••Solving inequalities

•Inequality symbols

Just as we use "$=$" to replace the words "is equal to," we will use the **inequality symbols** "$<$" and "$>$" to represent "is less than" and "is greater than," respectively. Thus, we can write the following:

$a < b$	a is less than b
$a > b$	a is greater than b
$a \leq b$	a is less than or equal to b
$a \geq b$	a is greater than or equal to b

It no doubt seems obvious to you that

$5 < 8$

but does it seem equally obvious that

$$-8 < -5 \qquad 0 > -10 \qquad -30,000 < -1$$

To make the inequality relation precise so that we can interpret it relative to *all* real numbers, we need a careful definition of the concept.

Definition of < and >

If a and b are real numbers, then we write

$a < b$

if there is a positive real number p such that $a + p = b$. We write

$c > d$

if there is a positive real number q such that

$c - q = d$

Certainly, one would expect that if a positive number were added to *any* real number it would make it larger and if it were subtracted from *any* real number it would make it smaller. That is essentially what the definition states. Note that if

$$a > b \qquad \text{then} \qquad b < a$$

and vice versa.

EXAMPLE 14

(A) $5 < 8$ since $5 + 3 = 8$

(B) $-8 < -5$ since $-8 + 3 = -5$

(C) $0 > -10$ since $0 - 10 = -10$

(D) $-1 > -1,000$ since $-1 - 999 = -1,000$

Important Real Number Property

Given any two real numbers a and b, then either

$$a < b \qquad a > b \qquad \text{or} \qquad a = b$$

Work Problem 14 and check solution in Solutions to Matched Problems on page 181.

PROBLEM 14 Insert $<$, $>$, or $=$ in each square as appropriate.

(A) $4 \square 6$ (B) $6 \square 4$ (C) $-6 \square -4$

(D) $-8 \square 8$ (E) $9 \square 9$ (F) $3 \square -9$

(G) $0 \square -1$ (H) $-500 \square -3$ (I) $0 \square -30$

••Inequalities and line graphs

The inequality symbols have a very clear geometric interpretation on the real number line. If $a < b$, then a is to the left of b; if $c > d$, then c is to the right of d (Figure 1).

Figure 1

EXAMPLE 15
Refer to Figure 1.

(A) $a < d$ since a is to the left of d
(B) $c > 0$ since c is to the right of 0
(C) $d < c$ since d is to the left of c
(D) $a < 0$ since a is to the left of 0

Work Problem 15 and check solution in Solutions to Matched Problems on page 181.

PROBLEM 15 Referring to Figure 1, insert $<$ or $>$ in each square.	
(A) $b \square d$	**(B)** $0 \square b$
(C) $a \square c$	**(D)** $d \square 0$

Now let us turn to simple inequality statements of the form

$x > 2$ $-2 < x \leq 3$

$x \leq -3$ $0 \leq x < 5$

We are interested in graphing such statements on a real number line. In general, to **graph an inequality statement** in one variable on a real number line is to graph the set of all real number replacements of the variable that make the statement true. This set is called the **solution set** of the inequality statement.

EXAMPLE 16
Graph $x \geq -3$ on a real number line.

Solution
The solution set for

$x \geq -3$

is the set of *all* real numbers greater than or equal to -3. Graphically, this includes *all* the points from -3 to the right—a solid line.

Solid dot indicates
-3 is included.

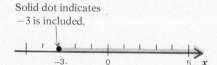

Work Problem 16 and check solution in Solutions to Matched Problems on page 181.

PROBLEM 16   Graph $x \leq 1$ on a real number line.

Solution

EXAMPLE 17
Graph $-4 \leq x < 2$ on a real number line.

Solution
The solution set for

$$-4 \leq x < 2 \qquad x \text{ a real number}$$

is the set of *all* real numbers between -4 and 2, including -4, but excluding 2. The graph is a solid line including the left endpoint, -4, and excluding the right endpoint, 2.

Note that a hollow circle indicates that an endpoint is not included and a solid circle indicates that an endpoint is included.

Work Problem 17 and check solution in Solutions to Matched Problems on page 181.

PROBLEM 17   Graph $-2 < x \leq 3$ on a real number line.

Solution

•••Solving inequalities

The solution sets for inequalities such as

$$x > 2 \qquad -4 < \leq 3 \qquad x \leq -1$$

are obvious. We will now consider inequality statements that do not have obvious solution sets. Can you, for example, guess the real number solutions for

$$3(x - 2) + 1 < 3x - (x + 7)$$

By the end of this section you will be able to solve this type of inequality almost as easily as you solved first-degree equations.

When solving equations we made considerable use of the addition, subtraction, multiplication, and division properties of equality. We can use similar properties to help us solve inequalities. Look over the following examples carefully and notice what happens to the sense of (direction) of the inequality in each case.

$$-4 < 2$$
$$-4 + 3 \; ? \; 2 + 3$$
$$-1 < 5$$

$$-4 < 2$$
$$-4 - 3 \; ? \; 2 - 3$$
$$-7 < -1$$

$$-4 < 2$$
$$3(-4) \; ? \; 3(2)$$
$$-12 < 6$$

$$-4 < 2$$
$$(-3)(-4) \; ? \; (-3)(2)$$
$$12 > -6$$
$$\uparrow$$
sense
reverses

$$-4 < 2$$
$$\frac{-4}{2} \; ? \; \frac{2}{2}$$
$$-2 < 1$$

$$-4 < 2$$
$$\frac{-4}{-2} \; ? \; \frac{2}{-2}$$
$$2 > -1$$
$$\uparrow$$
sense
reverses

Each example above illustrates one of the following general inequality properties.

Inequality Properties

1. *Addition property.* The sense of an inequality remains unchanged if the same quantity is added to each side.

   $$a < b$$
   $$a + c < b + c$$

2. *Subtraction property.* The sense of an inequality remains unchanged if the same quantity is subtracted from each side.

   $$a < b$$
   $$a - c < b - c$$

3. *Multiplication property.* The sense of an inequality remains unchanged if each side is multiplied by the same positive quantity. The sense reverses if each side is multiplied by the same negative quantity.

   $$a < b$$
   $$ca < cb \quad \text{if } c > 0$$
   $$ca > cb \quad \text{if } c < 0$$

4. *Division property.* The sense of an inequality remains unchanged if each side is divided by the same positive quantity. The sense reverses if each side is divided by the same negative quantity.

   $$a < b$$
   $$\frac{a}{c} < \frac{b}{c} \quad \text{if } c > 0$$
   $$\frac{a}{c} > \frac{b}{c} \quad \text{if } c < 0$$

NOTE: If $c > 0$, then c is positive. If $c < 0$, then c is negative.

Similar properties hold if $<$ is replaced with $\leq$ and $>$ is replaced with $\geq$. Thus, we find that we can perform essentially the same operations on inequality statements to produce equivalent inequality statements that we performed on equations to produce equivalent equations, with the exception that:

The sense of the inequality sign reverses if we multiply or divide both sides of an inequality by a negative number.

EXAMPLE 18
Solve and graph $3(x - 1) + 5 < 5(x + 2)$.

Solution

$3(x - 1) + 5 < 5(x + 2)$ Simplify left and right sides.

$3x - 3 + 5 < 5x + 10$

$3x + 2 < 5x + 10$

$3x + 2 - 2 < 5x + 10 - 2$ Sense unchanged (property 2).

$3x < 5x + 8$

$3x - 5x < 5x + 8 - 5x$ Sense unchanged (property 2).

$-2x < 8$

$\dfrac{-2x}{-2} > \dfrac{8}{-2}$ Sense reverses (property 4).

$x > -4$

Work Problem 18 and check solution in Solutions to Matched Problems on page 181.

PROBLEM 18   Solve and graph:

$$2(2x + 3) \geq 6(x - 2) + 10$$

EXAMPLE 19
Solve and graph $\dfrac{2x - 3}{4} + 6 \geq 2 + \dfrac{4x}{3}$.

Solution

$$\frac{2x - 3}{4} + 6 \geq 2 + \frac{4x}{3}$$ Multiply both sides by 12, the lcm of 4 and 3, to clear fractions.

$$12 \cdot \frac{(2x - 3)}{4} + 12 \cdot 6 \geq 12 \cdot 2 + 12 \cdot \frac{4x}{3}$$ Sense unchanged (property 3). Cancel denominators.

$$3(2x - 3) + 72 \geq 24 + 4 \cdot 4x$$

$$6x - 9 + 72 \geq 24 + 16x$$

$$6x + 63 \geq 24 + 16x$$

$$-10x + 63 \geq 24$$

$$-10x \geq -39$$

$$x \leq 3.9$$ Sense reverses (why?)

Work Problem 19 and check solution in Solutions to Matched Problems on page 181.

PROBLEM 19   Solve and graph:
$$\frac{4x - 3}{3} + 8 > 6 + \frac{3x}{2}$$

EXAMPLE 20

Solve and graph $-2 < 5 - 7x \leq 19$.

Solution

We proceed as above, except we try to isolate x in the middle with a coefficient of 1.

$$-2 < 5 - 7x \leq 19$$ Subtract 5 from each member.

$$-2 - 5 < 5 - 7x - 5 \leq 19 - 5$$ Sense unchanged.

$$-7 < -7x \leq 14$$ Divide each member by -7.

$$\frac{-7}{-7} > \frac{-7x}{-7} \geq \frac{14}{-7}$$ Sense reversed.

$$1 > x \geq -2 \quad \text{or} \quad -2 \leq x < 1$$

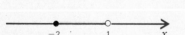

Work Problem 20 and check solution in Solutions to Matched Problems on page 182.

PROBLEM 20   Solve and graph:

$$-3 \le 7 - 2x < 7$$

EXAMPLE 21

Solve $30 \le \frac{5}{9}(F - 32) \le 35$

Solution

$$30 \le \frac{5}{9}(F - 32) \le 35$$

Multiply each member by $\frac{9}{5}$, the reciprocal of $\frac{5}{9}$.

$$\frac{9}{5} \cdot 30 \le \frac{9}{5} \cdot \frac{5}{9}(F - 32) \le \frac{9}{5} \cdot 35$$

Sense does not change (why?)

$$54 \le F - 32 \le 63$$

Add 32 to each member.

$$54 + 32 \le F - 32 + 32 \le 63 + 32$$

$$86 \le F \le 95$$

Work Problem 21 and check solution in Solutions to Matched Problems on page 182.

PROBLEM 21   Solve:

$$68 \le \frac{9}{5}C + 32 \le 77$$

SOLUTIONS TO MATCHED PROBLEMS

14. **(A)** $4 < 6$ **(B)** $6 > 4$ **(C)** $-6 < -4$ **(D)** $-8 < 8$ **(E)** $9 = 9$ **(F)** $3 > -9$
   (G) $0 > -1$ **(H)** $-500 < -3$ **(I)** $0 > -30$

15. **(A)** $b > d$ **(B)** $0 > b$ **(C)** $a < c$ **(D)** $d < 0$

16.   **17.**

18. $2(2x + 3) \ge 6(x - 2) + 10$
    $4x + 6 \ge 6x - 12 + 10$
    $4x + 6 \ge 6x - 2$
    $-2x \ge -8$
    $x \le 4$

19. $\dfrac{4x - 3}{3} + 8 > 6 + \dfrac{3x}{2}$

$6 \cdot \dfrac{4x - 3}{3} + 6 \cdot 8 > 6 \cdot 6 + 6 \cdot \dfrac{3x}{2}$

$2(4x - 3) + 48 > 36 + 3 \cdot 3x$
$8x - 6 + 48 > 36 + 9x$
$8x + 42 > 36 + 9x$
$-x > -6$
$x < 6$

20.
$$-3 \leq 7 - 2x < 7$$
$$-3 - 7 \leq 7 - 2x - 7 < 7 - 7$$
$$-10 \leq -2x < 0$$
$$\frac{-10}{-2} \geq \frac{-2x}{-2} > \frac{0}{-2}$$
$$5 \geq x > 0$$
or $\quad 0 < x \leq 5$

21.
$$68 \leq \tfrac{9}{5}C + 32 \leq 77$$
$$68 - 32 \leq \tfrac{9}{5}C + 32 - 32 \leq 77 - 32$$
$$36 \leq \tfrac{9}{5}C \leq 45$$
$$\tfrac{5}{9} \cdot 36 \leq \tfrac{5}{9} \cdot \tfrac{9}{5}C \leq \tfrac{5}{9} \cdot 45$$
$$20 \leq C \leq 25$$

PRACTICE EXERCISE 4.4

Work odd-numbered problems first, check answers, and then work even-numbered problems in areas of weakness. Answers to all problems are in the back of the book. Make every effort to work a problem yourself before you look at an answer.

A *Replace each question mark with $<$ or $>$.*

1. $6 \,?\, 3$ _____

2. $5 \,?\, 7$ _____

3. $-3 \,?\, -6$ _____

4. $-7 \,?\, -5$ _____

5. $-6 \,?\, -3$ _____

6. $-5 \,?\, -7$ _____

7. $5 \,?\, 0$ _____

8. $0 \,?\, 8$ _____

9. $-5 \,?\, 0$ _____

10. $0 \,?\, -8$ _____

11. $-8 \,?\, -4$ _____

12. $-7 \,?\, 5$ _____

Referring to

replace each question mark in Problems 13 to 18 with either $<$ or $>$.

13. $e \,?\, a$ _____

14. $a \,?\, d$ _____

15. $c \,?\, b$ _____

16. $e \,?\, f$ _____

17. $0 \,?\, d$ _____

18. $0 \,?\, a$ _____

Graph Problems 19 to 24 on a real number line.

19. $x > -5$ _____

20. $x \leq 2$ _____

21. $-5 < x \leq -1$ _____

22. $-4 \leq x < 1$ _____

23. $-1 < x < 3$ _____

24. $-2 \leq x \leq 2$ _____

Solve (do not graph).

25. $x - 4 < -1$ _____

26. $x - 2 > 5$ _____

27. $x + 3 > -4$ _____

28. $x + 5 < -2$ _____

29. $3x < 6$ _____

30. $2x > 8$ _____

31. $-3x < 6$ _____

32. $-2x > 8$ _____

33. $\dfrac{x}{3} < -7$ _____

34. $\dfrac{x}{5} > -2$ _____

35. $\dfrac{x}{-5} > -2$ _____

36. $\dfrac{x}{-3} < -7$ _____

37. $2x - 3 > 5$ _____

38. $3x + 7 < 13$ _____

39. $12 - 7x > 5$ _____

40. $8 - 5x < -2$ _____

41. $-3x \le -2x + 1$ _____

42. $-5x > -3x - 2$ _____

43. $-3 \le x - 5 < 8$ _____

44. $2 < x + 3 < 5$ _____

45. $-24 \le 6x \le 6$ _____

46. $-6 < 3x < 9$ _____

47. $-2 < 6 - x < 3$ _____

48. $-4 < 3 - x \le 4$ _____

B *Solve and graph.*

49. $3 - m < 4(m - 3)$ _____

50. $2(1 - u) \ge 5u$ _____

51. $3 - x \ge 5(3 - x)$ _____

52. $3 - (2 + x) > -9$ _____

53. $3(x - 5) - 2(x + 1) \ge 2(x - 3)$ _____

54. $4(2u - 3) < 2(3u + 1) - (5 - 3u)$ _____

55. $-4 < 5t + 6 \le 21$ _____

56. $2 \le 3m - 7 \le 14$ _____

57. $-3 \le 3 - 2x < 7$ _____

58. $-5 < 7 - 4x \le 15$ _____

59. $-11 < 5 - 4x \le 9$ _____

60. $-5 \le 3 - 2x < 9$ _____

61. $x - \dfrac{2}{3} > \dfrac{x}{3} + 2$ _____

62. $\dfrac{x}{5} - 3 < \dfrac{3}{5} - x$ _____

63. $\dfrac{x - 3}{2} - 1 > \dfrac{x}{4}$ _____

64. $-2 - \dfrac{x}{4} < \dfrac{1 + x}{3}$ _____

65. $-2 - \dfrac{B}{4} \le \dfrac{1 + B}{3}$ _____

66. $\dfrac{y - 3}{4} - 1 > \dfrac{y}{2}$ _____

67. $\dfrac{p}{3} - \dfrac{p - 2}{2} \le \dfrac{p}{4} - 4$ _____

68. $\dfrac{3q}{7} - \dfrac{q - 4}{3} > 4 + \dfrac{2q}{7}$ _____

69. $-4 \le \tfrac{9}{5}C + 32 \le 68$ _____

70. $-1 \le \tfrac{2}{3}m + 5 \le 11$ _____

71. $-5 \le \tfrac{5}{9}(F - 32) \le 10$ _____

72. $-10 \le \tfrac{5}{9}(F - 32) \le 25$ _____

73. If we add a positive real number to any real number:

(**A**) Will the sum be greater than or less than the original number? _____

(**B**) Will the sum be to the right or left of the original number on a number line? _____

74. If we subtract a positive real number from any real number:

(**A**) Will the difference be greater than or less than the original number? _____

(**B**) Will the difference be to the right or left of the original number on a number line? _____

C *Solve and graph.*

75. $\dfrac{2x}{5} - \dfrac{1}{2}(x - 3) \le \dfrac{2x}{3} - \dfrac{3}{10}(x + 2)$ _____

76. $\dfrac{2}{3}(x + 7) - \dfrac{x}{4} > \dfrac{1}{2}(3 - x) + \dfrac{x}{6}$ _____

77. $-6 < \frac{2}{3}(2 - x) < 8$ _____

78. $-9 \le \frac{3}{4}(3 - x) \le 12$ _____

The Check Exercise for this section is on page 199.

4.5 ABSOLUTE VALUE IN EQUATIONS AND INEQUALITIES

•**Absolute value**
••**Equations involving absolute value**
•••**Inequalities involving absolute value**
••••**Summary**

•**Absolute value**

Recall that the absolute value of x, denoted by $|x|$, can be thought of as the distance that x is from the origin 0 on a number line (see Figure 2). Absolute value is also defined nongeometrically as follows:

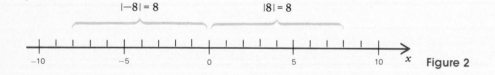

Figure 2

$$|x| = \begin{cases} x & \text{if } x > 0 \quad (x \text{ is positive}) \\ 0 & \text{if } x = 0 \quad (x \text{ is zero}) \\ -x & \text{if } x < 0 \quad (x \text{ is negative}) \end{cases}$$

Absolute Value

Thus,

$$|8| = 8 \qquad |0| = 0 \qquad |-8| = -(-8) = 8$$

••Equations involving absolute value

If we want to solve an equation such as

$$|x| = 6$$

then it follows from the definition of absolute value that x can be replaced with either $+6$ or -6, since $|+6| = 6$ and $|-6| = 6$; that is, both $+6$ and -6 are 6 units from the origin. We can write the solution in the compact form

$$x = \pm 6$$

Graphing the solution set, we obtain

Similarly, to solve the slightly more complicated equation

$$|2x - 3| = 5$$

we can replace $2x - 3$ with either $+5$ or -5. Thus, the single equation $|2x - 3| = 5$ is equivalent to the two equations

$$2x - 3 = 5 \quad \text{or} \quad 2x - 3 = -5$$

which are often written more compactly as

$$2x - 3 = \pm 5$$

and both are solved simultaneously as follows:

$$2x = 3 \pm 5$$

$$x = \frac{3 \pm 5}{2}$$

$$x = -1, 4$$

CHECK

$x = 4 \qquad |2(4) - 3| = |5| = 5$

$x = -1 \qquad |2(-1) - 3| = |-5| = 5$

EXAMPLE 22

Solve and graph:

(A) $|x - 7| = 4$

Solution
$$x - 7 = \pm 4$$
$$x = 7 \pm 4$$
$$x = 3, 11$$

(B) $|3x + 8| = 4$

Solution
$$3x + 8 = \pm 4$$
$$3x = -8 \pm 4$$
$$x = \frac{-8 \pm 4}{3}$$
$$x = -\tfrac{4}{3}, -4$$

Work Problem 22 and check solution in Solutions to Matched Problems on page 189.

PROBLEM 22   Solve and graph:

   (A) $|x| = 3$

   (B) $|m + 3| = 9$

   (C) $|2x - 5| = 7$

•••Inequalities involving absolute value

Inequalities such as

$$|x| < 5 \qquad |x - 3| \le 4 \qquad |x + 2| > 3$$

are encountered fairly frequently and are compact ways of writing double-inequality statements. Theorem 1 below shows how we can transform such statements into double-inequality forms that can then be solved by methods considered in the preceding section.

THEOREM 1

For p a positive number

(A) $|x| < p$ is equivalent to $-p < x < p$
(B) $|x| > p$ is equivalent to $x < -p$ or $x > p$

Interpreted geometrically, $|x| < p$ states that x cannot be any further from the origin than p units in either direction; that is, x must be between $-p$ and p. On the other hand, $|x| > p$ states that x must be more than p units from the origin (in either direction); thus x must be less than $-p$ or x must be greater than p.

EXAMPLE 23
Solve and graph:

(A) $|x| < 5$

Solution
$|x| < 5$ is equivalent to

$-5 < x < 5$

(B) $|x| > 5$

Solution
$|x| > 5$ is equivalent to

$x < -5$ or $x > 5$

COMMON ERROR
Don't write $-5 > x > 5$ for part **B**. It is impossible for x to be greater than 5 and less than -5 at the same time.

Work Problem 23 and check solution in Solutions to Matched Problems on page 189.

PROBLEM 23   Solve and graph:

(A) $|x| < 8$

(B) $|x| > 8$

EXAMPLE 24
Solve and graph:

(A) $|2x - 3| \le 5$

Solution
$|2x - 3| \le 5$ is equivalent to

$-5 \le 2x - 3 \le 5$ Now solve for x.

$-5 + 3 \le 2x - 3 + 3 \le 5 + 3$

$-2 \le 2x \le 8$

$-1 \le x \le 4$

(B) $|2x - 3| > 5$

Solution

$|2x - 3| > 5$ is equivalent to

$2x - 3 < -5$ or $2x - 3 > 5$ Solve each for x.

$\qquad 2x < -2 \qquad\qquad\qquad 2x > 8$

$\qquad\quad x < -1 \qquad\qquad\qquad x > 4$

Thus, $x < -1$ or $x > 4$.

Work Problem 24 and check solution in Solutions to Matched Problems on page 189.

PROBLEM 24   Solve and graph:

   (A) $|2x + 1| < 9$

   (B) $|2x + 1| \geq 9$

•••• Summary

We compare each of the forms discussed above for convenient reference:

Summary of Absolute-Value Forms

For p a positive number

$|x| = p \qquad x = \pm p$

$|x| < p \qquad -p < x < p$

$|x| > p \qquad x < -p$ or $x > p$

REMARK:

What if p were negative? For example, what are the solution sets for

$|x| < -4$ and $|x| > -4$

The first is the null set and the latter is the set of *all* real numbers. Think about this.

SOLUTIONS TO MATCHED PROBLEMS

22. **(A)** $|x| = 3$
$x = \pm 3$

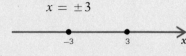

(B) $|m + 3| = 9$
$m + 3 = \pm 9$
$m = -3 \pm 9$
$m = -12, 6$

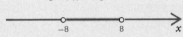

(C) $|2x - 5| = 7$
$2x - 5 = \pm 7$
$2x = 5 \pm 7$
$x = \dfrac{5 \pm 7}{2}$
$x = -1, 6$

23. **(A)** $|x| < 8$
$-8 < x < 8$

(B) $|x| > 8$
$x < -8 \qquad \text{or} \qquad x > 8$

24. **(A)** $|2x + 1| < 9$
$-9 < 2x + 1 < 9$
$-10 < 2x < 8$
$-5 < x < 4$

(B) $|2x + 1| \geq 9$
$2x + 1 \leq -9 \qquad \text{or} \qquad 2x + 1 \geq 9$
$2x \leq -10 \qquad\qquad 2x \geq 8$
$x \leq -5 \qquad\qquad x \geq 4$

PRACTICE EXERCISE 4.5

Work odd-numbered problems first, check answers, and then work even-numbered problems in areas of weakness. Answers to all problems are in the back of the book. Make every effort to work a problem yourself before you look at an answer.

Solve and graph:

A **1.** $|x| = 5$ _____

2. $|x| = 7$ _____

3. $|t - 3| = 4$ _____

4. $|y - 5| = 3$ _____

5. $|x + 1| = 5$ _____

6. $|u + 8| = 3$ _____

7. $|t| \leq 5$ _____

8. $|x| \leq 7$ _____

9. $|t - 3| < 4$ _____

10. $|y - 5| < 3$ _____

11. $|x + 1| \leq 5$ _____

12. $|u + 8| \leq 3$ _____

B **13.** $|2x - 3| = 5$ _____

14. $|3x + 4| = 8$ _____

15. $|2x - 3| \leq 5$ _____

16. $|5x - 3| \leq 12$ _____

17. $|6m + 9| - 6 = 7$ _____

18. $|5t - 7| - 8 = 3$ _____

19. $|9M - 7| < 15$ _____

20. $|7u + 9| < 14$ _____

21. $|x| \geq 7$ _____

22. $|x| \geq 5$ _____

C 23. $|t - 3| > 4$ _____

24. $|y - 5| > 3$ _____

25. $|x + 1| \geq 5$ _____

26. $|u + 8| \geq 3$ _____

27. $|3u + 4| \geq 3$ _____

28. $|2y - 8| > 2$ _____

29. $|3 - 2x| < 5$ _____

30. $|5 - 2x| \leq 1$ _____

The Check Exercise for this section is on page 201.

4.6 Chapter Review

Important terms and symbols

4.1 EQUATIONS WITH INTEGER COEFFICIENTS   solution, root, solving an equation, equivalent equations, properties of equality, first-degree equation in one variable, $ax + b = 0, a \neq 0$

4.2 EQUATIONS INVOLVING FRACTIONAL FORMS   equations with constants in denominators, equations with variables in denominators

4.3 SOLVING FOR A PARTICULAR VARIABLE

4.4 INEQUALITIES   inequality symbols, inequalities and line graphs, solving inequalities, properties of inequalities, $a < b, a > b, a \leq b, a \geq b$

4.5 ABSOLUTE VALUE IN EQUATIONS AND INEQUALITIES   absolute value, equations involving absolute value, inequalities involving absolute value, $|x|, |x| = p, |x| < p, |x| > p$

DIAGNOSTIC (REVIEW) EXERCISE 4.6 _____

Work through all the problems in this chapter review and check answers in the back of the book. (Answers to all problems are there, and following each answer is a number in italics indicating the section in which that type of problem is discussed.) Where weaknesses show up, review appropriate sections in the text. When you are satisfied that you know the material, take the practice test following this review.

A *Solve (do not graph).*

1. $4x - 9 = x - 15$ _____

2. $2x + 3(x - 1) = 5 - (x - 4)$ _____

3. $4x - 9 < x - 15$ _____

4. $-3 < 2x - 5 < 7$ _____

5. $|x| = 6$ _____

6. $|x| < 6$ _____

7. $|x| > 6$ _____

8. $|y + 9| = 5$ _____

9. $|y + 9| < 5$ _____

10. $|y + 9| > 5$ _____

11. $0.4x + 0.3x = 6.3$ _____

12. $-\frac{3}{5}y = \frac{2}{3}$ _____

13. $\dfrac{x}{4} - 3 = \dfrac{x}{5}$ _____

14. $\dfrac{x}{4} - 1 \geq \dfrac{x}{3}$ _____

15. Solve $W = I^2R$ for R _____ *(Electric circuits)*

16. Solve $A = \dfrac{bh}{2}$ for b _____ *(Area of a triangle)*

B *Solve each and graph each inequality or absolute-value statement.*

17. $-14 \leq 3x - 2 < 7$ _____

18. $-3 \leq 5 - 2x < 3$ _____

19. $3(2 - x) - 2 \leq 2x - 1$ _____

20. $0.4x - 0.3(x - 3) = 5$ _____

21. $\dfrac{x}{4} - \dfrac{x - 3}{3} = 2$ _____

22. $\dfrac{2}{3m} - \dfrac{1}{4m} = \dfrac{1}{12}$ _____

23. $\dfrac{3x}{x - 5} - 8 = \dfrac{15}{x - 5}$ _____

24. $|4x - 7| = 5$ _____

25. $|4x - 7| \leq 5$ _____

26. $|4x - 7| > 5$ _____

27. $0.05n + 0.1(n - 3) = 1.35$ _____

28. $\dfrac{x + 3}{8} \leq 5 - \dfrac{2 - x}{3}$ _____

29. $-6 < \frac{3}{5}(x - 4) \leq -3$ _____

30. $\dfrac{5}{2x + 3} - 5 = \dfrac{-5x}{2x + 3}$ _____

31. $\dfrac{3}{x} - \dfrac{2}{x + 1} = \dfrac{1}{2x}$ _____

32. $\dfrac{11}{9x} - \dfrac{1}{6x^2} = \dfrac{3}{2x}$ _____

33. $\dfrac{u - 3}{2u - 2} = \dfrac{1}{6} - \dfrac{1 - u}{3u - 3}$ _____

34. $\dfrac{x}{x^2 - 6x + 9} - \dfrac{1}{x^2 - 9} = \dfrac{1}{x + 3}$ _____

35. Solve $S = \dfrac{n(a + L)}{2}$ for L _____ *(Arithmetic progression)*

36. Solve $Q = M + MT$ for M _____

37. Indicate true (T) or false (F):

    (A) If $x < y$ and $a > 0$, then $ax < ay$. _____

    (B) If $x < y$ and $a < 0$, then $ax > ay$. _____

C *Solve and graph each inequality or absolute-value statement.*

38. $3 - 2\{x - 2[3x - 2(x + 1)]\} = 5(x - 2) + 2$ _____

39. $\dfrac{x - 3}{12} - \dfrac{x + 2}{9} = \dfrac{1 - x}{6} - 1$ _____   **40.** $\dfrac{3x}{5} - \dfrac{1}{2}(x - 3) \leq \dfrac{1}{3}(x + 2)$ _____

41. $-4 \leq \frac{2}{3}(6 - 2x) \leq 8$ _____   **42.** $\dfrac{7}{2 - x} = \dfrac{10 - 4x}{x^2 + 3x - 10}$ _____

43. $|2x - 3| < -2$ _____   **44.** Solve $y = \dfrac{4x + 3}{2x - 5}$ for x in terms of y. _____

45. Solve $\dfrac{1}{f} = \dfrac{1}{f_1} + \dfrac{1}{f_2}$ for f_1 _____   *(Optics)*

PRACTICE TEST CHAPTER 4 _____

Take this as if it were a graded test. Allow yourself up to 50 minutes. Work the problems without looking back in the chapter. Correct your work, using the answers (keyed to appropriate sections) in the back of the book.

Solve (do not graph).

1. $2(x - 3) - 3(2x + 1) = 2 - (x - 3)$ _____   **2.** $3 - (x - 1) \leq 5 + 3(3x - 7)$ _____

3. $-8 \leq 3x - 5 \leq 1$ _____   **4.** $|x| = 2$ _____

5. $|x| < 2$ _____   **6.** $|x| > 2$ _____

7. $|x| = -2$ _____   **8.** $\dfrac{2x}{9} - \dfrac{x - 1}{3} = \dfrac{1}{6}$ _____

9. $\dfrac{3}{x} - \dfrac{1}{4} = \dfrac{2}{x}$ _____   **10.** $\dfrac{3}{2x - 3} - 2 = \dfrac{2x}{2x - 3}$ _____

11. $0.05(x - 4) + 0.25x = 3.40$ _____   **12.** $\dfrac{3}{x^2 - 4} - \dfrac{1}{x - 2} = \dfrac{3}{x + 2}$ _____

13. $P = 2a + 2b$ for b _____   **14.** $S = \dfrac{n}{2}(A + L)$ for n _____

Solve and graph.

15. $\dfrac{x}{18} - \dfrac{1}{6} > \dfrac{x}{12} + \dfrac{1}{4}$ _____   **16.** $-7 < 3 - 2x \leq 11$ _____

17. $-3 \leq \dfrac{3}{4}(x + 2) \leq 9$ _____   **18.** $|2x + 3| \leq 7$ _____

19. $|2x - 3| = 9$ _____   **20.** $|2x - 3| > 5$ _____

CHECK EXERCISE 4.1

Work the following problems without looking at any text examples. Show your work in the space provided. Write the letter that best indicates your answer in the answer column.

Solve and check.

1. $-2x + 8 = 2x - 8$

    (A) $x = 0$
    (B) $x = 4$
    (C) No solution
    (D) None of these

2. $5(20 - x) + 10x = 360$

    (A) $x = 92$
    (B) $x = 52$
    (C) No solution
    (D) None of these

3. $2(3x - 5) - 8x = x - (6 - x)$

    (A) $x = -1$
    (B) $x = 8$
    (C) No solution
    (D) None of these

4. $2[x - (2 - x)] = 3(x - 2) - (2 - x)$

    (A) $x = -12$
    (B) $x = -4$
    (C) No solution
    (D) None of these

5. $2x(x - 3) - (x^2 - 16) = x[5 - 3(x + 1)] + 4x^2$

    (A) $x = 2$
    (B) $x = -2$
    (C) No solution
    (D) None of these

CHECK EXERCISE 4.2

Work the following problems without looking at any text examples. Show your work in the space provided. Write the letter that best indicates your answer in the answer column.

Solve each equation.

1. $\dfrac{2x - 8}{6} - \dfrac{x - 1}{3} = \dfrac{x}{2}$

    (A) $x = -\frac{1}{2}$
    (B) 2
    (C) No solution
    (D) None of these

2. $0.25(20 - x) + 0.05x = 3.20$

    (A) $x = -9$
    (B) $x = 9$
    (C) No solution
    (D) None of these

3. $2 - \dfrac{4x}{3x - 2} = \dfrac{2}{3x - 2}$

    (A) $x = 3$
    (B) $x = -1$
    (C) No solution
    (D) None of these

4. $\dfrac{y-4}{10y-20} = \dfrac{1}{6} - \dfrac{y+1}{5y-10}$

    (A) $y = -1$
    (B) $y = -7$
    (C) No solution
    (D) None of these

5. $3 - \dfrac{2x}{3-x} = \dfrac{6}{x-3}$

    (A) $x = 3$
    (B) $x = -\frac{3}{5}$
    (C) No solution
    (D) None of these

CHECK EXERCISE 4.3

Work the following problems without looking at any text examples. Show your work in the space provided. Write the letter that best indicates your answer in the answer column.

Solve each formula or equation for the indicated letter.

1. $r = \dfrac{d}{t}$ for t

    (A) $t = rd$

    (B) $t = \dfrac{r}{d}$

    (C) $t = \dfrac{d}{r}$

    (D) None of these

2. $Q = \dfrac{100M}{C}$ for M

    (A) $M = \dfrac{CQ}{100}$

    (B) $M = \dfrac{100Q}{C}$

    (C) $M = CQ - 100$

    (D) None of these

3. $y = 3x - 4$ for x

    (A) $x = \dfrac{y - 4}{3}$

    (B) $x = \dfrac{y}{3} + 4$

    (C) $x = \dfrac{y + 4}{3}$

    (D) None of these

4. $3x + 2y = 6$ for y

    (A) $2y = 6 - 3x$

    (B) $y = \dfrac{6}{3x} - 2$

    (C) $y = \dfrac{3x - 6}{2}$

    (D) None of these

5. $y = mx + b$ for m

    (A) $m = \dfrac{y + b}{x}$

    (B) $m = \dfrac{y - b}{x}$

    (C) $m = \dfrac{x}{y - b}$

    (D) None of these

6. $A = \frac{4}{3}\pi R^3$ for π

    (A) $\pi = \dfrac{4R^3}{3A}$

    (B) $\pi = \dfrac{4A}{3R^3}$

    (C) $\pi = \dfrac{3A}{4R^3}$

    (D) None of these

7. $A = P(1 + rt)$ for r

    (A) $r = \dfrac{A - P}{Pt}$

    (B) $r = \dfrac{A}{Pt} - 1$

    (C) $r = \dfrac{Pt}{A - P}$

    (D) None of these

8. $U = R + RST$ for R

    (A) $R = \dfrac{U}{1 + ST}$

    (B) $R = U - RST$

    (C) $R = \dfrac{1 + ST}{U}$

    (D) None of these

9. $\dfrac{1}{a} = \dfrac{1}{b} + \dfrac{1}{c}$ for b

    (A) $b = \dfrac{ac + ab}{c}$

    (B) $b = \dfrac{ac}{c - a}$

    (C) $bc = ac + ab$

    (D) None of these

10. $y = \dfrac{5x - 3}{3x - 1}$ for x

    (A) $x = \dfrac{y - 3}{3y - 5}$

    (B) $y = 3xy - 5x + 3$

    (C) $x = \dfrac{y - 3 + 5x}{3y}$

    (C) None of these

CHECK EXERCISE 4.4

NAME _____

CLASS _____

SCORE _____

*Work the following problems without looking at any text examples. Show
your work in the space provided. Write the letter that best indicates your
answer in the answer column.*

ANSWER
COLUMN

1. _____
2. _____
3. _____
4. _____
5. _____
6. _____
7. _____
8. _____
9. _____
10. _____

1. One of the statements on the right is false.
Indicate which one.

    (A) $\quad 0 > -11$
    (B) $\quad 0 < -11$
    (C) $\quad 10 > -10$
    (D) $\quad -27 < -1$

2. One of the statements on the right is false relative
to the figure below. Indicate which one.

    (A) $\quad a < b$
    (B) $\quad 0 > b$
    (C) $\quad b < c$
    (D) $\quad 0 < a$

Solve (do not graph).

3. $-6x < 24$

    (A) $\quad x < -4$
    (B) $\quad x > -4$
    (C) $\quad x = -4$
    (D) None of these

4. $\dfrac{y}{-3} \geq -2$

    (A) $\quad y \geq 6$
    (B) $\quad y \geq -5$
    (C) $\quad y \leq -5$
    (D) None of these

5. $x + 4 > -2$

    (A) $\quad x > -6$
    (B) $\quad x < -6$
    (C) $\quad x > 2$
    (D) None of these

199

6. $0 \le 7 - x \le 9$

    (A) $7 \le x \le 2$
    (B) $-7 \le -x \le 2$
    (C) $-2 \le x \le 7$
    (D)   None of these

Solve and graph.

7. $5 - (3 + x) \le x + 6$

    (A) $x \le -2$

    (B) $x \ge -2$
    (C)   No solution
    (D)   None of these

8. $-11 \le 3x - 2 < 1$

    (A) $\dfrac{-13}{2} \le x < -\dfrac{1}{3}$

    (B) $-12 \le x < 0$

    (C) $-3 \le x < 1$

    (D)   None of these

9. $\dfrac{3m}{5} - \dfrac{2m - 2}{15} > \dfrac{2}{3} + m$

    (A) $m > -1$

    (B) $m < -1$

    (C) $m < -\frac{3}{2}$
    (D)   None of these

10. $-3 < \frac{3}{4}u - 6 < 3$

    (A) $4 < u < 12$

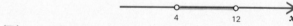

    (B) $-12 < u < -4$

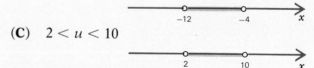

    (C) $2 < u < 10$

    (D)   None of these

CHECK EXERCISE 4.5

Work the following problems without looking at any text examples. Show your work in the space provided. Write the letter that best indicates your answer in the answer column.

Solve and graph.

1. $|x - 3| = 5$

    (A) $x = 8$

    (B) $x = -2$

    (C) $x = -2, 8$

    (D)   None of these

2. $|x| \leq 4$

    (A) $x \leq 4$

    (B) $-4 \leq x \leq 4$

    (C) $x \leq -4 \text{ or } x \geq 4$

    (D)   None of these

3. $|x| > 4$

    (A) $x > 4$

    (B) $x < -4 \text{ or } x > 4$

    (C) $-4 < x < 4$

    (D)   None of these

201

4. $|2x - 5| \leq 7$

(A) $-6 \leq x \leq 1$

(B) $-1 \leq x \leq 6$

(C) $x \leq -1$ or $x \geq 6$

(D) None of these

5. $|2x - 5| > 7$

(A) $-1 < x < 6$

(B) $x < -1$ or $x > 6$

(C) $x < -6$ or $x > 1$

(D) None of these

Chapter Five

APPLICATIONS

CONTENTS

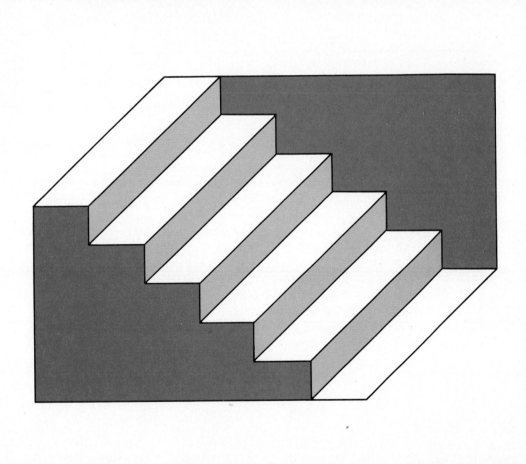

Are you looking down from above or up from below at the "stairs" in the figure? Slowly rotate the figure 180°.

INSTRUCTIONS FOR STUDENTS IN A
SELF-PACED CLASS OR LAB

yes — **HAVE YOU HAD INTERMEDIATE ALGEBRA BEFORE THIS COURSE?** — no

1. Work Diagnostic (Review) Exercise 5.7 on page 241. Check answers in back of book; then work through text sections corresponding to problems missed. (Section numbers are in italics following each answer.)
2. When finished with step 1, take Practice Test: Chapter 5 on page 243 as a final check of your understanding of the chapter. Check answers in the back of the book; then review sections where weakness still prevails. (Corresponding section numbers are in italics following each answer.)
3. When you think you are ready, ask your instructor for a graded test for Chapter 5.
4. If your instructor approves, after the test is corrected, go to the next chapter.

1. Work through each section in the chapter as follows:
   (a) Read discussion.
   (b) Read each example and work the corresponding matched problem. Check your solution to the matched problem in Solutions to Matched Problems on the indicated page.
   (c) At the end of a section work the odd-numbered problems in the Practice Exercise and check answers; then work even-numbered problems in areas of weakness. (Answers to *all* Practice Exercise sets are in the back of the book.)
   (d) Work Check Exercise as instructed. Tear out and turn in as directed by your instructor. (Answers are not in the text.)
2. Repeat each step in item 1 for each section in the chapter.
3. After the instructional part of the chapter is completed, proceed with steps 1 to 4 in the box above.

A strategy for solving word problems is introduced in Section 5.1 below. This strategy is put to immediate use in Section 5.2, where relatively easy number and geometric problems are presented to provide practice in translating words into symbolic forms. The sections following include significant problems of a slightly more difficult nature from many different areas of applications.

5.1 A STRATEGY FOR SOLVING WORD PROBLEMS

A great many practical problems can be solved using algebraic techniques—so many, in fact, there is no one method of attack that will work for all. However, we can formulate a strategy that will help you organize your approach.

A Strategy for Solving Word Problems

1. Read the problem carefully—several times if necessary—that is, until you understand the problem, know what is to be found, and know what is given.
2. Draw figures or diagrams and label known and unknown parts.
3. Look for formulas connecting the known quantities with the unknown quantities.
4. Let one of the unknown quantities be represented by a variable, say x, and try to represent all other unknown quantities in terms of x. This is an important step and must be done carefully.
5. Form an equation or inequality relating the unknown quantities with the known quantities.
6. Solve the equation or inequality and write answers to *all* parts of the problem requested.
7. Check and interpret all solutions in terms of the original problem and not just in the equation or inequality found in step 5 (a mistake might have been made in setting up the equation or inequality in step 5).

5.2 NUMBER AND GEOMETRIC PROBLEMS

- •Number problems
- ••Geometric problems

•Number problems

In earlier sections you had experience in translating verbal forms into symbolic forms. We now take advantage of that experience to solve a variety of number problems.

EXAMPLE 1

Find a number such that 6 more than one-half the number is two-thirds the number.

Solution

Let x = the number.

We symbolize each part of the problem as follows:

6 (more than) (one-half the number) (is) (two-thirds the number*)

6   + $\dfrac{x}{2}$ = $\dfrac{2x}{3}$

We now solve the equation as in the last chapter.

$$6 + \frac{x}{2} = \frac{2x}{3}$$

$$6 \cdot 6 + 6 \cdot \frac{x}{2} = 6 \cdot \frac{2x}{3}$$

$$36 + 3x = 4x$$

$$-x = -36$$

$$x = 36$$

CHECK

6 more than one-half the number: $6 + \frac{36}{2} = 6 + 18 = 24$

Two-thirds the number: $\frac{2}{3}(36) = 2 \cdot 12 = 24$

Work Problem 1 and check solution in Solutions to Matched Problems on page 211.

PROBLEM 1   Find a number such that 10 less than two-thirds the number is one-fourth the number. Write an equation and solve.

Solution

EXAMPLE 2

Find a number such that 2 less than twice the number is 5 times the quantity that is 2 more than the number.

---

*If x is a number, then two-thirds x can be written $\frac{2}{3}x$ or $\frac{2x}{3}$, since $\frac{2}{3}x = \frac{2}{3} \cdot \frac{x}{1} = \frac{2x}{3}$. The latter form will be more convenient for our purposes.

Solution
Let $x =$ the number.
Symbolize each part:

2 less than twice the number: $2x - 2$ not $2 - 2x$

is: $=$

5 times the quantity
that is 2 more than the number: $5(x + 2)$ Why would $5x + 2$ be incorrect?

Write an equation and solve:

$2x - 2 = 5(x + 2)$
$2x - 2 = 5x + 10$
$-3x = 12$
$x = -4$

Check is left to the reader.

Work Problem 2 and check solution in Solutions to Matched Problems on page 211.

PROBLEM 2 Find a number such that 4 times the quantity that is 2 less than the number is 1 more than 3 times the number. Write an equation and solve.

Solution

EXAMPLE 3
Find three consecutive even numbers such that twice the first plus the third is 10 more than the second.

Solution
Let $x =$ first of three consecutive even numbers
        $x + 2 =$ second of three consecutive even numbers
        $x + 4 =$ third of three consecutive even numbers

Form an equation and solve:

$$\underset{2x}{\boxed{\text{Twice the first}}} \quad \underset{+}{\boxed{\text{plus}}} \quad \underset{(x+4)}{\boxed{\text{the third}}} \quad \underset{=}{\boxed{\text{is}}} \quad 10 \quad \underset{+}{\boxed{\text{more than}}} \quad \underset{(x+2)}{\boxed{\text{the second}}}$$

$$2x + x + 4 = 10 + x + 2$$
$$3x + 4 = x + 12$$
$$2x = 8$$
$$\left.\begin{array}{l} x = 4 \\ x + 2 = 6 \\ x + 4 = 8 \end{array}\right\} \text{three consecutive even numbers}$$

Thus, the three consecutive even numbers are 4, 6, and 8.

CHECK
Twice the first plus the third: $2 \cdot 4 + 8 = 8 + 8 = 16$

10 more than the second: $10 + 6 = 16$

Work Problem 3 and check solution in Solutions to Matched Problems on page 211.

PROBLEM 3 Find three consecutive odd numbers such that 4 times the first minus the third is the same as the second. Write an equation and solve.

Solution

••Geometric problems

Recall that the perimeter of a rectangle or triangle is the distance around the figure. Symbolically:

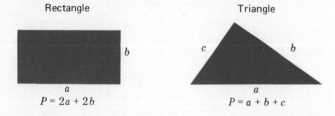

Rectangle
$P = 2a + 2b$

Triangle
$P = a + b + c$

EXAMPLE 4
If one side of a triangle is one-fourth the perimeter, the second side is 7 meters, and the third side is two-fifths the perimeter, what is the perimeter?

Solution

Let P = the perimeter. Draw a triangle and label sides:

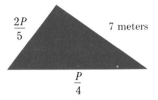

Thus,

$$P = a + b + c$$

$$P = \frac{P}{4} + 7 + \frac{2P}{5}$$

$$20 \cdot P = 20 \cdot \frac{P}{4} + 20 \cdot 7 + 20 \cdot \frac{2P}{5}$$

$$20P = 5P + 140 + 8P$$

$$7P = 140$$

$$P = 20 \text{ meters}$$

CHECK

Side 1 $= \dfrac{P}{4} = \dfrac{20}{4}$ $= \;\; 5 \text{ meters}$

Side 2 $= 7 \text{ meters}$ $= \;\; 7 \text{ meters}$

Side 3 $= \dfrac{2P}{5} = \dfrac{2 \cdot 20}{5} = \;\; \underline{8} \text{ meters}$

$\qquad\qquad\qquad\qquad\quad 20 \text{ meters (perimeter)}$

Work Problem 4 and check solution in Solutions to Matched Problems on page 211.

PROBLEM 4 If one side of a triangle is one-third the perimeter, the second side is 7 cm, and the third side is one-fifth the perimeter, what is the perimeter of the triangle? Set up an equation and solve.

Solution

EXAMPLE 5

Find the dimensions of a rectangle with perimeter 84 cm if its width is two-fifths the length.

Solution

Draw a rectangle and label sides. Let x = length.

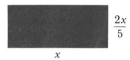

$\dfrac{2x}{5}$

x

Begin with the formula for the perimeter of a rectangle:

$$2a + 2b = P$$

$$2x + 2 \cdot \frac{2x}{5} = 84$$

$$2x + \frac{4x}{5} = 84$$

$$5\left[2x + 5 \cdot \frac{4x}{5}\right] = 5 \cdot 84$$

$$10x + 4x = 420$$

$$14x = 420$$

$$x = 30 \text{ cm} \qquad \text{(length)}$$

$$\frac{2x}{5} = 12 \text{ cm} \qquad \text{(width)}$$

Check is left to the reader.

Work Problem 5 and check solution in Solutions to Matched Problems on page 211.

PROBLEM 5 Find the dimensions of a rectangle with perimeter 176 cm if its width is three-eighths its length. Write an equation and solve.

Solution

SOLUTIONS TO MATCHED PROBLEMS

Checks are left to the reader.

1. Let x = the number

$$\frac{2x}{3} - 10 = \frac{x}{4}$$

$$12 \cdot \frac{2x}{3} - 12 \cdot 10 = 12 \cdot \frac{x}{4}$$

$$8x - 120 = 3x$$
$$5x = 120$$
$$x = 24$$

2. Let x = the number
$$4(x - 2) = 3x + 1$$
$$4x - 8 = 3x + 1$$
$$x = 9$$

3. Let $\left.\begin{array}{l} x \\ x + 2 \\ x + 4 \end{array}\right\}$ represent three consecutive odd numbers

$$4x - (x + 4) = x + 2$$
$$4x - x - 4 = x + 2$$
$$3x - 4 = x + 2$$
$$2x = 6$$
$$\left.\begin{array}{l} x = 3 \\ x + 2 = 5 \\ x + 4 = 7 \end{array}\right\}$$ the three consecutive odd numbers

4.

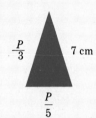

$$\frac{P}{3} + 7 + \frac{P}{5} = P$$

$$15 \cdot \frac{P}{3} + 15 \cdot 7 + 15 \cdot \frac{P}{5} = 15P$$

$$5P + 105 + 3P = 15P$$
$$8P + 105 = 15P$$
$$-7P = -105$$
$$P = 15 \text{ cm}$$

5.

$$2x + 2\left(\frac{3x}{8}\right) = 176$$

$$2x + \frac{3x}{4} = 176$$

$$4 \cdot 2x + 4 \cdot \frac{3x}{4} = 4 \cdot 176$$

$$8x + 3x = 704$$
$$11x = 704$$
$$x = 64 \text{ cm (length)}$$
$$\frac{3x}{8} = 24 \text{ cm (width)}$$

PRACTICE EXERCISE 5.2

Work odd-numbered problems first, check answers, and then work even-numbered problems in areas of weakness. Answers to all problems are in the back of the book. Make every effort to work a problem yourself before you look at an answer.

A *If x represents a number, write an algebraic expression for each of the following numbers.*

1. one-third x _____

2. one-fifth x _____

3. three-fourths x _____

4. 5 more than three-eighths x _____

5. two-thirds x _____

6. 7 more than two-fifths x _____

7. 9 less than one-third x _____

8. 11 less than one-fourth x _____

9. two-thirds the number that is 5 less than 3 times x _____

10. three-fourths the number that is 7 less than twice x _____

Find numbers meeting each of the indicated conditions. Write an equation, using x, and solve.

11. 3 more than one-sixth the number is $\frac{2}{3}$ _____

12. 2 more than one-fourth the number is $\frac{1}{2}$ _____

13. Three consecutive integers whose sum is 96 _____

14. Three consecutive integers whose sum is 78 _____

15. Three consecutive even numbers whose sum is 42 _____

16. Three consecutive even numbers whose sum is 54 _____

B 17. 3 less than one-third the number is one-fourth the number _____

18. 2 less than one-half the number is one-third the number _____

19. 2 less than one-sixth the number is 1 more than one-fourth the number _____

20. 5 less than half the number is 3 more than one-third the number _____

21. 4 less than three-fifths the number is 8 more than one-third the number _____

22. 5 more than two-thirds the number is 10 less than one-fourth the number _____

23. Three consecutive odd numbers such that the sum of the first and second is 5 more than the third

_____

24. Three consecutive odd numbers such that the sum of the second and third is 1 more than 3 times the first

_____

Set up equations and solve.

25. Find the dimensions of a rectangle with perimeter 66 cm if its length is 3 cm more than twice its width.

_____

26. Find the dimensions of a rectangle with perimeter 128 meters, if its length is 6 meters less than 4 times the width.

_____

27. Find the dimensions of a rectangle with perimeter 84 meters, if its width is one-sixth its length.

_____

28. Find the dimensions of a rectangle with perimeter 72 cm, if its width is one-third its length.

_____

29. Find the dimensions of a rectangle with perimeter 264 cm, if its width is 11 cm less than three-eighths its length.

   _____

30. Find the dimensions of a rectangle with perimeter 112 cm, if its width is 7 cm less than two-fifths its length.

   _____

C **31.** If one side of a triangle is two-fifths the perimeter P, the second side is 70 cm, and the third side is one-fourth the perimeter, what is the perimeter?

   _____

32. If one side of a triangle is one-fourth the perimeter P, the second side is 3 meters, and the third side is one-third the perimeter, what is the perimeter?

   _____

33. On a trip across the Grand Canyon in Arizona, a group traveled one-third the distance by mule, 6 km by boat, and one-half the distance by foot. How long was the trip?

   _____

34. A high diving tower is located in a lake. If one-fifth the height of the tower is in sand, 6 meters in water, and one-half the total height in air, what is the total height of the tower?

   _____

The Check Exercise for this section is on page 245.

5.3 RATIO AND PROPORTION

•Ratios
••Proportion
•••Metric conversion

Many problems in science and technology courses can be solved using ratio and proportion methods. In addition, we will find a convenient way of converting metric units into English units and vice versa.

•Ratios

The ratio of two quantities is the first divided by the second. Symbolically:

> *The Ratio of a to b*
>
> The ratio of a to b, $b \neq 0$, is $\dfrac{a}{b}$.

EXAMPLE 6

If a parking meter has 45 nickels, 30 dimes, and 15 quarters, then the ratio of nickels to quarters is

$$\frac{45}{15} = \frac{3}{1}$$

(which is also written 3:1 or 3/1 and is read 3 to 1).

Work Problem 6 and check solution in Solutions to Matched Problems on page 217.

PROBLEM 6   In Example 6:

(A)   What is the ratio of quarters to nickels? _____

(B)   What is the ratio of dimes to nickels? _____

••Proportion

In addition to providing a way of comparing known quantities, ratios also provide a way of finding unknown quantities.

EXAMPLE 7

Suppose you are told that the ratio of quarters to dimes in a parking meter is 3/5 and that there are 40 dimes in the meter. How many quarters are in the meter?

Solution

Let q = the number of quarters; then the ratio of quarters to dimes is $q/40$. Thus,

$$\frac{q}{40} = \frac{3}{5}$$

To isolate q, multiply both sides by 40—we do not need to use the lcm of 40 and 5.

$$q = 40 \cdot \frac{3}{5}$$

$$q = 24 \text{ quarters}$$

Work Problem 7 and check solution in Solutions to Matched Problems on page 217.

PROBLEM 7   If the ratio of dimes to quarters in a parking meter is 3/2 and there are 24 quarters in the meter, how many dimes are there?

Solution

The equation in Example 7 is called a proportion. In general, a statement of equality between two ratios is called **a proportion.** That is,

$$
\boxed{
\begin{array}{c}
\textit{A Proportion} \\[6pt]
\hline
\dfrac{a}{b} = \dfrac{c}{d} \qquad b, d \neq 0
\end{array}
}
$$

EXAMPLE 8
If a car can travel 192 km on 32 liters of gas, how far will it go on 60 liters?

Solution
Let x = distance traveled on 60 liters. Then,

$$\frac{x}{60} = \frac{192}{32} \qquad \frac{km}{l} = \frac{km}{l} \text{(kilometers per liter)}$$

$$x = 60 \cdot \frac{192}{32} \qquad \text{We isolate } x \text{ by multiplying both sides by } 60 \text{—we do not need to use the lcm of 60 and 32.}$$

$$= 360 \text{ km}$$

Work Problem 8 and check solution in Solutions to Matched Problems on page 217.

PROBLEM 8   If there are 24 milliliters (ml) of sulfuric acid in 64 ml of solution, how many milliliters of acid are in 48 ml of the same solution? Set up a proportion and solve.

Solution

•••Metric conversion

A summary of metric units is located inside the back cover of the text. We will show how the concept of proportion can be used to convert metric units to English units and vice versa.

EXAMPLE 9

If there is 0.45 kilogram (kg) in 1 pound, how many pounds are in 90 kg?

Solution

Let $x =$ number of pounds in 90 kg. Set up a proportion (preferably with x in the numerator on the left side, since you have a choice). That is, set up a proportion of the form

$$\frac{\text{Pounds}}{\text{Kilograms}} = \frac{\text{pounds}}{\text{kilograms}}$$ Each ratio represents pounds per kilogram.

Thus,

$$\frac{x}{90} = \frac{1}{0.45}$$

$$x = 90 \cdot \frac{1}{0.45}$$

$$x = 200 \text{ lb}$$

We could also have set up the proportion as

$$\frac{\text{Kilograms}}{\text{Pounds}} = \frac{\text{kilograms}}{\text{pounds}}$$

$$\frac{90}{x} = \frac{0.45}{1}$$

but not as

$$\frac{\text{Pounds}}{\text{Kilograms}} = \frac{\text{kilograms}}{\text{pounds}} \qquad \text{or} \qquad \frac{\text{kilograms}}{\text{pounds}} = \frac{\text{pounds}}{\text{kilograms}}$$

$$\frac{x}{90} = \frac{0.45}{1} \qquad\qquad\qquad \frac{90}{x} = \frac{1}{0.45}$$

The important thing to remember is that no matter how the proportion is set up it is valid only as long as the denominate* numbers appearing on one side of the proportion are placed in the same relative position as those on the other side—the arrangement of units must be consistent on both sides of the proportion.

Work Problem 9 and check solution in Solutions to Matched Problems on page 217.

PROBLEM 9   If there are 2.2 pounds in 1 kilogram, how many kilograms are in 100 lb? Set up a proportion and solve to two decimal places.

Solution

---

*A denominate number is a number whose unit represents a unit of measure—such as 3 lb, 5 meters, or 10 grams.

EXAMPLE 10

If there is 0.94 liter in 1 quart, how many quarts are in 50 liters? Set up a proportion (with the variable in the numerator on the left side) and solve to two decimal places.

Solution

Let x = number of quarts in 50 liters. We set up a proportion of the form

$$\frac{\text{Quarts}}{\text{Liters}} = \frac{\text{quarts}}{\text{liters}} \qquad \text{Each side gives quarts per liter.}$$

Thus,

$$\frac{x}{50} = \frac{1}{0.94}$$

$$x = 50 \cdot \frac{1}{0.94}$$

$$x = 53.19 \text{ qt}$$

Work Problem 10 and check solution in Solutions to Matched Problems on page 217.

PROBLEM 10   If there are 1.09 yards in 1 meter, how many meters are in 80 yd? Set up a proportion and solve to two decimal places.

Solution

SOLUTIONS TO MATCHED PROBLEMS

6. (A) 1/3   **(B)** 2/3

7. Let x = number of dimes

$$\frac{x}{24} = \frac{3}{2}$$

$$x = 24 \cdot \frac{3}{2} = 36 \text{ dimes}$$

8. Let x = amount of acid

$$\frac{x}{48} = \frac{24}{64}$$

$$x = 48 \cdot \frac{24}{64} = 18 \text{ ml}$$

9. Let x = number of kilograms

$$\frac{x}{100} = \frac{1}{2.2}$$

$$x = 100 \cdot \frac{1}{2.2} \approx 45.45 \text{ kg}$$

10. Let x = number of meters

$$\frac{x}{80} = \frac{1}{1.09}$$

$$x = 80 \frac{1}{1.09} \approx 73.39 \text{ meters}$$

PRACTICE EXERCISE 5.3 _____

Work odd-numbered problems first, check answers, and then work even-numbered problems in areas of weakness. Answers to all problems are in the back of the book. Make every effort to work a problem yourself before you look at an answer.

A *Write as a ratio.*

1. 33 dimes to 22 nickels _____

2. 17 quarters to 51 dimes _____

3. 25 cm to 10 cm _____

4. 30 meters to 18 meters _____

5. 300 km to 24 liters _____

6. 320 mi to 12 gal _____

Solve each proportion.

7. $\dfrac{m}{16} \dfrac{5}{4}$

8. $\dfrac{n}{12} = \dfrac{2}{3}$ _____

9. $\dfrac{x}{13} \dfrac{21}{39}$

10. $\dfrac{x}{12} = \dfrac{27}{18}$ _____

11. $\dfrac{350}{560} = \dfrac{x}{32}$ _____

12. $\dfrac{180}{270} = \dfrac{h}{6}$ _____

Set up proportions and solve. Compute decimal answers to two decimal places.

13. If in a vending machine the ratio of quarters to dimes is 5/8 and there are 96 dimes, how many quarters are there?

_____

14. If in a parking meter the ratio of pennies to nickels is 13/6 and there are 78 nickels, how many pennies are there?

_____

15. If the ratio of the length of a rectangle to its width is 5/3 and its width is 24 meters, how long is it?

_____

16. If the ratio of the width of a rectangle to its length is 4/7 and its length is 56 cm, how wide is it?

_____

17. If a car can travel 108 km on 12 liters of gas, how far will it go on 18 liters? _____

18. If a boat can travel 72 mi on 18 gal of diesel fuel, how far will it travel on 15 gal? _____

B 19. STOCK COMMISSION   If a commission of $27 is charged on the purchase of 600 shares of a stock, how much commission would be charged for 840 shares of the same stock?

_____

20. COMMON STOCKS   If the price-earning ratio of a common stock is 8.4 and the stock earns $23.50 per share, what is the price of the stock per share? NOTE: Ratios are often written as decimal fractions.

In this case 8.4 is the ratio of price per share to earnings per share. Thus, if x is the price per share, we obtain the proportion $x/23.5 = 8.4$.

_____

21. MIXTURE   If there are 9 ml of hydrochloric acid in 46 ml of solution, how many milliliters will be in 52 ml of solution?

_____

22. MIXTURE   If 3.6 cups of flour are needed in a recipe that will feed six people, how much flour will be needed in the recipe that will feed nine people?

_____

23. PHOTOGRAPHY   A 35- by 23-millimeter (mm) colored slide is used to make an enlargement whose longest side is 10 in. How wide will the enlargement be if all of the slide is used?

_____

24. PHOTOGRAPHY   A 3.25- by 4.25-in. negative is used to produce an enlargement whose shortest side is 12 in. How long will the enlargement be if all of the negative is used?

_____

25. TECHNOLOGY   A welder knows that a 3.4-meter piece of steel rod weighs 20.2 kg. How much will 2.7 meters of the same rod weigh?

_____

26. TECHNOLOGY   If 5 kg of pressure on a hydraulic lift can lift 1,200 kg, how much pressure would be required to lift 1,600 kg?

_____

27. METRIC CONVERSION   If there are 1.06 quarts in 1 liter, how many liters are in 1 gal (4 qt)?

_____

28. METRIC CONVERSION   If there are 2.2 pounds in 1 kilogram, how many kilograms are in 10 lb?

_____

29. METRIC CONVERSION   If there are 39.37 inches in 1 meter, how many inches are in 10 cm?

_____

30. METRIC CONVERSION   If there are 28.57 grams in 1 ounce, how many ounces are in 1 kg (1,000 grams)?

_____

31. METRIC CONVERSION   If there is 0.62 mile in 1 kilometer, how many kilometers are in 1 mi?

_____

32. **METRIC CONVERSION** If there is 0.91 meter in 1 yard, how many yards are in 1 meter?

_____

33. **METRIC CONVERSION** If there is 0.26 gallon in 1 liter, how many liters are in 5 gal?

_____

34. **METRIC CONVERSION** If there is 0.94 liter in 1 quart, how many quarts are in 10 liters?

_____

C 35. **WILDLIFE MANAGEMENT** Estimate the total number of trout in a lake if a sample of 300 are netted, marked, and released, and after a suitable period for mixing, a second sample of 250 produces 25 marked ones. (Assume that the ratio of the marked trout in the second sample to the total number of the sample is the same as the ratio of those marked in the first sample to the total lake population.)

36. **WILDLIFE MANAGEMENT** Repeat the preceding problem, with a first (marked) sample of 400 and a second sample of 264 with only 24 marked ones.

_____

37. **ENGINEERING** If in Figure 1 the diameter of the smaller pipe is 12 mm and the diameter of the larger pipe is 24 cm, how much force, f, would be required to lift a 1,200-kg car ($F = 1,200$)? (Neglect weight of lift equipment and use the proportion in Figure 1.)

_____

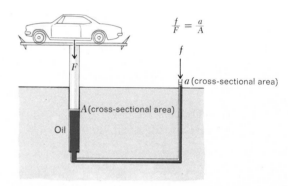

$$\frac{f}{F} = \frac{a}{A}$$

Figure 1
Hydraulic Lift

38. **ASTRONOMY** Do you have any idea how one might measure the circumference of the earth? In 240 B.C. Eratosthenes measured the size of the earth from its curvature. At Syene, Egypt (lying on the Tropic of Cancer), the sun was directly overhead at noon on June 21. At the same time in Alexandria, a town 500 mi directly north, the sun's rays fell at an angle of 7.5° to the vertical. Using this information and a little knowledge of geometry (see Figure 2), Eratosthenes was able to approximate the circumference of the earth using the following proportion: 7.5 is to 360 as 500 is to the circumference of the earth. Compute Eratosthenes' estimate.

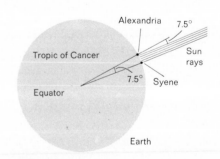

Figure 2

The Check Exercise for this section is on page 247.

5.4 RATE-TIME PROBLEMS

There are many types of rate-time problems in addition to the distance-rate-time problems with which you are probably familiar. In this section we will look at rate-time problems as a general class of problems that include distance-rate-time problems as one of many special cases.

If a runner travels 20 km in 2 hr, then the ratio

$$\frac{20 \text{ km}}{2 \text{ hr}} \qquad \text{or} \qquad 10 \text{ km/hr}$$

(read 10 kilometers per hour) is called a **rate.** It is the number of kilometers produced (covered) in each unit of time (each hour). Similarly, if an automatic bottling machine bottles 3,200 bottles of soft drinks in 20 minutes, then the ratio

$$\frac{3,200 \text{ bottles}}{20 \text{ min}} = 160 \text{ bottles/min}$$

(read 160 bottles per minute) is also a rate. It is the number of bottles produced in each unit of time (each minute).

In general, if Q is the quantity of something produced (kilometers, words, parts, and so on) in T units of time (hours, years, minutes, seconds, and so on), then

---

Quantity-Rate*-Time Formulas

$$R = \frac{Q}{T} \qquad \text{Rate} = \frac{\text{quantity}}{\text{time}}$$

$$Q = RT \qquad \text{Quantity} = (\text{rate})(\text{time})$$

$$T = \frac{Q}{R} \qquad \text{Time} = \frac{\text{quantity}}{\text{rate}}$$

If Q is distance D, then

$$R = \frac{D}{T} \qquad D = RT \qquad T = \frac{D}{R}$$

---

*These are also referred to as **average rates.**

EXAMPLE 11

A gas pump in a service station can deliver 36 liters in 4 min.

(A) What is its rate (liters per minute)?

Solution

$$R = \frac{Q}{T} = \frac{36 \text{ liters}}{4 \text{ min}} = 9 \text{ liters/min}$$

(B) How much gas can be delivered in 5 min?

Solution

$Q = RT = (9\ \text{liters/min})(5\ \text{min}) = 45\ \text{liters}$

(C) How long will it take to deliver 54 liters?

Solution

$T = \dfrac{Q}{R} = \dfrac{54\ \text{liters}}{9\ \text{liters/min}} = 6\ \text{min}$

Work Problem 11 and check solution in Solutions to Matched Problems on page 226.

PROBLEM 11 A person jogs 18 km in 2 hr.

(A) What is his or her rate (kilometers per hour)?

$R = \dfrac{D}{T} =$

(B) How far will the person jog in 1.5 hr?

$D = RT =$

(C) How long will it take to jog 12 km?

$T = \dfrac{D}{R} =$

EXAMPLE 12

A jet plane leaves San Francisco and travels at 650 km/hr toward Los Angeles. At the same time another plane leaves Los Angeles and travels at 800 km/hr toward San Francisco. If the cities are 570 km apart, how long will it take the jets to meet, and how far from San Francisco will they be at that time?

Solution

Let T = number of hours until both planes meet. Draw a diagram and label known and unknown parts. Both planes will have traveled the same amount of time when they meet.

650 km/hr   800 km/hr

Meeting

SF $D_1 = 650\,T$ point $D_2 = 800\,T$ LA

$$\begin{pmatrix} \text{Distance plane} \\ \text{from SF travels} \\ \text{to meeting point} \end{pmatrix} + \begin{pmatrix} \text{distance plane} \\ \text{from LA travels} \\ \text{to meeting point} \end{pmatrix} = \begin{pmatrix} \text{total distance} \\ \text{from} \\ \text{SF to LA} \end{pmatrix}$$

$$\begin{array}{ccccc} D_1 & + & D_2 & = & 570 \\ & 650T + 800T & = & 570 \\ & 1{,}450T & = & 570 \\ & T & = & \dfrac{570}{1{,}450} \approx 0.39 \text{ hr} \end{array}$$

Distance from SF $\approx$ (rate from SF)(time from SF)

$$= (650)(0.39) = 253.5 \text{ km}$$

Work Problem 12 and check solution in Solutions to Matched Problems on page 226.

PROBLEM 12 If an older printing press can print 45 fliers per minute and a newer press can print 80, how long will it take both presses together to print 4,500 fliers? How many will the older press have printed by then? (Note that mathematically Problem 12 and Example 12 are the same.)

Solution

EXAMPLE 13

Find the total amount of time to print the fliers in Problem 12 if the newer press is brought on the job 10 min later than the older press and both continue until the job is completed.

Solution

Let x = time to complete whole job
Then x = time old press is on the job
$x - 10$ = time new press is on the job

$$\left(\begin{array}{c}\text{Quantity}\\\text{printed}\\\text{by old press}\end{array}\right) + \left(\begin{array}{c}\text{quantity}\\\text{printed}\\\text{by new press}\end{array}\right) = \text{total needed}$$

$$45x \quad + \quad 80(x - 10) \quad = 4{,}500$$

$$45x + 80x - 800 = 4{,}500$$

$$125x = 5{,}300$$

$$x = 42.4 \text{ min}$$

Work Problem 13 and check solution in Solutions to Matched Problems on page 226.

PROBLEM 13 A car leaves a city traveling at 60 km/hr. How long will it take a second car traveling at 80 km/hr to catch up to the first car if it leaves 2 hr later? Set up an equation and solve. HINT: Both cars will have traveled the same distance when the second car catches up to the first.

Solution

EXAMPLE 14

A speedboat takes 1.5 times longer to go 120 mi up a river than to return. If the boat cruises at 25 mi/hr in still water, what is the rate of the current?

Solution

Let
$$x = \text{rate of current in miles/hour}$$
$$25 - x = \text{rate of boat upstream}$$
$$25 + x = \text{rate of boat downstream}$$

$$\text{Time upstream} = (1.5)(\text{time downstream})$$

$$\frac{\text{Distance upstream}}{\text{Rate upstream}} = (1.5)\left(\frac{\text{distance downstream}}{\text{rate downstream}}\right) \qquad \text{Recall } T = D/R \text{ from } D = RT$$

$$\frac{120}{25 - x} = (1.5)\left(\frac{120}{25 + x}\right)$$

$$\frac{120}{25 - x} = \frac{180}{25 + x} \qquad \text{Multiply both sides by } (25 - x)(25 + x) \text{ to clear fractions.}$$

$$(25 + x)(120) = (25 - x)(180)$$

$$3{,}000 + 120x = 4{,}500 - 180x$$

$$300x = 1{,}500$$

$$x = 5 \text{ mi/hr} \qquad \text{(rate of current)}$$

CHECK

Time upstream $= \dfrac{\text{distance upstream}}{\text{rate upstream}} = \dfrac{120}{20} = 6 \text{ hr}$

Time downstream $= \dfrac{\text{distance downstream}}{\text{rate downstream}} = \dfrac{120}{30} = 4 \text{ hr}$

Thus, time upstream is 1.5 times longer than time downstream.

Work Problem 14 and check solution in Solutions to Matched Problems on page 227.

PROBLEM 14 A fishing boat takes twice as long to go 25 mi up a river than to return. If the boat cruises at 9 mi/hr in still water, what is the rate of the current?

Solution

EXAMPLE 15

In an electronic computer center a card-sorter operator is given the job of alphabetizing a given quantity of data cards. The operator knows that an older sorter can do the job by itself in 6 hr. With the help of a newer machine the job is completed in 2 hr. How long would it take the new machine to do the job alone?

Solution

Let $x =$ time for new machine to do the whole job alone

Part of job completed $=$ rate $\times$ time

Rate of old machine $= \frac{1}{6}$ job per hour

Rate of new machine $= 1/x$ job per hour

$$\left(\begin{array}{c}\text{Part of job}\\\text{completed by}\\\text{old machine}\\\text{in 2 hr}\end{array}\right) + \left(\begin{array}{c}\text{Part of job}\\\text{completed by}\\\text{new machine}\\\text{in 2 hr}\end{array}\right) = 1 \text{ whole job}$$

$$\text{(Rate)(time)} + \text{(rate)(time)} = 1$$

$$\frac{1}{6}(2) + \frac{1}{x}(2) = 1$$

$$\frac{1}{3} + \frac{2}{x} = 1$$

$$x + 6 = 3x$$

$$2x = 6$$

$$x = 3 \text{ hr}$$

CHECK

$\frac{1}{6}(2) + \frac{1}{3}(2) = \frac{1}{3} + \frac{2}{3} = 1$ whole job

Work Problem 15 and check solution in Solutions to Matched Problems on page 227.

PROBLEM 15   At a family cabin, water is pumped and stored in a large water tank. Two pumps are used for this purpose. One can fill the tank by itself in 6 hr, and the other can do the job in 9 hr. How long will it take both pumps operating together to fill the tank?

Solution

SOLUTIONS TO MATCHED PROBLEMS

11. **(A)** $R = \frac{D}{T} = \frac{18\text{ km}}{2\text{ hr}} = 9\text{ km/hr}$

(B) $D = RT = (9\text{ km/hr})(1.5\text{ hr}) = 13.5\text{ km}$

(C) $T = \frac{D}{R} = \frac{12\text{ km}}{9\text{ km/hr}} = 1\frac{1}{3}\text{ hr}$

12. Let T = number of hours for both presses together to print the 4,500 fliers

$$\begin{pmatrix} \text{Number of} \\ \text{fliers printed} \\ \text{by first press} \\ \text{in } T \text{ min} \end{pmatrix} + \begin{pmatrix} \text{number of} \\ \text{fliers printed} \\ \text{by second press} \\ \text{in } T \text{ min} \end{pmatrix} = (\text{total printing})$$

$$45T \qquad + \qquad 80T \qquad = 4,500 \qquad\qquad Q = RT$$
$$125T = 4,500$$
$$T = 36\text{ min} \qquad (\text{total time})$$

Old press: $Q = RT = (45)(36) = 1,620$ fliers in 36 min

13. Let x = time for second car to catch up to first car

1st car — Time = $x + 2$ hr, Rate = 60 km/hr

2nd car — Time = x hr, Rate = 80 km/hr

When second car catches up to first car, both will have traveled the same distance.

(Distance first car travels) = (distance second car travels)
$$60(x + 2) = 80x \qquad D = RT$$
$$60x + 120 = 80x$$
$$-20x = -120$$
$$x = 6\text{ hr}$$

14. Let x = rate of current in miles per hour

$9 - x$ = rate of boat upstream

$9 + x$ = rate of boat downstream

(Time upstream) = 2(time downstream)

$$\frac{25}{9 - x} = 2\left(\frac{25}{9 + x}\right) \qquad T = \frac{D}{R}$$

$$\frac{25}{9 - x} = \frac{50}{9 + x}$$

$$25(9 + x) = 50(9 - x)$$

$$225 + 25x = 450 - 50x$$

$$75x = 225$$

$$x = 3 \text{ mi/hr} \qquad \text{(rate of current)}$$

15.   Let x = time for both pumps together to fill tank

Rate of faster pump = $\frac{1}{6}$ tank per hour

Rate of slower pump = $\frac{1}{9}$ tank per hour

Amount of tank filled = rate × time

$$\begin{pmatrix} \text{Amount filled} \\ \text{by faster pump} \\ \text{in } x \text{ hr} \end{pmatrix} + \begin{pmatrix} \text{amount filled} \\ \text{by slower pump} \\ \text{in } x \text{ hr} \end{pmatrix} = 1 \text{ full tank}$$

$$\frac{1}{6}x \qquad + \qquad \frac{1}{9}x \qquad = 1 \qquad Q = RT$$

$$\frac{x}{6} + \frac{x}{9} = 1$$

$$18 \cdot \frac{x}{6} + 18 \cdot \frac{x}{9} = 18 \cdot 1$$

$$3x + 2x = 18$$

$$5x = 18$$

$$x = 3.6 \text{ hr} \qquad \text{(time for both to fill tank)}$$

PRACTICE EXERCISE 5.4

Work odd-numbered problems first, check answers, and then work even-numbered problems in areas of weakness. Answers to all problems are in the back of the book. Make every effort to work a problem yourself before you look at an answer.

In each problem set up an equation and solve. Compute all decimal answers to two decimal places.

A   **1.** If a car travels 264 kilometers in 4 hr, what is its rate? _____

     2. If a printing press prints 14,000 evening newspapers in 2.5 hr, what is its rate? _____

     3. If the labor costs for a plumber are $78, how long did she stay on the job if she receives $12 per hour?

     _____

     4. If a bilge pump on a boat can pump 36 liters per minute, how long will it take it to pump out 81 liters of bilge water?

     _____

5. Two cars leave Chicago at the same time and travel in opposite directions. If one travels at 62 km/hr and the other at 88 km/hr, how long will it take them to be 750 km apart?

   _____

6. Two airplanes leave Miami at the same time and fly in opposite directions. If one flies at 840 km/hr and the other at 510 km/hr, how long will it take them to be 3,510 km apart?

   _____

B 7. The distance between towns A and B is 750 km. If a passenger train leaves town A and travels toward town B at 90 km/hr at the same time a freight train leaves town B and travels toward A at 35 km/hr, how long will it take the two trains to meet?

   _____

8. Repeat Problem 7, using 630 km for the distance between the two towns, 100 km/hr as the rate for the passenger train, and 40 km/hr as the rate for the freight train.

   _____

9. An office worker can fold and stuff 14 envelopes per minute. If another office worker can do 10, how long will it take them working together to fold and stuff 1,560 envelopes?

   _____

10. One file clerk can file 12 folders per minute and a second clerk 9. How long will it take them working together to file 672 folders?

    _____

11. A car leaves a town traveling at 50 km/hr. How long will it take a second car traveling at 60 km/hr to catch up to the first car if it leaves 1 hr later?

    _____

12. Repeat Problem 11 if the first car travels at 45 km/hr and the second car leaves 2 hr later traveling at 75 km/hr.

    _____

13. Find the total time to complete the job in Problem 9 if the second (slower) office worker is brought on the job 15 min after the first person has started.

    _____

14. Find the total time to complete the job in Problem 10 if the faster file clerk is brought on the job 14 min after the slower clerk has started.

    _____

15. Pipe A can fill a tank in 8 hr and pipe B can fill the same tank in 6 hr. How long will it take both pipes together to fill the tank?

    _____

16. A typist can complete a mailing in 5 hr. If another typist requires 7 hr, how long will it take both working together to complete the mailing?

_____

17. A painter can paint a house in 5 days. With the help of another painter, the house can be painted in 3 days. How long would it take the second painter to paint the house alone?

_____

18. You are at a river resort and rent a motorboat for 5 hr at 7 A.M. You are told that the boat will travel at 8 km/hr upstream and 12 km/hr returning. You decide that you would like to go as far up the river as you can and still be back at noon. At what time should you turn back, and how far from the resort will you be at that time?

_____

C 19. In an electronic computer center a card-sorter operator is given the job of alphabetizing a given quantity of data cards. The operator knows that an older sorter can do the job by itself in 3 hr. With the help of a newer machine the job is completed in 1 hr. How long would it take the new machine to do the job alone?

_____

20. Repeat Problem 18 if the boat travels 30 km/hr upstream and 45 km/hr downstream.

_____

21. Three seconds after a person fires a rifle at a target, she hears the sound of impact. If sound travels at 335 meters/sec and the bullet at 670 meters/sec, how far away is the target?

_____

22. An explosion is set off on the surface of the water 11,000 ft from a ship. If the sound reaches the ship through the water 7.77 sec before it arrives through the air and if sound travels through water 4.5 times faster than through air, how fast (to the nearest foot) does sound travel in air and in water? (See Figure 3.)

_____

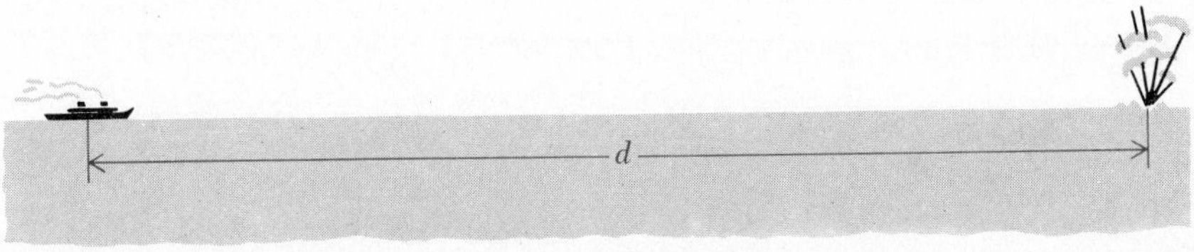

Figure 3

The Check Exercise for this section is on page 249.

5.5 MIXTURE PROBLEMS

A variety of applications can be classified as mixture problems. And even though the problems come from different areas, their mathematical treatment is essentially the same.

EXAMPLE 16

A concert brought in $27,200 on the sale of 4,000 tickets. If tickets sold for $5 and $8, how many of each were sold?

Solution

Let

$x =$ the *number* of $5 tickets sold

then

$4,000 - x =$ the *number* of $8 tickets sold

We now form an equation using the value of the tickets before and after mixing:

Value before mixing   *Value after mixing*

$$\begin{pmatrix} \text{Value of} \\ \text{\$5 tickets} \\ \text{sold} \end{pmatrix} + \begin{pmatrix} \text{value of} \\ \text{\$8 tickets} \\ \text{sold} \end{pmatrix} = \begin{pmatrix} \text{total value} \\ \text{of all} \\ \text{tickets sold} \end{pmatrix}$$

$$5x + 8(4,000 - x) = 27,200$$
$$5x + 32,000 - 8x = 27,200$$
$$-3x = -4,800$$
$$x = 1,600 \quad (\$5 \text{ tickets})$$
$$4,000 - x = 2,400 \quad (\$8 \text{ tickets})$$

Work Problem 16 and check solution in Solutions to Matched Problems on page 233.

PROBLEM 16 Suppose you receive 40 nickels and quarters in change worth $4. How many of each type of coin do you have?

Solution

Let us now consider some mixture problems involving percent. Recall 23 percent in decimal form is 0.23, 6.5 percent is 0.065, and so on.

EXAMPLE 17

How many milliliters (ml) of distilled water must be added to 60 ml of 70% acid solution to obtain a 60% solution?

Solution

Let $x =$ the number of milliliters of distilled water added. We illustrate the situation before and after mixing, keeping in mind that the amount of acid present before mixing must equal the amount of acid present after mixing.

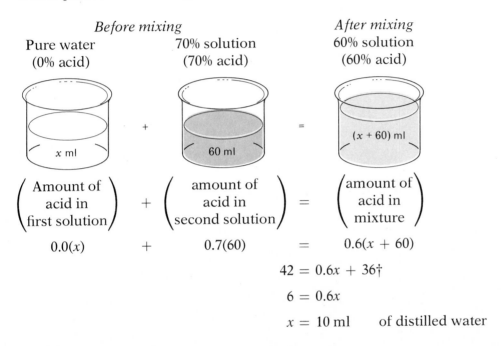

$$
\begin{pmatrix} \text{Amount of} \\ \text{acid in} \\ \text{first solution} \end{pmatrix} + \begin{pmatrix} \text{amount of} \\ \text{acid in} \\ \text{second solution} \end{pmatrix} = \begin{pmatrix} \text{amount of} \\ \text{acid in} \\ \text{mixture} \end{pmatrix}
$$

$$0.0(x) \quad + \quad 0.7(60) \quad = \quad 0.6(x + 60)$$

$$42 = 0.6x + 36†$$

$$6 = 0.6x$$

$$x = 10 \text{ ml} \quad \text{of distilled water}$$

Work Problem 17 and check solution in Solutions to Matched Problems on page 233.

PROBLEM 17   How many centiliters (cl) of pure alcohol must be added to 35 cl of a 20% solution to obtain a 30% solution?

Solution

---

†At this point we can multiply both sides by 10 to eliminate decimals (see Section 4.2). Eliminating decimals at an early stage in an equation-solving process is usually a matter of personal preference. Use the device if you wish, or carry through with decimal arithmetic, whichever is easier for you.

EXAMPLE 18

A chemical storeroom has a 20% alcohol solution and a 50% alcohol solution. How many centi-liters (cl) must be taken from each to obtain 24 cl of a 30% solution?

Solution

Let $\quad x$ = amount of 20% solution used

Then $24 - x$ = amount of 50% solution used

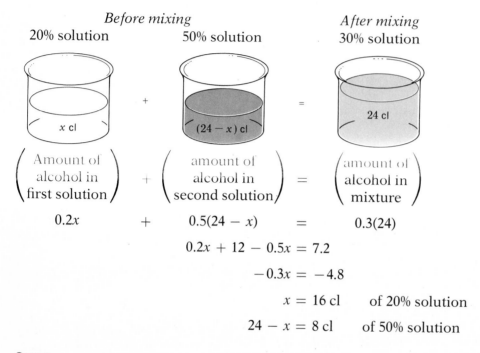

Before mixing   *After mixing*

20% solution   50% solution   30% solution

$$
\begin{pmatrix} \text{Amount of} \\ \text{alcohol in} \\ \text{first solution} \end{pmatrix} + \begin{pmatrix} \text{amount of} \\ \text{alcohol in} \\ \text{second solution} \end{pmatrix} = \begin{pmatrix} \text{amount of} \\ \text{alcohol in} \\ \text{mixture} \end{pmatrix}
$$

$$0.2x \quad + \quad 0.5(24 - x) \quad = \quad 0.3(24)$$

$$0.2x + 12 - 0.5x = 7.2$$

$$-0.3x = -4.8$$

$$x = 16 \text{ cl} \qquad \text{of 20\% solution}$$

$$24 - x = 8 \text{ cl} \qquad \text{of 50\% solution}$$

CHECK:

$$(20\% \text{ of } 16 \text{ cl}) + (50\% \text{ of } 8 \text{ cl}) \overset{?}{=} 30\% \text{ of } 24 \text{ cl}$$

$$(0.2)(16) + (0.5)(8) \overset{?}{=} (0.3)(24)$$

$$3.2 + 4.0 \overset{?}{=} 7.2$$

$$7.2 \overset{\checkmark}{=} 7.2$$

Work Problem 18 and check solution in Solutions to Matched Problems on page 234.

PROBLEM 18   Repeat Example 18, using 10% and 40% stockroom solutions.

Solution

EXAMPLE 19

A coffee shop wishes to blend $6-per-kilogram coffee with $8.50-per-kilogram coffee to produce a blend selling for $7 per kilogram. How much of each should be used to produce 50 kg of the new blend?

Solution

This problem is mathematically very close to Example 18.

Let x = amount of $6-per-kilogram coffee used

Then $50 - x$ = amount of $8.50-per-kilogram coffee used

$$\underbrace{\begin{pmatrix} \text{Value of} \\ \text{\$6-per-kilogram} \\ \text{coffee used} \end{pmatrix} + \begin{pmatrix} \text{value of} \\ \text{\$8.50-per-kilogram} \\ \text{coffee used} \end{pmatrix}}_{\textit{Value before blending}} = \underbrace{\begin{pmatrix} \text{total value} \\ \text{of 50 kg} \\ \text{of the blend} \end{pmatrix}}_{\textit{Value after blending}}$$

$$6x \qquad + \qquad 8.5(50 - x) \qquad = \qquad 7(50)$$

$$6x + 425 - 8.5x = 350$$

$$-2.5x = -75$$

$$x = 30 \text{ kg of \$6 coffee}$$

$$50 - x = 20 \text{ kg of \$8.50 coffee}$$

Work Problem 19 and check solution in Solutions to Matched Problems on page 234.

PROBLEM 19 Repeat Example 19, using $5.50-per-kilogram coffee and $8-per-kilogram coffee.

Solution

SOLUTIONS TO MATCHED PROBLEMS

Checks are left to the reader.

16. Let x = number of nickels

$40 - x$ = number of quarters

$$\underbrace{\begin{pmatrix} \text{Value of} \\ \text{nickels} \\ \text{in cents} \end{pmatrix} + \begin{pmatrix} \text{value of} \\ \text{quarters} \\ \text{in cents} \end{pmatrix}}_{\textit{Value of coins before mixing}} = \underbrace{\begin{pmatrix} \text{total value} \\ \text{of all coins} \\ \text{in cents} \end{pmatrix}}_{\textit{Value of coins after mixing}}$$

$$5x \quad + \quad 25(40 - x) \quad = \quad 400 \qquad \text{not 4}$$

$$5x + 1,000 - 25x = 400$$

$$-20x = -600$$

$$x = 30 \text{ nickels}$$

$$40 - x = 10 \text{ quarters}$$

17. Let x = amount of pure alcohol to be added

$$\underbrace{\begin{pmatrix} \text{Amount of} \\ \text{alcohol in 35 cl} \\ \text{of 20\% solution} \end{pmatrix} + \begin{pmatrix} \text{amount of} \\ \text{pure alcohol} \\ \text{added} \end{pmatrix}}_{\textit{Amount of alcohol before mixing}} = \underbrace{\begin{pmatrix} \text{amount of} \\ \text{alcohol in new} \\ \text{30\% solution} \end{pmatrix}}_{\textit{Amount of alcohol after mixing}}$$

$$(0.2)(35) \qquad + \qquad x \qquad = \qquad 0.3(35 + x)$$

$$7 + x = 10.5 + 0.3x$$

$$0.7x = 3.5$$

$$x = 5 \text{ cl of pure alcohol must be added}$$

18. Let x = amount of 10% solution used
    $24 - x$ = amount of 40% solution used

Amount of alcohol before mixing   *Amount of alcohol after mixing*

$\left(\begin{array}{c}\text{Amount of}\\\text{alcohol in } x \text{ cl}\\\text{of 10\% solution}\end{array}\right) + \left(\begin{array}{c}\text{amount of}\\\text{alcohol in } (24-x) \text{ cl}\\\text{of 40\% solution}\end{array}\right) = \left(\begin{array}{c}\text{amount of}\\\text{alcohol in 24 cl}\\\text{of 30\% solution}\end{array}\right)$

$$0.1x \quad + \quad 0.4(24-x) \quad = \quad 0.3(24)$$
$$x + 4(24-x) = 3(24) \quad \text{Multiply both sides by 10.}$$
$$x + 96 - 4x = 72$$
$$-3x = -24$$
$$x = 8 \text{ cl of 10\% solution}$$
$$24 - x = 16 \text{ cl of 40\% solution}$$

19. Let x = amount of $5.50 coffee used
    $50 - x$ = amount of $8 coffee used

Value before mixing   *Value after mixing*

$\left(\begin{array}{c}\text{Value of}\\\text{\$5.50-per-kilogram}\\\text{coffee used}\end{array}\right) + \left(\begin{array}{c}\text{value of}\\\text{\$8-per-kilogram}\\\text{coffee used}\end{array}\right) = \left(\begin{array}{c}\text{total value of}\\\text{50 kg of \$7-per-kilogram}\\\text{blend}\end{array}\right)$

$$5.5x \quad + \quad 8(50-x) \quad = \quad 7(50)$$
$$5.5x + 400 - 8x = 350$$
$$-2.5x = -50$$
$$x = 20 \text{ kg of \$5.50 coffee}$$
$$50 - x = 30 \text{ kg of \$8 coffee}$$

PRACTICE EXERCISE 5.5

Work odd-numbered problems first, check answers, and then work even-numbered problems in areas of weakness. Answers to all problems are in the back of the book. Make every effort to work a problem before you look at an answer.

Set up appropriate equations and solve.

A 1. A vending machine takes only dimes and quarters. If it contains 100 coins with a total value of $14.50, how many of each type of coin is in the machine?

2. A parking meter contains only nickels and dimes. If it contains 50 coins at a total value of $3.50, how many of each type of coin is in the meter?

3. A school musical production brought in $12,600 on the sale of 3,500 tickets. If the tickets sold for $2 and $4, how many of each type was sold?

4. A concert brought in $60,000 on the sale of 8,000 tickets. If tickets sold for $6 and $10, how many of each type of ticket was sold?

B 5. How many deciliters (dl) of alcohol must be added to 100 dl of a 40% alcohol solution to obtain a 50% solution?

_____

6. How many milliliters (ml) of hydrochloric acid must be added to 12 ml of a 30% solution to obtain a 40% solution?

_____

7. How many liters of distilled water must be added to 140 liters of a 80% alcohol solution to obtain a 70% solution?

_____

8. How many centiliters (cl) of distilled water must be added to 500 cl of a 60% acid solution to obtain a 50% solution?

_____

9. A chemical stockroom has a 20% alcohol solution and a 50% solution. How many deciliters of each should be used to obtain 90 dl of a 30% solution?

_____

10. A chemical supply company has a 30% sulfuric acid solution and a 70% sulfuric acid solution. How many liters of each should be used to obtain 100 liters of a 40% solution?

_____

11. A tea shop wishes to blend a $5-per-kilogram tea with a $6.50-per-kilogram tea to produce a blend selling for $6 per kilogram. How much of each should be used to obtain 75 kg of the new blend?

_____

12. A gourmet food store wishes to blend a $7-per-kilogram coffee with a $9.50-per-kilogram coffee to produce a blend selling for $8 per kilogram. How much of each should be used to obtain 100 kg of the new blend?

_____

13. You have inherited $20,000 and wish to invest it. If part is to be invested at 8 percent and the rest at 12 percent, how much should be invested at each rate to yield the same amount as if all had been invested at 11 percent?

_____

14. An investor has $10,000 to invest. If part is invested at 8 percent and the rest at 12 percent, how much should be invested at each rate to produce the same yield as if all had been invested at 9 percent?

_____

C 15. A 9-liter radiator contains a 50% solution of antifreeze in distilled water. How much should be drained and replaced with pure antifreeze to obtain a 70% solution?

_____

16. A 12-liter radiator contains a 60% solution of antifreeze and distilled water. How much should be drained and replaced with pure antifreeze to obtain an 80% solution?

   _____

17. A box contains 140 packages and the contents weigh 1.92 kg. If some of the packages weigh 12 grams each and the rest 16 grams each, how many of each type of package are in the box?

   _____

18. A newly hired laboratory assistant is asked (as a test) to prepare 80 centiliters (cl) of a 70% solution of hydrochloric acid from stock solutions of 20% and 50%. How much of each should be used? Interpret your answer.

   _____

The Check Exercise for this section is on page 251.

5.6 INEQUALITY APPLICATIONS

Inequality forms are encountered frequently in applied mathematics. In this section we will indicate a few ways in which they are used.

EXAMPLE 20

What numbers satisfy the condition that 5 less than 3 times the number is less than 9 more than 5 times the number?

Solution

Let $x =$ a number satisfying the given conditions

$$3x - 5 < 5x + 9$$

$$-2x < 14$$

$$x > -7$$

Work Problem 20 and check solution in Solutions to Matched Problems on page 238.

PROBLEM 20 What numbers satisfy the condition that 6 more than twice the number is more than 8 less than 4 times the number?

Solution

EXAMPLE 21

If the perimeter of a rectangle with one side 50 cm must be greater than 140 cm, how may the other side vary?

Solution

Draw a figure and label known and unknown parts. Write down any relevant formulas.

$$P > 140 \text{ cm}$$
$$2a + 2b > 140$$
$$2 \cdot 50 + 2x > 140$$
$$100 + 2x > 140$$
$$2x > 40$$
$$x > 20 \text{ cm}$$

x $P = 2a + 2b$

50 cm

Thus, the other side must be greater than 20 cm.

Work Problem 21 and check solution in Solutions to Matched Problems on page 238.

PROBLEM 21 If the area of a rectangle with one side 20 meters must be less than 360 square meters, how may the other side vary?

Solution

EXAMPLE 22

If the temperature for a 24-hr period in Antarctica ranged between -49 and $14°F$ (that is, $-49 \leq F \leq 14$), what was the range in Celsius* degrees? (Recall $F = \frac{9}{5}C + 32$.)

Solution

Since $F = \frac{9}{5}C + 32$, we replace F in $-49 \leq F \leq 14$ with $\frac{9}{5}C + 32$ and solve the double inequality for C:

$$-49 \leq \tfrac{9}{5}C + 32 \leq 14$$

$$-49 - 32 \leq \tfrac{9}{5}C + 32 - 32 \leq 14 - 32$$

$$-81 \leq \tfrac{9}{5}C \leq -18$$

$$(\tfrac{5}{9})(-81) \leq (\tfrac{5}{9})(\tfrac{9}{5}C) \leq (\tfrac{5}{9})(-18)$$

$$-45 \leq C \leq -10$$

Work Problem 22 and check solution in Solutions to Matched Problems on page 238.

PROBLEM 22   Repeat Example 22 for $-31 \leq F \leq 5$.

Solution

SOLUTIONS TO MATCHED PROBLEMS

20. Let x = a number satisfying the given conditions
$$2x + 6 > 4x - 8$$
$$-2x > -14$$
$$x < 7$$

21. Let x = the length of the other side
$$20x < 360$$
$$x < 18 \text{ meters} \quad \text{(side assumed positive)}$$

22. Replace F in $-31 \leq F \leq 5$ with $\frac{9}{5}C + 32$ and solve for C:
$$-31 \leq \tfrac{9}{5}C + 32 \leq 5$$
$$-31 - 32 \leq \tfrac{9}{5}C + 32 - 32 \leq 5 - 32$$
$$-63 \leq \tfrac{9}{5}C \leq -27$$
$$\tfrac{5}{9}(-63) \leq \tfrac{5}{9} \cdot \tfrac{9}{5}C \leq \tfrac{5}{9}(-27)$$
$$-35 \leq C \leq -15$$

---

*The Celsius scale is widely known as the centigrade scale because the temperature between freezing water and boiling water is divided into 100°. This temperature scale is named after the Swedish astronomer Anders Celsius (1701–1744), who established it in 1742.

PRACTICE EXERCISE 5.6 _____

Work odd-numbered problems first, check answers, and then work even-numbered problems in areas of weakness. Answers to all problems are in the back of the book. Make every effort to work a problem yourself before you look at an answer.

A *What numbers satisfy the given conditions? Solve, using inequality methods.*

1. 3 less than twice the number is greater than or equal to −6. _____

2. 5 more than twice the number is less than or equal to 7. _____

3. 15 reduced by 3 times the number is less than 6. _____

4. 5 less than 3 times the number is less than or equal to 4 times the number. _____

Set up inequalities and solve.

5. GEOMETRY   If the area of a rectangle with one side 10 cm must be greater than 65 cm² (square centi¡meters), how large may the other side be?

_____

6. GEOMETRY   If the perimeter of a rectangle with one side 10 cm must be smaller than 30 cm, how large may the other side be?

_____

B 7. EARTH SCIENCE   As dry air moves upward it expands and in so doing cools at a rate of about 5.5°F for each 1,000 ft up to about 40,000 ft. If the ground temperature is 70°F, then the temperature T at height h is given approximately by $T = 70 - 0.0055h$. For what range in altitude will the temperature range from −40 to 26°F?

_____

8. EARTH SCIENCE   Repeat Problem 7 for a temperature range from −95 to 37°F. _____

9. BUSINESS   For a business to make a profit it is clear that revenue R must be greater than costs C; in short, a profit will result only if $R > C$. If a company manufactures records and its cost equation for a week is $C = 300 + 1.5x$, where x is the number of records manufactured in a week, and its revenue equation is $R = 2x$, where x is the number of records sold in a week, how many records must be sold for the company to realize a profit?

_____

10. BUSINESS   Repeat Problem 9 for $C = 490 + 1.8x$ and $R = 2.5x$. _____

11. ELECTRIC CIRCUITS   If the power demand in an 110-volt electric circuit in a home varies from 220 to 2,750 watts, what is the range of current flowing through the circuit? ($W = EI$, where W = power in watts, E = pressure in volts, and I = current in amperes.)

_____

12. **ELECTRIC CIRCUITS**   In an 110-volt electric house circuit a 30-ampere fuse is used. In order to keep the fuse from "blowing," the total wattages in all of the appliances on that circuit must be kept below what figure? ($W = EI$, where W = power in watts, E = pressure in volts, and I = current in amperes.) HINT: $I = W/E$ and $I \leq 30$.

_____

13. **CHEMISTRY**   In a chemistry experiment the solution of hydrochloric acid is to be kept between 30 and 35°C, inclusive. What would the range of temperature be in Fahrenheit degrees [$C = \frac{5}{9}(F - 32)$]?

_____

14. **PHOTOGRAPHY**   A photographic developer is to be kept between 68 and 77°F, inclusive. What is the range of temperature in degrees Celsius ($F = \frac{9}{5}C + 32$)? Solve, using inequality methods.

_____

15. **AERONAUTICS**   It is customary in supersonic studies to specify the velocity of an object relative to the velocity of sound. The ratio between these two velocities is called the Mach number, and it is given by the formula

$$M \text{ (Mach number)} = \frac{V \text{ (speed of object)}}{S \text{ (speed of sound)}}$$

If a supersonic transport is designed to operate from Mach 1.7 to 2.4, what is the speed range of the transport in miles per hour? (Assume that the speed of sound is 740 mi/hr.) Solve, using inequality methods.

_____

16. **AERONAUTICS**   Repeat Problem 15 for Mach numbers 1.2 and 1.8. _____

17. **PSYCHOLOGY**   A person's IQ is found by dividing mental age, MA, as indicated by standard tests, by chronological age, CA, and then multiplying this ratio by 100. In terms of a formula,

$$IQ = \frac{MA \cdot 100}{CA}$$

If the IQ range of a group of 12-year-olds is $70 \leq IQ \leq 120$, what is the mental-age range of this group? Solve, using inequality methods.

_____

18. **PSYCHOLOGY**   Repeat Problem 17 for an IQ range of 80 to 160. _____

19. **ANTHROPOLOGY**   In the study of race and human genetic groupings, anthropologists use a ratio called the cephalic index. This is the ratio of the breadth of the head to its length (looking down from the top), expressed as a percent. Thus,

$$C = \frac{100B}{L}$$

(long-headed, $C < 75$; intermediate, $75 \leq C \leq 80$; round-headed, $C > 80$). If the length of a person's head is 20 cm, how may the breadth vary for the person to be classified in the intermediate category?

20. **ANTHROPOLOGY** For the person in Problem 19, how may the breadth of a head vary for them to be classified as round-headed?

C 21. **PUZZLE** A railroad worker is walking through a train tunnel (Figure 4) when he notices an unscheduled train approaching him. If he is three-quarters of the way through the tunnel and the train is one tunnel length ahead of him, which way should he run to maximize his chances of escaping?

Figure 4
Worker in a Train Tunnel

The Check Exercise for this section is on page 253.

5.7 CHAPTER REVIEW

DIAGNOSTIC (REVIEW) EXERCISE 5.7

Work through all the problems in this chapter review and check answers in the back of the book. (Answers to all problems are there, and following each answer is a number in italics indicating the section in which that type of problem is discussed.) Where weaknesses show up, review appropriate sections in the text. When you are satisfied that you know the material, take the practice test following this review.

Set up an appropriate equation, proportion, or inequality and solve. Round decimal answers to one decimal place.

1. Find a number such that 3 less than one-half the number is 2 more than one-third the number.

2. A parking meter contains 50 nickels and dimes worth $3.50. How many of each type of coin is in the meter?

3. Find the dimensions of a rectangle with perimeter 200 meters, if the width is two-thirds the length.

4. If a car can travel 320 kilometers (km) on 35 liters of gas, how far will it go on 42 liters? Set up a proportion and solve.

5. What numbers satisfy the condition that 6 less than 3 times the number is greater than 2 more than 4 times the number? Set up an inequality and solve.

6. A coffee shop wishes to mix a $4-per-pound coffee with a $7-per-pound coffee to obtain a blend worth $5 per pound. How much of each should be used to obtain 60 lb of the new blend?

7. Find three consecutive even numbers such that the sum of the first two minus the third is 30.

8. If the perimeter of a rectangle with one side 25 cm must be greater than 66 cm, how large must the other side be? Set up an inequality and solve.

9. A fishing boat takes twice as long to go 16 kilometers (km) up a river than to return. If the boat cruises at 6 km/hr in still water, what is the rate of the current?

10. If there are 40 milliliters (ml) of sulfuric acid in 90 ml of solution, how many milliliters of sulfuric acid would be in 54 ml of the same solution? Set up a proportion and solve.

11. A chemical storeroom has a 30% sulfuric acid solution and a 70% sulfuric acid solution. How many milliliters (ml) of each should be used to obtain 100 ml of a 40% solution?

12. A printing press can print 70 leaflets per minute. A newer model can print 100 per minute. If both presses are used, how long will it take to print 8,500 leaflets?

13. If in the preceding problem the newer press is brought on the job 10 min after the first press, what will be the total time to complete the job?

14. If there are 2.54 centimeters (cm) in 1 inch, how many inches are in 1 meter (100 cm)? Set up a proportion and solve.

   _____

15. A chemical must be kept between 77 and 86°F, inclusive. What would be the range of temperature in Celsius degrees ($F = \frac{9}{5}C + 32$)? Set up an inequality and solve.

   _____

PRACTICE TEST CHAPTER 5 _____

Take this as if it were a graded test. Allow yourself up to 50 minutes. Work the problems without looking back in the chapter. Correct your work, using the answers (keyed to appropriate sections) in the back of the book.

Set up equations or inequalities and solve. Round decimal answers to one decimal place.

1. Find three consecutive odd numbers such that 3 times their sum is 261. _____

2. If the width of a rectangle with perimeter 76 cm is 2 cm less than three-fifths the length, what are the dimensions of the rectangle?

   _____

3. If 50 ml of a solution contains 18 ml of alcohol, how many milliliters of alcohol are in 70 ml of the same solution?

   _____

4. If one printing press can print 45 brochures per minute and a newer press can print 55, how long will it take to print 3,000 brochures, if the newer press is brought on the job 10 minutes after the older press has started and both continue until finished?

   _____

5. If there are 2.54 centimeters in 1 inch, how many inches are in 127 cm? _____

6. A candy store wishes to mix $3-per-pound candy with $6-per-pound candy to obtain a mix selling for $5 per pound. How much of each should be used to obtain 90 pounds of the new mix?

   _____

7. If the cost equation for manufacturing x units is $C = 240 + 2.5x$ and the revenue equation is $R = 4x$, how many units must be manufactured and sold for the company to make a profit? Set up an inequality and solve.

   _____

8. If one car leaves a town traveling at 56 km/hr, how long will it take a second car traveling at 76 km/hr to catch up, if the second car leaves 1.5 hr later?

   _____

9. A 50% alcohol solution and an 80% alcohol solution are in a stockroom. How many deciliters (dl) of each should a chemist take to obtain 36 dl of a 60% solution?

   _____

10. A chemical solution is to be kept between 10 and 15°C, inclusive; that is, $10 \leq C \leq 15$. What is the temperature range in Fahrenheit degrees? $[C = \frac{5}{9}(F - 32)]$

    _____

CHECK EXERCISE 5.2

NAME _____

CLASS _____

SCORE _____

Work the following problems without looking at any text examples. Show your work in the space provided. Write the letter that best indicates your answer in the answer column.

ANSWER
COLUMN

1. _____

2. _____

3. _____

4. _____

5. _____

For each problem set up an appropriate equation and solve.

1. Find a number such that 5 less than one-third the number is 3 less than one-fourth the number.

   (A) 24
   (B) −24
   (C) −96
   (D) None of these

2. Find a number such that one-third the number is one-half the quantity that is 6 less than the number.

   (A) 36
   (B) 18
   (C) 24
   (D) None of these

3. Find three consecutive odd numbers such that twice the sum of the first two is 21 more than 3 times the third.

   (A) 37, 39, 41
   (B) 38, 39, 40
   (C) 29, 31, 33
   (D) None of these

4. Find the dimensions of a rectangle with perimeter 44 cm, if the width is five-sixths the length.

(**A**) 24 cm by 20 cm
(**B**) 12 cm by 10 cm
(**C**) 14 cm by 8 cm
(**D**) None of these

5. If one side of a triangle is three-fifths the perimeter P, the second side is 21 meters, and the third side is one-sixth the perimeter, what is the perimeter?

(**A**) 90 meters
(**B**) 60 meters
(**C**) 120 meters
(**D**) None of these

CHECK EXERCISE 5.3

Work the following problems without looking at any text examples. Show your work in the space provided. Write the letter that best indicates your answer in the answer column.

Set up a proportion and solve.

1. If a car can travel 260 kilometers (km) on 39 liters of gas, how far will it go on 24 liters?

    (A) 36 km
    (B) 160 km
    (C) 0.20 km
    (D) None of these

2. If $3,000 earns $330 interest, how much would $5,000 earn invested at the same rate?

    (A) $45,454.55
    (B) $650.00
    (C) $550.00
    (D) None of these

3. If there is 0.62 mile in 1 kilometer (km), how many kilometers are in 124 mi?

    (A) 200 km
    (B) 0.01 km
    (C) 76.88 km
    (D) None of these

4. If there are 2.2 pounds in 1 kilogram (kg), how many kilograms are in 22 lb?

    (A) 48.4 kg
    (B) 0.1 kg
    (C) 10 kg
    (D) None of these

5. A 35- by 23-millimeter (mm) colored slide is used to make an enlargement whose longest side is 16 in. How wide (to two decimal places) will the enlargement be if all the slide is used?

    (A) 10.51 in
    (B) 24.35 in
    (C) 14.72 in
    (D) None of these

CHECK EXERCISE 5.4

Work the following problems without looking at any text examples. Show your work in the space provided. Write the letter that best indicates your answer in the answer column.

Set up an equation and solve. Compute all decimal answers to two decimal places.

1. An older capping machine can fill and cap 15 bottles per minute, while a newer machine can fill and cap 20 bottles per minute. How long will it take both machines working together to fill and cap anorder of 2,450 bottles?

   (A) 285.83 min
   (B) 70 min
   (C) 60.5 min
   (D) None of these

2. Towns *A* and *B* are 3,600 kilometers (km) apart. If a truck leaves town *A* and travels toward town *B* at 82 km/hr at the same time a truck leaves town *B* and travels toward town *A* at 38 km/hr, how long will it take the trucks to meet?

   (A) 30 hr
   (B) 43.90 hr
   (C) 94.74 hr
   (D) None of these

3. A person starts jogging around a large lake at 7 A.M., moving at 200 meters/min. How long will it take a friend to catch up if the friend starts at 7:10 A.M. and jogs at 300 meters/min?

   (A) 10 min
   (B) 20 min
   (C) 30 min
   (D) None of these

249

4. A speedboat takes 1.25 as long to go 36 kilometers (km) up a river than to return. If the boat cruises at 30 km/hr in still water, what is the rate of the current?

    (A)   30 km/hr
    (B) -3.33 km/hr
    (C)   3.33 km/hr
    (D)   None of these

5. If an automated typewriter can type all of the "personalized letters" for an advertising campaign in 15 hr, while a later model can do the job in 10 hr, how long will it take both models together to do the whole job?

    (A)   6 hr
    (B)   5 hr
    (C)   25 hr
    (D)   None of these

CHECK EXERCISE 5.5

Work the following problems without looking at any text examples. Show your work in the space provided. Write the letter that best indicates your answer in the answer column.

1. A vending machine contains 30 nickels and dimes worth $2.35. How many of each type of coin is in the machine?

    (**A**) 15 nickels, 16 dimes
    (**B**) 11 nickels, 18 dimes
    (**C**) 9 nickels, 19 dimes
    (**D**) None of these

2. A chemical storeroom has a 20% hydrochloric acid solution and a 60% hydrochloric acid solution. How many milliliters (ml) of each should be used to obtain 60 ml of a 50% solution?

    (**A**) 15 ml of 20% solution
            45 ml of 60% solution
    (**B**) 10 ml of 20% solution
            50 ml of 60% solution
    (**C**) 18 ml of 20% solution
            42 ml of 60% solution
    (**D**) None of these

3. A candy store wishes to mix $2-per-pound candy with $3.50-per-pound candy to obtain a mix selling for $3 per pound. How much of each should be used to obtain 60 lb of the mix?

    (**A**) 15 lb of $2 candy
            45 lb of $3.50 candy
    (**B**) 25 lb of $2 candy
            35 lb of $3.50 candy
    (**C**) 20 lb of $2 candy
            40 lb of $3.50 candy
    (**D**) None of these

4. How much pure acid must be added to 60 centiliters (cl) of a 60% solution to obtain a 70% solution?

   (**A**)   15 cl
   (**B**)   25 cl
   (**C**)   20 cl
   (**D**)   None of these

5. You have $1,000 to invest. If a part is invested at 10 percent and the rest at 18 percent, how much should be invested at each rate to produce the same yield as if all had been invested at 12 percent?

   (**A**) $700 at 10 percent
             $300 at 18 percent
   (**B**) $750 at 10 percent
             $250 at 18 percent
   (**C**) $250 at 10 percent
             $750 at 18 percent
   (**D**)   None of these

CHECK EXERCISE 5.6

NAME _____

CLASS _____

SCORE _____

Work the following problems without looking at any text examples. Show your work in the space provided. Write the letter that best indicates your answer in the answer column.

ANSWER
COLUMN
1. _____
2. _____
3. _____
4. _____
5. _____

Set up appropriate inequalities and solve.

1. What numbers satisfy the condition that 7 less than 5 times a number is more than 6 times the number?

    (A) $x < 7$
    (B) $x < -7$
    (C) $x > -7$
    (D) None of these

2. The area of a rectangle with one side 20 meters must be between 100 and 150 square meters. How may the other side vary?

    (A) $50 < x < 75$
    (B) $5 < x < 7.5$
    (C) $30 < x < 55$
    (D) None of these

3. If the perimeter of a rectangle with one side 30 cm must be greater than 100 cm, how large must the other side be?

    (A) $x > 30$ cm
    (B) $x > 20$ cm
    (C) $x > 70$ cm
    (D) None of these

4. If the cost equation for a manufacturing company is $C = 4x + 150$ and the revenue equation is $R = 6x$, where x is the number of units manufactured, how many units must be produced and sold for the company to make a profit?

    (A) $x > 75$
    (B) $x < 75$
    (C) $x = 75$
    (D) None of these

5. A chemical must be kept between 20 and 30°C, inclusive. What would be the range of temperature in Fahrenheit degrees $[C = \frac{5}{9}(F - 32)]$?

    (A) $4 \le F \le 22$
    (B) $60 \le F \le 80$
    (C) $68 \le F \le 86$
    (D) None of these

Chapter Six

EXPONENTS, RADICALS, AND COMPLEX NUMBERS

CONTENTS

Can you ever reach the top or bottom of the stairs in the figure?

INSTRUCTIONS FOR STUDENTS IN A
SELF-PACED CLASS OR LAB

(yes)　**HAVE YOU HAD INTERMEDIATE ALGEBRA BEFORE THIS COURSE?**　(no)

1. Work Diagnostic (Review) Exercise 6.8 on page 301. Check answers in back of book; then work through text sections corresponding to problems missed. (Section numbers are in italics following each answer.)
2. When finished with step 1, take Practice Test: Chapter 6 on page 304 as a final check of your understanding of the chapter. Check answers in the back of the book; then review sections where weakness still prevails. (Corresponding section numbers are in italics following each answer.)
3. When you think you are ready, ask your instructor for a graded test for Chapter 6.
4. If your instructor approves, after the test is corrected, go to the next chapter.

1. Work through each section in the chapter as follows:
   (a) Read discussion.
   (b) Read each example and work the corresponding matched problem. Check your solution to the matched problem in Solutions to Matched Problems on the indicated page.
   (c) At the end of a section work the odd-numbered problems in the Practice Exercise and check answers; then work even-numbered problems in areas of weakness. (Answers to *all* Practice Exercise sets are in the back of the book.)
   (d) Work Check Exercise as instructed. Tear out and turn in as directed by your instructor. (Answers are not in the text.)
2. Repeat each step in item 1 for each section in the chapter.
3. After the instructional part of the chapter is completed, proceed with steps 1 to 4 in the box above.

6.1 Integer Exponents

- •Positive-integer exponents
- ••Zero exponents
- •••Negative-integer exponents
- ••••Common errors

•Positive-integer exponents

In Section 1.6 we introduced the concept of natural number (positive-integer) exponents and five basic properties that control their use. Recall

$$a^5 = a \cdot a \cdot a \cdot a \cdot a \qquad \text{(five factors of } a\text{)}$$

and, in general,

> **Definition of Positive-Integer Exponent**
>
> For n a positive integer
>
> $$a^n = a \cdot a \cdot \cdots \cdot a \qquad n \text{ factors of } a$$

As a consequence of this definition we stated the following five important properties of exponents:

> **Exponent Properties**
>
> For m and n positive integers
>
> 1. $a^m a^n = a^{m+n}$
> 2. $(a^n)^m = a^{mn}$
> 3. $(ab)^m = a^m b^m$
> 4. $\left(\dfrac{a}{b}\right)^m = \dfrac{a^m}{b^m}$
> 5. $\dfrac{a^m}{a^n} = \begin{cases} a^{m-n} & \text{if } m > n \\ 1 & \text{if } m = n \\ \dfrac{1}{a^{n-m}} & \text{if } n > m \end{cases}$

Now let us turn to other types of exponents. For example, how should symbols such as

$$8^0 \qquad \text{and} \qquad 7^{-3}$$

be defined? In this section we will extend the meaning of exponent to include 0 and negative integers. Thus, typical scientific expressions such as

The diameter of a red corpuscle is approximately 8×10^{-5} cm.
The amount of water found in the air as vapor is about 9×10^{-6} times that found in seas.
The focal length of a thin lens is given by $f^{-1} = a^{-1} + b^{-1}$.

will then make sense.

In extending the concept of exponent beyond the natural numbers, we will require that any new exponent symbol be defined in such a way that all five laws of exponents for natural numbers continue to hold. Thus, we will need only one set of laws for all types of exponents rather than a new set for each new exponent.

••Zero exponents

We will start by defining the zero exponent. If all the exponent laws must hold even if some of the exponents are 0, then a^0 ($a \neq 0$) should be defined so that when the first law of exponents is applied,

$$a^0 \cdot a^2 = a^{0+2} = a^2$$

This suggests that a^0 should be defined as 1 for all nonzero real numbers a, since 1 is the only real number that gives a^2 when multiplied by a^2. If we let $a = 0$ and follow the same reasoning, we find that

$$0^0 \cdot 0^2 = 0^{0+2} = 0^2 = 0$$

and 0^0 could be any real number, since $0^2 = 0$; hence 0^0 is not uniquely determined. For this reason and others, we choose not to define 0^0.

> *Definition of Zero Exponent*
>
> For all real numbers $a \neq 0$
>
> $a^0 = 1$
>
> 0^0 is not defined

EXAMPLE 1
(A) $5^0 = 1$
(B) $325^0 = 1$
(C) $(\frac{1}{3})^0 = 1$
(D) $t^0 = 1$ ($t \neq 0$)
(E) $(x^2 y^3)^0 = 1$ ($x \neq 0, y \neq 0$)

Work Problem 1 and check solution in Solutions to Matched Problems on page 263.

PROBLEM 1 (A) $12^0 =$
(B) $999^0 =$
(C) $x^0 = $; $x \neq 0$
(D) $(m^3 n^3)^0 = $; $m, n \neq 0$

•••Negative-integer exponents

To get an idea of how a negative-integer exponent should be defined, we can proceed as above. If the first law of exponents is to hold, then a^{-2} ($a \neq 0$) must be defined so that

$$a^{-2} \cdot a^2 = a^{-2+2} = a^0 = 1$$

Thus a^{-2} must be the reciprocal of a^2; that is,

$$a^{-2} = \frac{1}{a^2}$$

This kind of reasoning leads us to the following general definition.

Definition of Negative-Integer Exponents

If n is a positive integer and a is a nonzero real number, then

$$a^{-n} = \frac{1}{a^n} \qquad a^{-4} = \frac{1}{a^4}$$

Of course, it follows, using equality properties,* that

$$a^n = \frac{1}{a^{-n}} \qquad \frac{1}{x^{-3}} = x^3$$

---

*Multiply both sides of $a^{-n} = \dfrac{1}{a^n}$ by $\dfrac{a^n}{a^{-n}}$ to obtain $\dfrac{a^n}{a^{-n}} \cdot a^{-n} = \dfrac{a^n}{a^{-n}} \cdot \dfrac{1}{a^n}$

$$a^n = \frac{1}{a^{-n}}$$

EXAMPLE 2

(A) $a^{-7} = \dfrac{1}{a^7}$

(B) $\dfrac{1}{x^{-8}} = x^8$

(C) $10^{-3} = \dfrac{1}{10^3}$ or $\dfrac{1}{1,000}$ or 0.001

(D) $\dfrac{x^{-3}}{y^{-5}} = \dfrac{x^{-3}}{1} \cdot \dfrac{1}{y^{-5}} = \dfrac{1}{x^3} \cdot \dfrac{y^5}{1} = \dfrac{y^5}{x^3}$

Work Problem 2 and check solution in Solutions to Matched Problems on page 263.

PROBLEM 2 Write, using positive exponents or no exponents:

(A) $x^{-5} =$

(B) $\dfrac{1}{y^{-4}} =$

(C) $10^{-2} =$

(D) $\dfrac{m^{-2}}{n^{-3}} =$

With the definition of negative exponent and zero exponent behind us, we can now replace the fifth law of exponents with a simpler form that does not have any restrictions on the relative size of the exponents. Thus

$$\frac{a^m}{a^n} = a^{m-n} = \frac{1}{a^{n-m}}$$

EXAMPLE 3

Simplify, leaving answers with negative exponents.

(A) $\dfrac{2^5}{2^8} = 2^{5-8} = 2^{-3}$ **(B)** $\dfrac{10^{-3}}{10^6} = 10^{-3-6} = 10^{-9}$

Simplify, leaving answers with positive exponents.

(C) $\dfrac{2^5}{2^8} = \dfrac{1}{2^{8-5}} = \dfrac{1}{2^3}$ **(D)** $\dfrac{10^{-3}}{10^6} = \dfrac{1}{10^{6-(-3)}} = \dfrac{1}{10^9}$

Work Problem 3 and check solution in Solutions to Matched Problems on page 263.

PROBLEM 3   Simplify, leaving answers with negative exponents.

   (A) $\dfrac{3^4}{3^9} =$ **(B)** $\dfrac{x^{-2}}{x^3} =$

Simplify, leaving answers with positive exponents.

   (C) $\dfrac{3^4}{3^9} =$ **(D)** $\dfrac{x^{-2}}{x^3} =$

Table 1 provides a summary of all of our work on exponents to this point:

TABLE 1
Integer exponents and their laws (summary)

Definition of a^p p an integer and a a real number	Laws of exponents n and m integers, a and b real numbers
1. *If p is a positive integer, then* $a^p = a \cdot a \cdot \cdots \cdot a$ *p factors of a* EXAMPLE: $3^5 = 3 \cdot 3 \cdot 3 \cdot 3 \cdot 3$	1. $a^m a^n = a^{m+n}$
2. *If p = 0, then* $a^p = 1$ $a \neq 0$ EXAMPLE: $3^0 = 1$	2. $(a^n)^m = a^{mn}$ 3. $(ab)^m = a^m b^m$
3. *If p is a negative integer, then* $a^p = \dfrac{1}{a^{-p}}$ $a \neq 0$ EXAMPLE: $3^{-4} = \dfrac{1}{3^{-(-4)}} = \dfrac{1}{3^4}$	4. $\left(\dfrac{a}{b}\right)^m = \dfrac{a^m}{b^m}$ 5. $\dfrac{a^m}{a^n} = a^{m-n} = \dfrac{1}{a^{n-m}}$

EXAMPLE 4

Simplify and express answers using positive exponents* only.

(A) $a^5a^{-2} = a^{5-2} = a^3$

(B) $(a^{-3}b^2)^{-2} = (a^{-3})^{-2}(b^2)^{-2} = a^6b^{-4} = \dfrac{a^6}{b^4}$

(C) $\left(\dfrac{a^{-5}}{a^{-2}}\right)^{-1} = \dfrac{(a^{-5})^{-1}}{(a^{-2})^{-1}} = \dfrac{a^5}{a^2} = a^3$

(D) $\dfrac{4x^{-3}y^{-5}}{6x^{-4}y^3} = \dfrac{2x^{-3-(-4)}}{3y^{3-(-5)}} = \dfrac{2x^{-3+4}}{3y^{3+5}} = \dfrac{2x}{3y^8}$

or, changing to positive exponents first,

$$\frac{4x^{-3}y^{-5}}{6x^{-4}y^3} = \frac{2x^4}{3x^3y^3y^5} = \frac{2x}{3y^8}$$

(E) $\dfrac{10^{-4} \cdot 10^2}{10^{-3} \cdot 10^5} = \dfrac{10^{-4+2}}{10^{-3+5}} = \dfrac{10^{-2}}{10^2} = \dfrac{1}{10^4} = \dfrac{1}{10,000} = 0.0001$

(F) $\left(\dfrac{m^{-3}m^3}{n^{-2}}\right)^{-2} = \left(\dfrac{m^{-3+3}}{n^{-2}}\right)^{-2} = \left(\dfrac{m^0}{n^{-2}}\right)^{-2}$

$$= \left(\frac{1}{n^{-2}}\right)^{-2} = \frac{1^{-2}}{(n^{-2})^{-2}} = \frac{1}{n^4}$$

Work Problem 4 and check solution in Solutions to Matched Problems on page 263.

PROBLEM 4 Simplify and express answers using positive exponents only.

(A) $x^{-2}x^6 =$

(B) $(x^3y^{-2})^{-2} =$

(C) $\left(\dfrac{x^{-6}}{x^{-2}}\right)^{-1} =$

(D) $\dfrac{8m^{-2}n^{-4}}{6m^{-5}n^2} =$

(E) $\left(\dfrac{6m^2n^{-3}}{8m^{-1}n^3}\right)^{-2} =$

(F) $\dfrac{10^{-3} \cdot 10^5}{10^{-2} \cdot 10^6} =$

••••Common errors

As was stated earlier, laws of exponents involve products and quotients, not sums and differences. Consider:

Correct   *Common error*

$\dfrac{a^{-2}y}{b} = \dfrac{y}{a^2b}$ $\dfrac{a^{-2}+y}{b} \neq \dfrac{y}{a^2b}$

The plus sign in the numerator of the second illustration makes a big difference. Actually, $\dfrac{a^{-2} + y}{b}$

---

*It is important to realize that there are situations where it is desirable to allow negative exponents in an answer (see Section 6.2, for example). In this section we ask you to write answers using positive exponents only so that problems in the exercise set will have unique answer forms.

represents a compact way of writing a complex fraction. To simplify, we replace a^{-2} with $1/a^2$, then proceed as in Section 3.4.

$$\frac{a^{-2} + y}{b} = \frac{\frac{1}{a^2} + y}{b} = \frac{a^2\left(\frac{1}{a^2} + y\right)}{a^2 \cdot b}$$

$$= \frac{1 + a^2 y}{a^2 b}$$

Also, consider the following:

Correct	*Common error*
$(a^{-1}b^{-1})^2 = a^{-2}b^{-2}$	$(a^{-1} + b^{-1})^2 = a^{-2} + b^{-2}$
$= \frac{1}{a^2 b^2}$	$= \frac{1}{a^2 + b^2}$

The second illustration contains two errors:

$(a^{-1} + b^{-1})^2 = a^{-2} + b^{-2}$ Wrong!

and

$a^{-2} + b^{-2} = \frac{1}{a^2 + b^2}$ Wrong!

The problem is worked correctly in Example 5**B**.

EXAMPLE 5
Simplify and express answers using positive exponents only.

(A) $\frac{3^{-2} + 2^{-1}}{11} = \frac{\frac{1}{3^2} + \frac{1}{2}}{11} = \frac{\frac{2}{18} + \frac{9}{18}}{11} = \frac{11}{18} \div 11 = \frac{11}{18} \cdot \frac{1}{11} = \frac{1}{18}$

(B) $(a^{-1} + b^{-1})^2 = \left(\frac{1}{a} + \frac{1}{b}\right)^2 = \left(\frac{b + a}{ab}\right)^2 = \frac{(b + a)^2}{(ab)^2} = \frac{b^2 + 2ab + a^2}{a^2 b^2}$

Work Problem 5 and check solution in Solutions to Matched Problems on page 263.

PROBLEM 5   Simplify and express answers using positive exponents only:

(A) $\frac{2^{-2} + 3^{-1}}{5} =$

(B) $(x^{-1} + y^{-1})^2 =$

SOLUTIONS TO MATCHED PROBLEMS

1. All are equal to 1

2. **(A)** $x^{-5} = \dfrac{1}{x^5}$ **(B)** $\dfrac{1}{y^{-4}} = y^4$ **(C)** $10^{-2} = \dfrac{1}{10^2} = \dfrac{1}{100} = 0.01$ **(D)** $\dfrac{m^{-2}}{n^{-3}} = \dfrac{n^3}{m^2}$

3. **(A)** $\dfrac{3^4}{3^9} = 3^{4-9} = 3^{-5}$ **(B)** $\dfrac{x^{-2}}{x^3} = x^{-2-3} = x^{-5}$ **(C)** $\dfrac{3^4}{3^9} = \dfrac{1}{3^{9-4}} = \dfrac{1}{3^5}$

   (D) $\dfrac{x^{-2}}{x^3} = \dfrac{1}{x^{3-(-2)}} = \dfrac{1}{x^5}$

4. **(A)** $x^{-2}x^6 = x^4$ **(B)** $(x^3y^{-2})^{-2} = x^{-6}y^4 = \dfrac{y^4}{x^6}$ **(C)** $\left(\dfrac{x^{-6}}{x^{-2}}\right)^{-1} = \dfrac{x^6}{x^2} = x^4$

   (D) $\dfrac{8m^{-2}n^{-4}}{6m^{-5}n^2} = \dfrac{4m^{-2-(-5)}}{3n^{2-(-4)}} = \dfrac{4m^3}{3n^6}$ **(E)** $\left(\dfrac{6m^2n^{-3}}{8m^{-1}n^3}\right)^{-2} = \left(\dfrac{3n^{-6}}{4m^{-3}}\right)^{-2} = \dfrac{3^{-2}n^{12}}{4^{-2}m^6} = \dfrac{4^2n^{12}}{3^2m^6} = \dfrac{16n^{12}}{9m^6}$

   (F) $\dfrac{10^{-3}\cdot 10^5}{10^{-2}\cdot 10^6} = \dfrac{10^2}{10^4} = \dfrac{1}{10^2} = \dfrac{1}{100} = 0.01$

5. **(A)** $\dfrac{2^{-2} + 3^{-1}}{5} = \dfrac{\dfrac{1}{2^2} + \dfrac{1}{3}}{5} = \dfrac{\dfrac{1}{4} + \dfrac{1}{3}}{5} = \dfrac{12\cdot\dfrac{1}{4} + 12\cdot\dfrac{1}{3}}{12\cdot 5} = \dfrac{3+4}{60} = \dfrac{7}{60}$

   (B) $(x^{-1} + y^{-1})^2 = \left(\dfrac{1}{x} + \dfrac{1}{y}\right)^2 = \left(\dfrac{y+x}{xy}\right)^2$ or $\dfrac{x^2 + 2xy + y^2}{x^2y^2}$

PRACTICE EXERCISE 6.1

Work odd-numbered problems first, check answers, and then work even-numbered problems in areas of weakness. Answers to all problems are in the back of the book. Make every effort to work a problem yourself before you look at an answer.

Simplify and write answers using positive exponents only.

A 1. 23^0 _____   2. 10^0 _____   3. y^0 _____   4. x^0 _____

5. 3^{-3} _____   6. 2^{-2} _____   7. m^{-7} _____   8. x^{-4} _____

9. $\dfrac{1}{4^{-3}}$ _____   10. $\dfrac{1}{3^{-2}}$ _____   11. $\dfrac{1}{y^{-5}}$ _____   12. $\dfrac{1}{x^{-3}}$ _____

13. $10^7\cdot 10^{-5}$ _____   14. $10^{-4}\cdot 10^6$ _____   15. $y^{-3}y^4$ _____

16. x^6x^{-2} _____   17. u^5u^{-5} _____   18. $m^{-3}m^3$ _____   19. $\dfrac{10^3}{10^{-7}}$ _____

20. $\dfrac{10^8}{10^{-3}}$ _____   21. $\dfrac{x^9}{x^{-2}}$ _____   22. $\dfrac{a^8}{a^{-4}}$ _____   23. $\dfrac{z^{-2}}{z^3}$ _____

24. $\dfrac{b^{-3}}{b^5}$ _____   25. $\dfrac{10^{-1}}{10^6}$ _____   26. $\dfrac{10^{-4}}{10^2}$ _____   27. $(10^{-4})^{-3}$ _____

28. $(2^{-3})^{-2}$ _____   29. $(y^{-2})^{-4}$ _____   30. $(x^{-5})^{-2}$ _____

31. $(u^{-5}v^{-3})^{-2}$ _____   32. $(x^{-3}y^{-2})^{-1}$ _____   33. $(x^2y^{-3})^2$ _____

34. $(x^{-2}y^3)^2$ _____

35. $(x^{-2}y^3)^{-1}$ _____

36. $(x^2y^{-3})^{-1}$ _____

B 37. $(m^2)^0$ _____

38. $1,231^0$ _____

39. $\dfrac{10^{-3}}{10^{-5}}$ _____

40. $\dfrac{10^{-2}}{10^{-4}}$ _____

41. $\dfrac{y^{-2}}{y^{-3}}$ _____

42. $\dfrac{x^{-3}}{x^{-2}}$ _____

43. $\dfrac{10^{-13} \cdot 10^{-4}}{10^{-21} \cdot 10^3}$ _____

44. $\dfrac{10^{23} \cdot 10^{-11}}{10^{-3} \cdot 10^{-2}}$ _____

45. $\dfrac{18 \times 10^{12}}{6 \times 10^{-4}}$ _____

46. $\dfrac{8 \times 10^{-3}}{2 \times 10^{-5}}$ _____

47. $\left(\dfrac{y}{y^{-2}}\right)^3$ _____

48. $\left(\dfrac{x^2}{x^{-1}}\right)^2$ _____

49. $\dfrac{1}{(3mn)^{-2}}$ _____

50. $(2cd^2)^{-3}$ _____

51. $(2mn^{-3})^3$ _____

52. $(3x^3y^{-2})^2$ _____

53. $(m^4n^{-5})^{-3}$ _____

54. $(x^{-3}y^2)^{-2}$ _____

55. $(2^2 3^{-3})^{-1}$ _____

56. $(2^{-3}3^2)^{-2}$ _____

57. $(10^{12} \cdot 10^{-12})^{-1}$ _____

58. $(10^2 \cdot 3^0)^{-2}$ _____

59. $\dfrac{8x^{-3}y^{-1}}{6x^2y^{-4}}$ _____

60. $\dfrac{9m^{-4}n^3}{12m^{-1}n^{-1}}$ _____

61. $\dfrac{2a^6b^{-2}}{16a^{-3}b^2}$ _____

62. $\dfrac{4x^{-2}y^{-3}}{2x^{-3}y^{-1}}$ _____

63. $\left(\dfrac{x^{-1}}{x^{-8}}\right)^{-1}$ _____

64. $\left(\dfrac{n^{-3}}{n^{-2}}\right)^{-2}$ _____

65. $\left(\dfrac{m^{-2}n^3}{m^4n^{-1}}\right)^2$ _____

66. $\left(\dfrac{x^4y^{-1}}{x^{-2}y^3}\right)^2$ _____

67. $\left(\dfrac{6nm^{-2}}{3m^{-1}n^2}\right)^{-3}$ _____

68. $\left(\dfrac{2x^{-3}y^2}{4xy^{-1}}\right)^{-2}$ _____

69. $\left[\left(\dfrac{x^{-2}y^3t}{x^{-3}y^{-2}t^2}\right)^2\right]^{-1}$ _____

70. $\left[\left(\dfrac{u^3v^{-1}w^{-2}}{u^{-2}v^{-2}w}\right)^{-2}\right]^2$ _____

71. $\left(\dfrac{2^2x^2y^0}{8x^{-1}}\right)^{-2}\left(\dfrac{x^{-3}}{x^{-5}}\right)^3$ _____

72. $\left(\dfrac{3^3x^0y^{-2}}{2^3x^3y^{-5}}\right)^{-1}\left(\dfrac{3^3x^{-1}y}{2^2x^2y^{-2}}\right)^2$ _____

C 73. $(a^2 - b^2)^{-1}$ _____

74. $(x + 2)^{-2}$ _____

75. $\dfrac{x^{-1} + y^{-1}}{x + y}$ _____

76. $\dfrac{2^{-1} + 3^{-1}}{25}$ _____

77. $\dfrac{c - d}{c^{-1} - d^{-1}}$ _____

78. $\dfrac{12}{2^{-2} + 3^{-1}}$ _____

79. $(x^{-1} + y^{-1})^{-1}$ _____

80. $(2^{-2} + 3^{-2})^{-1}$ _____

81. $(x^{-1} - y^{-1})^2$ _____

82. $(10^{-2} + 10^{-3})^{-1}$ _____

The Check Exercise for this section is on page 305.

6.2 SCIENTIFIC NOTATION

Work in science and engineering often involves the use of very, very large numbers:

The estimated free oxygen of the earth weighs approximately 1,500,000,000,000,000,000,000 grams.

Also involved is the use of very, very small numbers:

The probable mass of a hydrogen atom is 0.000 000 000 000 000 000 000 000 001 7 gram.

Writing and working with numbers of this type in standard decimal notation is generally awkward. It is often convenient to represent numbers of this type in **scientific notation;** that is, as the product of a number between 1 and 10 and a power of 10. Any decimal fraction, however large or small, can be represented in scientific notation.

EXAMPLE 6
DECIMAL FRACTIONS AND SCIENTIFIC NOTATION

$$5 = 5 \times 10^0 \qquad\qquad 0.7 = 7 \times 10^{-1}$$

$$35 = 3.5 \times 10 \qquad\qquad 0.083 = 8.3 \times 10^{-2}$$

$$430 = 4.3 \times 10^2 \qquad\qquad 0.0043 = 4.3 \times 10^{-3}$$

$$5,870 = 5.87 \times 10^3 \qquad\qquad 0.000687 = 6.87 \times 10^{-4}$$

$$8,910,000 = 8.91 \times 10^6 \qquad 0.00000036 = 3.6 \times 10^{-7}$$

Can you discover a simple mechanical rule that relates the number of decimal places the decimal is moved with the power of 10 that is used?

$$7,320,000 = 7.320\,000. \times 10^6 = 7.32 \times 10^6$$

6 places left

positive exponent

$$0.000\,000\,54 = 0.000\,000\,5.4 \times 10^{-7} = 5.4 \times 10^{-7}$$

7 places right

negative exponent

Work Problem 6 and check solution in Solutions to Matched Problems on page 267.

PROBLEM 6 Write in scientific notation.

(A) $720 =$ (B) $43,000 =$

(C) $0.08 =$ (D) $0.000\,057 =$

Figure 1 shows the relative size of a number of familiar objects on a power-of-ten scale. Note that 10^{10} is not just double 10^5. (Why?)

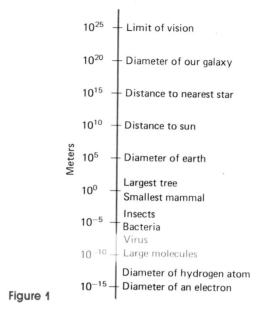

Figure 1

EXAMPLE 7

Evaluate the following complicated arithmetic problem by first converting to scientific notation.

$$\frac{(0.000\ 000\ 000\ 000\ 026)(720)}{(48,000,000,000)(0.0013)}$$

Write each number in scientific notation.

$$= \frac{(2.6 \times 10^{-14})(7.2 \times 10^2)}{(4.8 \times 10^{10})(1.3 \times 10^{-3})}$$

Collect all powers of 10 together.

$$= \frac{\overset{2}{\cancel{(2.6)}}\overset{6}{\cancel{(7.2)}}}{\underset{4}{\cancel{(4.8)}}\underset{1}{\cancel{(1.3)}}} \cdot \frac{(10^{-14})(10^2)}{(10^{10})(10^{-3})}$$

Simplify each group and write the result in scientific notation.

$$= 3 \times 10^{-19}$$

NOTE: If you try to work this problem directly using a hand calculator, you will find that some of the numbers will not fit unless they are first converted to scientific notation. If you have a calculator, try it. Some calculators can compute directly in scientific notation and read out in scientific notation.

Work Problem 7 and check solution in Solutions to Matched Problems on page 267.

PROBLEM 7 Convert to scientific notation and evaluate:

$$\frac{(42,000)(0.000\ 000\ 000\ 09)}{(600,000,000,000)(0.000\ 21)} =$$

We are able to look back into time by looking out into space. Since light travels at a fast but finite rate, we see heavenly bodies not as they exist now, but as they existed at some time in the past. If the distance between the sun and the earth is approximately 9.3×10^7 mi and if light travels at the rate of approximately 1.86×10^5 mi/sec, we see the sun as it was how many minutes ago?

$$t = \frac{d}{r} = \frac{9.3 \times 10^7}{1.86 \times 10^5} = 5 \times 10^2 = 500 \text{ sec} \qquad \text{or} \qquad \frac{500}{60} \approx 8.3 \text{ min}$$

Hence, we always see the sun as it was 8.3 min ago.

SOLUTIONS TO MATCHED PROBLEMS

6. (A) $720 = 7.2 \times 10^2$ (B) $43,000 = 4.3 \times 10^4$ (C) $0.08 = 8 \times 10^{-2}$
(D) $0.000\ 057 = 5.7 \times 10^{-5}$

7. $\dfrac{(42,000)(0.000\ 000\ 000\ 09)}{(600,000,000,000)(0.000\ 21)} = \dfrac{(4.2 \times 10^4)(9 \times 10^{-11})}{(6 \times 10^{11})(2.1 \times 10^{-4})} = \dfrac{(4.2)(9)}{(6)(2.1)} \cdot \dfrac{(10^4)(10^{-11})}{(10^{11})(10^{-4})} = 3 \times 10^{-14}$

PRACTICE EXERCISE 6.2

Work odd-numbered problems first, check answers, and then work even-numbered problems in areas of weakness. Answers to all problems are in the back of the book. Make every effort to work a problem yourself before you look at an answer.

A *Write in scientific notation:*

1. 70 _____

2. 50 _____

3. 800 _____

4. 600 _____

5. 80,000 _____

6. 600,000 _____

7. 0.008 _____

8. 0.06 _____

9. 0.00000008 _____

10. 0.00006 _____

11. 52 _____

12. 35 _____

13. 0.63 _____

14. 0.72 _____

15. 340 _____

16. 270 _____

17. 0.085 _____

18. 0.032 _____

19. 6,300 _____

20. 5,200 _____

21. 0.0000068 _____

22. 0.00072 _____

Write as a decimal fraction:

23. 8×10^2 _____

24. 5×10^2 _____

25. 4×10^{-2} _____

26. 8×10^{-2} _____

27. 3×10^5 _____

28. 6×10^6 _____

29. 9×10^{-4} _____

30. 2×10^{-5} _____

31. 5.6×10^4 _____

32. 7.1×10^3 _____

33. 9.7×10^{-3} _____

34. 8.6×10^{-4} _____

35. 4.3×10^5 _____

36. 8.8×10^6 _____

37. 3.8×10^{-7} _____

38. 6.1×10^{-6} _____

B *Write in scientific notation:*

39. 5,460,000,000 _____   40. 42,700,000 _____

41. 0.000 000 072 9 _____   42. 0.000 072 3 _____

43. The energy of a laser beam can go as high as 10,000,000,000,000 watts. _____

44. The distance that light travels in 1 year is called a light-year. It is approximately 5,870,000,000,000 mi.

_____

45. The nucleus of an atom has a diameter of a little more than 1/100,000 that of the whole atom.

_____

46. The mass of one water molecule is 0.000 000 000 000 000 000 000 000 03 gram.

Write as a decimal fraction:

47. 8.35×10^{10} _____   48. 3.46×10^9 _____

49. 6.14×10^{-12} _____   50. 6.23×10^{-7} _____

51. The diameter of the sun is approximately 8.65×10^5 mi. _____

52. The distance from the earth to the sun is approximately 9.3×10^7 mi. _____

53. The approximate mass of a hydrogen atom is 1.7×10^{-24} gram. _____

54. The diameter of a red corpuscle is approximately 7.5×10^{-5} cm. _____

Simplify and express answer in scientific notation.

55. $(3 \times 10^{-6})(3 \times 10^{10})$ _____

56. $(4 \times 10^5)(2 \times 10^{-3})$ _____

57. $(2 \times 10^3)(3 \times 10^{-7})$ _____

58. $(4 \times 10^{-8})(2 \times 10^5)$ _____

59. $\dfrac{6 \times 10^{12}}{2 \times 10^7}$ _____

60. $\dfrac{9 \times 10^8}{3 \times 10^5}$ _____

61. $\dfrac{15 \times 10^{-2}}{3 \times 10^{-6}}$ _____

62. $\dfrac{12 \times 10^3}{4 \times 10^{-4}}$ _____

Convert each numeral to scientific notation and simplify. Express answer in scientific notation and as a decimal fraction.

63. $\dfrac{(90,000)(0.000\ 002)}{0.006}$ _____

64. $\dfrac{(0.000\ 6)(4,000)}{0.000\ 12}$ _____

65. $\dfrac{(60,000)(0.000\ 003)}{(0.000\ 4)(1,500,000)}$ _____

66. $\dfrac{(0.000\ 039)(140)}{(130,000)(0.000\ 21)}$ _____

C 67. If the mass of the earth is 6×10^{27} grams and each gram is 1.1×10^{-6} ton, find the mass of the earth in tons.

_____

68. In 1929 Vernadsky, a biologist, estimated that all of the free oxygen of the earth is 1.5×10^{21} grams and that it is produced by life alone. If one gram is approximately 2.2×10^{-3} lb, what is the amount of free oxygen in pounds?

_____

69. Some of the designers of high-speed computers are currently thinking of single-addition times of 10^{-7} sec (100 nanosec). How many additions would such a computer be able to perform in 1 sec? In 1 min?

_____

70. If electricity travels in a computer circuit at the speed of light (1.86×10^5 mi/sec), how far will it travel in the time it takes the computer in the preceding problem to complete a single addition? (Size of circuits is becoming a critical problem in computer design.) Give the answer in miles and in feet.

_____

OPTIONAL CALCULATOR EXERCISE
Use a calculator and work Problems 55 to 70 using scientific notation.

The Check Exercise for this section is on page 307.

6.3 RATIONAL EXPONENTS

•**Roots of real numbers**
••**Rational exponents**

•Roots of real numbers

What do we mean by a root of a number? Perhaps you recall that a square root of a number b is a number a such that $a^2 = b$ and a cube root of a number b is a number a such that $a^3 = b$.
   What are the square roots of 4?

   2 is a square root of 4, since $2^2 = 4$

   -2 is a square root of 4, since $(-2)^2 = 4$

Thus, 4 has two real square roots, one the negative of the other.
    What are the cube roots of 8?

2 is a cube root of 8, since $2^3 = 8$

and 2 is the only real number with this property. In general,

> *Definition of an nth Root*
> _____
>
> For n a natural number
>
> a is an nth root of b if $a^n = b$
> 2 is a fourth root of 16, since $2^4 = 16$

    How many real square roots of 9 exist? Of 7? Of -4? How many real fourth roots of 7 exist?
Of -7? How many real cube roots of 27 are there? The following important theorem (which we
state without proof) answers these questions completely.

THEOREM 1
Number of real nth roots of a real number b*

	n even	*n* odd
b **positive**	Two real nth roots -2 and 2 are both fourth roots of 16	One real nth root 2 is the only real cube root of 8
b **negative**	No real nth root -4 has no real square roots	One real nth root -2 is the only real cube root of -8

Thus,

5 has two real square roots, two real fourth roots, and so on.

7 has one real cube root, one real fifth root, and so on.

What symbols do we use to represent these roots? We turn to this question now.

••Rational exponents

If all exponent laws are to continue to hold even if some of the exponents are not integers; then

$$(5^{1/2})^2 = 5^{2/2} = 5 \qquad \text{and} \qquad (7^{1/3})^3 = 7^{3/3} = 7$$

Hence, $5^{1/2}$ must name a square root of 5, since $(5^{1/2})^2 = 5$. Similarly, $7^{1/3}$ must name the cube
root of 7, since $(7^{1/3})^3 = 7$.

---

*In this section we limit our discussion to real roots of real numbers. After the real numbers are extended to the com-
plex numbers (see Section 6.7), then additional roots can be considered. For example, it turns out that 8 has three
cube roots: in addition to the real number 2, there are two other cube roots in the complex number system. A systematic discussion of roots in the complex number system is reserved for more advanced courses on the subject.

In general, for n a natural number and b not negative when n is even,

$$(b^{1/n})^n = b^{n/n} = b$$

Thus, $b^{1/n}$ must name an nth root of b. Which real nth root of b does $b^{1/n}$ represent if n is even and b is positive? (According to Theorem 1 there are two real nth roots.) We answer this question in the following definition:

Definition of $b^{1/n}$

For n a natural number

$b^{1/n}$ is an nth root of b

If n is even and b is positive, then $b^{1/n}$ represents the positive real nth root of b (sometimes called the **principal nth root of** b), $-b^{1/n}$ represents the negative real nth root of b, and $(-b)^{1/n}$ does not represent a real number. If n is odd, then $b^{1/n}$ represents *the* real nth root of b. $0^{1/n} = 0$.

$16^{1/2} = 4$ $-16^{1/2} = -4$
(not -4 and 4) $[-16^{1/2}$ and $(-16)^{1/2}$ are not the same]

$(-16)^{1/2}$ is not real $32^{1/5} = 2$ $(-32)^{1/5} = -2$ $0^{1/9} = 0$

EXAMPLE 8
(A) $4^{1/2} = 2$
(B) $-4^{1/2} = -2$
(C) $(-4)^{1/2}$ is not a real number
(D) $8^{1/3} = 2$
(E) $(-8)^{1/3} = -2$
(F) $0^{1/5} = 0$

Note carefully the difference between parts **B** and **C**.

Work Problem 8 and check solution in Solutions to Matched Problems on page 273.

PROBLEM 8   Find each of the following:

(A) $9^{1/2} =$ (B) $-9^{1/2} =$

(C) $(-9)^{1/2} =$ (D) $27^{1/3} =$

(E) $(-27)^{1/3} =$ (F) $0^{1/4} =$

How should an expression such as $5^{2/3}$ be defined? If the properties of exponents are to continue to hold for all rational exponents, then $5^{2/3} = (5^{1/3})^2$; that is, $5^{2/3}$ must represent the square of the cube root of 5. Thus, we are lead to the following general definition:

Definition of $b^{m/n}$ and $b^{-m/n}$

For m and n natural numbers and b any real number, except b cannot be negative when n is even,

$$b^{m/n} = (b^{1/n})^m \qquad \text{and} \qquad b^{-m/n} = \frac{1}{b^{m/n}}$$

$4^{3/2} = (4^{1/2})^3 = 2^3 = 8$ $\qquad (-32)^{3/5} = [(-32)^{1/5}]^3 = (-2)^3 = -8$

$4^{-3/2} = \dfrac{1}{4^{3/2}} = \dfrac{1}{8}$ $\qquad (-4)^{3/2}$ is not real

---

We have now discussed $b^{m/n}$ for all rational numbers m/n and real numbers b. It can be shown, though we will not do so, that all five properties of exponents discussed in Section 6.1 continue to hold for rational exponents so long as we avoid even roots of negative numbers. With the latter restriction in effect, the following useful relationship is an immediate consequence of the exponent properties:

$$b^{m/n} = (b^{1/n})^m = (b^m)^{1/n}$$

EXAMPLE 9
(A) $8^{2/3} = (8^{1/3})^2 = 2^2 = 4 \qquad$ or $\qquad 8^{2/3} = (8^2)^{1/3} = 64^{1/3} = 4$
(B) $(-8)^{5/3} = [(-8)^{1/3}]^5 = (-2)^5 = -32 \qquad$ Easier than computing $[(-8)^5]^{1/3}$
(C) $(3x^{1/3})(2x^{1/2}) = 6x^{1/3+1/2} = 6x^{5/6}$

(D) $(2x^{1/3}y^{-2/3})^3 = 8xy^{-2}$ or $\dfrac{8x}{y^2}$

(E) $\left(\dfrac{4x^{1/3}}{x^{1/2}}\right)^{1/2} = \dfrac{4^{1/2}x^{1/6}}{x^{1/4}} = \dfrac{2}{x^{1/4-1/6}} = \dfrac{2}{x^{1/12}} \qquad$ or $\qquad 2x^{-1/12}$
(F) $(2a^{1/2} + b^{1/2})(a^{1/2} + 3b^{1/2}) = 2a + 7a^{1/2}b^{1/2} + 3b$

Work Problem 9 and check solution in Solutions to Matched Problems on page 273.

PROBLEM 9 Simplify, and express answers using positive exponents only.

(A) $9^{3/2} =$

(B) $(-27)^{4/3} =$

(C) $(5y^{3/4})(2y^{1/3}) =$

(D) $(2x^{-3/4}y^{1/4})^4 =$

(E) $\left(\dfrac{8x^{1/2}}{x^{2/3}}\right)^{1/3} =$

(F) $(x^{1/2} - 2y^{1/2})(3x^{1/2} + y^{1/2}) =$

The properties of exponents can be used as long as we are dealing with symbols that name real numbers. Can you resolve the following contradiction?

$$-1 = (-1)^{2/2} = [(-1)^2]^{1/2} = 1^{1/2} = 1$$

The second member of the equality chain, $(-1)^{2/2}$, involves an even root of a negative number, which is not real. Thus we see that the properties of exponents do not necessarily hold when we are dealing with nonreal quantities unless further restrictions are imposed. One such restriction is to require all rational exponents to be reduced to lowest terms.

SOLUTIONS TO MATCHED PROBLEMS

8. **(A)** $9^{1/2} = 3$ **(B)** $-9^{1/2} = -3$ **(C)** $(-9)^{1/2}$ is not real   **(D)** $27^{1/3} = 3$
(E) $(-27)^{1/3} = -3$ **(F)** $0^{1/4} = 0$

9. **(A)** $9^{3/2} = (9^{1/2})^3 = 3^3 = 27$ **(B)** $(-27)^{4/3} = [(-27)^{1/3}]^4 = (-3)^4 = 81$

   (C) $(5y^{3/4})(2y^{1/3}) = 10y^{3/4+1/3} = 10y^{13/12}$ **(D)** $(2x^{-3/4}y^{1/4})^4 = 2^4x^{-3}y = \dfrac{16y}{x^3}$

   (E) $\left(\dfrac{8x^{1/2}}{x^{2/3}}\right)^{1/3} = \dfrac{8^{1/3}x^{1/6}}{x^{2/9}} = \dfrac{2}{x^{2/9-1/6}} = \dfrac{2}{x^{1/18}}$ **(F)** $(x^{1/2} - 2y^{1/2})(3x^{1/2} + y^{1/2}) = 3x - 5x^{1/2}y^{1/2} - 2y$

PRACTICE EXERCISE 6.3

Work odd-numbered problems first, check answers, and then work even-numbered problems in areas of weakness. Answers to all problems are in the back of the book. Make every effort to work a problem yourself before you look at an answer.

In Problems 1 to 72 all variables represent positive real numbers.

A *Most of the following are integers. Find them.*

1. $25^{1/2}$ _____   **2.** $36^{1/2}$ _____   **3.** $(-25)^{1/2}$ _____   **4.** $(-36)^{1/2}$ _____

5. $8^{1/3}$ _____   **6.** $27^{1/3}$ _____   **7.** $(-8)^{1/3}$ _____   **8.** $(-27)^{1/3}$ _____

9. $-8^{1/3}$ _____   **10.** $-27^{1/3}$ _____   **11.** $16^{3/2}$ _____   **12.** $25^{3/2}$ _____

13. $8^{2/3}$ _____   **14.** $27^{2/3}$ _____

Simplify, and express the answer using positive exponents only.

15. $x^{1/4}x^{3/4}$ _____   **16.** $y^{1/5}y^{2/5}$ _____   **17.** $\dfrac{x^{2/5}}{x^{3/5}}$ _____   **18.** $\dfrac{a^{2/3}}{a^{1/3}}$ _____

19. $(x^4)^{1/2}$ _____   **20.** $(y^{1/2})^4$ _____   **21.** $(a^3b^9)^{1/3}$ _____   **22.** $(x^4y^2)^{1/2}$ _____

23. $\left(\dfrac{x^9}{y^{12}}\right)^{1/3}$ _____   **24.** $\left(\dfrac{m^{12}}{n^{16}}\right)^{1/4}$ _____   **25.** $(x^{1/3}y^{1/2})^6$ _____   **26.** $\left(\dfrac{u^{1/2}}{v^{1/3}}\right)^{12}$ _____

B *Most of the following are rational numbers. Find them.*

27. $\left(\tfrac{4}{25}\right)^{1/2}$ _____   **28.** $\left(\tfrac{9}{4}\right)^{1/2}$ _____   **29.** $\left(\tfrac{4}{25}\right)^{3/2}$ _____   **30.** $\left(\tfrac{9}{4}\right)^{3/2}$ _____

31. $(\frac{1}{8})^{2/3}$ _____   **32.** $(\frac{1}{27})^{2/3}$ _____   **33.** $36^{-1/2}$ _____   **34.** $25^{-1/2}$ _____

35. $25^{-3/2}$ _____   **36.** $16^{-3/2}$ _____   **37.** $(-\frac{4}{9})^{3/2}$ _____   **38.** $(-\frac{25}{81})^{1/2}$ _____

39. $(3^6)^{-1/3}$ _____   **40.** $(4^{-8})^{3\sqrt{16}}$ _____

Simplify, and express the answer using positive exponents only.

41. $x^{1/4}x^{-3/4}$ _____   **42.** $\dfrac{d^{2/3}}{d^{-1/3}}$ _____   **43.** $n^{3/4}n^{-2/3}$ _____

44. $m^{1/2}m^{-1/3}$ _____   **45.** $(x^{-2/3})^{-6}$ _____   **46.** $(y^{-8})^{1/16}$ _____

47. $(4u^{-2}v^4)^{1/2}$ _____   **48.** $(8x^3y^{-6})^{1/3}$ _____   **49.** $(x^4y^6)^{-1/2}$ _____

50. $(4x^{1/2}y^{3/2})^2$ _____   **51.** $\left(\dfrac{x^{-2/3}}{y^{-1/2}}\right)^{-6}$ _____   **52.** $\left(\dfrac{m^{-3}}{n^2}\right)^{-1/6}$ _____

53. $\left(\dfrac{25x^5y^{-1}}{16x^{-3}y^{-5}}\right)^{1/2}$ _____   **54.** $\left(\dfrac{8a^{-4}b^3}{27a^2b^{-3}}\right)^{1/3}$ _____   **55.** $\left(\dfrac{8y^{1/3}y^{-1/4}}{y^{-1/12}}\right)^2$ _____

56. $\left(\dfrac{9x^{1/3}x^{1/2}}{x^{-1/6}}\right)^{1/2}$ _____

Multiply, and express answer using positive exponents only.

57. $3m^{3/4}(4m^{1/4} - 2m^8)$ _____   **58.** $2x^{1/3}(3x^{2/3} - x^6)$ _____

59. $(2x^{1/2} + y^{1/2})(x^{1/2} + y^{1/2})$ _____   **60.** $(x^{1/2} + y^{1/2})(x^{1/2} - y^{1/2})$ _____

61. $(x^{1/2} + y^{1/2})^2$ _____   **62.** $(x^{1/2} - y^{1/2})^2$ _____

C *Simplify, and express answer using positive exponents only.*

63. $(-16)^{-3/2}$ _____   **64.** $-16^{-3/2}$ _____

65. $(a^{-1/2} + 3b^{-1/2})(2a^{-1/2} - b^{-1/2})$ _____   **66.** $(x^{-1/2} - y^{-1/2})^2$ _____

67. $(a^{n/2}b^{n/3})^{1/n}, n > 0$ _____   **68.** $(a^{3/n}b^{3/m})^{1/3}, n > 0, m > 0$ _____

69. $\left(\dfrac{x^{m+2}}{x^m}\right)^{1/2}, m > 0$ _____   **70.** $\left(\dfrac{a^m}{a^{m-2}}\right)^{1/2}, m > 0$ _____

71. Find a value of x such that $(x^2)^{1/2} \neq x$. _____

72. For which real numbers does $(x^2)^{1/2} = |x|$? (More will be said about this form in Section 6.5.)

The Check Exercise for this section is on page 309.

6.4 RADICAL FORMS AND RATIONAL EXPONENTS

In the preceding section we introduced the symbol $b^{1/n}$ to represent an nth root of b, and found that the symbol could be combined with other exponent forms, using the properties of exponents. Another symbol is also used to represent an nth root, the radical sign. Both symbols are widely used, and you should become familiar with them and their respective properties.

---

nth-Root Radical

For n a natural number greater than 1 and b any real number

$$\sqrt[n]{b} = b^{1/n}$$

Thus, $\sqrt[n]{b}$ represents an nth root of b.

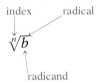

index   radical

$\sqrt[n]{b}$

radicand

If $n = 2$, we write

$$\sqrt{b} \qquad \text{and not} \qquad \sqrt[2]{b}$$

and refer to $\sqrt{b}$ as "the positive square root of b." Thus, it follows that

$$\sqrt{3} = 3^{1/2}$$
$$\sqrt[8]{5} = 5^{1/8}$$
$$\sqrt[3]{x^2} = (x^2)^{1/3} = x^{2/3}$$
$$(\sqrt[3]{x})^2 = (x^{1/3})^2 = x^{2/3}$$

There are occasions when it is more convenient to work with radicals than with rational exponents, and vice versa. It is often an advantage to be able to shift back and forth between the two forms. The following relationships, suggested by the above examples, are useful in this regard.

For b nonnegative when n is even

$$b^{m/n} = (b^m)^{1/n} = \sqrt[n]{b^m}$$
$$b^{m/n} = (b^{1/n})^m = (\sqrt[n]{b})^m$$

The following examples should make clear the process of changing from one form to the other. All variables represent *positive* real numbers.

EXAMPLE 10

Convert from rational exponent form to radical form.

(A) $5^{1/2} = \sqrt{5}$ Positive square root of 5.

(B) $x^{1/7} = \sqrt[7]{x}$

(C) $m^{2/3} = \sqrt[3]{m^2}$ or $(\sqrt[3]{m})^2$ First form is usually preferred.

(D) $(3u^2v^3)^{3/5} = \sqrt[5]{(3u^2v^3)^3}$ or $(\sqrt[5]{3u^2v^3})^3$ First form is usually preferred.

(E) $y^{-2/3} = \dfrac{1}{y^{2/3}} = \dfrac{1}{\sqrt[3]{y^2}}$ or $\dfrac{1}{(\sqrt[3]{y})^2}$ Index of radical cannot be negative.

(F) $(x^2 + y^2)^{1/2} = \sqrt{x^2 + y^2}$ $\neq x + y$ Why?

Work Problem 10 and check solution in Solutions to Matched Problems on page 277.

PROBLEM 10   Convert to radical form (do not simplify).

(A) $7^{1/2} =$ (B) $u^{1/5} =$

(C) $x^{3/5} =$ (D) $(2x^3y^2)^{2/3} =$

(E) $x^{-3/4} =$ (F) $(x^3 + y^3)^{1/3} =$

EXAMPLE 11

Convert from radical form to rational exponent form.

(A) $\sqrt{13} = 13^{1/2}$ Positive square root of 13.

(B) $\sqrt[5]{x} = x^{1/5}$

(C) $\sqrt[4]{w^3} = w^{3/4}$

(D) $\sqrt[5]{(3x^2y^2)^4} = (3x^2y^2)^{4/5}$

(E) $\dfrac{1}{\sqrt[3]{x^2}} = \dfrac{1}{x^{2/3}} = x^{-2/3}$

(F) $\sqrt[4]{x^4 + y^4} = (x^4 + y^4)^{1/4}$ $\neq x + y$ Why?

Work Problem 11 and check solution in Solutions to Matched Problems on page 277.

PROBLEM 11   Convert to rational exponent form.

(A) $\sqrt{17} =$ (B) $\sqrt[7]{m} =$

(C) $\sqrt[5]{x^2} =$ (D) $(\sqrt[7]{5m^3n^4})^3 =$

(E) $\dfrac{1}{\sqrt[6]{u^5}} =$ (F) $\sqrt[5]{x^5 - y^5} =$

SOLUTIONS TO MATCHED PROBLEMS

10. **(A)** $7^{1/2} = \sqrt{7}$ **(B)** $u^{1/5} = \sqrt[5]{u}$ **(C)** $x^{3/5} = \sqrt[5]{x^3}$ or $(\sqrt[5]{x})^3$

(D) $(2x^3y^2)^{2/3} = \sqrt[3]{(2x^3y^2)^2}$ or $(\sqrt[3]{2x^3y^2})^2$ **(E)** $x^{-3/4} = \dfrac{1}{\sqrt[4]{x^3}}$ or $\dfrac{1}{(\sqrt[4]{x})^3}$

(F) $(x^3 + y^3)^{1/3} = \sqrt[3]{x^3 + y^3}$

11. **(A)** $\sqrt{17} = 17^{1/2}$ **(B)** $\sqrt[7]{m} = m^{1/7}$ **(C)** $\sqrt[5]{x^2} = x^{2/5}$ **(D)** $(\sqrt[7]{5m^3n^4})^3 = (5m^3n^4)^{3/7}$

(E) $\dfrac{1}{\sqrt[6]{u^5}} = u^{-5/6}$ **(F)** $\sqrt[5]{x^5 - y^5} = (x^5 - y^5)^{1/5}$

PRACTICE EXERCISE 6.4

Work odd-numbered problems first, check answers, and then work even-numbered problems in areas of weakness. Answers to all problems are in the back of the book. Make every effort to work a problem yourself before you look at an answer.

All variables are restricted to avoid even roots of negative numbers.

A *Change to radical form. (Do not simplify.)*

1. $11^{1/2}$ _____   **2.** $7^{1/2}$ _____   **3.** $5^{1/3}$ _____   **4.** $6^{1/4}$ _____

5. $u^{3/5}$ _____   **6.** $x^{3/4}$ _____   **7.** $4y^{3/7}$ _____   **8.** $5m^{2/3}$ _____

9. $(4y)^{3/7}$ _____   **10.** $(5m)^{2/3}$ _____   **11.** $(4ab^3)^{2/5}$ _____

12. $(7x^2y)^{2/3}$ _____   **13.** $(a + b)^{1/2}$ _____   **14.** $(a^2 + b^2)^{1/2}$ _____

Change to rational exponent form. (Do not simplify.)

15. $\sqrt{6}$ _____   **16.** $\sqrt{3}$ _____   **17.** $\sqrt[4]{m}$ _____   **18.** $\sqrt[7]{m}$ _____

19. $\sqrt[5]{y^3}$ _____   **20.** $\sqrt[3]{a^2}$ _____   **21.** $\sqrt[4]{(xy)^3}$ _____

22. $\sqrt[5]{(7m^3n^3)^4}$ _____   **23.** $\sqrt{x^2 - y^2}$ _____   **24.** $\sqrt{1 + y^2}$ _____

B *Change to radical form. (Do not simplify.)*

25. $-5y^{2/5}$ _____   **26.** $-3x^{1/2}$ _____   **27.** $(1 + m^2n^2)^{3/7}$ _____

28. $(x^2y^2 - w^3)^{4/5}$ _____   **29.** $w^{-2/3}$ _____   **30.** $y^{-3/5}$ _____

31. $(3m^2n^3)^{-3/5}$ _____   **32.** $(2xy)^{-2/3}$ _____   **33.** $a^{1/2} + b^{1/2}$ _____

34. $x^{-1/2} + y^{-1/2}$ _____   **35.** $(a^3 + b^3)^{2/3}$ _____   **36.** $(x^{1/2} + y^{-1/2})^{1/3}$ _____

Change to rational exponent form. (Do not simplify.)

37. $\sqrt[3]{(a + b)^2}$ _____   **38.** $\sqrt[5]{(x - y)^2}$ _____   **39.** $-3x\sqrt[4]{a^3b}$ _____

40. $-5\sqrt[3]{2x^2y^2}$ _____   **41.** $\sqrt[9]{-2x^3y^7}$ _____   **42.** $\sqrt[5]{-4m^2n^3}$ _____

43. $\dfrac{3}{\sqrt[3]{y}}$ _____   **44.** $\dfrac{2x}{\sqrt{y}}$ _____   **45.** $\dfrac{-2x}{\sqrt{x^2 + y^2}}$ _____

46. $\dfrac{2}{\sqrt{x}} + \dfrac{3}{\sqrt{y}}$ _____

47. $\sqrt[3]{m^2} - \sqrt{n}$ _____

48. $\dfrac{-5u^2}{\sqrt{u} + \sqrt[3]{v^3}}$ _____

49. Show that $(x^2 + y^2)^{1/2} \neq x + y$ by evaluating each side for $x = 3$ and $y = 4$. _____

50. Show that $\sqrt{a^2 + b^2} \neq a + b$ by evaluating each side for $a = 4$ and $b = 3$. _____

51. Find a value of x such that $\sqrt{x^2} \neq x$. _____

52. For what real numbers does $\sqrt{x^2} = |x|$? _____

NOTE: More will be said about the material in Problems 51 and 52 in the next section.

The Check Exercise for this section is on page 311.

6.5 CHANGING AND SIMPLIFYING RADICAL EXPRESSIONS

- •Properties of radicals
- ••Simplest radical form
- •••Simplifying $\sqrt[n]{x^n}$ for all real x

•Properties of radicals

Changing and simplifying radical expressions is aided by the introduction of several properties of radicals that follow directly from the exponent properties considered earlier. To start, consider the following examples:

1. $\sqrt[5]{2^5} = (2^5)^{1/5} = 2^{5/5} = 2^1 = 2$

2. $\sqrt{4 \cdot 9} = \sqrt{36} = 6$ and $\sqrt{4}\sqrt{9} = 2 \cdot 3 = 6$

3. $\sqrt{\dfrac{36}{4}} = \sqrt{9} = 3$ and $\dfrac{\sqrt{36}}{\sqrt{4}} = \dfrac{6}{2} = 3$

4. $\sqrt[6]{2^4} = (2^4)^{1/6} = 2^{4/6} = 2^{2/3} = (2^2)^{1/3} = \sqrt[3]{2^2}$

These examples suggest the following general properties of radicals.

Properties of Radicals

In the following n, m, and k are natural numbers ≥ 2; x and y are positive real numbers.

1. $\sqrt[n]{x^n} = x \qquad (\sqrt[n]{x})^n = x$ $\qquad$ $\sqrt[3]{x^3} = x, (\sqrt[3]{x})^3 = x$

2. $\sqrt[n]{xy} = \sqrt[n]{x}\,\sqrt[n]{y}$ $\qquad$ $\sqrt[4]{xy} = \sqrt[4]{x}\,\sqrt[4]{y}$

3. $\sqrt[n]{\dfrac{x}{y}} = \dfrac{\sqrt[n]{x}}{\sqrt[n]{y}}$ $\qquad$ $\sqrt[5]{\dfrac{x}{y}} = \dfrac{\sqrt[5]{x}}{\sqrt[5]{y}}$

4. $\sqrt[kn]{x^{km}} = \sqrt[n]{x^m}$ $\qquad$ $\sqrt[12]{x^8} = \sqrt[4\cdot 3]{x^{4\cdot 2}} = \sqrt[3]{x^2}$

It is important to remember that the above properties hold in general only if x and y are restricted to positive numbers. **In this section, unless otherwise stated, all variables are assumed to represent positive numbers.** Near the end of the section we will discuss what happens if we relax this restriction.

The properties of radicals above are readily established using exponent properties:

1. $\sqrt[n]{x^n} = (x^n)^{1/n} = x^{n/n} = x$
2. $\sqrt[n]{xy} = (xy)^{1/n} = x^{1/n}y^{1/n} = \sqrt[n]{x}\,\sqrt[n]{y}$
3. $\sqrt[n]{\dfrac{x}{y}} = \left(\dfrac{x}{y}\right)^{1/n} = \dfrac{x^{1/n}}{y^{1/n}} = \dfrac{\sqrt[n]{x}}{\sqrt[n]{y}}$
4. $\sqrt[kn]{x^{km}} = (x^{km})^{1/kn} = x^{km/kn} = x^{m/n} = \sqrt[n]{x^m}$

The following example illustrates how these properties are used. Properties 2 and 3 are used from right to left as well as from left to right.

EXAMPLE 12

(A) $\sqrt[5]{(3x^2y)^5} = 3x^2y$ $\qquad\qquad$ Property 1

(B) $\sqrt{10}\sqrt{5} = \sqrt{50} = \sqrt{25\cdot 2} = \sqrt{25}\sqrt{2} = 5\sqrt{2}$ $\qquad$ Property 2

(C) $\sqrt[3]{\dfrac{x}{27}} = \dfrac{\sqrt[3]{x}}{\sqrt[3]{27}} = \dfrac{\sqrt[3]{x}}{3}$ $\quad$ or $\quad$ $\tfrac{1}{3}\sqrt[3]{x}$ $\qquad$ Property 3 (left to right)

(D) $\dfrac{\sqrt[3]{16x^5}}{\sqrt[3]{2x^2}} = \sqrt[3]{\dfrac{16x^5}{2x^2}} = \sqrt[3]{8x^3} = 2x$ $\qquad$ Property 3 (right to left)

(E) $\sqrt[6]{x^4} = \sqrt[2\cdot 3]{x^{2\cdot 2}} = \sqrt[3]{x^2}$ $\qquad$ Property 4

Work Problem 12 and check solution in Solutions to Matched Problems on page 284.

PROBLEM 12 Simplify as in Example 12.

(A) $\sqrt[7]{(u^2 + v^2)^7} =$

(B) $\sqrt[3]{4} \cdot \sqrt[3]{4} =$

(C) $\sqrt[3]{\dfrac{x^2}{8}} =$

(D) $\dfrac{\sqrt[3]{54x^8}}{\sqrt[3]{2x^2}} =$

(E) $\sqrt[8]{y^6} =$

••Simplest radical form

The laws of radicals provide us with the means for changing algebraic expressions containing radicals to a variety of equivalent forms. One form often useful is the simplest radical form. An algebraic expression that contains radicals is said to be in the **simplest radical form** if all four of the following conditions are satisfied:

Simplest Radical Form
_____

1. A radicand (the expression within the radical sign) contains no polynomial factor to a power greater than or equal to the index of the radical. ($\sqrt{x^3}$ violates this condition.)
2. The power of the radicand and the index of the radical have no common factor other than 1. ($\sqrt[6]{x^4}$ violates this condition.)
3. No radical appears in a denominator. ($3/\sqrt{5}$ violates this condition.)
4. No fraction appears within a radical. ($\sqrt{\frac{2}{3}}$ violates this condition.)

It should be understood that forms other than the simplest radical form may be more useful on occasion. The choice depends on the situation.

EXAMPLE 13
Change to simplest radical form:

(A) $\sqrt{8x^3} = \sqrt{(4x^2)(2x)}$

$\qquad = \sqrt{4x^2}\sqrt{2x} = 2x\sqrt{2x}$

Violates condition 1. Factor $8x^3$ into a perfect-square part, $4x^2$, and what is left, $2x$; then use multiplication property 2.

(B) $\sqrt[3]{54x^5} = \sqrt[3]{(27x^3)(2x^2)}$

$\qquad = \sqrt[3]{27x^3}\,\sqrt[3]{2x^2}$

$\qquad = 3x\sqrt[3]{2x^2}$

Violates condition 1. Factor $54x^5$ into a perfect-cube part, $27x^3$, and what is left, $2x^2$; then use multiplication property 2.

(C) $\sqrt[9]{x^6} = \sqrt[3]{x^2}$

Violates condition 2. Index 9 and power of radicand 6 have the common factor 3.

(D) $\dfrac{4x}{\sqrt{8x}} = \dfrac{4x}{\sqrt{8x}} \cdot \dfrac{\sqrt{2x}}{\sqrt{2x}}$

$\qquad = \dfrac{4x\sqrt{2x}}{\sqrt{16x^2}}$

$\qquad = \dfrac{4x\sqrt{2x}}{4x} = \sqrt{2x}$

Violates condition 3 (has a radical in the denominator). Multiply numerator and denominator by "smallest" or "simplest" expression that will make denominator the square root of a perfect square. Using $\sqrt{2x}$ rather than $\sqrt{8x}$ results in less work—try the latter to see why.

281

(E) $\sqrt[3]{\dfrac{y}{4x}} = \sqrt[3]{\dfrac{y}{4x}\cdot\dfrac{2x^2}{2x^2}}$

$= \sqrt[3]{\dfrac{2x^2y}{8x^3}} = \dfrac{\sqrt[3]{2x^2y}}{\sqrt[3]{8x^3}}$

$= \dfrac{\sqrt[3]{2x^2y}}{2x}$

Violates condition 4 (there is a fraction within the radical). Multiply numerator and denominator inside the radical by the "smallest" or "simplest" expression that will make the denominator a perfect cube. Using $2x^2$ rather than 4^2x^2 results in less work—try the latter to see why.

NOTE: The process of removing radicals from a denominator is called rationalizing the denominator.

Work Problem 13 and check solution in Solutions to Matched Problems on page 284.

PROBLEM 13 Change to simplest radical form:

(A) $\sqrt{18y^5} =$

(B) $\sqrt[3]{32m^8} =$

(C) $\sqrt[12]{y^8} =$

(D) $\dfrac{6u}{\sqrt[3]{4x}} =$

(E) $\sqrt{\dfrac{3y}{8x}} =$

EXAMPLE 14
Change to simplest radical form:

(A) $\sqrt{12x^3y^5z^2} = \sqrt{(2^2x^2y^4z^2)(3xy)} = \sqrt{2^2x^2y^4z^2}\sqrt{3xy} = 2xy^2z\sqrt{3xy}$
(B) $\sqrt[6]{16x^4y^2} = \sqrt[6]{(2^2x^2y)^2} = \sqrt[3]{4x^2y}$
(C) $\dfrac{3}{\sqrt{12}} = \dfrac{3}{\sqrt{12}}\cdot\dfrac{\sqrt{3}}{\sqrt{3}} = \dfrac{3\sqrt{3}}{\sqrt{36}} = \dfrac{3\sqrt{3}}{6} = \dfrac{\sqrt{3}}{2}$ or $\dfrac{1}{2}\sqrt{3}$
(D) $\dfrac{6x^2}{\sqrt[3]{9x}} = \dfrac{6x^2}{\sqrt[3]{9x}}\cdot\dfrac{\sqrt[3]{3x^2}}{\sqrt[3]{3x^2}} = \dfrac{6x^2\sqrt[3]{3x^2}}{\sqrt[3]{3^3x^3}} = \dfrac{6x^2\sqrt[3]{9x^2}}{3x} = 2x\sqrt[3]{3x^2}$
(E) $\sqrt[3]{\dfrac{2a^2}{3b^2}} = \sqrt[3]{\dfrac{(2a^2)(3^2b)}{(3b^2)(3^2b)}} = \sqrt[3]{\dfrac{18a^2b}{3^3b^3}} = \dfrac{\sqrt[3]{18a^2b}}{\sqrt[3]{3^3b^3}} = \dfrac{\sqrt[3]{18a^2b}}{3b}$

Work Problem 14 and check solution in Solutions to Matched Problems on page 284.

PROBLEM 14 Change to simplest radical form:

(A) $\sqrt[3]{16} =$

(B) $\sqrt[3]{16x^7y^4z^3}$

(C) $\sqrt[9]{8x^6y^3} =$

(D) $\dfrac{6}{\sqrt{2x}} =$

(E) $\dfrac{10x^3}{\sqrt[3]{4x^2}} =$

(F) $\sqrt[3]{\dfrac{3y^2}{2x^4}} =$

•••Simplifying $\sqrt[n]{x^n}$ for all real x

In the preceding discussion we restricted variables to nonnegative quantities. If we lift this restriction, then

$$\sqrt{x^2} = x$$

is true only for certain values of x and is not true for others. If x is positive or 0, then the equation is true; if x is negative, then the equation is false. For example, test the equation for $x = 2$ and for $x = -2$:

$x = 2$	$x = -2$
$\sqrt{x^2} = x$	$\sqrt{x^2} = x$
$\sqrt{2^2} \overset{?}{=} 2$	$\sqrt{(-2)^2} \overset{?}{=} (-2)$
$\sqrt{4} \overset{?}{=} 2$	$\sqrt{4} \overset{?}{=} -2$
$2 \overset{\checkmark}{=} 2$	$2 \neq -2$

Thus, we see that if x is negative, then we must write

$$\sqrt{x^2} = -x$$

Now both sides represent positive numbers. In summary, for x any real number

$$\sqrt{x^2} = \begin{cases} x & \text{if } x \text{ is positive} \\ 0 & \text{if } x \text{ is } 0 \\ -x & \text{if } x \text{ is negative} \end{cases}$$

Also, recall the definition of absolute value from Chapter 1:

$$|x| = \begin{cases} x & \text{if } x \text{ is positive} \\ 0 & \text{if } x \text{ is } 0 \\ -x & \text{if } x \text{ is negative} \end{cases}$$

We see that $\sqrt{x^2}$ and $|x|$ actually are the same, and we can write

> For x *any* real number
>
> $$\sqrt{x^2} = |x|$$

Thus, only if x is restricted to nonnegative real numbers can we drop the absolute-value sign.

Now let us consider $\sqrt[3]{x^3}$. Here we do not have the same kind of problem that we had above. It turns out that

> For x *any* real number
>
> $$\sqrt[3]{x^3} = x$$

We do not need the absolute-value sign on the right. As above, let us evaluate the equation for $x = 2$ and $x = -2$.

$x = 2$	$x = -2$
$\sqrt[3]{x^3} = x$	$\sqrt[3]{x^3} = x$
$\sqrt[3]{2^3} \overset{?}{=} 2$	$\sqrt[3]{(-2)^3} \overset{?}{=} (-2)$
$\sqrt[3]{8} \overset{?}{=} 2$	$\sqrt[3]{-8} \overset{?}{=} -2$
$2 \overset{\checkmark}{=} 2$	$-2 \overset{\checkmark}{=} -2$

If asked to simplify $\sqrt[3]{x^3} + \sqrt{x^2}$, many students would write

$$\sqrt[3]{x^3} + \sqrt{x^2} = x + x = 2x$$

and not think any more about it. But, if we evaluate each side for $x = -2$, we find that

$$\sqrt[3]{(-2)^3} + \sqrt{(-2)^2} = \sqrt[3]{-8} + \sqrt{4} = -2 + 2 = 0 \qquad \text{Left side}$$

and

$$2(-2) = -4 \qquad \text{Right side}$$

Both sides are not equal! What is wrong? When x is not restricted to positive values or zero, we should write

$$\sqrt[3]{x^3} + \sqrt{x^2} = x + |x|$$

then the right side will equal the left side for *all* real numbers. Consider the following example.

EXAMPLE 15
For x a positive number

$$\sqrt[3]{x^3} + \sqrt{x^2} = x + |x| = x + x = 2x \qquad \text{Remember: } |x| = x \text{ if } x \geq 0.$$

For x a negative number

$$\sqrt[3]{x^3} + \sqrt{x^2} = x + |x| = x - x = 0 \qquad \text{Remember: } |x| = -x \text{ if } x < 0.$$

Work Problem 15 and check solution in Solutions to Matched Problems on page 284.

PROBLEM 15 (A) For x a positive number

$$2\sqrt[3]{x^3} - \sqrt{x^2} =$$

(B) For x a negative number

$$2\sqrt[3]{x^3} - \sqrt{x^2} =$$

Following the same reasoning as above, we can obtain the more general result:

In general, for x *any* real number and n a positive integer > 1

$$\sqrt[n]{x^n} = \begin{cases} |x| & \text{if } n \text{ is even} \\ x & \text{if } n \text{ is odd} \end{cases} \qquad \begin{array}{l} \sqrt[4]{x^4} = |x| \\ \sqrt[5]{x^5} = x \end{array}$$

SOLUTIONS TO MATCHED PROBLEMS

12. (A) $\sqrt[7]{(u^2 + v^2)^7} = u^2 + v^2$ (B) $\sqrt[3]{4} \cdot \sqrt[3]{4} = \sqrt[3]{4 \cdot 4} = \sqrt[3]{16} = \sqrt[3]{8 \cdot 2} = \sqrt[3]{8} \cdot \sqrt[3]{2} = 2\sqrt[3]{2}$

(C) $\sqrt[3]{\dfrac{x^2}{8}} = \dfrac{\sqrt[3]{x^2}}{\sqrt[3]{8}} = \dfrac{\sqrt[3]{x^2}}{2}$ or $\dfrac{1}{2}\sqrt[3]{x^2}$ (D) $\dfrac{\sqrt[3]{54x^8}}{\sqrt[3]{2x^2}} = \sqrt[3]{\dfrac{54x^8}{2x^2}} = \sqrt[3]{27x^6} = 3x^2$ (E) $\sqrt[8]{y^6} = \sqrt[4]{y^3}$

13. (A) $\sqrt{18y^5} = \sqrt{(9y^4)(2y)} = \sqrt{9y^4}\sqrt{2y} = 3y^2\sqrt{2y}$

(B) $\sqrt[3]{32m^8} = \sqrt[3]{(8m^6)(4m^2)} = \sqrt[3]{8m^6}\sqrt[3]{4m^2} = 2m^2\sqrt[3]{4m^2}$ (C) $\sqrt[12]{y^8} = \sqrt[3]{y^2}$

(D) $\dfrac{6u}{\sqrt[3]{4x}} = \dfrac{6u}{\sqrt[3]{4x}} \cdot \dfrac{\sqrt[3]{2x^2}}{\sqrt[3]{2x^2}} = \dfrac{6u\sqrt[3]{2x^2}}{\sqrt[3]{2^3x^3}} = \dfrac{6u\sqrt[3]{2x^2}}{2x} = \dfrac{3u\sqrt[3]{2x^2}}{x}$

(E) $\sqrt{\dfrac{3y}{8x}} = \sqrt{\dfrac{3y \cdot 2x}{8x \cdot 2x}} = \sqrt{\dfrac{6xy}{4^2x^2}} = \dfrac{\sqrt{6xy}}{\sqrt{4^2x^2}} = \dfrac{\sqrt{6xy}}{4x}$

14. (A) $\sqrt[3]{16} = \sqrt[3]{8 \cdot 2} = \sqrt[3]{8} \cdot \sqrt[3]{2} = 2\sqrt[3]{2}$

(B) $\sqrt[3]{16x^7y^5z^3} = \sqrt[3]{(8x^6y^3z^3)(2xy^2)} = \sqrt[3]{8x^6y^3z^3}\sqrt[3]{2xy^2} = 2x^2yz\sqrt[3]{2xy^2}$

(C) $\sqrt[9]{8x^6y^3} = \sqrt[9]{(2x^2y)^3} = \sqrt[3]{2x^2y}$ (D) $\dfrac{6}{\sqrt{2x}} = \dfrac{6\sqrt{2x}}{\sqrt{2x}\sqrt{2x}} = \dfrac{6\sqrt{2x}}{2x} = \dfrac{3\sqrt{2x}}{x}$

(E) $\dfrac{10x^3}{\sqrt[3]{4x^2}} = \dfrac{10x^3\sqrt[3]{2x}}{\sqrt[3]{4x^2}\sqrt[3]{2x}} = \dfrac{10x^3\sqrt[3]{2x}}{\sqrt[3]{8x^3}} = \dfrac{10x^3\sqrt[3]{2x}}{2x}$

(F) $\sqrt[3]{\dfrac{3y^2}{2x^4}} = \sqrt[3]{\dfrac{3y^2 \cdot 2^2x^2}{2x^4 \cdot 2^2x^2}} = \dfrac{\sqrt[3]{12x^2y^2}}{\sqrt[3]{2^3x^6}} = \dfrac{\sqrt[3]{12x^2y^2}}{2x^2} = 5x^2\sqrt[3]{2x}$

15. (A) $2\sqrt[3]{x^3} - \sqrt{x^2} = 2x - x = x$ (B) $2\sqrt[3]{x^3} - \sqrt{x^2} = 2x - (-x) = 2x + x = 3x$

PRACTICE EXERCISE 6.5

Work odd-numbered problems first, check answers, and then work even-numbered problems in areas of weakness. Answers to all problems are in the back of the book. Make every effort to work a problem yourself before you look at an answer.

In Problems 1 to 76, simplify and write answers in simplest radical form. All variables represent positive real numbers.

A 1. $\sqrt{y^2}$ _____ 2. $\sqrt{x^2}$ _____ 3. $\sqrt{4u^2}$ _____ 4. $\sqrt{9m^2}$ _____

5. $\sqrt{49x^4y^2}$ _____ 6. $\sqrt{25x^2y^4}$ _____ 7. $\sqrt{18}$ _____ 8. $\sqrt{8}$ _____

9. $\sqrt{m^3}$ _____ 10. $\sqrt{x^3}$ _____ 11. $\sqrt{8x^3}$ _____ 12. $\sqrt{18y^3}$ _____

13. $\sqrt{\dfrac{1}{9}}$ _____ 14. $\sqrt{\dfrac{1}{4}}$ _____ 15. $\dfrac{1}{\sqrt{y^2}}$ _____ 16. $\dfrac{1}{\sqrt{x^2}}$ _____

17. $\dfrac{1}{\sqrt{5}}$ _____ 18. $\dfrac{1}{\sqrt{3}}$ _____ 19. $\sqrt{\dfrac{1}{5}}$ _____ 20. $\sqrt{\dfrac{1}{3}}$ _____

21. $\dfrac{1}{\sqrt{y}}$ _____ 22. $\dfrac{1}{\sqrt{x}}$ _____ 23. $\sqrt{\dfrac{1}{y}}$ _____ 24. $\sqrt{\dfrac{1}{x}}$ _____

25. $\sqrt{9x^3y^5}$ _____ 26. $\sqrt{4x^5y^3}$ _____ 27. $\sqrt{18x^8y^5}$ _____ 28. $\sqrt{8x^7y^6}$ _____

29. $\dfrac{1}{\sqrt{2x}}$ _____ 30. $\dfrac{1}{\sqrt{3y}}$ _____ 31. $\dfrac{6x^2}{\sqrt{3x}}$ _____ 32. $\dfrac{4xy}{\sqrt{2y}}$ _____

33. $\dfrac{3a}{\sqrt{2ab}}$ _____ 34. $\dfrac{2x^2y}{\sqrt{3xy}}$ _____ 35. $\sqrt{\dfrac{6x}{7y}}$ _____ 36. $\sqrt{\dfrac{3m}{2n}}$ _____

B 37. $\sqrt{\dfrac{9m^5}{2n}}$ _____ 38. $\sqrt{\dfrac{4a^3}{3b}}$ _____ 39. $\sqrt[4]{16x^8y^4}$ _____

40. $\sqrt[5]{32m^5n^{15}}$ _____ 41. $\sqrt[3]{2^4x^4y^7}$ _____ 42. $\sqrt[4]{2^4a^5b^8}$ _____

43. $\sqrt[4]{x^2}$ _____ 44. $\sqrt[10]{x^6}$ _____ 45. $\sqrt{2}\sqrt{8}$ _____

46. $\sqrt[3]{3}\,\sqrt[3]{9}$ _____ 47. $\sqrt{18m^3n^4}\,\sqrt{2m^3n^2}$ _____ 48. $\sqrt[3]{9x^2y}\,\sqrt[3]{3xy^2}$ _____

49. $\dfrac{6}{\sqrt[3]{3}}$ _____ 50. $\dfrac{2}{\sqrt[3]{2}}$ _____ 51. $\dfrac{\sqrt{4a^3}}{\sqrt{3b}}$ _____

52. $\dfrac{\sqrt{9m^5}}{\sqrt{2n}}$ _____ 53. $\sqrt{a^2+b^2}$ _____ 54. $\sqrt[3]{x^3+y^3}$ _____

55. $\sqrt[3]{\dfrac{8x^3}{27y^6}}$ _____ 56. $\sqrt[4]{\dfrac{a^8b^4}{16c^{12}}}$ _____ 57. $-m\sqrt[5]{3^6m^7n^{11}}$ _____

58. $-2x\sqrt[3]{8x^8y^{13}}$ _____ 59. $\sqrt[6]{x^4(x-y)^2}$ _____ 60. $\sqrt[8]{2^6(x+y)^6}$ _____

61. $\sqrt[3]{2x^2y^3}\,\sqrt[3]{3x^5y}$ _____ 62. $\sqrt[4]{6u^3v^4}\,\sqrt[4]{4u^5v}$ _____ 63. $\dfrac{4x^3y^2}{\sqrt[3]{2xy^2}}$ _____

64. $\dfrac{8u^3v^5}{\sqrt[3]{4u^2v^2}}$ _____   **65.** $-2x\sqrt[3]{\dfrac{3y^2}{4x}}$ _____   **66.** $6c\sqrt[3]{\dfrac{2ab}{9c^2}}$ _____

C 67. $\dfrac{x-y}{\sqrt[3]{x-y}}$ _____   **68.** $\dfrac{1}{\sqrt[3]{(x-y)^2}}$ _____   **69.** $\sqrt[4]{\dfrac{3y^3}{4x}}$ _____

70. $\sqrt[5]{\dfrac{4n^2}{16m^3}}$ _____   **71.** $-\sqrt{x^4+2x^2}$ _____   **72.** $\sqrt[4]{m^4+4m^6}$ _____

73. $\sqrt[4]{16x^4}\sqrt[3]{16x^{24}y^4}$ _____   **74.** $\sqrt[3]{8\sqrt{16x^6y^4}}$ _____

In Problems 75 to 82, simplify for (A) x a positive number and for (B) x a negative number.

75. $5\sqrt[3]{x^3}+2\sqrt{x^2}$ _____   **76.** $2\sqrt[3]{x^3}+4\sqrt{x^2}$ _____   **77.** $6\sqrt[3]{x^3}-7\sqrt{x^2}$ _____

78. $4\sqrt[3]{x^3}-6\sqrt{x^2}$ _____   **79.** $\sqrt[5]{x^5}+\sqrt[4]{x^4}$ _____   **80.** $\sqrt[6]{x^6}+\sqrt[3]{x^3}$ _____

81. $3\sqrt[4]{x^4}-2\sqrt[5]{x^5}$ _____   **82.** $5\sqrt[7]{x^7}-3\sqrt[6]{x^6}$ _____

The Check Exercise for this section is on page 313.

6.6 BASIC OPERATIONS ON RADICALS

- •Sums and differences
- ••Products
- •••Quotients—rationalizing denominators

•Sums and differences

Algebraic expressions involving radicals can often be simplified by adding and subtracting terms that contain exactly the same radical expressions. We proceed in essentially the same way as we do when we combine like terms in polynomials. You will recall that the distributive property of real numbers played a central role in this process. All variables represent positive real numbers.

EXAMPLE 16
Combine as many terms as possible.

(A) $5\sqrt{3}+4\sqrt{3} = (5+4)\sqrt{3} = 9\sqrt{3}$

(B) $2\sqrt[3]{xy^2}-7\sqrt[3]{xy^2} = (2-7)\sqrt[3]{xy^2} = -5\sqrt[3]{xy^2}$

(C) $3\sqrt{xy}-2\sqrt[3]{xy}+4\sqrt{xy}-7\sqrt[3]{xy} = 3\sqrt{xy}+4\sqrt{xy}-2\sqrt[3]{xy}-7\sqrt[3]{xy}$
$$= 7\sqrt{xy}-9\sqrt[3]{xy}$$

Work Problem 16 and check solution in Solutions to Matched Problems on page 290.

PROBLEM 16 Combine as many terms as possible.

(A) $6\sqrt{2} + 2\sqrt{2} =$

(B) $3\sqrt[5]{2x^2y^3} - 8\sqrt[5]{2x^2y^3} =$

(C) $5\sqrt[3]{mn} - 3\sqrt{mn} - 2\sqrt[3]{mn} + 7\sqrt{mn} =$

Thus we see that if two terms contain exactly the same radical—having the same index and the same radicand—they can be combined into a single term. Occasionally, terms containing radicals can be combined after they have been expressed in simplest radical form.

EXAMPLE 17

Express terms in simplest radical form and combine where possible.

(A) $4\sqrt{8} - 2\sqrt{18} = 4\sqrt{4 \cdot 2} - 2\sqrt{9 \cdot 2}$

$$= 8\sqrt{2} - 6\sqrt{2}$$

$$= 2\sqrt{2}$$

(B) $2\sqrt{12} - \dfrac{1}{\sqrt{3}} = 2 \cdot \sqrt{4} \cdot \sqrt{3} - \dfrac{1 \cdot \sqrt{3}}{\sqrt{3} \cdot \sqrt{3}}$

$$= 4\sqrt{3} - \dfrac{\sqrt{3}}{3}$$

$$= (4 - \tfrac{1}{3})\sqrt{3}$$

$$= \dfrac{11}{3}\sqrt{3} \quad \text{or} \quad \dfrac{11\sqrt{3}}{3}$$

(C) $\sqrt[3]{81} - \sqrt[3]{\dfrac{1}{9}} = \sqrt[3]{3^3 \cdot 3} - \sqrt[3]{\dfrac{3}{3^3}} = 3\sqrt[3]{3} - \tfrac{1}{3}\sqrt[3]{3}$

$$= (3 - \tfrac{1}{3})\sqrt[3]{3} = \tfrac{8}{3}\sqrt[3]{3}$$

Work Problem 17 and check solution in Solutions to Matched Problems on page 290.

PROBLEM 17 Express terms in simplest radical form and combine where possible.

(A) $\sqrt{12} - \sqrt{48} =$

(B) $3\sqrt{8} - \dfrac{1}{\sqrt{2}} =$

(C) $\sqrt[3]{\tfrac{1}{4}} - \sqrt[3]{16} =$

••Products

We will now consider several types of special products that involve radicals. The distributive property of real numbers plays a central role in our approach to these problems. In the discussion that follows, all variables represent positive real numbers.

EXAMPLE 18

Multiply and simplify:

(A) $\sqrt{2}(\sqrt{10} - 3) = \sqrt{2}\sqrt{10} - \sqrt{2} \cdot 3 = \sqrt{20} - 3\sqrt{2} = 2\sqrt{5} - 3\sqrt{2}$

(B) $(\sqrt{2} - 3)(\sqrt{2} + 5) = \sqrt{2}\sqrt{2} - 3\sqrt{2} + 5\sqrt{2} - 15$

$$= 2 + 2\sqrt{2} - 15$$
$$= 2\sqrt{2} - 13$$

(C) $(\sqrt{x} - 3)(\sqrt{x} + 5) = \sqrt{x}\sqrt{x} - 3\sqrt{x} + 5\sqrt{x} - 15$

$$= x + 2\sqrt{x} - 15$$

(D) $(\sqrt{x} - 3)^2 = (\sqrt{x})^2 - 2 \cdot 3\sqrt{x} + 3^2 = x - 6\sqrt{x} + 9$

(E) $(\sqrt{x - 3})^2 = x - 3 \qquad x \geq 3$

(F) $(\sqrt[3]{m} + \sqrt[3]{n^2})(\sqrt[3]{m^2} - \sqrt[3]{n}) = \sqrt[3]{m^3} + \sqrt[3]{m^2 n^2} - \sqrt[3]{mn} - \sqrt[3]{n^3}$

$$= m + \sqrt[3]{m^2 n^2} - \sqrt[3]{mn} - n$$

Work Problem 18 and check solution in Solutions to Matched Problems on page 290.

PROBLEM 18   Multiply and simplify.

(A) $\sqrt{3}(\sqrt{6} - 4) =$

(B) $(\sqrt{3} - 2)(\sqrt{3} + 4) =$

(C) $(\sqrt{y} - 2)(\sqrt{y} + 4) =$

(D) $(\sqrt{m} - \sqrt{n})^2 =$

(E) $(\sqrt{m - n})^2 =$

(F) $(\sqrt[3]{x^2} - \sqrt[3]{y^2})(\sqrt[3]{x} + \sqrt[3]{y}) =$

EXAMPLE 19

Show that $(2 - \sqrt{3})$ is a solution of the equation $x^2 - 4x + 1 = 0$.

Solution

$$x^2 - 4x + 1 = 0$$
$$(2 - \sqrt{3})^2 - 4(2 - \sqrt{3}) + 1 \stackrel{?}{=} 0$$
$$4 - 4\sqrt{3} + 3 - 8 + 4\sqrt{3} + 1 \stackrel{?}{=} 0$$
$$0 \stackrel{\checkmark}{=} 0$$

Work Problem 19 and check solution in Solutions to Matched Problems on page 290.

PROBLEM 19   Show that $(2 + \sqrt{3})$ is a solution of $x^2 - 4x + 1 = 0$.

Solution

•••Quotients—rationalizing denominators

Recall that to express $\sqrt{2}/\sqrt{3}$ in simplest radical form, we multiplied the numerator and denominator by $\sqrt{3}$ to clear the denominator of the the radical:

$$\frac{\sqrt{2}}{\sqrt{3}} = \frac{\sqrt{2} \cdot \sqrt{3}}{\sqrt{3} \cdot \sqrt{3}} = \frac{\sqrt{6}}{3}$$

The denominator is thus converted to a rational number. Also recall that the process of converting irrational denominators to rational forms is called **rationalizing the denominator.**
How can we rationalize the binomial denominator in

$$\frac{1}{\sqrt{3} - \sqrt{2}}$$

Multiplying the numerator and denominator by $\sqrt{3}$ or $\sqrt{2}$ does not help. Try it! Recalling the special product

$$(a - b)(a + b) = a^2 - b^2$$

suggests that we multiply the numerator and denominator by the denominator, only with the middle sign changed. Thus,

$$\frac{1}{\sqrt{3} - \sqrt{2}} = \frac{1(\sqrt{3} + \sqrt{2})}{(\sqrt{3} - \sqrt{2})(\sqrt{3} + \sqrt{2})} = \frac{\sqrt{3} + \sqrt{2}}{3 - 2} = \sqrt{3} + \sqrt{2}$$

EXAMPLE 20
Rationalize denominators and simplify:

(A) $\dfrac{\sqrt{2}}{\sqrt{6} - 2} = \dfrac{\sqrt{2}(\sqrt{6} + 2)}{(\sqrt{6} - 2)(\sqrt{6} + 2)} = \dfrac{\sqrt{12} + 2\sqrt{2}}{6 - 4}$

$\qquad = \dfrac{2\sqrt{3} + 2\sqrt{2}}{2} = \dfrac{2(\sqrt{3} + \sqrt{2})}{2} = \sqrt{3} + \sqrt{2}$

(B) $\dfrac{\sqrt{x} - \sqrt{y}}{\sqrt{x} + \sqrt{y}} = \dfrac{(\sqrt{x} - \sqrt{y})(\sqrt{x} - \sqrt{y})}{(\sqrt{x} + \sqrt{y})(\sqrt{x} - \sqrt{y})} = \dfrac{x - 2\sqrt{xy} + y}{x - y}$

Work Problem 20 and check solution in Solutions to Matched Problems on page 290.

PROBLEM 20   Rationalize denominators and simplify:

(A) $\dfrac{\sqrt{2}}{\sqrt{2}+3} =$

(B) $\dfrac{\sqrt{x}+\sqrt{y}}{\sqrt{x}-\sqrt{y}} =$

SOLUTIONS TO MATCHED PROBLEMS

16. (A) $6\sqrt{2}+2\sqrt{2}=8\sqrt{2}$ (B) $3\sqrt[6]{2x^2y^3}-8\sqrt[6]{2x^2y^3}=-5\sqrt[6]{2x^2y^3}$
(C) $5\sqrt[3]{mn}-3\sqrt{mn}-2\sqrt[3]{mn}+7\sqrt{mn}=3\sqrt[3]{mn}+4\sqrt{mn}$
17. (A) $\sqrt{12}-\sqrt{48}=\sqrt{4\cdot3}-\sqrt{16\cdot3}=2\sqrt{3}-4\sqrt{3}=-2\sqrt{3}$
(B) $3\sqrt{8}-\dfrac{1}{\sqrt{2}}=3\sqrt{4\cdot2}-\dfrac{1\sqrt{2}}{\sqrt{2}\sqrt{2}}=6\sqrt{2}-\dfrac{\sqrt{2}}{2}=(6-\frac12)(\sqrt{2})=\frac{11}{2}\sqrt{2}$ or $\dfrac{11\sqrt{2}}{2}$
(C) $\sqrt[3]{\frac14}-\sqrt[3]{16}=\sqrt[3]{\frac{1\cdot2}{4\cdot2}}-\sqrt[3]{8\cdot2}=\dfrac{\sqrt[3]{2}}{\sqrt[3]{8}}-\sqrt[3]{8}\sqrt[3]{2}=\dfrac{\sqrt[3]{2}}{2}-2\sqrt[3]{2}=(\frac12-2)(\sqrt[3]{2})=-\frac32\sqrt[3]{2}$
18. (A) $\sqrt{3}(\sqrt{6}-4)=\sqrt{18}-4\sqrt{3}=\sqrt{9\cdot2}-4\sqrt{3}=3\sqrt{2}-4\sqrt{3}$
(B) $(\sqrt{3}-2)(\sqrt{3}+4)=3+2\sqrt{3}-8=-5+2\sqrt{3}$ (C) $(\sqrt{y}-2)(\sqrt{y}+4)=y+2\sqrt{y}-8$
(D) $(\sqrt{m}-\sqrt{n})^2=(\sqrt{m})^2-2\sqrt{m}\sqrt{n}+(\sqrt{n})^2=m-2\sqrt{mn}+n$
(E) $(\sqrt{m-n})^2=m-n, m\ge n$ (F) $\sqrt[3]{x^2}-\sqrt[3]{y^2})(\sqrt[3]{x}+\sqrt[3]{y})=x+\sqrt[3]{x^2y}-\sqrt[3]{xy^2}-y$
19. $(2+\sqrt{3})^2-4(2+\sqrt{3})+1=4+4\sqrt{3}+3-8-4\sqrt{3}+1=0$
20. (A) $\dfrac{\sqrt{2}}{\sqrt{2}+3}=\dfrac{\sqrt{2}(\sqrt{2}-3)}{(\sqrt{2}+3)(\sqrt{2}-3)}=\dfrac{2-3\sqrt{2}}{2-9}=\dfrac{2-3\sqrt{2}}{-7}$ or $\dfrac{3\sqrt{2}-2}{7}$
(B) $\dfrac{\sqrt{x}+\sqrt{y}}{\sqrt{x}-\sqrt{y}}=\dfrac{(\sqrt{x}+\sqrt{y})(\sqrt{x}+\sqrt{y})}{(\sqrt{x}-\sqrt{y})(\sqrt{x}+\sqrt{y})}=\dfrac{x+2\sqrt{xy}+y}{x-y}$

PRACTICE EXERCISE 6.6

Work odd-numbered problems first, check answers, and then work even-numbered problems in areas of weakness. Answers to all problems are in the back of the book. Make every effort to work a problem yourself before you look at an answer.

A *Combine where possible.*

1. $7\sqrt{3}+2\sqrt{3}$ _____
2. $5\sqrt{2}+3\sqrt{2}$ _____
3. $2\sqrt{a}-7\sqrt{a}$ _____
4. $\sqrt{y}-4\sqrt{y}$ _____
5. $\sqrt{n}-4\sqrt{n}-2\sqrt{n}$ _____
6. $2\sqrt{x}-\sqrt{x}+3\sqrt{x}$ _____
7. $\sqrt{5}-2\sqrt{3}+3\sqrt{5}$ _____
8. $3\sqrt{2}-2\sqrt{3}-\sqrt{2}$ _____
9. $\sqrt{m}-\sqrt{n}-2\sqrt{n}$ _____
10. $2\sqrt{x}-\sqrt{y}+3\sqrt{y}$ _____

11. $\sqrt{18} + \sqrt{2}$ _____   **12.** $\sqrt{8} - \sqrt{2}$ _____

13. $\sqrt{8} - 2\sqrt{32}$ _____   **14.** $\sqrt{27} - 3\sqrt{12}$ _____

Multiply and simplify where possible.

15. $\sqrt{7}(\sqrt{7} - 2)$ _____   **16.** $\sqrt{5}(\sqrt{5} - 2)$ _____   **17.** $\sqrt{2}(3 - \sqrt{2})$ _____

18. $\sqrt{3}(2 - \sqrt{3})$ _____   **19.** $\sqrt{y}(\sqrt{y} - 8)$ _____   **20.** $\sqrt{x}(\sqrt{x} - 3)$ _____

21. $\sqrt{n}(4 - \sqrt{n})$ _____   **22.** $\sqrt{m}(3 - \sqrt{m})$ _____   **23.** $\sqrt{3}(\sqrt{3} + \sqrt{6})$ _____

24. $\sqrt{5}(\sqrt{10} + \sqrt{5})$ _____   **25.** $(2 - \sqrt{3})(3 + \sqrt{3})$ _____

26. $(\sqrt{2} - 1)(\sqrt{2} + 3)$ _____   **27.** $(\sqrt{5} + 2)^2$ _____

28. $(\sqrt{3} - 3)^2$ _____   **29.** $(\sqrt{m} - 3)(\sqrt{m} - 4)$ _____

30. $(\sqrt{x} + 2)(\sqrt{x} - 3)$ _____

Rationalize denominators and simplify.

31. $\dfrac{1}{\sqrt{5} + 2}$ _____   **32.** $\dfrac{1}{\sqrt{11} - 3}$ _____   **33.** $\dfrac{2}{\sqrt{5} + 1}$ _____

34. $\dfrac{4}{\sqrt{6} - 2}$ _____   **35.** $\dfrac{\sqrt{2}}{\sqrt{10} - 2}$ _____   **36.** $\dfrac{\sqrt{2}}{\sqrt{6} + 2}$ _____

37. $\dfrac{\sqrt{y}}{\sqrt{y} + 3}$ _____   **38.** $\dfrac{\sqrt{x}}{\sqrt{x} - 2}$ _____

B *Express in simplest radical form and combine where possible.*

39. $\sqrt{8mn} + 2\sqrt{18mn}$ _____   **40.** $\sqrt{4x} - \sqrt{9x}$ _____

41. $\sqrt{8} - \sqrt{20} + 4\sqrt{2}$ _____   **42.** $\sqrt{24} - \sqrt{12} + 3\sqrt{3}$ _____

43. $\sqrt[5]{a} - 4\sqrt[5]{a} + 2\sqrt[5]{a}$ _____   **44.** $3\sqrt[3]{u} - 2\sqrt[3]{u} - 2\sqrt[3]{u}$ _____

45. $2\sqrt[3]{x} + 3\sqrt[3]{x} - \sqrt{x}$ _____   **46.** $5\sqrt[5]{y} - 2\sqrt[5]{y} + 3\sqrt[4]{y}$ _____

47. $\sqrt{\frac{1}{8}} + \sqrt{8}$ _____   **48.** $\sqrt{\frac{2}{3}} - \sqrt{\frac{3}{2}}$ _____

49. $\sqrt{\dfrac{3uv}{2}} - \sqrt{24uv}$ _____   **50.** $\sqrt{\dfrac{xy}{2}} + \sqrt{8xy}$ _____

In Problems 51 to 58, multiply and simplify where possible.

51. $(4\sqrt{3} - 1)(3\sqrt{3} - 2)$ _____   **52.** $(2\sqrt{7} - \sqrt{3})(2\sqrt{7} + \sqrt{3})$ _____

53. $(\sqrt{x} - \sqrt{y})(\sqrt{x} + \sqrt{y})$ _____   **54.** $(2\sqrt{x} + 3)(2\sqrt{x} - 3)$ _____

55. $(5\sqrt{m} + 2)(2\sqrt{m} - 3)$ _____   **56.** $(3\sqrt{u} - 2)(2\sqrt{u} + 4)$ _____

57. $(\sqrt[3]{4} + \sqrt[3]{9})(\sqrt[3]{2} + \sqrt[3]{3})$ _____

58. $\sqrt[3]{4}(\sqrt[3]{2} - \sqrt[3]{16})$ _____

59. Show that $2 - \sqrt{3}$ is a solution to $x^2 - 4x + 1 = 0$. _____

60. Show that $2 + \sqrt{3}$ is a solution to $x^2 - 4x + 1 = 0$. _____

Rationalize denominators and simplify.

61. $\dfrac{\sqrt{3} + 2}{\sqrt{3} - 2}$ _____

62. $\dfrac{\sqrt{2} - 1}{\sqrt{2} + 1}$ _____

63. $\dfrac{\sqrt{2} + \sqrt{3}}{\sqrt{3} - \sqrt{2}}$ _____

64. $\dfrac{3 - \sqrt{a}}{\sqrt{a} - 2}$ _____

65. $\dfrac{2 + \sqrt{x}}{\sqrt{x} - 3}$ _____

66. $\dfrac{\sqrt{5} - \sqrt{2}}{\sqrt{5} + \sqrt{2}}$ _____

67. $\dfrac{3\sqrt{x}}{2\sqrt{x} - 3}$ _____

68. $\dfrac{5\sqrt{a}}{3 - 2\sqrt{a}}$ _____

C *Express in simplest radical form and combine where possible.*

69. $\dfrac{\sqrt{3}}{3} + 2\sqrt{\dfrac{1}{3}} + \sqrt{12}$ _____

70. $\sqrt{\dfrac{1}{2}} + \dfrac{\sqrt{2}}{2} + \sqrt{8}$ _____

71. $\sqrt[3]{\dfrac{1}{3}} + \sqrt[3]{3^5}$ _____

72. $\sqrt[4]{32} - \sqrt[4]{\dfrac{1}{8}}$ _____

Multiply and simplify where possible.

73. $(\sqrt[3]{x} - \sqrt[3]{y^2})(\sqrt[3]{x^2} + 2\sqrt[3]{y})$ _____

74. $(\sqrt[4]{u^2} - \sqrt[5]{v^3})(\sqrt[5]{u^3} + \sqrt[5]{v^2})$ _____

Rationalize denominators and simplify.

75. $\dfrac{2\sqrt{x} - 3\sqrt{y}}{4\sqrt{x} + 5\sqrt{y}}$ _____

76. $\dfrac{3\sqrt{x} + 2\sqrt{y}}{2\sqrt{x} - 5\sqrt{y}}$ _____

The Check Exercise for this section is on page 315.

6.7 COMPLEX NUMBERS

- •Introductory remarks
- ••The complex number system
- •••Complex numbers and radicals
- ••••Concluding remarks

•Introductory remarks

The Pythagoreans (500–275 B.C.) found that the simple equation

$$x^2 = 2 \tag{1}$$

had no rational number solutions. If (1) were to have a solution, then a new kind of number had to be invented—the irrational numbers. The irrational numbers $\sqrt{2}$ and $-\sqrt{2}$ are both solutions to (1). Irrational numbers were not put on a firm mathematical foundation until the last century. The rational and irrational numbers together constitute the real number system (see Chapter 1).

Is there any need to extend the real number system further? Yes, since we find that another simple equation

$$x^2 = -1$$

has no real solutions (what number squared is negative?). Once again, we are forced to invent a new kind of number, a number that has the possibility of being negative when it is squared. These new numbers are called **complex numbers.** The complex numbers evolved over a long period of time, dating back to Cardono (1545). But, like the real numbers, it was not until the last century that they were finally put on a firm mathematical foundation.

••The complex number system

A **complex number** is any number of the form

$$\boxed{a + bi}$$

where a and b are real numbers; i is called the **imaginary unit.** Thus

$$5 + 2i \qquad \tfrac{1}{4} + 2i \qquad \sqrt{2} - \tfrac{1}{3}i \qquad\qquad 0 + 5i \qquad 6 + 0i \qquad 0 + 0i$$

are all complex numbers. Particular kinds of complex numbers are given special names:

$a + 0i = a$	real number
$0 + bi = bi$	pure imaginary number
$0 + 0i = 0$	zero
$1i = i$	imaginary unit
$a - bi$	conjugate of $a + bi$

Thus, we see that just as every integer is a rational number, every real number is a complex number; that is, the real numbers form a subset of the set of complex numbers.

To use complex numbers we must know how to add, subtract, multiply, and divide them. We start by defining equality, addition, and multiplication.

EQUALITY	$a + bi = c + di$ if and only if $a = c$ and $b = d$	
ADDITION	$(a + bi) + (c + di) = (a + c) + (b + d)i$	
MULTIPLICATION	$(a + bi)(c + di) = (ac - bd) + (ad + bc)i$	

These definitions, particularly the one for multiplication, may seem a little strange to you. But it turns out that if we want the real number properties discussed in Chapter 1 to continue to hold for complex numbers, and if we also want the possibility of having the square of a number negative, then we must define addition and multiplication as above. Let us use the definition of multiplication to see what happens to i when it is squared:

$$i^2 = \overset{a}{(0} + \overset{b}{1i)}\overset{c}{(0} + \overset{d}{1i)}$$

$$= \overset{a\ c}{(0 \cdot 0} - \overset{b\ d}{1 \cdot 1)} + \overset{a\ d}{(0 \cdot 1} + \overset{b\ c}{1 \cdot 0)}i$$

$$= -1 + 0i$$

$$= -1$$

Thus,

$$\boxed{i^2 = -1}$$

and we have a number whose square is negative (and a solution to $x^2 = -1$). We choose to let

$$\boxed{i = \sqrt{-1} \quad \text{and} \quad -i = -\sqrt{-1}}$$

Fortunately, you do not have to memorize the definitions of addition and multiplication above. We can show that the complex numbers under these definitions are associative and commutative, and that multiplication distributes over addition. As a consequence, we can manipulate complex numbers as if they were binomial forms in real number algebra, with the exception that i^2 is to be replaced with -1. The following example illustrates the mechanics of carrying out addition, subtraction, multiplication, and division.

EXAMPLE 21

Carry out the indicated operations and write each answer in the form $a + bi$.

(A) $(3 + 2i) + (2 - i)$

Solution

$(3 + 2i) + (2 - i) = 3 + 2i + 2 - i$ Remove parentheses
and combine like terms.

$\qquad\qquad\qquad = 5 + i$

(B) $(3 + 2i) - (2 - i)$

Solution

$$(3 + 2i) - (2 - i) = 3 + 2i - 2 + i \qquad \text{Remove parentheses and combine like terms.}$$

$$= 1 + 3i$$

(C) $(3 + 2i)(2 - i)$

Solution

$$(3 + 2i)(2 - i) = 6 + i - 2i^2 \qquad \text{Multiply and replace } i^2 \text{ with } -1.$$

$$= 6 + i - 2(-1)$$

$$= 6 + i + 2$$

$$= 8 + i$$

(D) $\dfrac{3 + 2i}{2 - i}$

Solution

In order to eliminate i from the denominator, we multiply the numerator and denominator by the conjugate of $2 - i$, that is, by $2 + i$:

$$\frac{3 + 2i}{2 - i} \cdot \frac{2 + i}{2 + i} = \frac{6 + 7i + 2i^2}{4 - i^2} = \frac{6 + 7i + 2(-1)}{4 - (-1)}$$

$$= \frac{4 + 7i}{5} = \frac{4}{5} + \frac{7}{5}i$$

Work Problem 21 and check solution in Solutions to Matched Problems on page 299.

PROBLEM 21   Carry out the indicated operations and write each answer in the form $a + bi$.

(A) $(3 + 2i) + (6 - 4i) =$

(B) $(3 - 5i) - (1 - 3i) =$

(C) $(2 - 4i)(3 + 2i) =$

(D) $\dfrac{2 + 4i}{3 + 2i} =$

EXAMPLE 22

Carry out the indicated operations and write each answer in the form $a + bi$.

(A) $(2 - 3i)^2 - (4i)^2$

Solution

$$(2 - 3i)^2 - (4i)^2 = 4 - 12i + 9i^2 - 16i^2$$

$$= 4 - 12i + 9(-1) - 16(-1)$$

$$= 4 - 12i - 9 + 16$$

$$= 11 - 12i$$

(B) $\dfrac{2 + i}{3i}$

Solution

$$\frac{2 + i}{3i} = \frac{2 + i}{3i} \cdot \frac{i}{i}$$

$$= \frac{2i + i^2}{3i^2}$$

$$= \frac{2i + (-1)}{3(-1)}$$

$$= \frac{2i - 1}{-3}$$

$$= \frac{-1}{-3} + \frac{2}{-3}i$$

$$= \frac{1}{3} - \frac{2}{3}i$$

Work Problem 22 and check solution in Solutions to Matched Problems on page 299.

PROBLEM 22 Carry out the indicated operations and write each answer in the form $a + bi$.

(A) $(3i)^2 - (3 - 2i)^2 =$

(B) $\dfrac{3 + i}{2i} =$

•••Complex numbers and radicals

Recall that we say y is a square root of x if $y^2 = x$. It can be shown that if x is a positive real number, then x has two real square roots, one the negative of the other; if x is negative, then x has two complex square roots, one also the negative of the other. In particular, if we let $x = -a$, $a > 0$, then one of the square roots of x is given by*

---

*Note that if in $a + bi$, $b = \sqrt{k}$, then we often write $a + i\sqrt{k}$ instead of $a + \sqrt{k}i$ so that i will not accidentally end up under the radical sign.

$$\sqrt{-a} = i\sqrt{a} \qquad a > 0 \qquad \sqrt{-9} = i\sqrt{9} = 3i$$

To check this, we square $i\sqrt{a}$ and obtain $-a$:

$$(i\sqrt{a})^2 = i^2(\sqrt{a})^2 = (-1)a = -a$$

EXAMPLE 23
Write in the form $a + bi$:

(A) $\sqrt{-4}$

Solution

$$\sqrt{-4} = i\sqrt{4} = 2i \qquad \text{or} \qquad 0 + 2i$$

(B) $4 + \sqrt{-40}$

Solution

$$4 + i\sqrt{40} = 4 + i\sqrt{4\cdot 10} = 4 + 2i\sqrt{10}$$

(C) $\dfrac{-3 - \sqrt{-7}}{2}$

Solution

$$\frac{-3 - \sqrt{-7}}{2} = \frac{-3 - i\sqrt{7}}{2} = -\frac{3}{2} - \frac{\sqrt{7}}{2}i$$

Work Problem 23 and check solution in Solutions to Matched Problems on page 299.

PROBLEM 23   Write in the form $a + bi$.

    (A) $\sqrt{-16} =$

    (B) $5 - \sqrt{-36} =$

    (C) $\dfrac{-5 - \sqrt{-2}}{2} =$

EXAMPLE 24
Convert square roots of negative numbers to complex form, perform the indicated operations, and express answers in $a + bi$ form.

(A) $(3 + \sqrt{-4})(2 - \sqrt{-9})$

Solution

$$(3 + \sqrt{-4})(2 - \sqrt{-9}) = (3 + i\sqrt{4})(2 - i\sqrt{9})$$

Note that

$$= (3 + 2i)(2 - 3i)$$

$$\sqrt{-4}\sqrt{-9} \neq \sqrt{(-4)(-9)}$$

$$= 6 - 5i - 6i^2$$

since
$$\sqrt{-4}\sqrt{-9} = (2i)(3i) = 6i^2$$
$$= -6$$

$$= 6 - 5i - 6(-1)$$

while
$$\sqrt{(-4)(-9)} = \sqrt{36} = 6$$

$$= 6 - 5i + 6$$

$$= 12 - 5i$$

(B) $\dfrac{1}{3 - \sqrt{-4}}$

Solution

$$\frac{1}{3 - \sqrt{-4}} = \frac{1}{3 - i\sqrt{4}}$$

$$= \frac{1}{3 - 2i}$$

$$= \frac{1}{3 - 2i} \cdot \frac{3 + 2i}{3 + 2i}$$

$$= \frac{3 + 2i}{9 - 4i^2}$$

$$= \frac{3 + 2i}{9 - 4(-1)}$$

$$= \frac{3 + 2i}{9 + 4}$$

$$= \frac{3 + 2i}{13} = \frac{3}{13} + \frac{2}{13}i$$

Work Problem 24 and check solution in Solutions to Matched Problems on page 299.

PROBLEM 24 Convert square roots of negative numbers to complex form, perform the indicated operations, and express answers in the form $a + bi$.

(A) $(4 - \sqrt{-25})(3 + \sqrt{-49}) =$

(B) $\dfrac{1}{2 + \sqrt{-9}} =$

••••Concluding remarks

You have just been introduced to a number system that is used extensively by electrical, aeronautical, and space scientists, as well as chemists and physicists. The interpretation of complex numbers $a + bi$ relative to the real world is not readily seen until you have had more experience in some of the above fields. We state only that a and b in the complex number $a + bi$ often represent real-world quantities. Our use of the complex numbers will be in connection with solutions to second-degree equations such as

$$x^2 - 4x + 5 = 0$$

which we will study in the next chapter.

SOLUTIONS TO MATCHED PROBLEMS _____

21. **(A)** $(3 + 2i) + (6 - 4i) = 3 + 2i + 6 - 4i = 9 - 2i$
(B) $(3 - 5i) - (1 - 3i) = 3 - 5i - 1 + 3i = 2 - 2i$
(C) $(2 - 4i)(3 + 2i) = 6 - 8i - 8i^2 = 6 - 8i + 8 = 14 - 8i$
(D) $\dfrac{2 + 4i}{3 + 2i} = \dfrac{(2 + 4i)}{(3 + 2i)} \cdot \dfrac{(3 - 2i)}{(3 - 2i)} = \dfrac{6 + 8i - 8i^2}{9 - 4i^2} = \dfrac{6 + 8i + 8}{9 + 4} = \dfrac{14 + 8i}{13} = \dfrac{14}{13} + \dfrac{8}{13}i$

22. **(A)** $(3i)^2 - (3 - 2i)^2 = 9i^2 - (9 - 12i + 4i^2) = 9i^2 - 9 + 12i - 4i^2 = -9 + 12i + 5i^2$
$= -9 + 12i - 5 = -14 + 12i$
(B) $\dfrac{3 + i}{2i} = \dfrac{3 + i}{2i} \cdot \dfrac{i}{i} = \dfrac{3i + i^2}{2i^2} = \dfrac{3i - 1}{-2} = \dfrac{-1}{-2} + \dfrac{3}{-2}i = \dfrac{1}{2} - \dfrac{3}{2}i$

23. **(A)** $\sqrt{-16} = i\sqrt{16} = 4i$ or $0 + 4i$ **(B)** $5 - \sqrt{-36} = 5 - i\sqrt{36} = 5 - 6i$
(C) $\dfrac{-5 - \sqrt{-2}}{2} = \dfrac{-5 - i\sqrt{2}}{2} = \dfrac{-5}{2} - \dfrac{\sqrt{2}}{2}i$ or $-\dfrac{5}{2} - \dfrac{\sqrt{2}}{2}i$

24. **(A)** $(4 - \sqrt{-25})(3 + \sqrt{-49}) = (4 - i\sqrt{25})(3 + i\sqrt{49}) = (4 - 5i)(3 + 7i) = 12 + 13i - 35i^2$
$= 12 + 13i + 35 = 47 + 13i$
(B) $\dfrac{1}{2 + \sqrt{-9}} = \dfrac{1}{2 + i\sqrt{9}} = \dfrac{1}{2 + 3i} = \dfrac{1}{(2 + 3i)} \cdot \dfrac{(2 - 3i)}{(2 - 3i)} = \dfrac{2 - 3i}{4 - 9i^2} = \dfrac{2 - 3i}{4 + 9} = \dfrac{2 - 3i}{13} = \dfrac{2}{13} - \dfrac{3}{13}i$

PRACTICE EXERCISE 6.7 _____

Work odd-numbered problems first, check answers, and then work even-numbered problems in areas of weakness. Answers to all problems are in the back of the book. Make every effort to work a problem yourself before you look at an answer.

A *Perform the indicated operations and write each answer in the form $a + bi$.*

1. $(5 + 2i) + (3 + i)$ _____

2. $(6 + i) + (2 + 3i)$ _____

3. $(-8 + 5i)(3 - 2i)$ _____

4. $(2 - 3i) + (5 - 2i)$ _____

5. $(8 + 5i) - (3 + 2i)$ _____

6. $(9 + 7i) - (2 + 5i)$ _____

7. $(4 + 7i) - (-2 - 6i)$ _____

8. $(9 - 3i) - (12 - 5i)$ _____

9. $(3 - 7i) + 5i$ _____

10. $12 + (5 - 2i)$ _____

11. $(5i)(3i)$ _____

12. $(2i)(4i)$ _____

13. $-2i(5 - 3i)$ _____

14. $-3i(2 - 4i)$ _____

15. $(2 - 3i)(3 + 3i)$ _____

16. $(3 - 5i)(-2 - 3i)$ _____

17. $(7 - 6i)(2 - 3i)$ _____

18. $(2 - i)(3 + 2i)$ _____

19. $(7 + 4i)(7 - 4i)$ _____

20. $(5 - 3i)(5 + 3i)$ _____

21. $\dfrac{1}{2 + i}$ _____

22. $\dfrac{1}{3 - i}$ _____

23. $\dfrac{3 + i}{2 - 3i}$ _____

24. $\dfrac{2 - i}{3 + 2i}$ _____

25. $\dfrac{13 + i}{2 - i}$ _____

26. $\dfrac{15 - 3i}{2 - 3i}$ _____

B *Convert square roots of negative numbers to complex form, perform the indicated operations, and express answers in the form a + bi.*

27. $(5 - \sqrt{-9}) + (2 - \sqrt{-4})$ _____

28. $(-8 + \sqrt{-25}) + (3 - \sqrt{-4})$ _____

29. $(9 - \sqrt{-9}) - (12 - \sqrt{-25})$ _____

30. $(4 + \sqrt{-49}) - (-2 - \sqrt{-36})$ _____

31. $(-2 + \sqrt{-49})(3 - \sqrt{-4})$ _____

32. $(5 + \sqrt{-9})(2 - \sqrt{-1})$ _____

33. $\dfrac{5 - \sqrt{-4}}{3}$ _____

34. $\dfrac{6 - \sqrt{-64}}{2}$ _____

35. $\dfrac{1}{2 - \sqrt{-9}}$ _____

36. $\dfrac{1}{3 - \sqrt{-16}}$ _____

37. $\dfrac{2}{5i}$ _____

38. $\dfrac{1}{3i}$ _____

39. $\dfrac{1 + 3i}{2i}$ _____

40. $\dfrac{2 - i}{3i}$ _____

41. $(2 - i)^2 + 3(2 - i) - 5$ _____

42. $(2 - 3i)^2 - 2(2 - 3i) + 9$ _____

43. Evaluate $x^2 - 2x + 2$ for $x = 1 - i$. _____

44. Evaluate $x^2 - 2x + 2$ for $x = 1 + i$. _____

45. Evaluate $x^2 - 4x + 5$ for $x = 2 + i$. _____

46. Evaluate $x^2 - 4x + 5$ for $x = 2 - i$. _____

C 47. For what values of x will $\sqrt{x - 10}$ be real? _____

48. When will $\dfrac{-b \pm \sqrt{b^2 - 4ac}}{2a}$ represent a complex number, assuming a, b, and c are all real numbers $(a \neq 0)$?

_____

The Check Exercise for this section is on page 317.

6.8 CHAPTER REVIEW

Important terms and symbols

6.1 INTEGER EXPONENTS positive-integer exponents, zero exponents, negative-integer exponents, exponent properties a^n, a^0, a^{-n}

6.2 SCIENTIFIC NOTATION $23{,}500{,}000 = 2.35 \times 10^7$, $0.000\,000\,001\,35 = 1.35 \times 10^{-9}$

6.3 RATIONAL EXPONENTS real roots of real numbers, rational exponents, properties of rational exponents, $b^{1/n}$, $b^{m/n}$

6.4 RADICAL FORMS AND RATIONAL EXPONENTS $\sqrt[n]{b} = b^{1/n}$, $b^{m/n} = \sqrt[n]{b^m}$

6.5 CHANGING RADICAL EXPRESSIONS properties of radicals, simplest radical form, simplifying $\sqrt[n]{x^n}$ for all real x, $\sqrt{x^2} = |x|$ for all real x, $\sqrt[3]{x^3} = x$ for all real x

6.6 BASIC OPERATIONS ON RADICALS sums and differences, special products, special quotients—rationalizing denominators

6.7 COMPLEX NUMBERS complex number system, imaginary unit, pure imaginary number, conjugate, basic operations, complex numbers and radicals, $a + bi$, i, bi, $i^2 = -1$, $i = \sqrt{-1}$, $\sqrt{-a} = i\sqrt{a}\,(a > 0)$

DIAGNOSTIC (REVIEW) EXERCISE 6.8

Work through all the problems in this chapter review and check answers in the back of the book. (Answers to all problems are there, and following each answer is a number in italics indicating the section in which that type of problem is discussed.) Where weaknesses show up, review appropriate sections in the text. When you are satisfied that you know the material, take the practice test following this review.

Unless otherwise stated, all variables represent positive real numbers.

A *Evaluate Problems 1 to 6, if possible.*

1. $\left(\dfrac{1}{3}\right)^0$ _____

2. 3^{-2} _____

3. $\dfrac{1}{2^{-3}}$ _____

4. $4^{-1/2}$ _____

5. $(-9)^{3/2}$ _____

6. $(-8)^{2/3}$ _____

7. Write in scientific notation: (**A**) 4,280,000,000 (**B**) 0.000 031 8

_____

8. Write as a decimal fraction: (**A**) 7.29×10^5 (**B**) 6.03×10^{-4}

_____

Simplify Problems 9 to 20, and write answers using positive exponents only.

9. $(3x^3y^2)(2xy^5)$ _____

10. $\dfrac{9u^8v^6}{3u^4v^8}$ _____

11. $6(xy^3)^5$ _____

12. $\left(\dfrac{c^2}{d^5}\right)^3$ _____

13. $\left(\dfrac{2x^2}{3y^3}\right)^2$ _____

14. $(x^{-3})^{-4}$ _____

15. $\dfrac{y^{-3}}{y^{-5}}$ _____

16. $(x^2y^{-3})^{-1}$ _____

17. $(x^9)^{1/3}$ _____

18. $(x^4)^{-1/2}$ _____

19. $x^{1/3}x^{-2/3}$ _____

20. $\dfrac{u^{5/3}}{u^{2/3}}$ _____

21. Change to radical form:

(**A**) $(3m)^{1/2}$ _____

(**B**) $3m^{1/2}$ _____

22. Change to rational exponent form:

(**A**) $\sqrt{2x}$ _____

(**B**) $\sqrt{a+b}$ _____

Simplify Problems 23 to 34, and write in simplest radical form. All variables represent positive real numbers.

23. $\sqrt{4x^2y^4}$ _____

24. $\sqrt{\dfrac{25}{y^2}}$ _____

25. $\sqrt{36x^4y^7}$ _____

26. $\dfrac{1}{\sqrt{2y}}$ _____

27. $\dfrac{6ab}{\sqrt{3a}}$ _____

28. $\sqrt{2x^2y^5}\,\sqrt{18x^3y^2}$ _____

29. $\sqrt{\dfrac{y}{2x}}$ _____

30. $4\sqrt{x}-7\sqrt{x}$ _____

31. $\sqrt{7}+2\sqrt{3}-4\sqrt{3}$ _____

32. $\sqrt{5}(\sqrt{5}+2)$ _____

33. $(\sqrt{3}-1)(\sqrt{3}+2)$ _____

34. $\dfrac{\sqrt{5}}{3-\sqrt{5}}$ _____

Perform the indicated operations in Problems 35 to 38, and write the answer in the form a + bi.

35. $(-3+2i)+(6-8i)$ _____

36. $(3-3i)(2+3i)$ _____

37. $\dfrac{13-i}{5-3i}$ _____

38. $\dfrac{2-i}{2i}$ _____

B 39. Convert each number to scientific notation, simplify, and write answer in scientific notation and as a decimal fraction:

$\dfrac{0.000\,052\,\cdot}{130(0.000\,2)}$ _____

Simplify Problems 40 to 53, and write answers using positive exponents only.

40. $\dfrac{3m^4n^{-7}}{6m^2n^{-2}}$ _____

41. $(x^{-3}y^2)^{-2}$ _____

42. $\dfrac{1}{(2x^2y^{-3})^{-2}}$ _____

43. $\left(-\dfrac{a^2b}{c}\right)^2\left(\dfrac{c}{b^2}\right)^3\left(\dfrac{1}{a^3}\right)^2$ _____

44. $\left(\dfrac{8u^{-1}}{2^2u^2v^0}\right)^{-2}\left(\dfrac{u^{-5}}{u^{-3}}\right)^3$ _____

45. $\left(\dfrac{9m^3n^{-3}}{3m^{-2}n^2}\right)^{-2}$ _____

46. $(x-y)^{-2}$ _____

47. $(9a^4b^{-2})^{1/2}$ _____

48. $\left(\dfrac{27x^2y^{-3}}{8x^{-4}y^3}\right)^{1/3}$ _____

49. $\dfrac{m^{-1/4}}{m^{3/4}}$ _____

50. $(2x^{1/2})(3x^{-1/3})$ _____

51. $\dfrac{3x^{-1/4}}{6x^{-1/3}}$ _____

52. $\dfrac{5^0}{3^2}+\dfrac{3^{-2}}{2^{-2}}$ _____

53. $(x^{1/2}+y^{1/2})^2$ _____

54. If a is a square root of b, then does $a^2=b$ or does $b^2=a$? _____

55. Change to radical form:

 (A) $(2mn)^{2/3}$ _____ **(B)** $3x^{2/5}$ _____

56. Change to rational exponent form:

 (A) $\sqrt[7]{x^5}$ _____ **(B)** $-4\sqrt[3]{(xy)^2}$ _____

Simplify Problems 57 to 67, and write in simplest radical form. All variables represent positive real numbers.

57. $\sqrt[3]{(2x^2y)^3}$ _____

58. $3x\sqrt[3]{x^5y^4}$ _____

59. $\dfrac{\sqrt{8m^3n^4}}{\sqrt{12m^2}}$ _____

60. $\sqrt[8]{y^6}$ _____

61. $-2x\sqrt[5]{3^6x^7y^{11}}$ _____

62. $\dfrac{2x^2}{\sqrt[3]{4x}}$ _____

63. $\sqrt[5]{\dfrac{3y^2}{8x^2}}$ _____

64. $(2\sqrt{x}-5\sqrt{y})(\sqrt{x}+\sqrt{y})$ _____

65. $\dfrac{\sqrt{x}-2}{\sqrt{x}+2}$ _____

66. $\dfrac{3\sqrt{x}}{2\sqrt{x}-\sqrt{y}}$ _____

67. $\sqrt{\tfrac{2}{3}}+\sqrt{\tfrac{3}{2}}$ _____

Perform the indicated operations in Problems 68 to 70 and write answers in the form $a+bi$.

68. $(2-2\sqrt{-4})-(3-\sqrt{-9})$ _____

69. $\dfrac{2-\sqrt{-1}}{3+\sqrt{-4}}$ _____

70. $(3+i)^2-2(3+i)+3$ _____

Simplify Problems 71 and 72, and write answers using positive exponents only.

71. $(x^{-1}+y^{-1})^{-1}$ _____

72. $\left(\dfrac{a^{-2}}{b^{-1}}+\dfrac{b^{-2}}{a^{-1}}\right)^{-1}$ _____

C *Simplify Problems 73 and 74, and write in simplest radical form. All variables represent positive real numbers.*

73. $\sqrt[9]{8x^6y^{12}}$ _____

74. $\sqrt[3]{3}-\dfrac{6}{\sqrt[3]{9}}+3\sqrt[3]{\dfrac{1}{9}}$ _____

75. Simplify $3\sqrt[3]{x^3}-2\sqrt{x^2}$

 (A) For x a positive number _____ **(B)** For x a negative number _____

PRACTICE TEST CHAPTER 6

Take this as if it were a graded test. Allow yourself up to 50 minutes. Work the problems without looking back in the chapter. Correct your work, using the answers (keyed to appropriate sections) in the back of the book.

Simplify Problems 1 to 9, and write answers using positive exponents only. All variables represent positive real numbers.

1. $\dfrac{3u^2v^5}{9u^3v^2}$ _____

2. $\left(\dfrac{2y^{-2}}{3x^{-3}}\right)^2$ _____

3. $(-8)^{-2/3}$ _____

4. $\dfrac{4x^{-2}y^3z^0}{8x^2y^{-2}}$ _____

5. $(4x^{-2}y^8)^{-1/2}$ _____

6. $\left(\dfrac{8m^6n^{-3}}{m^{-9}}\right)^{1/3}$ _____

7. $\left(\dfrac{x^{-1/2}y^{1/3}}{x^{-1/3}y^{1/2}}\right)^{-6}$ _____

8. $\dfrac{(3m^{1/3})(8m^{-1/2})}{12m^{1/6}}$ _____

9. $(x^{-1/2} + y^{-1/2})^2$ _____

10. Convert each number to scientific notation, simplify, and write answer in scientific notation:

$\dfrac{(0.000\ 075)(4,000)}{(2,000,000)(0.05)}$ _____

11. Change to rational exponent form: $4\sqrt[3]{(x-y)^2}$ _____

Perform any indicated operations in Problems 12 to 18, and express answers in simplest radical form.

12. $3x\sqrt[5]{2^6x^8y^{12}}$ _____

13. $\sqrt{\dfrac{3y^2}{2x}}$ _____

14. $\sqrt[3]{2} - \dfrac{6}{\sqrt[3]{4}}$ _____

15. $\sqrt[6]{(2xy)^4}$ _____

16. $(3 - \sqrt{2})^2 - 6(3 - \sqrt{2}) + 7$ _____

17. $\dfrac{\sqrt{x} + \sqrt{y}}{\sqrt{x} - \sqrt{y}}$ _____

18. Simplify for x a negative number: $4\sqrt[3]{x^3} + 5\sqrt{x^2}$ _____

Perform the indicated operations in Problems 19 and 20 (after converting square roots of negative numbers to complex form), and write answers in the form $a + bi$.

19. $(3 - 2i)^2 - 2(3 - 2i) + 3$ _____

20. $\dfrac{3 - \sqrt{-4}}{2 + \sqrt{-1}}$ _____

CHECK EXERCISE 6.1

ANSWER
COLUMN

1. _____
2. _____
3. _____
4. _____
5. _____
6. _____
7. _____
8. _____
9. _____
10. _____

Work the following problems without looking at any text examples. Show your work in the space provided. Write the letter that best indicates your answer in the answer column.

Simplify and write answers using positive exponents only.

1. $\dfrac{x^2 x^{-5}}{x^{-2}} =$

(**A**) x

(**B**) $\dfrac{1}{x}$

(**C**) x^5

(**D**) None of these

2. $(2^3 \cdot 3^0)^{-2} =$

(**A**) $\dfrac{-3}{2^6}$

(**B**) $\dfrac{1}{2^6}$

(**C**) -4^3

(**D**) None of these

3. $(2x^{-1})^{-2} =$

(**A**) $-4x^2$

(**B**) $\dfrac{-4}{x^3}$

(**C**) $-4x^{-3}$

(**D**) None of these

4. $(3x^{-3}y)^{-2} =$

(**A**) $\dfrac{x^6}{9y^2}$

(**B**) $\dfrac{-6x^6}{y^2}$

(**C**) $\dfrac{-6x^9}{y^2}$

(**D**) None of these

5. $\dfrac{8 \times 10^{-1}}{2 \times 10^{-5}} =$

(**A**) 4×10^4

(**B**) 4×10^6

(**C**) $\dfrac{4}{10^{-4}}$

(**D**) None of these

305

6. $\dfrac{9x^3y^{-2}}{12x^{-3}y} =$

   (A) $\dfrac{3y}{4x}$

   (B) $\dfrac{3x^5y^{-3}}{4}$

   (C) $\dfrac{3x^5}{4y^3}$

   (D) None of these

7. $\left(\dfrac{3^2x^3}{9x^{-3}y^0}\right)^{-2} =$

   (A) 1

   (B) -2

   (C) $\dfrac{1}{x^{12}}$

   (D) None of these

8. $\left|\left(\dfrac{3x^2y^{-1}}{x^{-1}y}\right)^{-2}\right|^{-1} =$

   (A) $\dfrac{9x^6}{y^4}$

   (B) $\dfrac{-6x^6}{y^4}$

   (C) $\dfrac{6x^6}{y^4}$

   (D) None of these

9. $(2^{-1} + 3^{-1})^0 =$

   (A) 0

   (B) 1

   (C) $\left(\frac{5}{6}\right)^0$

   (D) None of these

10. $(a^{-2} + b^{-2})^{-1} =$

   (A) $a^2 + b^2$

   (B) $\dfrac{1}{a^2 + b^2}$

   (C) $\dfrac{a^2b^2}{a^2 + b^2}$

   (D) None of these

CHECK EXERCISE 6.2

NAME _____

CLASS _____

SCORE _____

Work the following problems without looking at any text examples. Show your work in the space provided. Write the letter that best indicates your answer in the answer column.

ANSWER COLUMN

1. _____
2. _____
3. _____
4. _____
5. _____
6. _____
7. _____
8. _____
9. _____
10. _____

1. Write in scientific notation:

   $43{,}200{,}000 =$

   (A) 43×10^6
   (B) 4.3×10^7
   (C) 432×10^5
   (D) None of these

2. Write in scientific notation:

   $0.000\ 000\ 081 =$

   (A) 8.1×10^{-7}
   (B) 8.1×10^{-8}
   (C) 81×10^{-9}
   (D) None of these

3. Write in scientific notation:

   $0.6435 =$

   (A) $6{,}435 \times 10^{-4}$
   (B) 6.435×10^{-2}
   (C) 64.35×10^{-2}
   (D) None of these

4. Write as a decimal fraction:

   $5.03 \times 10^5 =$

   (A) $50{,}300{,}000$
   (B) $5{,}030{,}000$
   (C) $503{,}000$
   (D) None of these

5. Write as a decimal fraction:

   $5.07 \times 10^{-2} =$

   (A) 50.7
   (B) 0.507
   (C) 0.00507
   (D) None of these

6. Write as a decimal fraction:

$6.17 \times 10^{-6} =$

(A) 0.000 006 17
(B) 0.000 000 617
(C) 0.000 061 7
(D) None of these

7. Multiply and express answer in scientific notation:

$(3 \times 10^{-9})(2 \times 10^{4}) =$

(A) 6×10^{-5}
(B) $\dfrac{6}{10^5}$
(C) 6×10^{-36}
(D) None of these

8. Divide and express answer in scientific notation:

$\dfrac{8 \times 10^{-3}}{2 \times 10^{3}} =$

(A) 4
(B) $\dfrac{4}{10^6}$
(C) 4×10^{-6}
(D) None of these

9. Convert each numeral to scientific notation and simplify. Express answer in scientific notation:

$\dfrac{(0.000\ 000\ 032)(210)}{(0.000\ 24)(14)} =$

(A) 2×10^{-9}
(B) 2×10^{-3}
(C) 0.2×10^{-4}
(D) None of these

10. If the mass of the earth is 6×10^{27} grams and each gram is 3.53×10^{-2} ounce, find the mass of the earth in ounces. Express answer in scientific notation.

(A) 21.18×10^{25} oz
(B) 2.118×10^{26} oz
(C) 2.118×10^{24} oz
(D) None of these

CHECK EXERCISE 6.3

Work the following problems without looking at any text examples. Show your work in the space provided. Write the letter that best indicates your answer in the answer column.

ANSWER COLUMN

1. _____
2. _____
3. _____
4. _____
5. _____
6. _____
7. _____
8. _____
9. _____
10. _____

Evaluate each expression in Problems 1 to 5, if possible.

1. $4^{3/2} =$

   (A) 6
   (B) 8
   (C) 12
   (D) None of these

2. $(-8)^{4/3} =$

   (A) -8
   (B) -16
   (C) Not real
   (D) None of these

3. $(-16)^{3/4} =$

   (A) -8
   (B) -6
   (C) Not real
   (D) None of these

4. $64^{-1/2} =$

   (A) -8
   (B) $\frac{1}{8}$
   (C) -32
   (D) None of these

5. $\left(-\dfrac{8}{27}\right)^{2/3} =$

   (A) $\frac{4}{9}$
   (B) $-\frac{4}{9}$
   (C) Not real
   (D) None of these

Simplify Problems 6 to 10, and express answers using positive exponents only. All variables represent positive real numbers.

6. $(4x^4y^{16})^{3/2} =$

(A) $8x^6y^{24}$
(B) $6x^8y^{512}$
(C) $6x^6y^{24}$
(D) None of these

7. $(2m^{3/4})(3m^{-1/3}) =$

(A) $6m^{-1/4}$
(B) $6m^{5/12}$
(C) $\dfrac{6}{m^{1/4}}$
(D) None of these

8. $\left(\dfrac{27x^2y^{-3}}{8x^{-4}y^3}\right)^{1/3} =$

(A) $\dfrac{3x^2}{2y^2}$
(B) $\dfrac{9x^2}{8}$
(C) $\dfrac{3x^2}{2}$
(D) None of these

9. $\left(\dfrac{3x^{-1/2}}{x^{1/3}}\right)^{-2} =$

(A) $\dfrac{-6x}{x^{-2/3}}$
(B) $-6x^{5/3}$
(C) $\dfrac{x^{5/3}}{9}$
(D) None of these

10. $(2x^{1/2} - y^{1/2})(x^{1/2} + 2y^{1/2}) =$

(A) $2x^{1/2} - 3x^{1/2}y^{1/2} - 2y^{1/2}$
(B) $3x^{1/2} + y^{1/2}$
(C) $2x + 3x^{1/2}y^{1/2} - 2y$
(D) None of these

CHECK EXERCISE 6.4

Work the following problems without looking at any text examples. Show your work in the space provided. Write the letter that best indicates your answer in the answer column.

Change Problems 1 to 5 to radical form (do not simplify).

1. $5m^{3/4} =$

   (A) $(\sqrt[4]{5m})^3$
   (B) $\sqrt[4]{5m^3}$
   (C) $5\sqrt[4]{m^3}$
   (D) None of these

2. $(x^4 - y^4)^{1/4} =$

   (A) $x - y$
   (B) $\sqrt[4]{x^4} - \sqrt[4]{y^4}$
   (C) $\sqrt[4]{x^4 - y^4}$
   (D) None of these

3. $x^{-2/3} =$

   (A) $\sqrt[3]{x^2}^{-3}$
   (B) $\dfrac{1}{\sqrt[3]{x^2}}$
   (C) $-\frac{2}{3}x$
   (D) None of these

4. $\dfrac{3}{x^{1/2} + y^{1/2}} =$

   (A) $\sqrt{\dfrac{3}{x + y}}$
   (B) $\dfrac{3}{\sqrt{x + y}}$
   (C) $\dfrac{3}{x + y}$
   (D) None of these

5. $(x^3 - 2y^6)^{2/3} =$

   (A) $(\sqrt[3]{x^3 - 2y^6})^2$
   (B) $x^2 - 2y^4$
   (C) $\sqrt[3]{x^6 - 4y^{12}}$
   (D) None of these

311

Change Problems 6 to 10 to rational exponent form (do not simplify).

6. $\sqrt[3]{(2xy^2)^2} =$

(A) $(2xy^2)^{2/3}$
(B) $(2xy^2)^6$
(C) $(2xy^2)^{3/2}$
(D) None of these

7. $\dfrac{5}{\sqrt[7]{x^2}} =$

(A) $5x^{-7/2}$
(B) $5x^{-2/7}$
(C) $\dfrac{5}{y^{14}}$
(D) None of these

8. $7\sqrt[3]{x^2} =$

(A) $(7x)^{2/3}$
(B) $7x^{2/3}$
(C) $7x^{3/2}$
(D) None of these

9. $\sqrt[5]{x^5 + y^5} =$

(A) $x + y$
(B) $(x^5)^{1/5} + (y^5)^{1/5}$
(C) $(x^5 + y^5)^{1/5}$
(D) None of these

10. $\sqrt{x} - \sqrt[3]{x^4} =$

(A) $x^{1/2} - x^{3/4}$
(B) $x^2 - x^{12}$
(C) $x^{1/2} - x^{4/3}$
(D) None of these

CHECK EXERCISE 6.5

Work the following problems without looking at any text examples. Show your work in the space provided. Write the letter that best indicates your answer in the answer column.

In Problems 1 to 8, simplify and write in simplest radical form. All variables represent positive real numbers.

1. $\sqrt[3]{54x^{10}y^5z^3} =$

    (A) $3x^3yz\sqrt[3]{2xy^2}$

    (B) $z\sqrt[3]{54x^{10}y^5}$

    (C) $3z\sqrt{2x^{10}y^5}$

    (D) None of these

2. $\dfrac{12x^2}{\sqrt{3x}} =$

    (A) $12x^2$

    (B) $\dfrac{12x^2\sqrt{3x}}{\sqrt{9x^2}}$

    (C) $3x\sqrt{3x}$

    (D) None of these

3. $\sqrt{\dfrac{2u}{3v}} =$

    (A) $\dfrac{\sqrt{6uv}}{3v}$

    (B) $\sqrt{2u}$

    (C) $\sqrt{\dfrac{2u}{3v}}$

    (D) None of these

4. $\sqrt[12]{(a+b)^8} =$

    (A) $\sqrt[6]{(a+b)^4}$

    (B) $\sqrt[12]{(a+b)^8}$

    (C) $\sqrt[3]{(a+b)^2}$

    (D) None of these

5. $\dfrac{x}{\sqrt[4]{x^3}} =$

    (A) $\dfrac{\sqrt[4]{x^3}}{x^2}$

    (B) $\sqrt[4]{x}$

    (C) $x\sqrt[4]{x}$

    (D) None of these

313

6. $\sqrt[12]{16x^4y^8} =$

    (A) $\sqrt[12]{16x^4y^8}$
    (B) $\sqrt[3]{2xy^2}$
    (C) $\sqrt[3]{4xy^2}$
    (D) None of these

7. $\dfrac{6x^3y}{\sqrt[3]{9x^2y}} =$

    (A) $\dfrac{2x\sqrt[3]{9x^2y}}{3}$
    (B) $2x^2\sqrt[3]{3xy^2}$
    (C) $\dfrac{2x^2\sqrt[3]{81xy^2}}{3}$
    (D) None of these

8. $\sqrt[5]{\dfrac{3y}{16x^2}} =$

    (A) $\dfrac{1}{x}\sqrt{\dfrac{3x^3y}{16}}$
    (B) $\dfrac{\sqrt[5]{6x^3y}}{2x}$
    (C) $\sqrt[5]{3x^2y}$
    (D) None of these

Simplify Problems 9 and 10.

9. For x a positive number: $3\sqrt[3]{x^3} - 5\sqrt[4]{x^4} =$

    (A) $8x$
    (B) $-8x$
    (C) $-2x$
    (D) None of these

10. For x a negative number: $\sqrt[7]{x^7} + \sqrt{x^2} =$

    (A) $2x$
    (B) 0
    (C) $x^7 + x^2$
    (D) None of these

CHECK EXERCISE 6.6

Work the following problems without looking at any text examples. Show your work in the space provided. Write the letter that best indicates your answer in the answer column.

In Problems 1 to 4 combine terms where possible after expressing all terms in simplest radical form.

1. $3\sqrt{m} - 2\sqrt{n} - \sqrt{m} + \sqrt{n} =$

    (A) $\sqrt{mn}$
    (B) $2\sqrt{m} - \sqrt{n}$
    (C) $\sqrt{m}$
    (D) None of these

2. $\sqrt[3]{8x} + \sqrt[3]{27x} =$

    (A) $5\sqrt[3]{x}$
    (B) $\sqrt[3]{8x} + \sqrt[3]{27x}$
    (C) $5\sqrt{x}$
    (D) None of these

3. $\sqrt{18} - \dfrac{4}{\sqrt{2}} =$

    (A) $\sqrt{2}$
    (B) $\dfrac{2}{\sqrt{2}}$
    (C) $\sqrt{18} - 2\sqrt{2}$
    (D) None of these

4. $\sqrt{\dfrac{5}{3}} - \sqrt{\dfrac{3}{5}} =$

    (A) $\sqrt{\tfrac{5}{3}} - \sqrt{\tfrac{3}{5}}$
    (B) $\dfrac{\sqrt{15}}{3} - \dfrac{\sqrt{15}}{5}$
    (C) $\tfrac{2}{15}\sqrt{15}$
    (D) None of these

Multiply in Problems 5 to 8 and simplify where possible. Represent answers in simplest radical form.

5. $\sqrt{3}(\sqrt{6} - \sqrt{3}) =$

    (A) $\sqrt{18} - \sqrt{9}$
    (B) $\sqrt{18} - 3$
    (C) $3\sqrt{2} - 3$
    (D) None of these

315

6. $(\sqrt{m} - \sqrt{n})(\sqrt{m} + \sqrt{n}) =$

    (A) $m - n$
    (B) $\sqrt{m^2} - 2\sqrt{m}\sqrt{n} - \sqrt{n^2}$
    (C) $m - 2\sqrt{mn} - n$
    (D) None of these

7. $(2\sqrt{x} - 3)(3\sqrt{x} + 2) =$

    (A) $6\sqrt{x^2} - 5\sqrt{x} - 6$
    (B) $6x - 5\sqrt{x} - 6$
    (C) $\sqrt{x} - 6$
    (D) None of these

8. $(\sqrt[3]{m^2} - 2)(\sqrt[3]{m} + 2) =$

    (A) $m - 4$
    (B) $m + 2\sqrt[3]{m^2} - 2\sqrt[3]{m} - 4$
    (C) $\sqrt[3]{m^2}\sqrt[3]{m} - 4$
    (D) None of these

Rationalize denominators in Problems 9 and 10 and represent answers in simplest radical form.

9. $\dfrac{\sqrt{5}}{\sqrt{5} - 2} =$

    (A) $\dfrac{5 - 2\sqrt{5}}{3}$

    (B) $\dfrac{5}{5 - 2\sqrt{5}}$

    (C) $5 - 2\sqrt{5}$

    (D) None of these

10. $\dfrac{\sqrt{c} + \sqrt{d}}{\sqrt{c} - \sqrt{d}} =$

    (A) 1

    (B) $\dfrac{c + \sqrt{cd} + d}{c - d}$

    (C) $\dfrac{c + 2\sqrt{cd} + d}{c - d}$

    (D) None of these

CHECK EXERCISE 6.7

Work the following problems without looking at any text examples. Show your work in the space provided. Write the letter that best indicates your answer in the answer column.

Perform the indicated operations in Problems 1 to 6 and write each answer in the form $a + bi$.

1. $(-3 + 5i) - (2 - 4i) =$

    (A) $-1 - i$
    (B) $-5 - i$
    (C) $-5 + 9i$
    (D) None of these

2. $(7 + 3i) + (2 - i) - (1 - 2i) =$

    (A) 8
    (B) $8 - 3i$
    (C) $8 - 4i$
    (D) None of these

3. $(2 + 3i)(1 - 4i) =$

    (A) $2 - 5i - 12i^2$
    (B) $-10 - 5i$
    (C) $14 - 5i$
    (D) None of these

4. $\dfrac{3}{4 - 3i} =$

    (A) $\dfrac{12}{7} - \dfrac{9}{7}i$

    (B) $\dfrac{12}{25} - \dfrac{9}{25}i$

    (C) $\dfrac{12}{25} + \dfrac{9}{25}i$

    (D) None of these

5. $\dfrac{3 - i}{4i} =$

    (A) $-\dfrac{1}{4} - \dfrac{3}{4}i$

    (B) $\dfrac{3i + 1}{-4}$

    (C) $\dfrac{12i + 4}{-16}$

    (D) None of these

317

6. $(3 - 2i)^2 - 2(3 - 2i) + 4 =$

    (A) $3 - 8i$
    (B) $17 - 8i$
    (C) $17 - 16i$
    (D)   None of these

7.   Evaluate $x^2 - 2x + 5$ for $1 + 2i$.

    (A) $4 + 4i^2$
    (B) $8i$
    (C) 0
    (D)   None of these

In Problems 8 to 10 convert square roots of negative numbers to complex form, perform the indicated operations, and express answer in the form $a + bi$.

8. $(3 - \sqrt{-4})(2 + \sqrt{-25}) =$

    (A) $16 + 11i$
    (B) $-4 + 11i$
    (C) $6 + 11i - 10i^2$
    (D)   None of these

9. $\dfrac{-4 + \sqrt{-4}}{2} =$

    (A) $\dfrac{-4 + 2i}{2}$
    (B) $-2 + i$
    (C) $-2 - i$
    (D)   None of these

10. $\dfrac{2 + \sqrt{-1}}{1 - \sqrt{-4}} =$

    (A) $\dfrac{2 + i}{1 - 2i}$
    (B) $-\dfrac{1}{5} + i$
    (C) $-\dfrac{1}{3} + \dfrac{5}{3}i$
    (D)   None of these

Chapter Seven

SECOND-DEGREE EQUATIONS AND INEQUALITIES

CONTENTS

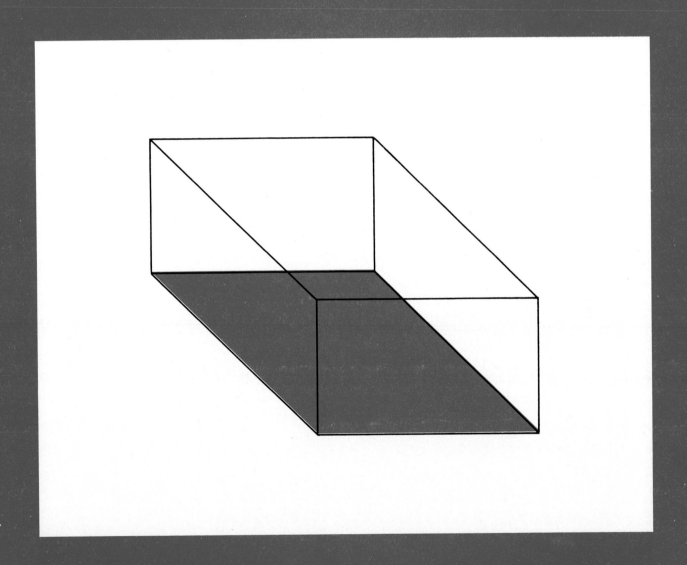

Are you looking down from above or up from below at the "box" in the figure?

INSTRUCTIONS FOR STUDENTS IN A
SELF-PACED CLASS OR LAB

(yes) **HAVE YOU HAD INTERMEDIATE ALGEBRA BEFORE THIS COURSE?** (no)

1. Work Diagnostic (Review) Exercise 7.8 on page 355. Check answers in back of book; then work through text sections corresponding to problems missed. (Section numbers are in italics following each answer.)
2. When finished with step 1, take Practice Test: Chapter 7 on page 357 as a final check of your understanding of the chapter. Check answers in the back of the book; then review sections where weakness still prevails. (Corresponding section numbers are in italics following each answer.)
3. When you think you are ready, ask your instructor for a graded test for Chapter 7.
4. If your instructor approves, after the test is corrected, go to the next chapter.

1. Work through each section in the chapter as follows:
   (a) Read discussion.
   (b) Read each example and work the corresponding matched problem. Check your solution to the matched problem in Solutions to Matched Problems on the indicated page.
   (c) At the end of a section work the odd-numbered problems in the Practice Exercise and check answers; then work even-numbered problems in areas of weakness. (Answers to *all* Practice Exercise sets are in the back of the book.)
   (d) Work Check Exercise as instructed. Tear out and turn in as directed by your instructor. (Answers are not in the text.)
2. Repeat each step in item 1 for each section in the chapter.
3. After the instructional part of the chapter is completed, proceed with steps 1 to 4 in the box above.

7.1 INTRODUCTION

The equation

$$\tfrac{1}{2}x - \tfrac{1}{3}(x + 3) = 2 - x$$

though complicated-looking, is actually a first-degree equation in one variable, since it can be transformed into the equivalent equation

$$7x - 18 = 0$$

which is a special case of

$$ax + b = 0 \qquad a \neq 0$$

We have solved many equations of this type and found that they always have a single solution. From a mathematical point of view we have essentially taken care of the problem of solving first-degree equations in one variable.

In this chapter we will consider the next class of polynomial equations called second-degree equations or quadratic equations. A **quadratic equation** in one variable is any equation that can be written in the form

---
Quadratic Equation

(Standard form)

$$ax^2 + bx + c = 0 \qquad a \neq 0 \qquad (1)$$

---

where x is a variable and a, b, and c are constants. We will refer to this form as the **standard form** for the quadratic equation. The equations

$$2x^2 - 3x + 5 = 0$$

$$15 = 180t - 16t^2$$

are both quadratic equations since they are either in the standard form or can be transformed into this form.

Applications that give rise to quadratic equations are many and varied. A brief glance at Section 7.5 will give you some indication of the variety.

7.2 SOLUTION BY SQUARE ROOT AND BY FACTORING

•Solution by square root
••Solution by factoring

•Solution by square root

The easiest type of quadratic equation to solve is the special form where the first-degree term is missing; that is, when (1) is of the form

$$ax^2 + c = 0 \qquad a \neq 0$$

The method of solution makes direct use of the definition of square root. The process is illustrated in the following example.

EXAMPLE 1
Solve by the square root method:

(A) $x^2 - 8 = 0$

Solution
$x^2 - 8 = 0$

$\qquad x^2 = 8$ What number squared is 8?

$\qquad x = \pm \sqrt{8}$ or $\pm 2\sqrt{2}$ $\pm 2\sqrt{2}$ is a short way of writing $-2\sqrt{2}$ or $+2\sqrt{2}$.

(B) $2x^2 - 3 = 0$

Solution
$2x^2 - 3 = 0$

$\qquad 2x^2 = 3$ Do not write $2x = \pm\sqrt{3}$ next (Why?).

$\qquad x^2 = \frac{3}{2}$ What number squared is $\frac{3}{2}$?

$\qquad x = \pm\sqrt{\dfrac{3}{2}}$ or $\pm\dfrac{\sqrt{6}}{2}$

(C) $3x^2 + 27 = 0$

Solution
$3x^2 + 27 = 0$

$\qquad 3x^2 = -27$ Do not write $3x = \pm\sqrt{-27}$ next (Why?).

$\qquad x^2 = -9$ What number squared is -9?

$\qquad x = \pm\sqrt{-9} = \pm 3i$

(D) $(x + \frac{1}{2})^2 = \frac{5}{4}$

Solution

$(x + \frac{1}{2})^2 = \frac{5}{4}$ Solve for $x + \frac{1}{2}$; then solve for x.

$x + \frac{1}{2} = \pm\sqrt{\frac{5}{4}}$

$x = -\frac{1}{2} \pm \frac{\sqrt{5}}{2}$

$= \frac{-1 \pm \sqrt{5}}{2}$ Short for $\frac{-1 + \sqrt{5}}{2}$ or $\frac{-1 - \sqrt{5}}{2}$.

Work Problem 1 and check solution in Solutions to Matched Problems on page 325.

PROBLEM 1 Solve by the square root method.

(A) $x^2 - 12 = 0$ **(B)** $3x^2 - 5 = 0$

(C) $2x^2 + 8 = 0$ **(D)** $(x + \frac{1}{3})^2 = \frac{2}{9}$

••Solution by factoring

If the coefficients a, b, and c in the quadratic equation

$$ax^2 + bx + c = 0$$

are such that $ax^2 + bx + c$ can be written as the product of two first-degree factors with integer coefficients, then the quadratic equation can be quickly and easily solved. The method of solution by factoring rests on the following property of the real numbers:

If a and b are real numbers, then

a · b = 0   if and only if   a = 0 or b = 0 (or both)

EXAMPLE 2
Solve by factoring, if possible.

(A) $x^2 + 2x - 15 = 0$

Solution

$x^2 + 2x - 15 = 0$

$(x - 3)(x + 5) = 0$ $(x - 3)(x + 5) = 0$ if and only if $(x - 3) = 0$ or $(x + 5) = 0$.

$x - 3 = 0$ or $x + 5 = 0$

$x = 3$ $x = -5$

(B) $4x^2 = 6x$

Solution

$$4x^2 = 6x$$

If both sides are divided by x, we lose one solution ($x = 0$). But we can simplify the equation by dividing both sides by 2, a common factor of both coefficients.

$$2x^2 = 3x$$

$$2x^2 - 3x = 0$$

$$x(2x - 3) = 0$$

$x(2x - 3) = 0$ if and only if $x = 0$ or $2x - 3 = 0$.

$$x = 0 \quad \text{or} \quad 2x - 3 = 0$$

$$x = 0 \qquad\qquad x = \tfrac{3}{2}$$

(C) $2x^2 - 8x + 3 = 0$

Solution

The polynomial cannot be factored using integer coefficients; hence, another method must be used to find the solution. This will be discussed later.

(D) $3 + \dfrac{5}{x} = \dfrac{2}{x^2}$

Solution

$$3 + \frac{5}{x} = \frac{2}{x^2}$$

Multiply both sides by x^2, the lcm of the denominators ($x \neq 0$).

$$3x^2 + 5x = 2$$

Write in standard form: $ax^2 + bx + c$

$$3x^2 + 5x - 2 = 0$$

Factor left side, if possible.

$$(3x - 1)(x + 2) = 0$$

$$3x - 1 = 0 \quad \text{or} \quad x + 2 = 0$$

$$3x = 1 \qquad\qquad x = -2$$

$$x = \tfrac{1}{3}$$

Work Problem 2 and check solution in Solutions to Matched Problems on page 326.

PROBLEM 2 Solve by factoring, if possible.

(A) $x^2 - 2x - 8 = 0$ **(B)** $9t^2 = 6t$

(C) $x^2 - 3x = 3$ **(D)** $x = \dfrac{3}{2x - 5}$

EXAMPLE 3

The length of a rectangle is 1 in. more than twice its width. If the area is 21 in.² (square inches), find its dimensions.

Solution

Draw a figure and label sides:

x

$2x + 1$

$$x(2x + 1) = 21$$

$$2x^2 + x - 21 = 0$$

$$(2x + 7)(x - 3) = 0$$

$2x + 7 = 0$ or $x - 3 = 0$

$\qquad 2x = -7 \qquad\qquad x = 3$ in.   (width)

$\qquad \cancel{x = -\tfrac{7}{2}} \qquad 2x + 1 = 7$ in.   (length)

Not possible, so
must be discarded.

NOTE: In practical problems involving quadratic equations one of two solutions must often be discarded because it will not make sense in the problem.

Work Problem 3 and check solution in Solutions to Matched Problems on page 326.

PROBLEM 3   The base of a triangle is 6 meters longer than its height. If the area is 20 square meters, find its dimensions ($A = \tfrac{1}{2}bh$).

Solution

SOLUTIONS TO MATCHED PROBLEMS

1. (A) $x^2 - 12 = 0$
$$x^2 = 12$$
$$x = \pm\sqrt{12} \text{ or } \pm 2\sqrt{3}$$

(B) $3x^2 - 5 = 0$
$$3x^2 = 5$$
$$x^2 = \tfrac{5}{3}$$
$$x = \pm\sqrt{\tfrac{5}{3}} \text{ or } \pm\frac{\sqrt{15}}{3}$$

(C) $2x^2 + 8 = 0$
$$2x^2 = -8$$
$$x^2 = -4$$
$$x = \pm\sqrt{-4}$$
$$x = \pm 2i$$

(D) $(x + \tfrac{1}{3})^2 = \tfrac{2}{9}$
$$x + \tfrac{1}{3} = \pm\sqrt{\tfrac{2}{9}}$$
$$x = -\frac{1}{3} \pm \frac{\sqrt{2}}{3} = \frac{-1 \pm \sqrt{2}}{3}$$

2. (A) $x^2 - 2x - 8 = 0$
$(x - 4)(x + 2) = 0$
$x - 4 = 0$ or $x + 2 = 0$
$x = 4$ $\qquad$ $x = -2$

(B) $9t^2 = 6t$
$3t^2 = 2t$
$3t^2 - 2t = 0$
$t(3t - 2) = 0$
$t = 0$ or $3t - 2 = 0$
$\qquad\qquad 3t = 2$
$\qquad\qquad t = \frac{2}{3}$

(C) $x^2 - 3x = 3$
$x^2 - 3x - 3 = 0$
Left side not factorable using integer coefficients.

(D) $\qquad x = \dfrac{3}{2x - 5}$
$x(2x - 5) = 3$
$2x^2 - 5x = 3$
$2x^2 - 5x - 3 = 0$
$(2x + 1)(x - 3) = 0$
$2x + 1 = 0$ or $x - 3 = 0$
$2x = -1$ $\qquad$ $x = 3$
$x = -\frac{1}{2}$

3.

$\frac{1}{2}bh = 20$
$\frac{1}{2}(x + 6)x = 20$
$(x + 6)x = 40$
$x^2 + 6x = 40$
$x^2 + 6x - 40 = 0$
$(x + 10)(x - 4) = 0$
$x + 10 = 0$ or $x - 4 = 0$
$\cancel{x = -10}$ $\qquad$ $x = 4$ meters (height)
$\qquad\qquad x + 6 = 10$ meters (base)

PRACTICE EXERCISE 7.2 ————————————————————————

Work odd-numbered problems first, check answers, and then work even-numbered problems in areas of weakness. Answers to all problems are in the back of the book. Make every effort to work a problem yourself before you look at an answer.

A *Solve by square root method.*

1. $x^2 - 16 = 0$ _____

2. $x^2 - 25 = 0$ _____

3. $x^2 + 16 = 0$ _____

4. $x^2 + 25 = 0$ _____

5. $y^2 - 45 = 0$ _____

6. $m^2 - 12 = 0$ _____

7. $4x^2 - 9 = 0$ _____

8. $9y^2 - 16 = 0$ _____

9. $16y^2 = 9$ _____

10. $9x^2 = 4$ _____

Solve by factoring.

11. $u^2 + 5u = 0$ _____

12. $v^2 - 3v = 0$ _____

13. $3A^2 = -12A$ _____

14. $4u^2 = 8u$ _____

15. $x^2 - 11x - 12 = 0$ _____

16. $y^2 - 6y + 5 = 0$ _____

17. $x^2 + 4x - 5 = 0$ _____

18. $x^2 - 4x - 12 = 0$ _____

19. $3Q^2 - 10Q - 8 = 0$ _____

20. $2d^2 + 15d - 8 = 0$ _____

B *Solve by square root method.*

21. $y^2 = 2$ _____

22. $x^2 = 3$ _____

23. $16a^2 + 9 = 0$ _____

24. $4x^2 + 25 = 0$ _____

25. $9x^2 - 7 = 0$ _____

26. $4t^2 - 3 = 0$ _____

27. $(m - 3)^2 = 25$ _____

28. $(n + 5)^2 = 9$ _____

29. $(t + 1)^2 = -9$ _____

30. $(d - 3)^2 = -4$ _____

31. $(x - \frac{1}{3})^2 = \frac{4}{9}$ _____

32. $(x - \frac{1}{2})^2 = \frac{9}{4}$ _____

Solve by factoring. (Write equations in standard form first.)

33. $u^2 = 2u + 3$ _____

34. $m^2 + 2m = 15$ _____

35. $3x^2 = x + 2$ _____

36. $2x^2 = 3 - 5x$ _____

37. $y^2 = 5y - 2$ _____

38. $3 = t^2 + 7t$ _____

39. $2x(x - 1) = 3(x + 1)$ _____

40. $3x(x - 2) = 2(x - 2)$ _____

41. $\dfrac{t}{2} = \dfrac{2}{t}$ _____

42. $y = \dfrac{9}{y}$ _____

43. $\dfrac{m}{4}(m + 1) = 3$ _____

44. $\dfrac{A^2}{2} = A + 4$ _____

45. $2y = \dfrac{2}{y} + 3$ _____

46. $L = \dfrac{15}{L - 2}$ _____

47. $2 + \dfrac{2}{x^2} = \dfrac{5}{x}$ _____

48. $1 - \dfrac{3}{x} = \dfrac{10}{x^2}$ _____

C *Solve by square root method.*

49. $(y + \frac{5}{2})^2 = \frac{5}{2}$ _____

50. $(x - \frac{3}{2})^2 = \frac{3}{2}$ _____

Solve for the indicated letters in terms of the other letters. Use positive square roots only.

51. $a^2 + b^2 = c^2$ (Solve for a.) _____

52. $s = \frac{1}{2}gt^2$ (Solve for t.) _____

APPLICATIONS

53. The width of a rectangle is 8 in. less than its length. If its area is 33 in.² (square inches), find its dimensions.

_____

54. Find the base and height of a triangle with area 2 ft² if its base is 3 ft longer than its height ($A = \frac{1}{2}bh$).

_____

55. In a given city on a given day, the demand equation for gasoline is $d = 900/p$ and the supply equation is $s = p - 80$, where $d = s$ denote the number of gallons demanded and supplied (in thousands), respectively, at a price of p cents per gallon. Find the price at which supply is equal to demand.

_____

56. To find the critical velocity at the top of the loop necessary to keep a steel ball on the track (see Figure 1), the centripetal force mv^2/r is equated to the force due to gravity mg. The mass m cancels out of the equation, and we are left with $v^2 = gr$. For a loop of radius 0.25 ft, find the critical velocity (in feet per second) at the top of the loop that is required to keep the ball on the track. Use $g = 32$ and compute your answer to two decimal places, using a square root table or a calculator.

Figure 1

The Check Exercise for this section is on page 359.

7.3 SOLUTION BY COMPLETING THE SQUARE

•Introduction
••Completing the square
•••Solution of quadratic equations by completing the square

•Introduction

The factoring and square root methods discussed in the last two sections are fast and easy to use when they apply. Unfortunately many quadratic equations will not yield directly to either method. For example, the very simple-looking polynomial in the equation

$$x^2 + 6x - 2 = 0$$

cannot be factored in the integers. The equation requires a new approach if it can be solved at all.

In this section we will discuss a method, called "solution by completing the square," that will work for all quadratic equations. In the next section we will use this method to develop a general formula that will be used in the future whenever the square root or factoring method fails.

The method of completing the square is based on the process of transforming the standard quadratic equation,

$$ax^2 + bx + c = 0$$

into the form

$$(x + A)^2 = B$$

where A and B are constants. This last equation can easily be solved by the square root method discussed in the preceding section. Thus,

$$(x + A)^2 = B$$
$$x + A = \pm \sqrt{B}$$
$$x = -A \pm \sqrt{B}$$

••Completing the square

Before considering how the first part is accomplished, let's pause for a moment and consider a related problem: What number must be added to $x^2 + 6x$ so that the result is the square of a linear expression? There is an easy mechanical rule for finding this number based on the squares of the following binomials:

$$(x + m)^2 = x^2 + 2mx + m^2$$
$$(x - m)^2 = x^2 - 2mx + m^2$$

In either case, we see that the third term on the right is the square of one-half of the coefficient of x in the second term on the right. This observation leads directly to the rule:

To **complete the square** of a quadratic of the form

$$x^2 + bx$$

add the square of one-half of the coefficient of x, that is

$$\left(\frac{b}{2}\right)^2 \qquad \text{or} \qquad \frac{b^2}{4}$$

Thus,

$$x^2 + bx + \left(\frac{b}{2}\right)^2 = \left(x + \frac{b}{2}\right)^2$$

EXAMPLE 4

(A) To complete the square of $x^2 + 6x$, add $(\frac{6}{2})^2$, that is, 9; thus

$$x^2 + 6x + 9 = (x + 3)^2$$

(B) To complete the square of $x^2 - 3x$, add $(-\frac{3}{2})^2$; that is $\frac{9}{4}$; thus

$$x^2 - 3x + \tfrac{9}{4} = (x - \tfrac{3}{2})^2$$

Work Problem 4 and check solution in Solutions to Matched Problems on page 331.

PROBLEM 4   **(A)** Complete the square of $x^2 + 10x$ and factor:

$$x^2 + 10x \qquad =$$

(B) Complete the square of $x^2 - 5x$ and factor:

$$x^2 - 5x \quad = $$

NOTE: It is important to note that the rule stated above applies only to quadratic forms where the coefficient of the second-degree term is 1.

We now use the method of completing the square to solve quadratic equations. In the next section we will use the method to develop a formula that will work for *all* quadratic equations.

•••Solution of quadratic equations by completing the square

Solving quadratic equations by the method of completing the square is best illustrated by examples.

EXAMPLE 5

Solve $x^2 + 6x - 2 = 0$ by the method of completing the square.

Solution

$x^2 + 6x - 2 = 0$	Add 2 to both sides of the equation to remove -2 from the left side.
$x^2 + 6x = 2$	To complete the square of the left side, add the square of one-half of the coefficient of x to each side of the equation.
$x^2 + 6x + 9 = 2 + 9$	Factor the left side.
$(x + 3)^2 = 11$	Solve by method of square root.
$x + 3 = \pm\sqrt{11}$	
$x = -3 \pm \sqrt{11}$	

Work Problem 5 and check solution in Solutions to Matched Problems on page 331.

PROBLEM 5   Solve by method of completing the square:

$$x^2 - 8x + 10 = 0$$

EXAMPLE 6

Solve $x^2 - 4x + 13 = 0$ by the method of completing the square.

Solution

$x^2 - 4x + 13 = 0$	
$x^2 - 4x = -13$	Add 4 to each side to complete the square on the left side.
$x^2 - 4x + 4 = 4 - 13$	
$(x - 2)^2 = -9$	
$x - 2 = \pm\sqrt{-9}$	
$x - 2 = \pm 3i$	
$x = 2 \pm 3i$	

Work Problem 6 and check solution in Solutions to Matched Problems on page 331.

PROBLEM 6   Solve by method of completing the square:

$$x^2 - 2x + 3 = 0$$

EXAMPLE 7

Solve $2x^2 - 4x - 3 = 0$ by the method of completing the square.

Solution

$$2x^2 - 4x - 3 = 0$$ Note that the coefficient of x^2 is not 1. Divide through by the leading coefficient and proceed as in the last example.

$$x^2 - 2x - \tfrac{3}{2} = 0$$

$$x^2 - 2x = \tfrac{3}{2}$$

$$x^2 - 2x + 1 = \tfrac{3}{2} + 1$$

$$(x - 1)^2 = \tfrac{5}{2}$$

$$x - 1 = \pm \sqrt{\tfrac{5}{2}}$$

$$x = 1 \pm \frac{\sqrt{10}}{2}$$

$$x = \frac{2 \pm \sqrt{10}}{2}$$

Work Problem 7 and check solution in Solutions to Matched Problems on page 332.

PROBLEM 7   Solve by method of completing the square:

$$2x^2 + 8x + 3 = 0$$

SOLUTIONS TO MATCHED PROBLEMS

4. **(A)** $x^2 + 10x + 25 = (x + 5)^2$ **(B)** $x^2 - 5x + (\tfrac{5}{2})^2 = (x - \tfrac{5}{2})^2$

5. $x^2 - 8x + 10 = 0$ **6.** $x^2 - 2x + 3 = 0$

$x^2 - 8x \quad\quad = -10$ $x^2 - 2x \quad\quad = -3$

$x^2 - 8x + 16 = 16 - 10$ $x^2 - 2x + 1 = 1 - 3$

$(x - 4)^2 = 6$ $(x - 1)^2 = -2$

$x - 4 = \pm \sqrt{6}$ $x - 1 = \pm \sqrt{-2}$

$x = 4 \pm \sqrt{6}$ $x = 1 \pm i\sqrt{2}$

7. $2x^2 + 8x + 3 = 0$

$2x^2 + 8x = -3$

$x^2 + 4x = -\frac{3}{2}$

$x^2 + 4x + 4 = 4 - \frac{3}{2}$

$(x + 2)^2 = \frac{5}{2}$

$x + 2 = \pm\sqrt{\frac{5}{2}}$

$x = -2 \pm \sqrt{\frac{5}{2}} \text{ or } \frac{-4 \pm \sqrt{10}}{2}$

PRACTICE EXERCISE 7.3

Work odd-numbered problems first, check answers, and then work even-numbered problems in areas of weakness. Answers to all problems are in the back of the book. Make every effort to work a problem yourself before you look at an answer.

A *Complete the square and factor.*

1. $x^2 + 4x$ _____   2. $x^2 + 8x$ _____   3. $x^2 - 6x$ _____

4. $x^2 - 10x$ _____   5. $x^2 + 12x$ _____   6. $x^2 + 2x$ _____

Solve by method of completing the square.

7. $x^2 + 4x + 2 = 0$ _____   8. $x^2 + 8x + 3 = 0$ _____

9. $x^2 - 6x - 3 = 0$ _____   10. $x^2 - 10x - 3 = 0$ _____

B *Complete the square and factor.*

11. $x^2 + 3x$ _____   12. $x^2 + x$ _____

13. $u^2 - 5u$ _____   14. $m^2 - 7m$ _____

Solve by method of completing the square.

15. $x^2 + x - 1 = 0$ _____   16. $x^2 + 3x - 1 = 0$ _____

17. $u^2 - 5u + 2 = 0$ _____   18. $n^2 - 3n - 1 = 0$ _____

19. $m^2 - 4m + 8 = 0$ _____   20. $x^2 - 2x + 3 = 0$ _____

21. $2y^2 - 4y + 1 = 0$ _____   22. $2x^2 - 6x + 3 = 0$ _____

23. $2u^2 + 3u - 1 = 0$ _____   24. $3x^2 + x - 1 = 0$ _____

C 25. $2u^2 - 3u + 2 = 0$ _____   26. $3x^2 - 5x + 3 = 0$ _____

27. Solve for x: $x^2 + mx + n = 0$ (treat m and n as constants). _____

The Check Exercise for this section is on page 361.

7.4 THE QUADRATIC FORMULA

• Quadratic formula
•• Which method?

• Quadratic formula

The method of completing the square can be used to solve any quadratic equation, but the process is often tedious. If you had a very large number of quadratic equations to solve by completing the square, before you finished you would probably ask yourself if the process could not be made more efficient. Why not take the general equation

$$ax^2 + bx + c = 0 \qquad a \neq 0$$

and solve it once and for all for x in terms of the coefficients a, b, and c by the method of completing the square, and thus obtain a formula that could be memorized and used whenever a, b, and c are known?

We start by making the leading coefficient 1. How? Multiply both sides of the equation by $1/a$. Thus

$$x^2 + \frac{b}{a}x + \frac{c}{a} = 0$$

Add $-c/a$ to both sides to clear c/a from the left side.

$$x^2 + \frac{b}{a}x = -\frac{c}{a}$$

Now complete the square on the left side by adding the square of one-half the coefficient of x to each side.

$$x^2 + \frac{b}{a}x + \frac{b^2}{4a^2} = \frac{b^2}{4a^2} - \frac{c}{a}$$

We now factor the left member and solve by the square root method.

$$\left(x + \frac{b}{2a}\right)^2 = \frac{b^2 - 4ac}{4a^2}$$

$$x + \frac{b}{2a} = \pm\sqrt{\frac{b^2 - 4ac}{4a^2}}$$

$$x = -\frac{b}{2a} \pm \frac{\sqrt{b^2 - 4ac}}{2a}$$

Thus,

$$\boxed{\begin{array}{c} \textit{Quadratic Formula} \\ \hline x = \dfrac{-b \pm \sqrt{b^2 - 4ac}}{2a} \qquad a \neq 0 \end{array}}$$

The preceding equation is called the **quadratic formula.** It should be memorized and used to solve quadratic equations when simpler methods fail. Note that $b^2 - 4ac$, called the **discriminant,** gives us the following useful information about roots.

$b^2 - 4ac$	$ax^2 + bx + c = 0$
Positive	Two real solutions
Zero	One real solution
Negative	Two complex solutions

EXAMPLE 8
Solve $2x^2 - 4x - 3 = 0$ by use of the quadratic formula.

Solution
$2x^2 - 4x - 3 = 0$

$x = \dfrac{-b \pm \sqrt{b^2 - 4ac}}{2a}$ $\begin{array}{l} a = 2 \\ b = -4 \\ c = -3 \end{array}$ Write down the quadratic formula, and identify a, b, and c.

$= \dfrac{-(-4) \pm \sqrt{(-4)^2 - 4(2)(-3)}}{2(2)}$ Substitute into formula and simplify. Be careful of sign errors here.

$= \dfrac{4 \pm \sqrt{40}}{4} = \dfrac{4 \pm 2\sqrt{10}}{4}$

$= \dfrac{2 \pm \sqrt{10}}{2}$

Work Problem 8 and check solution in Solutions to Matched Problems on page 337.

PROBLEM 8   Solve, using the quadratic formula:

$x^2 - 2x - 1 = 0$

EXAMPLE 9
Solve $x^2 + 11 = 6x$, using the quadratic formula.

Solution

$$x^2 + 11 = 6x \qquad \text{Write in standard form.}$$

$$x^2 - 6x + 11 = 0$$

$$x = \frac{-b \pm \sqrt{b^2 - 4ac}}{2a} \qquad \begin{aligned} a &= 1 \\ b &= -6 \\ c &= 11 \end{aligned}$$

$$= \frac{-(-6) \pm \sqrt{(-6)^2 - 4(1)(11)}}{2(1)} \qquad \text{Be careful of sign errors here.}$$

$$= \frac{6 \pm \sqrt{-8}}{2}$$

$$= \frac{6 \pm 2i\sqrt{2}}{2} = 3 \pm i\sqrt{2}$$

Work Problem 9 and check solution in Solutions to Matched Problems on page 337.

PROBLEM 9   Solve, using the quadratic formula:

$$2x^2 + 3 = 4x$$

••Which method?

In normal practice the quadratic formula is used whenever the square root method or the factoring method does not produce results easily. These latter methods are generally faster when they apply and should be used when possible.*
    Note that any equation of the form

$$ax^2 + c = 0$$

can always be solved by the square root method. And any equation of the form

$$ax^2 + bx = 0$$

can always be solved by factoring since $ax^2 + bx = x(ax + b)$.
    It is important to realize, however, that the quadratic formula can always be used and will produce the same results as any other method.

EXAMPLE 10

Solve $\dfrac{30}{8 + x} + 2 = \dfrac{30}{8 - x}$ by the most efficient method.

---

*The process of completing the square, in addition to producing the quadratic formula, is used in many other places in mathematics. See Sections 8.4 and 10.3, for example.

Solution

$$\frac{30}{8+x} + 2 = \frac{30}{8-x}$$

Multiply both sides by the lcm of the denominators $(8+x)(8-x)$. Note that $x \neq -8, 8$.

$$30(8-x) + 2(8+x)(8-x) = 30(8+x)$$

$$240 - 30x + 128 - 2x^2 = 240 + 30x$$

$$-2x^2 - 60x + 128 = 0$$

Divide both sides by -2.

$$x^2 + 30x - 64 = 0$$

Factor left side, if possible.

$$(x+32)(x-2) = 0$$

$$x + 32 = 0 \quad \text{or} \quad x - 2 = 0$$

$$x = -32 \qquad\qquad x = 2$$

We could have also solved $x^2 - 30x - 64 = 0$ using the quadratic formula.

$$x^2 + 30x - 64 = 0$$

$$x = \frac{-b \pm \sqrt{b^2 - 4ac}}{2a} \qquad \begin{aligned} a &= 1 \\ b &= 30 \\ c &= -64 \end{aligned}$$

$$= \frac{-30 \pm \sqrt{30^2 - 4(1)(-64)}}{2(1)}$$

$$= \frac{-30 \pm \sqrt{1,156}}{2}$$

$$= \frac{-30 \pm 34}{2}$$

Thus, $x = -32$ or $x = 2$.

It is clear that the factoring method was much easier. Nevertheless, we got the same result, as expected.

Work Problem 10 and check solution in Solutions to Matched Problems on page 337.

PROBLEM 10   Solve by the most efficient method:

$$\frac{6}{x-2} + 2 = \frac{4}{x}$$

SOLUTIONS TO MATCHED PROBLEMS

8. $x^2 - 2x - 1 = 0$

$$x = \frac{-b \pm \sqrt{b^2 - 4ac}}{2a} \qquad \begin{array}{l} a = 1 \\ b = -2 \\ c = -1 \end{array}$$

$$= \frac{-(-2) \pm \sqrt{(-2)^2 - 4(1)(-1)}}{2(1)}$$

$$= \frac{2 \pm \sqrt{8}}{2} = \frac{2 \pm 2\sqrt{2}}{2} = 1 \pm \sqrt{2}$$

9. $2x^2 + 3 = 4x$

$2x^2 - 4x + 3 = 0$

$$x = \frac{-b \pm \sqrt{b^2 - 4ac}}{2a} \qquad \begin{array}{l} a = 2 \\ b = -4 \\ c = 3 \end{array}$$

$$= \frac{-(-4) \pm \sqrt{(-4)^2 - 4(2)(3)}}{2(2)}$$

$$= \frac{4 \pm \sqrt{-8}}{4} = \frac{4 \pm 2i\sqrt{2}}{4} = 1 \pm \frac{\sqrt{2}}{2}i$$

10. $\qquad \dfrac{6}{x - 2} + 2 = \dfrac{4}{x} \qquad x \neq 0, 2$

$6x + 2x(x - 2) = 4(x - 2)$

$6x + 2x^2 - 4x = 4x - 8$

$2x^2 - 2x + 8 = 0$

$x^2 - x + 4 = 0$

Left side does not factor in the integers, so we go directly to the quadratic formula.

$$x = \frac{-b \pm \sqrt{b^2 - 4ac}}{2a} \qquad \begin{array}{l} a = 1 \\ b = -1 \\ c = 4 \end{array}$$

$$= \frac{-(-1) \pm \sqrt{(-1)^2 - 4(1)(4)}}{2(1)}$$

$$= \frac{1 \pm \sqrt{-15}}{2} = \frac{1 \pm i\sqrt{15}}{2} = \frac{1}{2} \pm \frac{\sqrt{15}}{2}i$$

PRACTICE EXERCISE 7.4

Work odd-numbered problems first, check answers, and then work even-numbered problems in areas of weakness. Answers to all problems are in the back of the book. Make every effort to work a problem yourself before you look at an answer.

A *Specify the constants a, b, and c for each quadratic equation when written in the the standard form $ax^2 + bx + c = 0$.*

1. $2x^2 - 5x + 3 = 0$ ＿＿＿＿＿＿＿

2. $3x^2 - 2x + 1 = 0$ ＿＿＿＿＿＿＿

3. $m = 1 - 3m^2$ ＿＿＿＿＿＿＿

4. $2u^2 = 1 - 3u$ ＿＿＿＿＿＿＿

5. $3y^2 - 5 = 0$ ＿＿＿＿＿＿＿

6. $2x^2 - 5x = 0$ ＿＿＿＿＿＿＿

Solve by use of the quadratic formula:

7. $x^2 + 8x + 3 = 0$ ＿＿＿＿＿＿＿

8. $x^2 + 4x + 2 = 0$ ＿＿＿＿＿＿＿

9. $y^2 - 10y - 3 = 0$ ＿＿＿＿＿＿＿

10. $y^2 - 6y - 3 = 0$ ＿＿＿＿＿＿＿

B 11. $u^2 = 1 - 3u$ ＿＿＿＿

12. $t^2 = 1 - t$ ＿＿＿＿

13. $y^2 + 3 = 2y$ ＿＿＿＿

14. $x^2 + 8 = 4x$ ＿＿＿＿

15. $2m^2 + 3 = 6m$ ＿＿＿＿

16. $2x^2 + 1 = 4x$ ＿＿＿＿

17. $p = 1 - 3p^2$ ＿＿＿＿

18. $3q + 2q^2 = 1$ ＿＿＿＿

Solve each of the following equations by any method, excluding completing the square.

19. $(x - 5)^2 = 7$ ＿＿＿＿

20. $(y + 4)^2 = 11$ ＿＿＿＿

21. $x^2 + 2x = 2$ ＿＿＿＿

22. $x^2 - 1 = 3x$ _____   **23.** $2u^2 + 3u = 0$ _____   **24.** $2n^2 = 4n$ _____

25. $x^2 - 2x + 9 = 2x - 4$ _____   **26.** $x^2 + 15 = 2 - 6x$ _____

27. $y^2 = 10y + 3$ _____   **28.** $3(2x + 1) = x^2$ _____   **29.** $2d^2 + 1 = 4d$ _____

30. $2y(3 - y) = 3$ _____   **31.** $\dfrac{2}{u} = \dfrac{3}{u^2} + 1$ _____   **32.** $1 + \dfrac{8}{x^2} = \dfrac{4}{x}$ _____

C 33. $\dfrac{1.2}{y - 1} + \dfrac{1.2}{y} = 1$ _____   **34.** $\dfrac{24}{10 + m} + 1 = \dfrac{24}{10 - m}$ _____

Solve for the indicated letter in terms of the other letters.

35. $d = \frac{1}{2}gt^2$ for t (positive) _____   **36.** $a^2 + b^2 = c^2$ for a (positive) _____

37. $A = P(1 + r)^2$ for (positive) _____   **38.** $P = EI - RI^2$ for I

The Check Exercise for this section is on page 363.

7.5 APPLICATIONS

We will now consider a number of applications from several fields that involve quadratic equations in their solutions. Since quadratic equations often have two solutions, it is important to check both solutions in the original problem to see if one or the other must be rejected. Also, a review of the strategy of solving word problems in Section 5.1 should prove helpful.

EXAMPLE 11
The sum of a number and its reciprocal is $\frac{5}{2}$. Find the number(s).

Solution

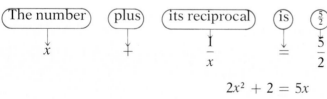

$$
\begin{array}{ll}
x + \dfrac{1}{x} = \dfrac{5}{2} & \text{Clear fractions. Note } x \neq 0. \\[2mm]
2x^2 + 2 = 5x & \text{Write in standard form.} \\[2mm]
2x^2 - 5x + 2 = 0 & \text{Solve by factoring.} \\[2mm]
(2x - 1)(x - 2) = 0 & \\[2mm]
x = \tfrac{1}{2} \quad \text{or} \quad 2 & \text{Both answers are good as can easily be checked.}
\end{array}
$$

Work Problem 11 and check solution in Solutions to Matched Problems on page 341.

PROBLEM 11   If the reciprocal of a number is subtracted from the original number, the difference is $\frac{8}{3}$. Find the number(s).

Solution

EXAMPLE 12

A tank can be filled in 4 hr by two pipes when both are used. How many hours are required for each pipe to fill the tank alone if the smaller pipe requires 3 hr more than the larger one?

Solution

Let $\;4\;$ = time for both pipes to fill the tank together
$\quad\;\; x\;$ = time for the larger pipe to fill the tank alone
$\;x + 3\;$ = time for the smaller pipe to fill the tank alone

Then $\frac{1}{4}$ = rate for both pipes together $\;\frac{1}{4}$ tank per hour

$\quad\;\frac{1}{x}$ = rate for larger pipe $\;\frac{1}{x}$ tank per hour

$\quad\;\frac{1}{x + 3}$ = rate for smaller pipe $\;\frac{1}{x + 3}$ tank per hour

$$\text{Sum of individual rates} = \text{rate together}$$

$$\frac{1}{x} + \frac{1}{x + 3} = \frac{1}{4}$$

$$4x(x + 3) \cdot \frac{1}{x} + 4x(x + 3) \cdot \frac{1}{x + 3} = 4x(x + 3) \cdot \frac{1}{4} \qquad \text{Clear fractions. Note that } x \neq -3, 0.$$

$$4(x + 3) + 4x = x(x + 3)$$

$$4x + 12 + 4x = x^2 + 3x$$

$$x^2 - 5x - 12 = 0 \qquad\qquad \text{Use quadratic formula.}$$

$$x = \frac{5 \pm \sqrt{73}}{2} \qquad\qquad \text{Why should we discard the negative answer?}$$

$$x = \frac{5 + \sqrt{73}}{2} \approx 6.77 \text{ hr (larger pipe)}$$

$$x + 3 \approx 9.77 \text{ hr (smaller pipe)}$$

Work Problem 12 and check solution in Solutions to Matched Problems on page 341.

PROBLEM 12   Two pipes can fill a tank in 3 hr when used together. Alone, one can fill the tank 2 hr faster than the other. How long will it take each pipe to fill the tank alone? Compute the answers to two decimal places, using the square root table at the end of the book or a calculator.

Solution

EXAMPLE 13

For a car traveling at a speed of v mi/hr, the least number of feet d under the best possible conditions that is necessary to stop a car (including a reaction time) is given approximately by the formula $d = 0.044v^2 + 1.1v$. Estimate the speed of a car requiring 200 ft to stop after danger is realized. A hand calculator will be useful for this problem. Compute the answer to two decimal places.

Solution

$$0.044v^2 + 1.1v = 200 \qquad \text{Write in standard form.}$$

$$0.044v^2 + 1.1v - 200 = 0 \qquad \text{Use the quadratic formula.}$$

$$v = \frac{-b \pm \sqrt{b^2 - 4ac}}{2a} \qquad \begin{aligned} a &= 0.044 \\ b &= 1.1 \\ c &= -200 \end{aligned}$$

$$v = \frac{-1.1 \pm \sqrt{1.1^2 - 4(0.044)(-200)}}{2(0.044)}$$

$$= \frac{-1.1 \pm \sqrt{36.41}}{0.088} \qquad \begin{aligned}&\text{Disregard the negative answer, since} \\ &\text{we are only interested in positive } v.\end{aligned}$$

$$= \frac{-1.1 + 6.03}{0.088} \quad = 56.02 \, \text{mi/hr} \qquad \text{Complete to two decimal places.}$$

NOTE: Example 13 is typical of most significant real-world problems in that decimal quantities rather than convenient small numbers are involved.

Work Problem 13 and check solution in Solutions to Matched Problems on page 341.

PROBLEM 13   Repeat Example 13 for a car requiring 300 ft to stop after danger is realized.

Solution

SOLUTIONS TO MATCHED PROBLEMS

11. Let x = the number
$$x - \frac{1}{x} = \frac{8}{3} \qquad x \neq 0$$
$$3x^2 - 3 = 8x$$
$$3x^2 - 8x - 3 = 0$$
$$(3x + 1)(x - 3) = 0$$
$$3x + 1 = 0 \qquad \text{or} \qquad x - 3 = 0$$
$$3x = -1 \qquad\qquad x = 3$$
$$x = -\tfrac{1}{3}$$

12. Let 3 = time for both pipes to fill tank together
x = time for faster pipe to fill tank alone
$x + 2$ = time for slower pipe to fill tank alone
Then $\frac{1}{3}$ = rate for both pipes together (1/3 tank per hour)
$$\frac{1}{x} = \text{rate for faster pipe}$$
$$\frac{1}{x+2} = \text{rate for slower pipe}$$
$$\frac{1}{x} + \frac{1}{x+2} = \frac{1}{3} \qquad x \neq -2, 0$$
$$3(x + 2) + 3x = x(x + 2)$$
$$3x + 6 + 3x = x^2 + 2x$$
$$-x^2 + 4x + 6 = 0$$
$$x^2 - 4x - 6 = 0$$
$$x = \frac{-(-4) \pm \sqrt{(-4)^2 - 4(1)(-6)}}{2(1)}$$
$$= \frac{4 \pm \sqrt{40}}{2}$$
$$= \frac{4 \pm 6.32}{2} \qquad \text{Keep positive answer only.}$$
$$x = 5.16 \text{ hr} \qquad \text{(faster pipe)}$$
$$x + 2 = 7.16 \text{ hr} \qquad \text{(slower pipe)}$$

13.
$$0.044v^2 + 1.1v = 300$$
$$0.044v^2 + 1.1v - 300 = 0$$
$$v = \frac{-1.1 \pm \sqrt{1.1^2 - 4(0.044)(-300)}}{2(0.044)}$$
$$= \frac{-1.1 \pm \sqrt{54.01}}{0.088}$$
$$= \frac{-1.1 + 7.35}{0.088}$$
$$= 71.02 \text{ mi/hr}$$

Thus, comparing this result with that obtained in Example 13, we see that a 27 percent increase in speed causes the car to go 50 percent farther before stopping.

PRACTICE EXERCISE 7.5

These problems are grouped according to type. The most difficult problems are double-starred (**), those of moderate difficulty single-starred (*), and the easier ones are not marked. A hand calculator with $\sqrt{\ }$ will prove useful in some cases. Compute decimal answers to two places.

NUMBER PROBLEMS

1. Find two consecutive positive even integers whose product is 168. _____

2. Find two positive numbers having a sum of 21 and a product of 104. _____

3. Find all numbers with the property that when the number is added to itself the sum is the same as when the number is multiplied by itself.

4. The sum of a number and its reciprocal is $\frac{10}{3}$. Find the number. _____

GEOMETRY

The following theorem may be used where needed:

PYTHAGOREAN THEOREM.   *A triangle is a right triangle if and only if the square of the longest side is equal to the sum of the squares of the two shorter sides.*

$$c^2 = a^2 + b^2$$

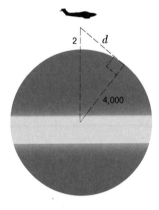

Figure 2

__*5.__ Approximately how far would the horizon be from an airplane 2 mi high? Assume the radius of the earth is 4,000 mi and use a square root table or a calculator to estimate the answer to the nearest mile (Figure 2).

_____

6. Find the base and height of a triangle with area 2 square meters if its base is 3 meters longer than its height. ($A = \frac{1}{2}bh$)

_____

*7. If the length and width of a 4- by 2-cm rectangle are each increased by the same amount, the area of the new rectangle will be twice the old. What are the dimensions of the new rectangle?

_____

8. The width of a rectangle is 2 meters less than its length. Find its dimensions if its area is 12 square meters.

_____

*9. A flag has a white cross of uniform width on a red background. Find the width of the cross so that it takes up exactly one-half the total area of a 4- by 3-ft flag.

_____

PHYSICS-ENGINEERING

10. The pressure p in pounds per square feet of wind blowing at v mi/hr is given by $p = 0.003v^2$. If a pressure gauge on a bridge registers a wind pressure of 14.7 lb/ft² what is the velocity of the wind?

_____

11. One method of measuring the velocity of water in a stream or river is to use an L-shaped tube as indicated in Figure 3. Torricelli's law in physics tells us that the height (in feet) that the water is pushed up into the tube above the surface is related to the water's velocity (in feet per second) by the formula $v^2 = 2gh$, where g is approximately 32 ft/sec² (feet per second per second). (NOTE: The device can also be used as a simple speedometer for a boat.) How fast is a stream flowing if $h = 0.5$ ft?

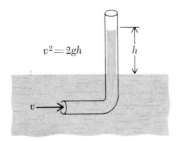

Figure 3

_____

12. At 20 mi/hr a car collides with a stationary object with the same force it would have if it had been dropped $13\frac{1}{2}$ ft, that is, if it had been pushed off the roof of an average one-story house. In general, a car moving at r mi/hr hits a stationary object with a force of impact that is equivalent to that force with which it would hit the ground when falling from a certain height h given by the formula $h = 0.0336r^2$. Approximately how fast would a car have to be moving if it crashed as hard as if it had been pushed off the top of a 12-story building 121 ft high?

_____

*13. For a car traveling at a speed of v mi/hr, the least number of feet d under the best possible conditions that is necessary to stop a car (including a reaction time) is given approximately by the formula $d = 0.044v^2 + 1.1v$. Estimate the speed of a car requiring 165 ft to stop after danger is realized.

_____

*14. If an arrow is shot vertically in the air (from the ground) with an initial velocity of 176 ft/sec its distance y above the ground t sec after it is released (neglecting air resistance) is given by $y = 176t - 16t^2$.
(A) Find the time when y is 0, and interpret physically.
(B) Find the times when the arrow is 16 ft off the ground.

_____

*15. A barrel 2 ft in diameter and 4 ft in height has a 1-in.-diameter drainpipe in the bottom. It can be shown that the height h of the surface of the water above the bottom of the barrel at time t min after the drain has been opened is given by the formula $h = (\sqrt{h_0} - \frac{5}{12}t)^2$, where h_0 is the water level above the drain at time $t = 0$. If the barrel is full and the drain opened, how long will it take to empty one-half of the contents? HINT: The problem is very easily solved if the right side of the equation is not squared.

_____

RATE-TIME
**16. One pipe can fill a tank in 5 hr less than another; together they fill the tank in 5 hr. How long would it take each alone to fill the tank? Compute the answer to two deciml places.

_____

**17. A new printing press can do a job in 1 hr less than an older press. Together they can do the same job in 1.2 hr. How long would it take each alone to do the job?

_____

*18. Two boats travel at right angles to each other after leaving the same dock at the same time; 1 hr later they are 13 km apart. If one travels 7 km/hr faster than the other, what is the rate of each? HINT: Use the pythagorean theorem near the beginning of this exercise set.

———————————————

*19. A speedboat takes 1 hr longer to go 24 km up a river than to return. If the boat cruises at 10 km/hr in still water, what is the rate of the current?

———————————————

ECONOMICS-BUSINESS

20. If P dollars is invested at r percent compounded annually, at the end of 2 years it will grow to $A = P(1 + r)^2$. At what interest rate will $100 grow to $144 in 2 years? NOTE: Find r for $A = 144$ and $P = 100$.

———————————————

*21. In a certain city the demand equation for popular records is $d = 3,000/p$, where d would be the quantity of records demanded on a given day if the selling price was p dollars per record. (Notice as the price goes up, the number of records the people are willing to buy goes down, and vice versa.) On the other hand, the supply equation is $s = 200p - 700$, where s is the quantity of records a supplier is willing to supply at p dollars per record. (Notice as the price goes up, the number of records a supplier is willing to sell goes up, and vice versa.) At what price will supply equal demand; that is, at what price will $d = s$? In economic theory the price at which supply equals demand is called the **equilibrium point**, the point at which the price ceases to change.

———————————————

> The Check Exercise for this section is on page 365.

7.6 EQUATIONS REDUCIBLE TO QUADRATIC FORM

•Equations involving radicals
••Other forms reducible to quadratic form

•Equations involving radicals

Consider the equation

$$x - 1 = \sqrt{x + 11}$$

What can we do to solve this equation? Perhaps doing something to the equation to eliminate the radical will help. What? Let us square both members to see what happens:

$$(x - 1)^2 = (\sqrt{x + 11})^2$$

$$x^2 - 2x + 1 = x + 11$$

$$x^2 - 3x - 10 = 0$$

$$(x + 2)(x - 5) = 0$$

$$x = -2, 5$$

CHECK

$x = -2$ $-2 - 1 \overset{?}{=} \sqrt{-2 + 11}$

$ -3 \overset{?}{=} \sqrt{9}$ Recall that $\sqrt{9}$ names the positive square root of 9.

$ -3 \neq 3$

Hence, $x = -2$ is not a solution.

$x = 5$ $5 - 1 \overset{?}{=} \sqrt{5 + 11}$

$ 4 \overset{?}{=} \sqrt{16}$

$ 4 \overset{\checkmark}{=} 4$

Hence, $x = 5$ is a solution.

     Therefore, 5 is a solution and -2 is not. The process of squaring introduced an "extraneous" solution. In general one can prove the following important theorem.

THEOREM 1

If both members of an equation are raised to a natural number power, then the solution set of the original equation is a subset of the solution set of the new equation.

Thus, any new equation obtained by raising both members of an equation to the same natural number power may have solutions (called **extraneous solutions**) that are not solutions of the original equation. On the other hand, any solution of the original equation must be among those of the new equation. We need to check all of the solutions at the end of the process to eliminate the so-called extraneous ones.

EXAMPLE 14

Solve: $x + \sqrt{x - 4} = 4$

Solution

$x + \sqrt{x - 4} = 4$ Isolate radical on one side.

$\sqrt{x - 4} = 4 - x$ Square both sides.

$\phantom{x + \sqrt{x - 4}} x - 4 = 16 - 8x + x^2$ Write in standard form.

$x^2 - 9x + 20 = 0$

$(x - 5)(x - 4) = 0$

$ x = 4, 5$

Checking, we find 4 is a solution and 5 is extraneous.

Work Problem 14 and check solution in Solutions to Matched Problems on page 348.

PROBLEM 14   Solve:

$$x = 5 + \sqrt{x - 3}$$

EXAMPLE 15
Solve: $\sqrt{2x + 3} - \sqrt{x - 2} = 2$

Solution

$\sqrt{2x + 3} - \sqrt{x - 2} = 2$	Easier to solve with a radical on each side.
$\sqrt{2x + 3} = \sqrt{x - 2} + 2$	Square both sides.
$2x + 3 = x - 2 + 4\sqrt{x - 2} + 4$	Isolate the radical on one side.
$x + 1 = 4\sqrt{x - 2}$	Square both sides again.
$x^2 + 2x + 1 = 16(x - 2)$	
$x^2 - 14x + 33 = 0$	
$(x - 11)(x - 3) = 0$	
$x = 3, 11$	Checking, we see that both numbers are solutions.

Work Problem 15 and check solution in Solutions to Matched Problems on page 348.

PROBLEM 15   Solve:

$$\sqrt{2x + 7} - \sqrt{x + 3} = 1$$

••Other forms reducible to quadratic form

Many equations that are not immediately recognizable as quadratic can be transformed into a quadratic form and then solved. Some examples are given below.

EXAMPLE 16
If asked to solve the equation

$$x^4 - x^2 - 12 = 0$$

you might at first have trouble. But if you recognize that the equation is quadratic in x^2, you can solve for x^2 first and then solve for x. You might find it convenient to make the substitution $u = x^2$, and then solve the equation

$$u^2 - u - 12 = 0$$
$$(u - 4)(u + 3) = 0$$
$$u = 4, -3$$

Replacing u with x^2, we obtain

$x^2 = 4 \qquad x^2 = -3$ Extraneous roots are not a problem here, since we have not
$x = \pm 2 \qquad x = \pm i\sqrt{3}$ performed operations that produce them.

Work Problem 16 and check solution in Solutions to Matched Problems on page 348.

PROBLEM 16   Solve for real solutions only:

$$x^6 + 6x^3 - 16 = 0$$

In general, if an equation that is not quadratic can be transformed into the form

$$au^2 + bu + c = 0$$

where u is an expression in some other variable, then the equation is said to be in **quadratic form.** Once recognized as a quadratic form, an equation can often be solved using quadratic methods.

EXAMPLE 17
Solve: $x^{2/3} - x^{1/3} - 6 = 0$

Solution
Let $u = x^{1/3}$; then $u^2 = x^{2/3}$. After substitution, the original equation becomes

$$u^2 - u - 6 = 0$$
$$(u - 3)(u + 2) = 0$$
$$u = 3, -2$$

Replacing u with $x^{1/3}$, we obtain

$x^{1/3} = 3 \qquad x^{1/3} = -2$ Now cube both sides of each equation to isolate x.
$x = 27 \qquad x = -8$ NOTE: $x \neq \sqrt[3]{3}$ and $x \neq \sqrt[3]{-2}$ (common errors)

Work Problem 17 and check solution in Solutions to Matched Problems on page 348.

PROBLEM 17 Solve:

$$x^{2/3} - x^{1/3} - 12 = 0$$

SOLUTIONS TO MATCHED PROBLEMS

14.
$$x = 5 + \sqrt{x - 3}$$
$$x - 5 = \sqrt{x - 3}$$
$$(x - 5)^2 = (\sqrt{x - 3})^2$$
$$x^2 - 10x + 25 = x - 3$$
$$x^2 - 11x + 28 = 0$$
$$(x - 4)(x - 7) = 0$$
$$x - 4 = 0 \quad \text{or} \quad x - 7 = 0$$
$$\cancel{x = 4} \qquad\qquad x = 7$$

15.
$$\sqrt{2x + 7} - \sqrt{x + 3} = 1$$
$$\sqrt{2x + 7} = \sqrt{x + 3} + 1$$
$$(\sqrt{2x + 7})^2 = (\sqrt{x + 3} + 1)^2$$
$$2x + 7 = x + 3 + 2\sqrt{x + 3} + 1$$
$$x + 3 = 2\sqrt{x + 3}$$
$$(x + 3)^2 = (2\sqrt{x + 3})^2$$
$$x^2 + 6x + 9 = 4(x + 3)$$
$$x^2 + 2x - 3 = 0$$
$$(x + 3)(x - 1) = 0$$
$$x + 3 = 0 \quad \text{or} \quad x - 1 = 0$$
$$x = -3 \qquad\qquad x = 1$$

Checking, we find both are solutions.

16. $x^6 + 6x^3 - 16 = 0$
Let $u = x^3$; then
$$u^2 + 6u - 16 = 0$$
$$(u + 8)(u - 2) = 0$$
$$u + 8 = 0 \quad \text{or} \quad u - 2 = 0$$
$$u = -8 \qquad\qquad u = 2$$
Replace u with x^3, and solve for x.
$$x^3 = -8 \qquad x^3 = 2$$
$$x = \sqrt[3]{-8} \qquad x = \sqrt[3]{2}$$
$$\quad = -2$$

17. $x^{2/3} - x^{1/3} - 12 = 0$
Let $u = x^{1/3}$; then
$$u^2 - u - 12 = 0$$
$$(u - 4)(u + 3) = 0$$
$$u - 4 = 0 \quad \text{or} \quad u + 3 = 0$$
$$u = 4 \qquad\qquad u = -3$$
Replace u with $x^{1/3}$, and solve for x.
$$x^{1/3} = 4 \qquad x^{1/3} = -3$$
$$(x^{1/3})^3 = 4^3 \qquad (x^{1/3})^3 = (-3)^3$$
$$x = 64 \qquad\qquad x = -27$$

PRACTICE EXERCISE 7.6

Work odd-numbered problems first, check answers, and then work even-numbered problems in areas of weakness. Answers to all problems are in the back of the book. Make every effort to work a problem yourself before you look at an answer.

Solve.

A **1.** $x - 2 = \sqrt{x}$ _____

2. $\sqrt{x} = x - 6$ _____

3. $m - 13 = \sqrt{m + 7}$ _____

4. $\sqrt{5n + 9} = n - 1$ _____

5. $x^4 - 10x^2 + 9 = 0$ _____

6. $x^4 - 13x^2 + 36 = 0$ _____

7. $x^4 - 7x^2 - 18 = 0$ _____

8. $y^4 - 2y^2 - 8 = 0$ _____

9. $\sqrt{x^2 - 3x} = 2$ _____

10. $\sqrt{x^2 - 8x} = 3$ _____

B **11.** $m - 7\sqrt{m} + 12 = 0$ _____

12. $t - 11\sqrt{t} + 18 = 0$ _____

13. $1 + \sqrt{x + 5} = x$ _____

14. $x - \sqrt{x + 10} = 2$ _____

15. $\sqrt{3x + 1} = \sqrt{x} - 1$ _____

16. $\sqrt{3x + 4} = 2 + \sqrt{x}$ _____

17. $\sqrt{3t + 4} + \sqrt{t} = -3$ _____

18. $\sqrt{3w - 2} - \sqrt{w} = 2$ _____

19. $\sqrt{u - 2} = 2 + \sqrt{2u + 3}$ _____

20. $\sqrt{3y - 2} = 3 - \sqrt{3y + 1}$ _____

21. $\sqrt{2x - 1} - \sqrt{x - 4} = 2$ _____

22. $\sqrt{y - 2} - \sqrt{5y + 1} = -3$ _____

23. $x^6 - 7x^3 - 8 = 0$ (real solutions only) _____

24. $x^6 + 3x^3 - 10 = 0$ (real solutions only) _____

25. $y^8 - 17y^4 + 16 = 0$ _____

26. $3m^4 - 4m^2 - 7 = 0$ _____

27. $x^{2/3} - 3x^{1/3} - 10 = 0$ _____

28. $2x^{2/3} + 3x^{1/3} - 2 = 0$ _____

29. $y^{1/2} - 3y^{1/4} + 2 = 0$ _____

30. $y^{1/2} - 5y^{1/4} + 6 = 0$ _____

31. $6x^{-2} - 5x^{-1} - 6 = 0$ _____

32. $3n^{-2} - 11n^{-1} - 20 = 0$ _____

C **33.** $4x^{-4} - 17x^{-2} + 4 = 0$ _____

34. $9y^{-4} - 10y^{-2} + 1 = 0$ _____

35. $(m^2 - m)^2 - 4(m^2 - m) = 12$ _____

36. $(x^2 + 2x)^2 - (x^2 + 2x) = 6$ _____

37. $(x - 3)^4 + 3(x - 3)^2 = 4$ _____

38. $(m - 5)^4 + 36 = 13(m - 5)^2$ _____

The Check Exercise for this section is on page 367.

7.7 NONLINEAR INQUALITIES

- •Quadratic inequalities
- ••Other inequalities

•Quadratic inequalities

You have now had quite a bit of experience solving first-degree inequalities in one variable such as

$2x - 3 \le 4(x - 4)$

But how do we solve second-degree inequalities such as

$x^2 + 2x < 8$

Using the quadratic formula directly doesn't work. However, if we move all terms to the left and factor, then we will be able to observe something that will lead to a solution.

$$x^2 + 2x - 8 < 0$$

$$(x + 4)(x - 2) < 0$$

We are looking for values of x that will make the left side less than 0, that is, negative. What will the signs of $(x + 4)$ and $(x - 2)$ have to be so that their product is negative? They must have opposite signs!

Let us see where each of the factors is positive, negative, and zero. The point at which either factor is zero is called a **critical point**. We will see why shortly.

Sign analysis for (x + 4):

Critical point	$(x + 4)$ is positive when	$(x + 4)$ is negative when
$x + 4 = 0$	$x + 4 > 0$	$x + 4 < 0$
$x = -4$	$x > -4$	$x < -4$

Geometrically:

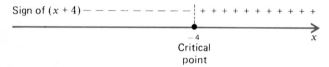

Thus, $(x + 4)$ is negative for values of x to the left of the critical point, and is positive for values of x to the right of the critical point.

Sign analysis for (x − 2):

Critical point	$(x - 2)$ is positive when	$(x - 2)$ is negative when
$x - 2 = 0$	$x - 2 > 0$	$x - 2 < 0$
$x = 2$	$x > 2$	$x < 2$

Geometrically:

Thus, $(x - 2)$ is negative for values of x to the left of the critical point, and is positive for values of x to the right of the critical point.

Combining the above results in a single geometric representation leads to a simple solution of the original problem.

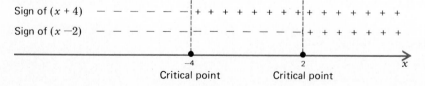

Now it is very easy to see that the factors have opposite signs for x between -4 and 2. Thus, the solution and graph of $x^2 + 2x < 8$ is

$-4 < x < 2$

The above discussion leads to the general result stated in Theorem 2, which is behind the sign-analysis method of solving quadratic inequalities.

THEOREM 2
The value of x at which $(ax + b)$ is zero is called a **critical point.** To the left of the critical point, on the real number line, $(ax + b)$ has one sign and to the right of the critical point, the opposite sign $(a \neq 0)$.

EXAMPLE 18
Solve and graph $x^2 \geq x + 6$

Solution

$$x^2 \geq x + 6$$

$$x^2 - x - 6 \geq 0$$

$$(x - 3)(x + 2) \geq 0$$

Critical points are: -2 and 3

Locate these points on a real number line and indicate the sign of each factor to the left and to the right of its critical point.

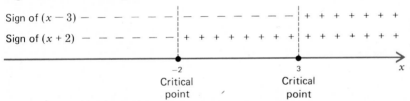

The inequality statement is satisfied when both factors have the same sign or when one or the other factor is 0. The former occurs when x is to the left of -2 or to the right of 3; the latter occurs at the critical points. Thus,

Solution: $x \leq -2$ or $x \geq 3$

Graph:

Work Problem 18 and check solution in Solutions to Matched Problems on page 353.

PROBLEM 18   Solve and graph:

     (A) $x^2 < x + 12$ **(B)** $x^2 \geq x + 12$

••Other inequalities

The procedures discussed above can also be used on some inequalities that are not quadratic.

EXAMPLE 19

Solve and graph: $\dfrac{x^2 - x + 1}{2 - x} \geq 1$

Solution

$\dfrac{x^2 - x + 1}{2 - x} \geq 1$ Since we do not know the sign of $2 - x$, we do not multiply both sides by it; instead, we subtract 1 from each side.

$\dfrac{x^2 - x + 1}{2 - x} - 1 \geq 0$ Combine terms on left side into a single fraction.

$\dfrac{x^2 - x + 1}{2 - x} - \dfrac{2 - x}{2 - x} \geq 0$

$\dfrac{x^2 - 1}{2 - x} \geq 0$ Factor numerator.

$\dfrac{(x - 1)(x + 1)}{2 - x} \geq 0$ Proceed as in Example 18. Critical points are $-1, 1, 2$.

Locate critical points on a real number line and indicate the sign of each first-degree form to the left and to the right of its critical point.

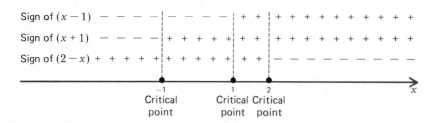

The inequality statement is satisfied when $(x - 1)$, $(x + 1)$, and $(2 - x)$ are all positive, two are negative and one is positive, or the numerator is 0. Two are negative to the left of -1 and all are positive between 1 and 2. The equality part of the inequality holds when x is 1 or -1, but not when $x = 2$. Thus,

Solution: $x \leq -1$ or $1 \leq x < 2$

Graph:

Work Problem 19 and check solution in Solutions to Matched Problems on page 353.

PROBLEM 19   Solve and graph:

$$\frac{3}{2 - x} \leq \frac{1}{x + 4}$$

SOLUTIONS TO MATCHED PROBLEMS

18. **(A)** $x^2 < x + 12$
$x^2 - x - 12 < 0$
$(x - 4)(x + 3) < 0$
Critical points: $-3, 4$

Sign of $(x - 4)$

Sign of $(x + 3)$

Solution and graph: $-3 < x < 4$

(B) $x^2 \geq x + 12$
$x^2 - x - 12 \geq 0$
$(x - 4)(x + 3) \geq 0$
Referring to part **A** we see that the solution and
graph are as follows:
$x \leq -3$ or $x \geq 4$

19. $\dfrac{3}{2 - x} \leq \dfrac{1}{x + 4}$ Move all terms to left side.

$\dfrac{3}{2 - x} - \dfrac{1}{x + 4} \leq 0$ Combine terms on left side.

$\dfrac{3(x + 4) - (2 - x)}{(2 - x)(x + 4)} \leq 0$ Simplify left side.

$\dfrac{4x + 10}{(2 - x)(x + 4)} \leq 0$ Find critical points.

Critical points: $-4, -\frac{5}{2}, 2$

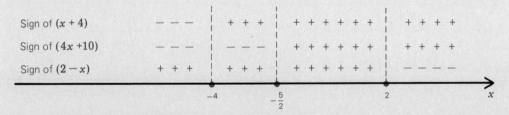

Sign of $(x + 4)$

Sign of $(4x + 10)$

Sign of $(2 - x)$

The inequality would be satisfied if $(4x + 10)$, $(2 - x)$, and $(x + 4)$ were all negative (which does not happen in this problem), or if one were negative and two were positive (which happens in two regions in this problem). The equality is satisfied only at $x = -\frac{5}{2}$. Thus, the solution and graph are as follows:
$-4 < x \leq -\frac{5}{2}$ or $x > 2$

PRACTICE EXERCISE 7.7

Work odd-numbered problems first, check answers, and then work even-numbered problems in areas of weakness. Answers to all problems are in the back of the book. Make every effort to work a problem yourself before you look at an answer.

Solve and graph (Problems 1 to 29).

A **1.** $(x - 3)(x + 4) < 0$ _____   **2.** $(x + 2)(x - 4) < 0$ _____

3. $(x - 3)(x + 4) \geq 0$ _____

4. $(x + 2)(x - 4) > 0$ _____

5. $x^2 + x < 12$ _____

6. $x^2 < 10 - 3x$ _____

7. $x^2 + 21 > 10x$ _____

8. $x^2 + 7x + 10 > 0$ _____

B **9.** $x(x + 6) \geq 0$ _____

10. $x(x - 8) \leq 0$ _____

11. $x^2 \geq 9$ _____

12. $x^2 > 4$ _____

13. $\dfrac{x - 5}{x + 2} \leq 0$ _____

14. $\dfrac{x + 2}{x - 3} < 0$ _____

15. $\dfrac{x - 5}{x + 2} > 0$ _____

16. $\dfrac{x + 2}{x - 3} \geq 0$ _____

17. $\dfrac{x - 4}{x(x + 2)} \; 0$ _____

18. $\dfrac{x(x + 5)}{x - 3} \; 0$ _____

19. $\dfrac{1}{x} < 4$ _____

20. $\dfrac{5}{x} > 3$ _____

21. $x^2 + 4 \geq 4x$ _____

22. $6x \leq x^2 + 9$ _____

23. $x^2 + 9 < 6x$ _____

24. $x^2 + 4 < 4x$ _____

25. $x^2 \geq 3$ _____

26. $x^2 < 2$ _____

27. $\dfrac{2}{x - 3} \leq -2$ _____

28. $\dfrac{2x}{x + 3} \geq 1$ _____

C **29.** $\dfrac{2}{x - 3} \leq \dfrac{2}{x + 2}$ _____

30. $\dfrac{2}{x + 1} \geq \dfrac{1}{x - 2}$ _____

31. For what values of x will $\sqrt{x^2 - 3x + 2}$ be a real number? _____

32. For what values of x will $\sqrt{\dfrac{x - 3}{x + 5}}$ be a real number? _____

33. If an object is shot straight up from the ground with an initial velocity of 160 ft/sec, its distance d in feet above the ground at the end of t sec (neglecting air resistance) is given by $d = 160t - 16t^2$. Find the duration of time for which $d \geq 256$.

_____

34. Repeat the preceding problem for $d \geq 0$. _____

> The Check Exercise for this section is on page 369.

7.8 Chapter Review

Important terms and symbols

7.1 INTRODUCTION   quadratic equation; standard form; $ax^2 + bx + c = 0, a \neq 0$

7.2 SOLUTION BY SQUARE ROOT AND BY FACTORING   solution by square root, solution by factoring

7.3 SOLUTION BY COMPLETING THE SQUARE   completing the square, solution by completing the square

$$x^2 + \underbrace{bx + \left(\frac{b}{2}\right)^2}_{} = \left(x + \frac{b}{2}\right)^2$$

Add the square of one-half of
the coefficient of x

7.4 THE QUADRATIC FORMULA   the quadratic formula, selection of method,

$$ax^2 + bx + c = 0 \qquad x = \frac{-b \pm \sqrt{b^2 - 4ac}}{2a}$$

7.6 EQUATIONS REDUCIBLE TO QUADRATIC FORM   equations involving radicals, other forms reducible to quadratic form

7.7 NONLINEAR INEQUALITIES   quadratic inequalities, other inequalities

DIAGNOSTIC (REVIEW) EXERCISE 7.8 _____

Work through all the problems in this chapter review and check answers in the back of the book. (Answers to all problems are there, and following each answer is a number in italics indicating the section in which that type of problem is discussed.) Where weaknesses show up, review appropriate sections in the text. When you are satisfied that you know the material, take the practice test following this review.

A *Find all solutions for Problems 1 to 5 by factoring or square root method.*

1. $x^2 - 3x = 0$ _____

2. $x^2 = 25$ _____

3. $x^2 - 5x + 6 = 0$ _____

4. $x^2 - 2x - 15 = 0$ _____

5. $x^2 - 7 = 0$ _____

6. Write $4x = 2 - 3x^2$ in standard form, $ax^2 + bx + c = 0$, and identify a, b, and c.

_____

7. Write down the quadratic formula associated with $ax^2 + bx + c = 0$ without looking at the text.

_____

8. Use the quadratic formula to solve $x^2 + 3x + 1 = 0$. _____

9. Find two positive numbers whose product is 27 if one is 6 more than the other.

_____

Solve and graph each inequality.

10. $x^2 + x < 20$ _____

11. $x^2 + x \geq 20$ _____

B *Find all solutions by factoring or square root method.*

12. $10x^2 = 20x$ _____

13. $3x^2 = 36$ _____

14. $3x^2 + 27 = 0$ _____

15. $(x - 2)^2 = 16$ _____

16. $3t^2 - 8t - 3 = 0$ _____

17. $2x = \dfrac{3}{x} - 5$ _____

Solve, using the quadratic formula.

18. $3x^2 = 2(x + 1)$ _____

19. $2x(x - 1) = 3$ _____

Solve, using any method.

20. $2x^2 - 2x = 40$ _____

21. $\dfrac{8m^2 + 15}{2m} = 13$ _____

22. $m^2 + m - 1 = 0$ _____

23. $u + \dfrac{3}{u} = 2$ _____

24. $\sqrt{5x - 6} - x = 0$ _____

25. $8\sqrt{x} = x + 15$ _____

26. $m^4 + 5m^2 - 36 = 0$ _____

27. $2x^{2/3} - 5x^{1/3} - 12 = 0$ _____

Solve and graph each inequality in Problems 28 to 31.

28. $x^2 \geq 4x + 21$ _____

29. $\dfrac{1}{x} < 2$ _____

30. $10x > x^2 + 25$ _____

31. $x^2 + 16 \geq 8x$ _____

32. The perimeter of a rectangle is 22 in. If its area is 30 in², find the length of each side.

_____

33. Solve $x^2 - 6x - 3 = 0$ by the completing-the-square method. _____

C *Solve Problems 34 to 40, using any method.*

34. $\left(t - \dfrac{3}{2}\right)^2 = -\dfrac{3}{2}$ _____

35. $3x - 1 = \dfrac{2(x + 1)}{x + 2}$ _____

36. $y^8 - 17y^4 + 16 = 0$ _____

37. $\sqrt{y - 2} - \sqrt{5y + 1} = -3$ _____

38. $\dfrac{3}{x-4} \leq \dfrac{2}{x-3}$ (also graph solution) _____

39. If the hypotenuse of a right triangle is 15 cm and its area is 54 cm², what are the lengths of the two sides? HINT: If x represents one side, use the pythagorean theorem to express the other side in terms of x; then use the formula for the area of a triangle, $A = \frac{1}{2}bh$.

_____

40. Cost equations for manufacturing companies are often quadratic in nature. (At very high or very low outputs the costs are more per unit because of inefficiency of plant operation at these extremes.) If the cost equation for manufacturing transistor radios is $C = x^2 - 10x + 31$, where C is the cost of manufacturing x units per week (both in thousands), find: (**A**) the output for a \$15,000 weekly cost? (**B**) the output for a \$6,000 weekly cost.

_____

PRACTICE TEST CHAPTER 7 _____

Take this as if it were a graded test. Allow yourself up to 50 minutes. Work the problems without looking back in the chapter. Correct your work, using the answers (keyed to appropriate sections) in the back of the book.

Solve Problems 1 to 7, using any method.

1. $2x^2 + 8 = 0$ _____   **2.** $2x^2 - 3x = 1$ _____

3. $x + \dfrac{13}{x} = 4$ _____   **4.** $m^4 + 3m^2 - 4 = 0$ _____

5. $x^{2/3} = x^{1/3} + 2$ _____   **6.** $x - \sqrt{12 - x} = 0$ _____

7. $\sqrt{x-2} = 2 + \sqrt{2x+3}$ _____

Solve and graph Problems 8 and 9.

8. $x^2 - x \leq 6$ _____   **9.** $\dfrac{3}{x-3} > \dfrac{2}{x+2}$ _____

10. The length of a rectangle is 1.5 meters longer than its width. If its area is 10 square meters, what are its dimensions?

_____

CHECK EXERCISE 7.2

NAME _____

CLASS _____

SCORE _____

Work the following problems without looking at any text examples. Show your work in the space provided. Write the letter that best indicates your answer in the answer column.

Solve Problems 1 to 5 by the square root method. By solve we mean to find all solutions.

ANSWER
COLUMN

1. _____
2. _____
3. _____
4. _____
5. _____
6. _____
7. _____
8. _____
9. _____
10. _____

1. $2x^2 - 32 = 0$

    (**A**) $x = \pm 2\sqrt{2}$
    (**B**) $x = 4$
    (**C**) $x = \pm 4$
    (**D**) None of these

2. $u^2 = 72$

    (**A**) $u = \pm 36$
    (**B**) $u = \sqrt{72}$
    (**C**) $u = 6\sqrt{2}$
    (**D**) None of these

3. $x^2 + 81 = 0$

    (**A**) $x = \pm 9i$
    (**B**) $x = \pm 9$
    (**C**) $x = 9i$
    (**D**) None of these

4. $(x - \frac{3}{2})^2 = \frac{3}{4}$

    (**A**) $\dfrac{-3 \pm \sqrt{3}}{2}$
    (**B**) $\dfrac{3 \pm \sqrt{3}}{2}$
    (**C**) $x = \pm \dfrac{\sqrt{3}}{2}$
    (**D**) None of these

5. $(x + 3)^2 = -4$

    (**A**) $x = -5, -1$
    (**B**) $x = -3 \pm 2i$
    (**C**) $x = 3 \pm 2i$
    (**D**) None of these

Solve Problems 6 to 9 by factoring, if possible.

6. $m^2 = 2m + 15$

    (A) $m = -5, 3$
    (B) $m = -3, 5$
    (C) $m = 3, 5$
    (D) None of these

7. $4x^2 + 23x = 6$

    (A) $x = -\frac{1}{4}, 6$
    (B) $x = -6, \frac{1}{4}$
    (C) Not solvable by factoring in the integers
    (D) None of these

8. $2x^2 = 4x + 3$

    (A) $x = -3, 2$
    (B) $x = -2, 3$
    (C) Not solvable by factoring in the integers
    (D) None of these

9. $6 = \dfrac{1}{x} + \dfrac{2}{x^2}$

    (A) $x = -\frac{1}{2}, \frac{2}{3}$
    (B) $x = -2, \frac{3}{2}$
    (C) $x = -\frac{2}{3}, \frac{1}{2}$
    (D) None of these

10. The height of a triangle is 2 centimeters (cm) less than its base. Find its base and height if its area is 12 cm². $(A = \frac{1}{2}bh)$

    (A) $b = 6\,\text{cm}, h = 4\,\text{cm}$
    (B) $b = 2\,\text{cm}, h = 12\,\text{cm}$
    (C) $b = 8\,\text{cm}, h = 3\,\text{cm}$
    (D) None of these

CHECK EXERCISE 7.3

NAME _____

CLASS _____

SCORE _____

Work the following problems without looking at any text examples. Show your work in the space provided. Write the letter that best indicates your answer in the answer column.

ANSWER COLUMN

1. _____

2. _____

3. _____

4. _____

5. _____

Solve by method of completing the square.

1. $x^2 + 2x - 1 = 0$

    (A) $x = -2, 0$
    (B) $x = -1 \pm \sqrt{2}$
    (C) $x = 1 \pm \sqrt{2}$
    (D) None of these

2. $x^2 - 2x - 8 = 0$

    (A) $x = -2, -4$
    (B) $x = -4, 2$
    (C) $x = 1 \pm \sqrt{7}$
    (D) None of these

3. $u^2 - 6u + 10 = 0$

    (A) $u = 3 \pm i$
    (B) $u = -3 \pm i$
    (C) $u = 3 \pm \sqrt{19}$
    (D) None of these

4. $m^2 + 3m - 1 = 0$

    (A) $m = -3 \pm \sqrt{10}$

    (B) $m = -\dfrac{3}{2} \pm \dfrac{\sqrt{13}}{2}i$

    (C) $m = \dfrac{-3 \pm \sqrt{13}}{2}$

    (D)   None of these

5. $2x^2 - 8x - 3 = 0$

    (A) $x = 2 \pm \sqrt{88}$

    (B) $x = \dfrac{4 \pm \sqrt{10}}{2}$

    (C) $x = \dfrac{4 \pm \sqrt{22}}{2}$

    (D)   None of these

CHECK EXERCISE 7.4

Work the following problems without looking at any text examples. Show your work in the space provided. Write the letter that best indicates your answer in the answer column.

Solve Problems 1 and 2, using the quadratic formula.

1. $x^2 = 6x - 7$

    (A) $x = -1, 7$
    (B) $x = 1, 5$
    (C) $x = 3 \pm \sqrt{2}$
    (D) None of these

2. $4x^2 + 3 = 4x$

    (A) $x = \dfrac{1}{2} \pm \dfrac{\sqrt{2}}{2}i$
    (B) $x = -2, 3$
    (C) $x = \dfrac{1 + \sqrt{2}}{2}$
    (D) None of these

363

, using the most efficient method.

(A) $m = \frac{1}{2}$
(B) $m = 0, \frac{1}{2}$
(C) $m = 0, 2$
(D) None of these

4. $(x - 3)^2 = -4$

(A) $x = 3 \pm 2i$
(B) $x = -3 \pm 2i$
(C) $x = 1, 5$
(D) None of these

5. $\dfrac{24}{7 + x} + 1 = \dfrac{24}{7 - x}$

(A) $x = 7$
(B) $x = -1, 49$
(C) $x = -49, 1$
(D) None of these

CHECK EXERCISE 7.5

Work the following problems without looking at any text examples. Show your work in the space provided. Write the letter that best indicates your answer in the answer column.

Set up an equation and solve. Use a square root table or a hand calculator to express answers to two decimal places.

1. Find all consecutive odd number pairs whose product is 143.

    (**A**) 11, 13
    (**B**) $-13, -11$
    (**C**) $-13, -11$ and 11, 13
    (**D**) None of these

2. The length of a rectangle is 3 centimeters (cm) less than twice the width. What are the rectangle's dimensions if its area is 10 cm²?

    (**A**) 2 by 5 cm
    (**B**) 3.11 by 3.22 cm
    (**C**) 2.99 by 2.97 cm
    (**D**) None of these

3. If *P* dollars is invested at *r* percent compounded annually, at the end of 2 years it will grow to $A = P(1 + r)^2$. At what rate of interest will $10 grow to $12.10?

    (**A**) -221 percent
    (**B**) 21 percent
    (**C**) 0.21 percent
    (**D**) None of these

365

4. If an arrow is shot vertically in the air (from the ground) with an initial velocity of 192 ft/sec its distance y above the ground t sec after it is released (neglecting air resistance) is given by $y = 192t - 16t^2$. Find the time at which y is 0.

(A) $t = 0$, 12 sec
(B) $t = 1$, 11 sec
(C) $t = 12$ sec
(D) None of these

5. One electronic typewriter can complete a personalized mailing in 1 hr less than an older model. Together they can complete the job in 3 hr. How long would it take each alone to complete the job?

(A) 4.3 and 5.3 hr
(B) 5.54 and 6.54 hr
(C) 0.7 and 1.7 hr
(D) None of these

CHECK EXERCISE 7.6

Work the following problems without looking at any text examples. Show your work in the space provided. Write the letter that best indicates your answer in the answer column.

Solve completely.

1. $\sqrt{x} = 2 - x$

      (A) $x = 4$
      (B) $x = -4$
      (C) $x = 1, 4$
      (D)   None of these

2. $x = \sqrt{x + 5} + 1$

      (A) $x = -1, 4$
      (B) $x = 4$
      (C) $x = -1$
      (D)   None of these

3. $\sqrt{m - 1} + \sqrt{3m + 3} = 4$

      (A) $m = 2, 26$
      (B) $m = 26$
      (C) $m = 2$
      (D)   None of these

367

4. $x^{-4} - 3x^{-2} - 4 = 0$

(A) $x = \pm\frac{1}{2}$
(B) $x = 1, 16$
(C) $x = \pm\frac{1}{2}, \pm i$
(D) None of these

5. $x^{2/3} + 7x^{1/3} - 8 = 0$

(A) $x = 1, 2$
(B) $x = -2, 1$
(C) $x = -512, 1$
(D) None of these

CHECK EXERCISE 7.7

Work the following problems without looking at any text examples. Show your work in the space provided. Write the letter that best indicates your answer in the answer column.

Solve and graph.

1. $(x - 3)(x + 1) < 0$

    (A) $-1 < x < 3$

    (B) $x < -1$ or $x > 3$

    (C) $x > -1$

    (D) None of these

2. $x^2 + 2x \geq 15$

    (A) $x < -5$ or $x > 3$

    (B) $x \leq -5$ or $x \geq 3$

    (C) $-5 \leq x \leq 3$

    (D) None of these

3. $\dfrac{x - 5}{x + 2} \leq 0$

    (A) $x < -2$ or $x \geq 5$

    (B) $-2 < x < 5$

    (C) $-2 \leq x \leq 5$

    (D) None of these

4. $\dfrac{2}{x} < 1$

(A) $x > 1$

(B) $0 < x < 2$

(C) $x < 0$ or $x > 2$

(D) None of these

5. $\dfrac{3}{x + 3} \geq \dfrac{1}{x - 1}$

(A) $x \geq 3$

(B) $x < -3$ or $1 < x \leq 3$

(C) $-3 < x < 1$ or $x \geq 3$

(D) None of these

Chapter Eight

GRAPHING INVOLVING TWO VARIABLES

CONTENTS

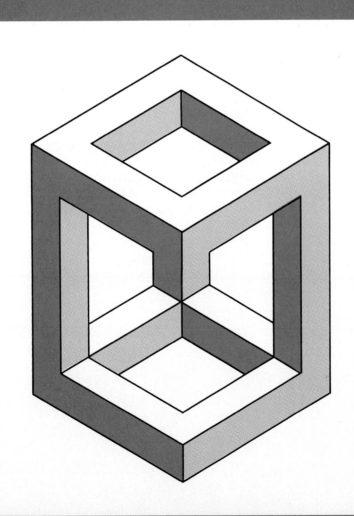

Could you build this "box"?

INSTRUCTIONS FOR STUDENTS IN A
SELF-PACED CLASS OR LAB

yes ← **HAVE YOU HAD INTERMEDIATE ALGEBRA BEFORE THIS COURSE?** → no

1. Work Diagnostic (Review) Exercise 8.6 on page 431. Check answers in back of book; then work through text sections corresponding to problems missed. (Section numbers are in italics following each answer.)
2. When finished with step 1, take Practice Test: Chapter 8 on page 434 as a final check of your understanding of the chapter. Check answers in the back of the book; then review sections where weakness still prevails. (Corresponding section numbers are in italics following each answer.)
3. When you think you are ready, ask your instructor for a graded test for Chapter 8.
4. If your instructor approves, after the test is corrected, go to the next chapter.

1. Work through each section in the chapter as follows:
   (a) Read discussion.
   (b) Read each example and work the corresponding matched problem. Check your solution to the matched problem in Solutions to Matched Problems on the indicated page.
   (c) At the end of a section work the odd-numbered problems in the Practice Exercise and check answers; then work even-numbered problems in areas of weakness. (Answers to *all* Practice Exercise sets are in the back of the book.)
   (d) Work Check Exercise as instructed. Tear out and turn in as directed by your instructor. (Answers are not in the text.)
2. Repeat each step in item 1 for each section in the chapter.
3. After the instructional part of the chapter is completed, proceed with steps 1 to 4 in the box above.

8.1 GRAPHING LINEAR EQUATIONS

- •Cartesian coordinate system
- ••Graphing a first-degree equation in two variables
- •••Vertical and horizontal lines
- ••••Use of different scales for each coordinate axis

In Chapters 4 and 7 we graphed equations and inequalities in one variable on a real number line. Recall,

Statement   *Graph*

$-3 < x \le 5$

Every real number can be associated with a unique point on a line, and, conversely, every point on a line can be associated with a unique real number. We now develop a system that will enable us to graph equations and inequalities in two variables such as

$$3x - 2y = 5 \qquad y \le 4x - 1$$

•Cartesian coordinate system

To form a cartesian coordinate system we select two real number lines, one vertical and one horizontal, and let them cross through their origins (zeros) as indicated in Figure 1. Up and to the right

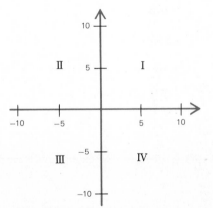

Figure 1

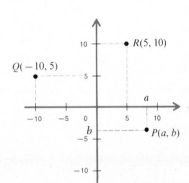

Figure 2

are the usual choices for the positive directions. These two number lines are called the **vertical axis** and the **horizontal axis** or (together) the **coordinate axes.** The coordinate axes divide the plane into four parts called **quadrants.** The quadrants are numbered counterclockwise from I to IV. All points in the plane lie in one of the four quadrants except for points on the coordinate axes.

Pick a point *P* in the plane at random (see Figure 2). Pass horizontal and vertical lines through the point. The vertical line will intersect the horizontal axis at a point with coordinate *a*, and the horizontal line will intersect the vertical axis at a point with coordinate *b*. These two numbers form the **coordinates**

$$(a, b)$$

of the point *P* in the plane. In particular, the coordinates of the point *Q* are $(-10, 5)$ and of the point *R* are $(5, 10)$.

The first coordinate a of the coordinates of point P is also called the **abscissa** of P; the second coordinate b of the coordinates of point P is also called the **ordinate** of P. The abscissa for Q in Figure 2 is -10 and the ordinate for Q is 5. The point with coordinates $(0, 0)$ is called the **origin.**

We know that coordinates (a, b) exist for each point in the plane since every point on each axis has a real number associated with it. Hence, by the procedure described, each point located in the plane can be labeled with a unique pair of real numbers. Conversely, by reversing the process, each pair of real numbers can be associated with a unique point in the plane.

The system that we have just defined is called a **cartesian coordinate system** (sometimes referred to as a **rectangular coordinate system**).

EXAMPLE 1

Find the coordinates of each of the points A, B, C, and D.

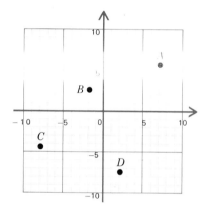

Solution

$A\,(7, 6)$ $B\,(-2, 3)$ $C\,(-8, -4)$ $D\,(2, -7)$

Work Problem 1 and check solution in Solutions to Matched Problems on page 381.

PROBLEM 1   Using the figure in Example 1, find the coordinates for the point:

     (A)   2 units to the right and 1 unit up from A _____

     (B)   2 units to the left and 2 units down from C _____

     (C)   1 unit up and 1 unit to the left of D _____

     (D)   2 units to the right of B _____

EXAMPLE 2

Graph (associate each ordered pair of numbers with a point in the cartesian coordinate system):

$(2, 7)$ $(7, 2)$ $(-8, 4)$ $(4, -8)$ $(-8, -4)$ $(-4, -8)$ $(3, 0)$ $(0, 3)$

Solution

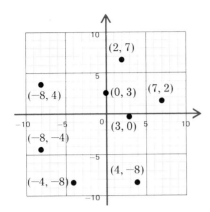

To plot $(-8, 4)$, for example, start at the origin and count 8 units to the left; then go straight up 4 units.

It is very important to note that the ordered pair $(2, 7)$ and the set $\{2, 7\}$ are not the same thing; $\{2, 7\} = \{7, 2\}$, but $(2, 7) \neq (7, 2)$.

Work Problem 2 and check solution in Solutions to Matched Problems on page 381.

PROBLEM 2   Graph $(3, 4), (-3, 2), (-2, -2), (4, -2), (0, 1)$, and $(-4, 0)$ in the same coordinate system.

Solution

Draw in coordinate axes and indicate coordinate scales on each axis first.

The development of the cartesian coordinate system represented a very important advance in mathematics. It was through the use of this system that René Descartes (1596–1650), a French philosopher-mathematician, was able to transform geometric problems requiring long tedious reasoning into algebraic problems that could be solved almost mechanically. This joining of algebra and geometry has now become known as **analytic geometry.**

Two fundamental problems of analytic geometry are the following:

1. Given an equation, find its graph.
2. Given a geometric figure, such as a straight line, circle, or ellipse, find its equation.

In this section we will be mainly interested in the first problem. In other parts of this chapter we will consider the second problem. Before we take up the first problem, let us refresh your memory on what is meant by "the graph of an equation." In general:

The Graph of an Equation

The graph of an equation in two variables in a rectangular coordinate system must meet the following two conditions:

1. If an ordered pair of numbers is a solution to the equation, the corresponding point must be on the graph of the equation.
2. If a point is on the graph of an equation, its coordinates must satisfy the equation.

••Graphing a first-degree equation in two variables

Suppose we are interested in graphing

$$y = 2x - 4$$

We start by finding some of its solutions. A **solution** of an equation in two variables is an ordered pair of real numbers that satisfies the equation. If we agree that the first element in the ordered pair will replace x and the second y, then

$$(0, -4)$$

is a solution of $y = 2x - 4$, as can easily be checked. How do we find other solutions? The answer is easy: We simply assign to x in $y = 2x - 4$ any convenient value and solve for y. For example, if $x = 3$, then

$$y = 2(3) - 4 = 2$$

Hence,

$$(3, 2)$$

is another solution of $y = 2x - 4$. It is clear that by proceeding in this manner, we can get solutions to this equation without end. Thus, the solution set is infinite. Let us make up a table of some solutions and graph them in a cartesian coordinate system, identifying the horizontal axis with x and the vertical axis with y.

Choose x	Compute $2x - 4 = y$	Write ordered pair (x, y)
-4	$2(-4) - 4 = -12$	$(-4, -12)$
-2	$2(-2) - 4 = -8$	$(-2, -8)$
0	$2(0) - 4 = -4$	$(0, -4)$
2	$2(2) - 4 = 0$	$(2, 0)$
4	$2(4) - 4 = 4$	$(4, 4)$
6	$2(6) - 4 = 8$	$(6, 8)$
8	$2(8) - 4 = 12$	$(8, 12)$

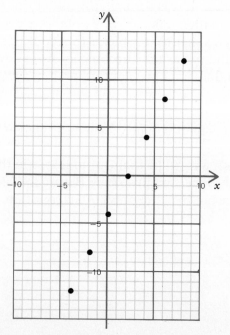

It appears that the graph of the equation is a straight line. If we knew this for a fact, then graphing $y = 2x - 4$ would be easy. We would simply find two solutions of the equation, plot them, and then plot as much of $y = 2x - 4$ as we liked by drawing a line through the two points, using a ruler. It turns out that it is true that the graph of $y = 2x - 4$ is a straight line. In fact, we have the following general theorem, which we state without proof.

THEOREM 1

> The graph of any equation of the form
>
> $$y = mx + b \quad \text{or} \quad Ax + By = C$$
>
> where $m, b, A, B,$ and C are constants (A and B both not 0) and x and y are variables, is a straight line.

Thus, the graphs of

$$y = \tfrac{2}{3}x - 5 \quad \text{and} \quad 2x - 3y = 12$$

are straight lines, since the first is of the form $y = mx + b$ and the second is of the form $Ax + By = C$.

> *Graphing Equations of the Form $y = mx + b$ or $Ax + By = C$*
>
> *Step 1.* Find two solutions of the equation. (A third solution is sometimes useful as a check point.)
> *Step 2.* Plot the solutions in a coordinate system.
> *Step 3.* Using a ruler, draw a straight line through the points plotted in step 2.
>
> NOTE: The third solution provides a check point, since if a ruler does not pass through all three points, a mistake has been made in finding the solutions.

EXAMPLE 3
(A) Graph $y = 2x - 4$.

Solution
Make up a table of at least two solutions (ordered pairs of numbers that satisfy the equation), plot these, and then draw a line through these points with a ruler.

x	2x − 4 = y	(x, y)
0	2(0) − 4 = −4	(0, −4)
2	2(2) − 4 = 0	(2, 0)
4	2(4) − 4 = 4	(4, 4)

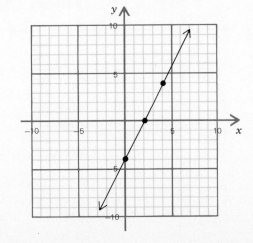

(B)   Graph $x + 3y = 6$.

Solution

To graph $x + 3y = 6$, assign to either x or y any convenient value and solve for the other variable. If we let $x = 0$, a convenient value, then

$$0 + 3y = 6$$

$$3y = 6$$

$$y = 2$$

Thus, $(0, 2)$ is a solution.

If we let $y = 0$, another convenient choice, then

$$x + 3(0) = 6$$

$$x + 0 = 6$$

$$x = 6$$

Thus, $(6, 0)$ is a solution.

To find a check point, choose another value for x or y, say $x = -6$; then

$$-6 + 3y = 6$$

$$3y = 12$$

$$y = 4$$

Thus, $(-6, 4)$ is also a solution.

We summarize the above results in a table and then draw the graph. The first two solutions indicate where the graph crosses the coordinate axes and are called the *y* **and** *x* **intercepts,** respectively. The intercepts are often the easiest points to find in an equation of the form $Ax + By = C$. To find the *y* intercept we let $x = 0$ and solve for *y*; to find the *x* intercept let $y = 0$ and solve for *x*. This is called the **intercept method** of graphing a straight line.

(x, y)	
$(0, 2)$	*y* intercept
$(6, 0)$	*x* intercept
$(-6, 4)$	Check point

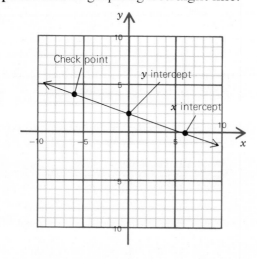

If a straight line does not pass through all three points, then we know we have made a mistake and must go back and check our work.

Work Problem 3 and check solution in Solutions to Matched Problems on page 381.

PROBLEM 3   Graph: (Be sure to draw in coordinate axes, label them, and indicate coordinate scales on each axis.)

$$\textbf{(A)} \quad y = -\tfrac{1}{3}x + 3 \qquad\qquad \textbf{(B)} \quad 3x - 2y = 15$$

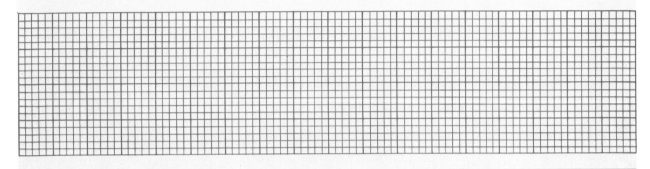

•••Vertical and horizontal lines

Vertical and horizontal lines in rectangular coordinate systems have particularly simple equations.

EXAMPLE 4

Graph $y = 4$ and $x = 3$ in a rectangular coordinate system.

Solution

To graph $y = 4$ or $x = 3$ in a rectangular coordinate system, each equation must be provided with the missing variable (usually done mentally) as follows:

$$y = 4 \qquad \text{is equivalent to} \qquad 0x + y = 4$$
$$x = 3 \qquad \text{is equivalent to} \qquad x + 0y = 3$$

In the first case, we see that no matter what value is assigned to x, $0x = 0$; thus, as long as $y = 4$, x can assume any value, and the graph of $y = 4$ is a horizontal line crossing the y axis at 4. Similarly, in the second case y can assume any value as long as $x = 3$, and the graph of $x = 3$ is a vertical line crossing the x axis at 3. Thus,

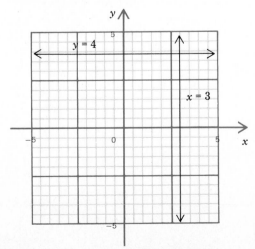

Work Problem 4 and check solution in Solutions to Matched Problems on page 381.

PROBLEM 4   Graph $y = -3$ and $x = -4$ in a rectangular coordinate system.

Solution

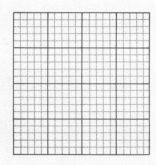

••••Use of different scales for each coordinate axis

Given the problem of graphing

$$A = 50 + 5t \qquad 0 \le t \le 10$$

we first note that t is restricted to values from 0 to 10. Let us find A for three values in this interval. We choose each end value and the middle value.

t	0	5	10
A	50	75	100

We see that as t varies from 0 to 10, A varies from 50 to 100. To keep the graph on the paper, we choose a different scale on the vertical axis (representing A) than that used on the horizontal axis (representing t). In addition, we note that the whole graph (under the restrictions for t) lies in the first quadrant. Thus, we obtain

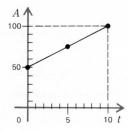

The graph is a line segment joining the two points (0, 50) and (10, 100). The line segment does not extend beyond these two points. The dotted lines are guide lines used to guide one's eyes to the endpoint, and are not part of the graph.

It should now be clear why equations of the form $Ax + By = C$ and $y = mx + b$ are called **linear equations:** their graphs are straight lines.

SOLUTIONS TO MATCHED PROBLEMS

1. **(A)** $(9, 7)$ **(B)** $(-10, -6)$ **(C)** $(1, -6)$ **(D)** $(0, 3)$

2.

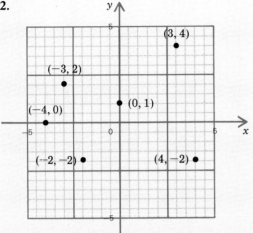

3. **(A)** $y = -\frac{1}{3}x + 3$

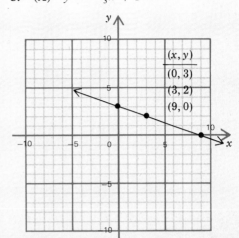

(x, y)
$(0, 3)$
$(3, 2)$
$(9, 0)$

4.

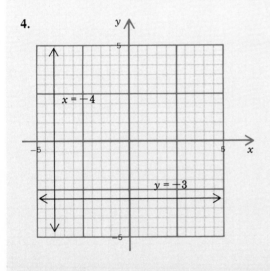

(B) $3x - 2y = 15$

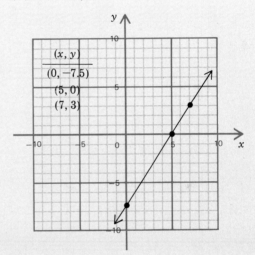

(x, y)
$(0, -7.5)$
$(5, 0)$
$(7, 3)$

PRACTICE EXERCISE 8.1

Work odd-numbered problems first, check answers, and then work even-numbered problems in areas of weakness. Answers to all problems are in the back of the book. Make every effort to work a problem yourself before you look at an answer.

Graph each equation in a cartesian coordinate system.

A **1.** $y = 2x$ **2.** $y = x$ **3.** $y = 2x - 3$

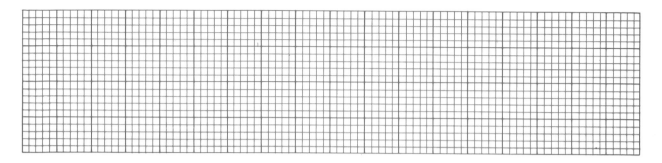

4. $y = x - 1$ **5.** $y = \dfrac{x}{3}$ **6.** $y = \dfrac{x}{2}$

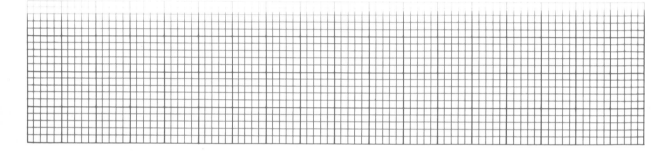

7. $y = \dfrac{x}{3} + 2$ **8.** $y = \dfrac{x}{2} + 1$ **9.** $x + y = -4$

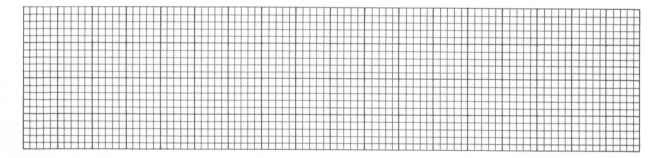

10. $x + y = 6$ **11.** $x - y = 3$ **12.** $x - y = 5$

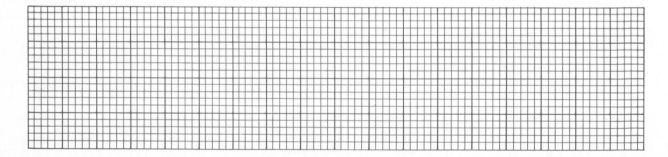

13. $3x + 4y = 12$ **14.** $2x + 3y = 12$ **15.** $8x - 3y = 24$

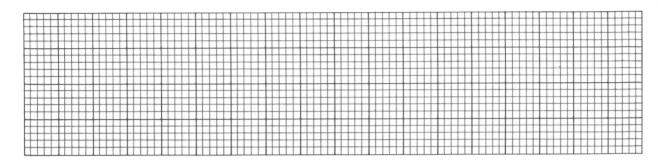

16. $3x - 5y = 15$ **17.** $y = 3$ **18.** $x = 2$

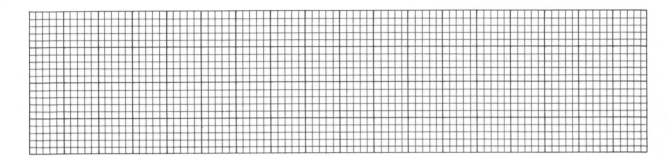

19. $x = -4$ **20.** $y = -3$

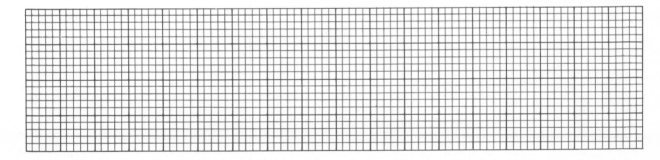

B 21. $y = \frac{1}{2}x$ **22.** $y = \frac{1}{4}x$ **23.** $y = \frac{1}{2}x - 1$

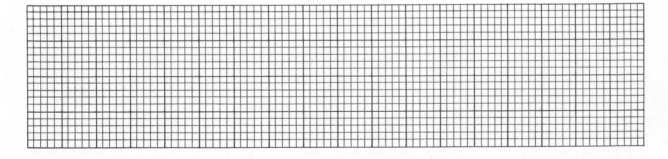

24. $y = \frac{1}{4}x + 1$ **25.** $y = -x + 2$ **26.** $y = -2x + 6$

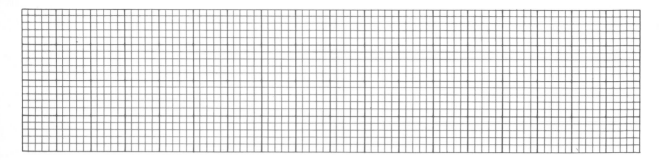

27. $y = \frac{1}{3}x - 1$ **28.** $y = -\frac{1}{2}x + 2$ **29.** $3x + 2y = 10$

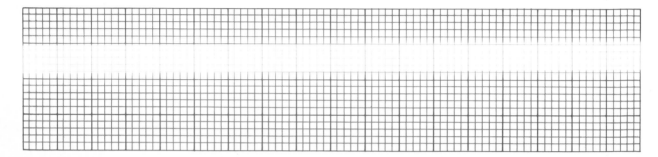

30. $2x + y = 7$ **31.** $5x - 6y = 15$ **32.** $7x - 4y = 21$

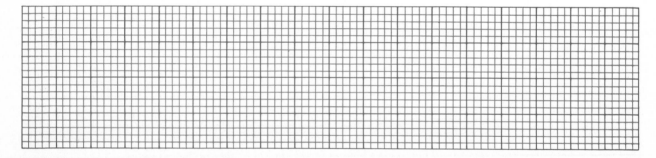

Graph Problems 33 to 42, using a different scale on the vertical axis from that on the horizontal axis to keep the size of the graph within reason.

33. $I = 6t, 0 \leq t \leq 10$ **34.** $d = 60t, 0 \leq t \leq 10$

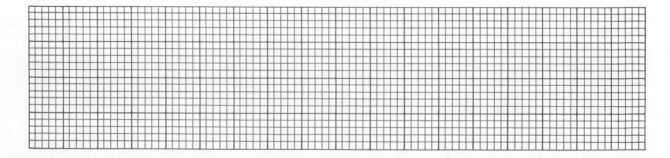

35. $v = 10 + 32t, 0 \le t \le 5$ **36.** $A = 100 + 10t, 0 \le t \le 10$

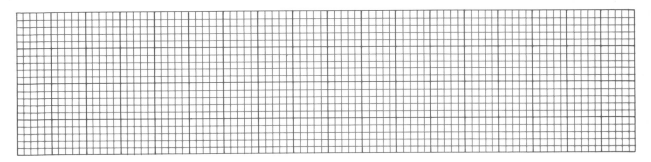

37. Graph $x + y = 3$ and $2x - y = 0$ on the same coordinate system. Determine by inspection the coordinates of the point where the graphs cross. Show that the coordinates of the point of intersection satisfy both equations.

38. Repeat the preceding problem with the equations $2x - 3y = -6$ and $x + 2y = 11$.

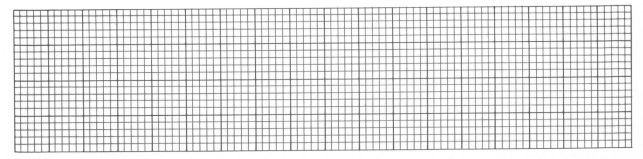

39. Graph $y = mx - 2$ for $m = 2, m = \frac{1}{2}, m = 0$, $m = -\frac{1}{2}$, and $m = -2$, all on the same coordinate system.

40. Graph $y = -\frac{1}{2}x + b$ for $b = -6, b = 0$, and $b = 6$, all on the same coordinate system.

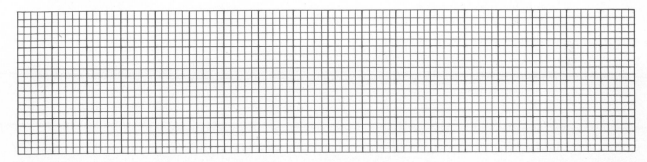

APPLICATION
41. ELECTRONICS In a simple electric circuit, such as found in a flashlight, if the resistance is 30 ohms, the current in the circuit I (in amperes) and the electromotive force E (in volts) are related by the equation $E = 30I$. Graph this equation for $0 \le I \le 1$.

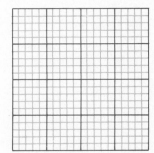

The Check Exercise for this section is on page 437.

8.2 SLOPE AND EQUATIONS OF A LINE

•Slope of a line
••Slope-intercept form
•••Point-slope form
••••Vertical and horizontal lines
•••••Parallel and perpendicular lines

In the preceding section we considered the problem: Given a linear equation of the form

$$Ax + By = C$$

or

$$y = mx + b$$

find its graph. Now we will consider the reverse problem: Given certain information about a straight line in a rectangular coordinate system, find its equation. We start by introducing a measure of the steepness of a line called slope.

•Slope of a line

If we take two points (x_1, y_1) and (x_2, y_2) on a line, then the ratio of the change in y to the change in x as we move from point P_1 to P_2 is called the **slope** of the line.

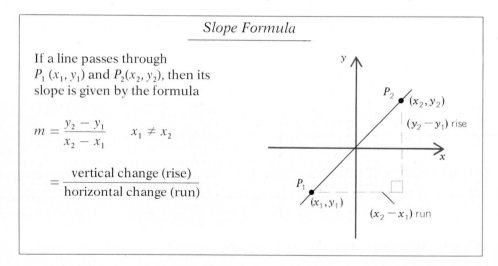

Slope Formula

If a line passes through $P_1 (x_1, y_1)$ and $P_2(x_2, y_2)$, then its slope is given by the formula

$$m = \frac{y_2 - y_1}{x_2 - x_1} \qquad x_1 \neq x_2$$

$$= \frac{\text{vertical change (rise)}}{\text{horizontal change (run)}}$$

EXAMPLE 5
Find the slope of the line passing through $(-3, -2)$ and $(3, 4)$.

Solution
Let $(x_1, y_1) = (-3, -2)$ and $(x_2, y_2) = (3, 4)$; then

$$m = \frac{y_2 - y_1}{x_2 - x_1} = \frac{4 - (-2)}{3 - (-3)} = \frac{6}{6} = 1$$

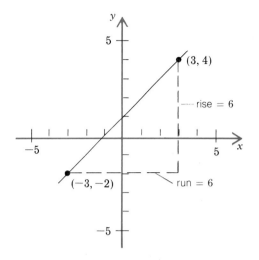

NOTE: It does not matter which point we call P_1 or P_2 so long as we stick to the choice once it is made. If we reverse the choice above, we obtain the same value for the slope, since the sign of the numerator and the denominator both change:

$$m = \frac{(-2) - 4}{(-3) - 3} = \frac{-6}{-6} = 1$$

Work Problem 5 and check solution in Solutions to Matched Problems on page 395.

PROBLEM 5 Graph the line passing through $(-2, 7)$ and $(3, -3)$; then compute its slope.

Solution

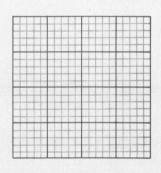

$$m =$$

For a horizontal line, y does not change as x changes; hence, its slope is zero. On the other hand, for a vertical line, x does not change as y changes; hence, $x_1 = x_2$, and

$$m = \frac{y_2 - y_1}{x_2 - x_1} = \frac{y_2 - y_1}{0}$$ Vertical-line slope is not defined.

In general, the slope of a line may be positive, negative, 0, or not defined. Each of these cases is interpreted geometrically as follows:

Going from left to right

Line	Slope	Example
Rising	Positive	
Falling	Negative	
Horizontal	0	
Vertical	Not defined	

••Slope-intercept form

Any equation of the form $Ax + By = C$, $B \neq 0$, can always be written in the form

$$y = mx + b$$

where m and b are constants. For example, starting with

$$2x + 3y = 6 \qquad \text{Form } Ax + By = C.$$

we solve for y to obtain

$$3y = -2x + 6$$
$$\tfrac{1}{3}(3y) = \tfrac{1}{3}(-2x + 6)$$
$$y = -\tfrac{2}{3}x + 2 \qquad \text{Form } y = mx + b.$$

The constants m and b in $y = mx + b$ have special geometric meaning. If we let $x = 0$, then

$$y = m \cdot 0 + b$$
$$= 0 + b$$
$$= b$$

Thus, **b is the y intercept,** the point where the graph crosses the y axis. In the example above, $y = -\tfrac{2}{3}x + 2$, the y intercept is 2.

Now let us determine the geometric significance of m in $y = mx + b$. We choose two points (x_1, y_1) and (x_2, y_2) on the graph of $y = mx + b$ (Figure 3).

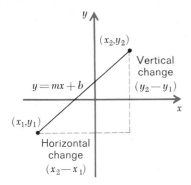

Figure 3

Since the two points are on the graph, they are solutions to the equation $y = mx + b$. Thus,

$$y_1 = mx_1 + b \quad \text{and} \quad y_2 = mx_2 + b$$

Solving both equations for b, we obtain

$$b = y_1 - mx_1 \quad \text{and} \quad b = y_2 - mx_2$$

Since $y_1 - mx_1$ and $y_2 - mx_2$ are both equal to b, they are equal to each other. Thus,

$$y_1 - mx_1 = y_2 - mx_2 \qquad \text{Now solve for } m.$$
$$mx_2 - mx_1 = y_2 - y_1$$
$$(x_2 - x_1)m = y_2 - y_1$$
$$m = \frac{y_2 - y_1}{x_2 - x_1} \qquad x_1 \neq x_2$$

Thus, m **is the slope** of the graph of $y = mx + b$.

In summary, if an equation of a line is written in the form $y = mx + b$, then b is the y intercept and m is the slope. Conversely, if we know the slope and y intercept of a line, we can write its equation in the form $y = mx + b$.

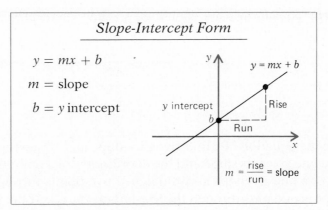

EXAMPLE 6

(A) Find the slope and y intercept of the line $y = \frac{1}{3}x + 2$. Graph the equation.

Solution

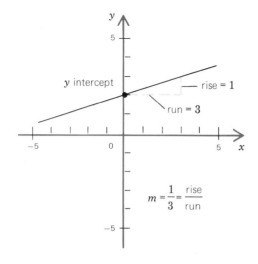

$$y = \tfrac{1}{3}x + 2$$

(slope) (y intercept)

(B) Find the equation of a line with slope -2 and y intercept 3.

Solution

Since $m = -2$ and $b = 3$, then, using $y = mx + b$, $y = -2x + 3$ is the equation.

Work Problem 6 and check solution in Solutions to Matched Problems on page 396.

PROBLEM 6   **(A)**   Find the slope and y intercept of the line $y = -\tfrac{1}{2}x + 4$. Graph the equation.

Solution

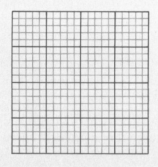

(B)   Find the equation of a line with slope $-\tfrac{1}{3}$ and y intercept 6.

Solution

•••Point-slope line

In Example 6 we found the equation of a line given its slope and y intercept. Often it is necessary to find the equation of a line given its slope and the coordinates of a point through which it passes, or to find the equation of a line given the coordinates of two points through which it passes.

Let a line have slope m and pass through the fixed point (x_1, y_1). If the variable point (x, y) is to be a point on the line, the slope of the line passing through (x, y) and (x_1, y_1) must be m (see Figure 4).

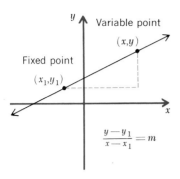

Figure 4

Thus the equation

$$\frac{y - y_1}{x - x_1} = m \qquad x \neq x_1$$

restricts the variable point (x, y) so that only those points in the plane lying on the line $(x \neq x_1)$ will have coordinates that satisfy the equation, and vice versa. This equation is usually written in the form

---

Point-Slope Form

The equation of a line passing through (x_1, y_1) with slope m is given by

$$y - y_1 = m(x - x_1) \qquad\qquad (1)$$

---

and is referred to as the **point-slope form of the equation of a line.** Note that (x_1, y_1) also satisfies equation (1). Hence, *any* point having coordinates that satisfy (1) is on the line that passes through (x_1, y_1) with slope m. Using equation (1) in conjunction with the slope formula, we can also find the equation of a line, knowing only the coordinates of two points through which it passes. The point-slope form should be memorized.

EXAMPLE 7

(A) Find an equation of a line with slope $-\frac{1}{3}$ that passes through $(6, -3)$. Write the resulting equation in the form $y = mx + b$.

Solution

$$y - y_1 = m(x - x_1)$$

$$y - (-3) = -\tfrac{1}{3}(x - 6)$$

$$y + 3 = -\tfrac{1}{3}(x - 6)$$

$$y + 3 = -\frac{x}{3} + 2$$

$$y = -\tfrac{1}{3}x - 1$$

(B) Find an equation of a line that passes through the two points $(-2, -6)$ and $(2, 2)$. Write the resulting equation in the form $y = mx + b$.

Solution

First find the slope of the line using the slope formula, then proceed as in part **A**, using the coordinates of either point for (x_1, y_1).

$$m = \frac{y_2 - y_1}{x_2 - x_1} = \frac{2 - (-6)}{2 - (-2)} = 2$$

Use $(x_1, y_1) = (-2, -6)$ or   use $(x_1, y_1) = (2, 2)$

$$y - y_1 = m(x - x_1) \qquad\qquad y - y_1 = m(x - x_1)$$
$$y - (-6) = 2[x - (-2)] \qquad\qquad y - 2 = 2(x - 2)$$
$$y + 6 = 2(x + 2) \qquad\qquad\quad y - 2 = 2x - 4$$
$$y + 6 = 2x + 4 \qquad\qquad\qquad y = 2x - 2$$
$$y = 2x - 2$$

Work Problem 7 and check solution in Solutions to Matched Problems on page 396.

PROBLEM 7   **(A)**   Find the equation of a line with slope $\frac{2}{3}$ that passes through $(-3, 4)$. Write the resulting equation in the form $y = mx + b$.

Solution

(B)   Find the equation of a line that passes through the two points $(6, -1)$ and $(-2, 3)$. Write the resulting equation in the form $y = mx + b$.

Solution

••••Vertical and horizontal lines

If a line is vertical, its slope is not defined. Since points on a vertical line have constant abscissas and arbitrary ordinates, the equation of a vertical line is of the form

$$x + 0y = c$$

or simply

$x = c$ vertical line

where c is the abscissa of each point on the line. Similarly, if a line is horizontal (slope 0), then every point on the line has constant ordinate and arbitrary abscissa. Thus, the equation of a horizontal line is of the form

$$0x + y = c$$

or simply

$$y = c \qquad \text{horizontal line}$$

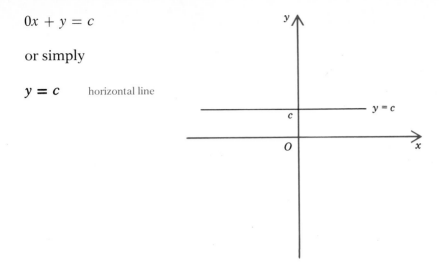

where c is the ordinate of each point on the line. Also, since a horizontal line has zero slope ($m = 0$), then, using the slope-intercept form, we obtain

$$y = mx + b$$
$$y = 0x + b$$
$$y = c$$

EXAMPLE 8
The equation of a vertical line through $(-2, -4)$ is $x = -2$, and the equation of a horizontal line through the same point is $y = -4$.

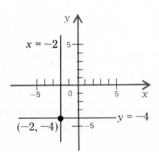

Work Problem 8 and check solution in Solutions to Matched Problems on page 396.

PROBLEM 8   Write the equations of the vertical and horizontal lines through $(3, -8)$.

Solution

•••••Parallel and perpendicular lines

It can be shown that if two nonvertical lines are parallel, then they have the same slope. And if two lines have the same slope, they are parallel. It can also be shown that if two nonvertical lines are perpendicular, then their slopes are the negative reciprocals of each other (that is, $m_2 = -1/m_1$, or, equivalently, $m_1 m_2 = -1$). And if the slopes of two lines are the negative reciprocals of each other, the lines are perpendicular. Symbolically,

---

Parallel and Perpendicular Lines

Given nonvertical lines L_1 and L_2 with slopes m_1 and m_2, respectively, then

$L_1 \parallel L_2$ if and only if $m_1 = m_2$

$L_1 \perp L_2$ if and only if $m_1 m_2 = -1$ or $m_2 = -\dfrac{1}{m_1}$

NOTE: $\parallel$ means "is parallel to" and $\perp$ means "is perpendicular to."

The lines determined by

$$y = \tfrac{2}{3}x - 5 \qquad \text{and} \qquad y = \tfrac{2}{3}x + 8$$

are parallel since both have the same slope, $\tfrac{2}{3}$. And the lines that are determined by

$$y = \frac{x}{3} + 5 \qquad \text{and} \qquad y = -3x - 7$$

are perpendicular, since the product of their slopes is -1; that is,

$$(\tfrac{1}{3})(-3) = -1$$

EXAMPLE 9

Given the line $x - 2y = 4$, find the equation of a line that passes through $(2, -3)$ and is

(A) Parallel to the given line.

(B) Perpendicular to the given line.

Write final equations in the form $y = mx + b$.

Solution

First find the slope of the given line by writing $x - 2y = 4$ in the form $y = mx + b$.

$$x - 2y = 4$$

$$-2y = 4 - x$$

$$-\tfrac{1}{2}(-2y) = -\tfrac{1}{2}(4 - x)$$

$$y = -2 + \tfrac{1}{2}x$$

$$y = \overset{\overset{\text{slope}}{\frown}}{\tfrac{1}{2}}x - 2$$

The slope of the given line is $\tfrac{1}{2}$.

(A) The slope of a line parallel to the given line is also $\frac{1}{2}$. We have only to find the equation of a line through $(2, -3)$ with slope $\frac{1}{2}$ to solve part **A**.

$$y - y_1 = m(x - x_1) \qquad m = \tfrac{1}{2} \text{ and } (x_1, y_1) = (2, -3)$$
$$y - (-3) = \tfrac{1}{2}(x - 2)$$
$$y + 3 = \tfrac{1}{2}x - 1$$
$$y = \tfrac{1}{2}x - 4$$

(B) The slope of the line perpendicular to the given line is the negative reciprocal of $\frac{1}{2}$, that is, -2. We have only to find the equation of a line through $(2, -3)$ with slope -2 to solve part **B**.

$$y - y_1 = m(x - x_1) \qquad m = -2 \text{ and } (x_1, y_1) = (2, -3)$$
$$y - (-3) = -2(x - 2)$$
$$y + 3 = -2x + 4$$
$$y = -2x + 1$$

Work Problem 9 and check solution in Solutions to Matched Problems on page 396.

PROBLEM 9 Given the line $2x = 6 - 3y$, find the equation of a line that passes through $(-3, 9)$ and is

    (A) Parallel to the given line.
    (B) Perpendicular to the given line.

    Write final equations in the form $y = mx + b$.

Solution

SOLUTIONS TO MATCHED PROBLEMS

5.

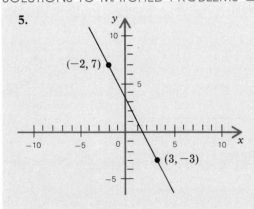

$$m = \frac{y_1 - y_2}{x_1 - x_2} = \frac{(-3) - 7}{3 - (-2)} = \frac{-10}{5} = -2$$

6. **(A)** $m = -\frac{1}{2}$; y intercept $= 4$ **(B)** $y = -\frac{1}{3}x + 6$

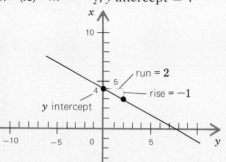

7. **(A)** $y - y_1 = m(x - x_1)$ **(B)** First find the slope; then use the point-slope formula.
$\quad\quad y - 4 = \frac{2}{3}(x + 3)$
$\quad\quad y - 4 = \frac{2}{3}x + 2$ $m = \dfrac{y_1 - y_2}{x_1 - x_2} = \dfrac{3 - (-1)}{(-2) - 6} = \dfrac{4}{-8} = -\dfrac{1}{2}$
$\quad\quad\quad y = \frac{2}{3}x + 6$
$\quad\quad\quad\quad\quad\quad\quad\quad y - y_1 = m(x - x_1)$
$\quad\quad\quad\quad\quad\quad\quad\quad y + 1 = -\frac{1}{2}(x - 6)$
$\quad\quad\quad\quad\quad\quad\quad\quad y + 1 = -\frac{1}{2}x + 3$
$\quad\quad\quad\quad\quad\quad\quad\quad\quad y = -\frac{1}{2}x + 2$

8. Vertical line through $(3, -8)$: $x = 3$
Horizontal line through $(3, -8)$: $y = -8$

9. First find the slope of the graph of $2x = 6 - 3y$:
$\quad 2x = 6 - 3y$
$\quad 3y = -2x + 6$
$\quad\; y = -\frac{2}{3}x + 2$
Slope of the given line is $-\frac{2}{3}$

(A) Find equation of a line with slope $-\frac{2}{3}$ that   **(B)** Find the equation of a line with slope $\frac{3}{2}$
passes through $(-3, 9)$:   that passes through $(-3, 9)$:
$\quad y - y_1 = m(x - x_1)$ $\quad y - y_1 = m(x - x_1)$
$\quad y - 9 = -\frac{2}{3}(x + 3)$ $\quad y - 9 = \frac{3}{2}(x + 3)$
$\quad y - 9 = -\frac{2}{3}x - 2$ $\quad y - 9 = \frac{3}{2}x + \frac{9}{2}$
$\quad\quad y = -\frac{2}{3}x + 7$ $\quad\quad y = \frac{3}{2}x + \frac{27}{2}$

PRACTICE EXERCISE 8.2 _____

Work odd-numbered problems first, check answers, and then work even-numbered problems in areas of weakness. Answers to all problems are in the back of the book. Make every effort to work a problem yourself before you look at an answer.

A *Find the slope and y intercept, and graph each equation.*

1. $y = 2x - 3$ _____   **2.** $y = x + 1$ _____

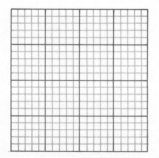   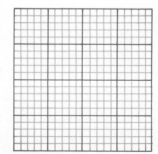

3. $y = -x + 2$ _____

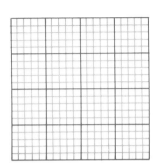

4. $y = -2x + 1$ _____

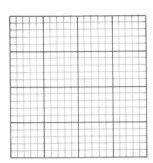

Write the equation of the line with slope and y intercept as indicated.

5. Slope $= 5$
   y intercept $= -2$ _____

6. Slope $= 3$
   y intercept $= -5$ _____

7. Slope $= -2$
   y intercept $= 4$ _____

8. Slope $= -1$
   y intercept $= 2$ _____

Write the equation of the line that passes through the given point with the indicated slope.

9. $m = 2,$ $(5, 4)$ _____

10. $m = 3,$ $(2, 5)$ _____

11. $m = -2,$ $(2, 1)$ _____

12. $m = -3,$ $(1, 3)$ _____

Find the slope of the line that passes through the given points.

13. $(3, 2)$ and $(5, 6)$ _____

14. $(1, 3)$ and $(2, 4)$ _____

15. $(2, 1)$ and $(10, 5)$ _____

16. $(1, 3)$ and $(7, 5)$ _____

Write the equation of the line through each indicated pair of points.

17. $(3, 2)$ and $(5, 6)$ _____

18. $(1, 3)$ and $(2, 4)$ _____

19. $(2, 1)$ and $(10, 5)$ _____

20. $(1, 3)$ and $(7, 5)$ _____

B *Find the slope and y intercept, and graph each equation.*

21. $y = -\dfrac{x}{3} + 2$ _____

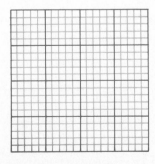

22. $y = -\dfrac{x}{4} - 1$ _____

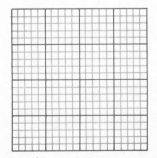

23. $x + 2y = 4$ _____

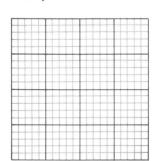

24. $x - 3y = -6$ _____

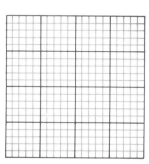

25. $2x + 3y = 6$ _____

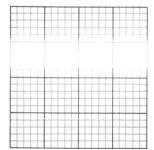

26. $3x + 4y = 12$ _____

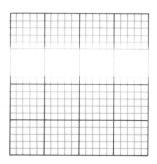

Write the equation of the line with slope and y intercept as given.

27. Slope $= -\frac{1}{2}$
y intercept $= -2$ _____

28. Slope $= -\frac{1}{3}$
y intercept $= -5$ _____

29. Slope $= \frac{2}{3}$
y intercept $= \frac{3}{2}$ _____

30. Slope $= -\frac{3}{2}$
y intercept $= \frac{5}{2}$ _____

Write the equation of the line that passes through the given point with the indicated slope. Transform the equation into the form $y = mx + b$.

31. $m = -2,$ $(-3, 2)$ _____

32. $m = -3,$ $(4, -1)$ _____

33. $m = \frac{1}{2},$ $(-4, 3)$ _____

34. $m = \frac{2}{3},$ $(-6, -5)$ _____

Find the slope of the line that passes through the given points.

35. $(3, 7)$ and $(-6, 4)$ _____

36. $(-5, -2)$ and $(-5, -4)$ _____

37. $(4, -2)$ and $(-4, 0)$ _____

38. $(-3, 0)$ and $(3, -2)$ _____

Write the equation of the line through each indicated pair of points. Transform the equation into the form $y = mx + b$.

39. $(3, 7)$ and $(-6, 4)$ _____

40. $(-5, -2)$ and $(5, -4)$ _____

41. $(4, -2)$ and $(-4, 0)$ _____

42. $(-3, 0)$ and $(3, -4)$ _____

Write the equations of the vertical and horizontal lines through each point.

43. $(-3, 5)$ —————————————

44. $(6, -2)$ —————————————

45. $(-1, 22)$ —————————————

46. $(5, 0)$ —————————————

C 47. Given the line $x - 4y = 8$, find the equation of a line that passes through $(4, 2)$ and is

(A) Parallel to the given line. —————————————

(B) Perpendicular to the given line. —————————————

Write final equations in the form $y = mx + b$.

48. Repeat Problem 47 for the given line $2x + y = 3$ and the point $(2, -3)$.

(A) ————————————— **(B)** —————————————

49. Repeat Problem 47 for the given line $2x = 6y + 6$ and the point $(4, -3)$.

(A) ————————————— **(B)** —————————————

50. Repeat Problem 47 for the given line $3y = 5 - 2x$ and the point $(-2, 2)$.

(A) ————————————— **(B)** —————————————

APPLICATION

51. BUSINESS A sporting goods store sells a pair of ski boots costing \$40 for \$66 and a pair of skis costing \$120 for \$186. **(A)** If the markup policy of the store for items costing over \$20 is assumed to be linear and is reflected in the pricing of these two items, write an equation that relates retail price R with cost C. [HINT: Find an equation of the form $R = mC + b$ having a graph that passes through $(40, 66)$ and $(120, 186)$.] **(B)** Use the equation from part **A** to find the retail price of a surfboard costing \$240.

(A) ————————————— **(B)** —————————————

The Check Exercise for this section is on page 439.

8.3 GRAPHING LINEAR INEQUALITIES

We know how to graph first-degree equations such as

$y = 2x - 3$

or

$2x - 3y = 5$

but how do we graph first-degree inequalities such as

$y \leq 2x - 3$

or

$2x - 3y > 5$

We will find that graphing inequalities is almost as easy as graphing equations. The following discussion leads to a simple solution to the problem.

A line in a cartesian coordinate system divides the plane into two **half planes.** A vertical line divides the plane into left and right half planes; a nonvertical line divides the plane into upper and lower half planes (Figure 5).

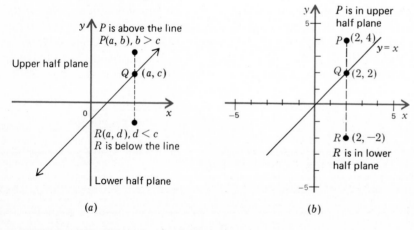

(a)   (b)   Figure 5

Now let us compare the graphs of

$y < 2x - 3$

$y = 2x - 3$

and

$y > 2x - 3$

Consider the vertical line $x = 3$, and ask what the relationship of y is to $2 \cdot 3 - 3$ as we move $(3, y)$ up and down this vertical line (see Figure 6a). If we are at point Q, a point on the graph of $y = 2x$

$- 3$, then $y = 2 \cdot 3 - 3$; if we move up the vertical line to P, the ordinate of $(3, y)$ increases and $y > 2 \cdot 3 - 3$; if we move down the line to R, the ordinate of $(3, y)$ decreases and $y < 2 \cdot 3 - 3$. Since the same results are obtained for each point x_0 on the x axis (see Figure 6b), we conclude that the graph of $y > 2x - 3$ is the upper half plane determined by $y = 2x - 3$, and $y < 2x - 3$ is the lower half plane.

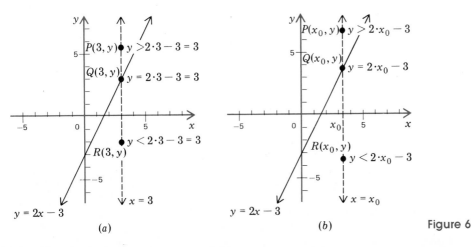

(a)   (b)   Figure 6

In graphing $y > 2x - 3$, we show the line $y = 2x - 3$ as a broken line, indicating that it is not part of the graph; in graphing $y \geq 2x - 3$, we show the line $y = 2x - 3$ as a solid line, indicating that it is part of the graph. Figure 7 illustrates four typical cases.

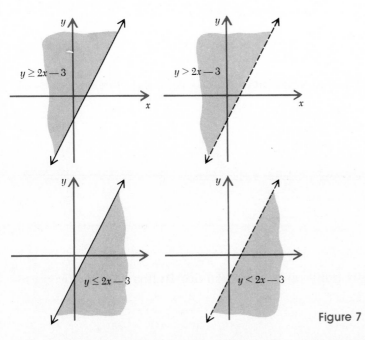

Figure 7

The preceding discussion suggests the following important theorem, which we state without proof.

THEOREM 2

The graph of a linear inequality

$$Ax + By < C \quad \text{or} \quad Ax + By > C$$

with $B \neq 0$, is either the upper half plane or the lower half plane (but not both) determined by the line $Ax + By = C$. If $B = 0$, the graph of

$$Ax < C \qquad \text{or} \qquad Ax > C$$

is either the left half plane or the right half plane (but not both) as determined by the line $Ax = C$.

This theorem leads to a simple, fast procedure for graphing linear inequalities in two variables:

Steps in Graphing Linear Inequalities

1. First, graph $Ax + By = C$:
   as a broken line if equality is not included in original statement, as a solid line if equality is included in original statement.
2. Choose a test point in the plane not on the line—the origin is the best choice if it is not on the line—and substitute the coordinates into the inequality.
3. The graph of the original inequality includes:
   - the half plane containing the test point, if the inequality is satisfied by that point, or
   - the half plane not containing the test point, if the inequality is not satisfied by that point.

EXAMPLE 10

Graph $3x - 4y \leq 12$.

Solution

Step 1. First graph the line $3x - 4y = 12$ as a solid line, since equality is included in $3x - 4y \leq 12$.

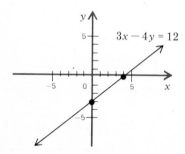

Step 2. Choose a convenient test point—any point on the line will do. In this case the origin results in the simplest computation.

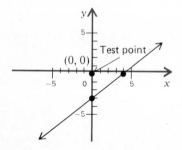

We see that the origin (0, 0) satisfies the original inequality:

$$3x - 4y \leq 12$$
$$3 \cdot 0 - 4 \cdot 0 \leq 12$$
$$0 \leq 12$$

Hence, all other points on the same side as the origin are also part of the graph. Thus, the graph is the upper half plane.

Step 3. The final graph is the upper half plane and the line:

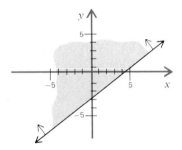

Work Problem 10 and check solution in Solutions to Matched Problems on page 405.

PROBLEM 10   Graph $2x + 2y \leq 5$.

Solution

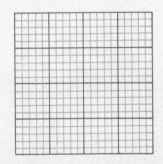

EXAMPLE 11

Graph in a rectangular coordinate system:

(A) $y > -3$

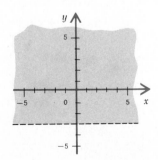

(B) $2x < 5$

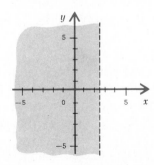

(C) $-2 \le x \le 4$

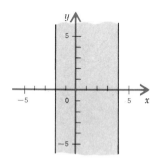

REMARK

If we are to graph an inequality involving only one variable, it is important to know the dimension of the coordinate system in which the graph is to occur. In Example 11 the graphs are to occur in a rectangular coordinate system; hence, in each case the presence of a second variable is assumed, but with a zero coefficient. For example, $y > -3$ is assumed to mean $0x + y > -3$. Thus, x can take on any value as long as y is greater than -3. The graph is the upper half plane above the line $y = -3$. If we were to graph $y > -3$ on a single number line, however, then we would obtain

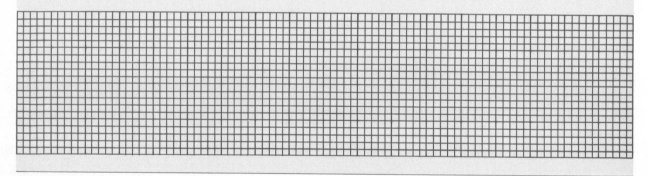

Work Problem 11 and check solution in Solutions to Matched Problems on page 405.

PROBLEM 11   Graph each inequality in a rectangular coordinate system:

   (A) $y < 2$ **(B)** $3x > -8$ **(C)** $1 \le y \le 5$

EXAMPLE 12

Graph $\{(x, y) \mid x \ge 0, y \ge 0, 2x + 3y \le 18\}$.

Solution

This is the set of all ordered pairs of real numbers (x, y) such that x and y are both nonnegative and satisfy $2x + 3y \le 18$. We graph each inequality in the same coordinate system and take the intersection (see Section 1.1) of the graphs of these solution sets.

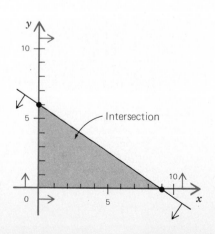

Work Problem 12 and check solution in Solutions to Matched Problems on page 405.

PROBLEM 12   Graph $\{(x, y) \mid 1 \leq x \leq 4,\ -2 \leq y \leq 2\}$.

Solution

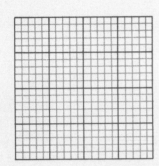

SOLUTIONS TO MATCHED PROBLEMS

10. $2x + 2y \leq 5$

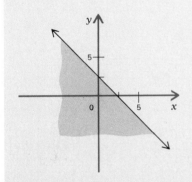

11. **(A)** $y < 2$

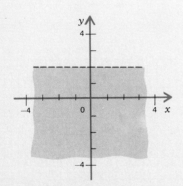

(B) $3x > -8$

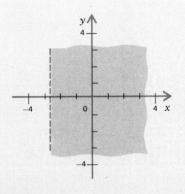

(C) $1 \leq y \leq 5$

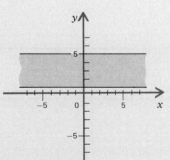

12. $\{(x, y) \mid 1 \leq x \leq 4,\ -2 \leq y \leq 2\}$

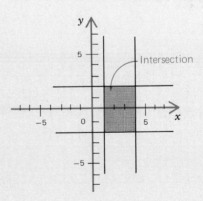

PRACTICE EXERCISE 8.3 ──────────────────────────────────────

Work odd-numbered problems first, check answers, and then work even-numbered problems in areas of weakness. Answers to all problems are in the back of the book. Make every effort to work a problem yourself before you look at an answer.

Graph each inequality in a rectangular coordinate system:

A **1.** $x + y \leq 6$ **2.** $x + y \geq 4$ **3.** $x - y > 3$

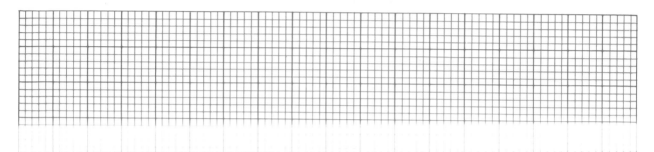

4. $x - y < 5$ **5.** $y \geq x - 2$ **6.** $y \leq x + 1$

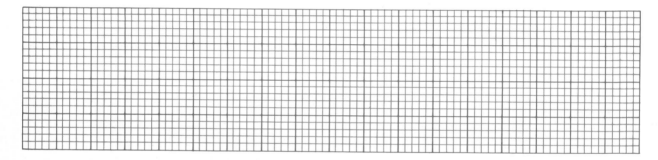

B **7.** $2x - 3y < 6$ **8.** $3x + 4y < 12$ **9.** $3y - 2x \geq 24$

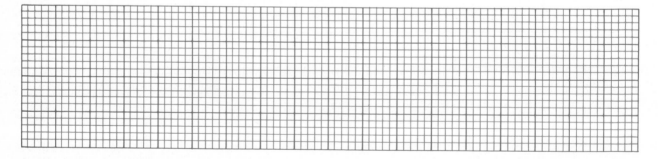

10. $3x + 2y \geq 18$ **11.** $y \geq \dfrac{x}{3} - 2$ **12.** $y \leq \dfrac{x}{2} - 4$

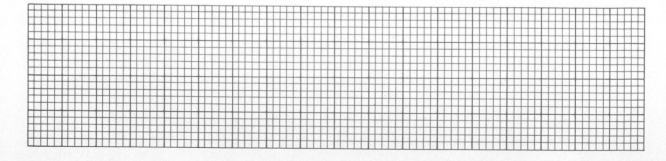

13. $y \leq \dfrac{2}{3}x + 5$ **14.** $y > \dfrac{x}{3} + 2$ **15.** $x \geq -5$

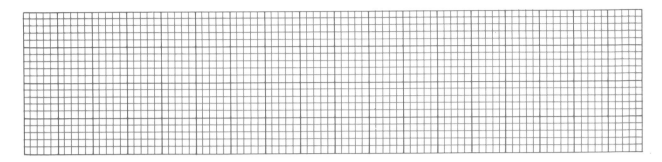

16. $y \leq 8$ **17.** $y < 0$ **18.** $x \geq 0$

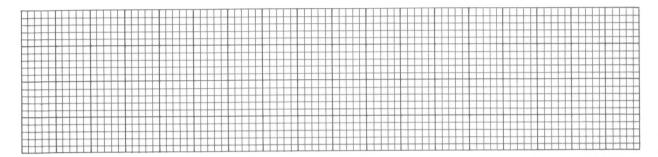

19. $-1 < x \leq 3$ **20.** $-3 \leq y < 2$ **21.** $-2 \leq y \leq 2$

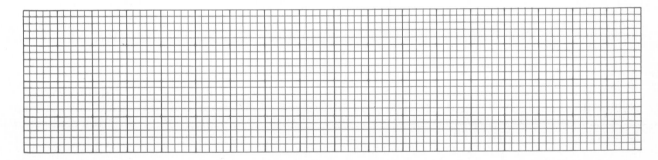

22. $-1 \leq x \leq 4$

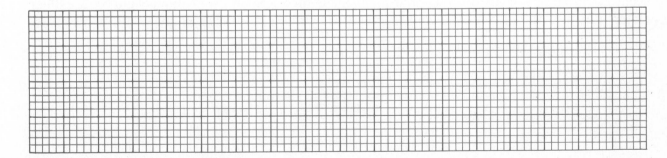

C *Graph each set.*

23. $\{(x, y) \mid x \geq 0, y \geq 0, \text{ and } 3x + 4y \leq 12\}$ **24.** $\{(x, y) \mid x \geq 0, y \geq 0, \text{ and } 3x + 2y \leq 18\}$

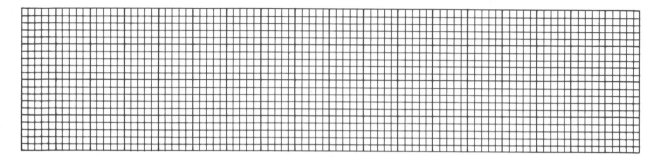

The Check Exercise for this section is on page 441.

8.4 CONIC SECTIONS, AN INTRODUCTION; CIRCLES

- •Introduction
- ••Distance-between-two-points formula
- •••Circles

•Introduction

In the first part of this chapter we discussed equations of a straight line. That is, equations such as

$$2x - 3y = 5 \qquad y = -\tfrac{1}{3}x + 2$$

These are first-degree equations in two variables. If we increase the degree of the equations by 1, what kind of graphs will we get? That is, what kind of graphs will second-degree equations such as

$$y = x^2 \qquad x^2 + y^2 = 25 \qquad \frac{x^2}{4} - \frac{y^2}{16} = 1 \qquad x^2 + 4y^2 - 3x + 7y = 4$$

produce? It can be shown that the graphs will be one of the plane curves you would get by intersecting a plane and a general cone—thus the name **conic sections.** Some typical curves are illustrated in Figure 8. The principal conic sections are circles, parabolas, ellipses, and hyperbolas.

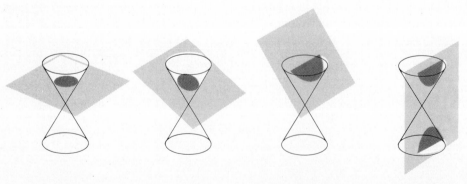

Circle   Ellipse   Parabola   Hyperbola   **Figure 8**
Conic Sections

In this book we will only consider a few simple, interesting, and useful second-degree equations in two variables. The subject is treated more thoroughly in a course in analytic geometry.

••Distance-between-two-points formula

A basic tool in determining equations of conics is the distance-between-two-points formula. The derivation of the formula is easy, making direct use of the pythagorean theorem.

Let P_1 and P_2 have coordinates as indicated in Figure 9.

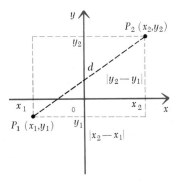

Figure 9

Using the pythagorean theorem, we can write

$$d^2 = |x_2 - x_1|^2 + |y_2 - y_1|^2 = (x_2 - x_1)^2 + (y_2 - y_1)^2 \qquad \text{Since } |N|^2 = N^2$$

Solving for d, a nonnegative value, we obtain the important formula:

> *Distance-between-Two-Points Formula*
>
> $$d = \sqrt{(x_2 - x_1)^2 + (y_2 - y_1)^2}$$

Note that it doesn't make any difference which point you call P_1 or P_2, since $(a - b)^2 = (b - a)^2$.

EXAMPLE 13
Find the distance between $(-3, 6)$ and $(4, -2)$.

Solution
Let $P_1 = (4, -2)$ and $P_2 = (-3, 6)$; then

$$d = \sqrt{(x_2 - x_1)^2 + (y_2 - y_1)^2}$$
$$= \sqrt{[(-3) - 4]^2 + [6 - (-2)]^2}$$
$$= \sqrt{(-7)^2 + (8)^2}$$
$$= \sqrt{113}$$

Or, let $P_1 = (-3, 6)$ and $P_2 = (4, -2)$; then

$$d = \sqrt{[4 - (-3)]^2 + [(-2) - 6]^2}$$
$$= \sqrt{(7)^2 + (-8)^2}$$
$$= \sqrt{113}$$

Work Problem 13 and check solution in Solutions to Matched Problems on page 413.

PROBLEM 13   Find the distance between $(4, -2)$ and $(3, 1)$.

Solution $d =$

•••Circles

We start with the definition of a circle; then we use the distance formula to find the standard equation of a circle.

Definition of a Circle
_____

A **circle** is the set of points equidistant from a fixed point. The fixed distance is called the **radius**, and the fixed point is called the **center**.

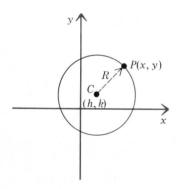

Figure 10
Circle

Let a circle have radius R and center at (h, k). Referring to Figure 10, we see that an arbitrary point $P(x, y)$ is on the circle if and only if

$$R = \sqrt{(x - h)^2 + (y - k)^2} \qquad \text{Distance from } C \text{ to } P \text{ is constant.}$$

or, equivalently,

$$(x - h)^2 + (y - k)^2 = R^2$$

Thus, we can state the following:

Equations of a Circle
_____

1. Radius R and center (h, k):

    $$(x - h)^2 + (y - k)^2 = R^2$$

2. Radius R and center at the origin:

    $$x^2 + y^2 = R^2 \qquad \text{since } (h, k) = (0, 0)$$

EXAMPLE 14

(A)   Graph $x^2 + y^2 = 25$.

Solution

This is a circle with radius 5 and center at the origin:

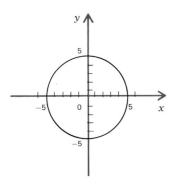

(B)   Graph $(x - 3)^2 + (y + 2)^2 = 9$.

Solution

Write $(x - 3)^2 + (y + 2)^2 = 9$ in the standard form $(x - h)^2 + (y - k)^2 = R^2$ to identify the center (h, k) and the radius R:

$$(x - 3)^2 + [y - (-2)]^2 = 3^2$$

with h, k, and R indicated.

This is an equation of a circle with radius 3 and center at $(h, k) = (3, -2)$. It is graphed as follows:

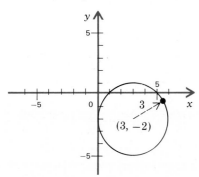

Work Problem 14 and check solution in Solutions to Matched Problems on page 413.

PROBLEM 14   Graph:

(A) $x^2 + y^2 = 4$

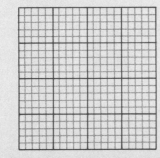

(B) $(x + 2)^2 + (y - 3)^2 = 16$

412

EXAMPLE 15

What is the equation of a circle with radius 8 and

(A) Center at the origin?
(B) Center at $(-5, 4)$?

Solution

(A) If the center is at the origin, then $(h, k) = (0, 0)$. Thus,

$$x^2 + y^2 = R^2 \qquad R = 8$$
$$x^2 + y^2 = 8^2 \quad \text{or} \quad x^2 + y^2 = 64$$

(B) If the center is at $(-5, 4)$, then $h = -5$ and $k = 4$. Thus,

$$(x - h)^2 + (y - k)^2 = R^2 \qquad h = -5, k = 4, R = 8$$
$$[x - (-5)]^2 + (y - 4)^2 = 8^2$$

or

$$(x + 5)^2 + (y - 4)^2 = 64$$

Work Problem 15 and check solution in Solutions to Matched Problems on page 414.

PROBLEM 15 What is the equation of a circle with radius $\sqrt{7}$ and

(A) Center at the origin?
(B) Center at $(6, -4)$?

Solutions **(A)** **(B)**

EXAMPLE 16

Graph $x^2 + y^2 - 6x + 8y + 9 = 0$.

Solution

We transform the equation into the form

$$(x - h)^2 + (y - k)^2 = R^2$$

by completing the square relative to x and relative to y (see Section 7.3).

$$x^2 - 6x + ? + y^2 + 8y + ? = -9 + ? + ?$$
$$x^2 - 6x + 9 + y^2 + 8y + 16 = -9 + 9 + 16$$
$$(x - 3)^2 + (y + 4)^2 = 16$$

or

$(x - 3)^2 + [y - (-4)]^2 = 16$ Thus, $(h, k) = (3, -4)$.

This is the equation of a circle with radius 4 and center at $(h, k) = (3, -4)$.

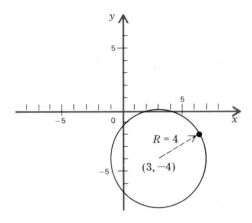

Work Problem 16 and check solution in Solutions to Matched Problems on page 414.

PROBLEM 16   Graph $x^2 + y^2 + 10x - 4y + 20 = 0$.

Solution

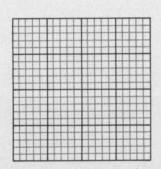

SOLUTIONS TO MATCHED PROBLEMS

13. $d = \sqrt{(x_2 - x_1)^2 + (y_2 - y_1)^2} = \sqrt{(3 - 4)^2 + [1 - (-2)]^2} = \sqrt{10}$

14. **(A)**

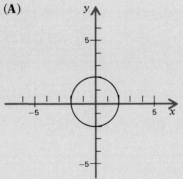

      (B)

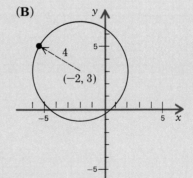

15. **(A)** $x^2 + y^2 = (\sqrt{7})^2$
$x^2 + y^2 = 7$

(B) $(x - 6)^2 + [y - (-4)]^2 = (\sqrt{7})^2$
$(x - 6)^2 + (y + 4)^2 = 7$

16. $(x + 5)^2 + (y - 2)^2 = 9$

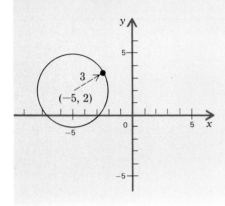

PRACTICE EXERCISE 8.4

Work odd-numbered problems first, check answers, and then work even-numbered problems in areas of weakness. Answers to all problems are in the back of the book. Make every effort to work a problem yourself before you look at an answer.

A *Find the distance between each pair of points. Leave answer in exact radical form.*

1. $(2, 4), (4, 5)$ _____

2. $(7, 3), (8, 6)$ _____

3. $(3, -7), (-2, 1)$ _____

4. $(-5, 3), (-1, -2)$ _____

5. $(-8, -2), (-5, 1)$ _____

6. $(2, -7), (-4, -3)$ _____

Graph each equation in a cartesian coordinate system.

7. $x^2 + y^2 = 16$

8. $x^2 + y^2 = 36$

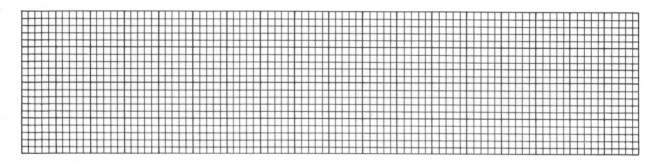

9. $x^2 + y^2 = 6$

10. $x^2 + y^2 = 10$

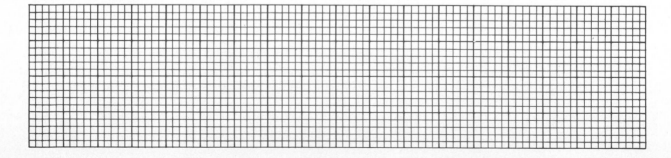

In Problems 11 to 14 write the equation of a circle with center at the origin and radius as given.

11. 7 _____   **12.** 8 _____

13. $\sqrt{5}$ _____   **14.** $\sqrt{10}$ _____

B 15. Is the triangle with vertices $(-1, 2), (2, -1), (3, 3)$ an isosceles triangle (a triangle with at least two sides of the same length)?

_____

16. Is the triangle in the preceding problem an equilateral triangle (a triangle with all sides the same length)?

_____

Graph each equation in a cartesian coordinate system.

17. $(x - 3)^2 + (y - 4)^2 = 16$ **18.** $(x - 4)^2 + (y - 2)^2 = 9$

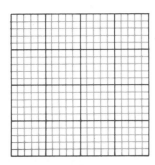   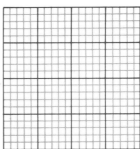

19. $(x - 4)^2 + (y + 3)^2 = 9$ **20.** $(x + 4)^2 + (y - 2)^2 = 25$

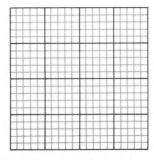   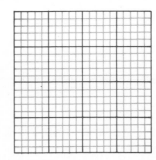

21. $(x + 3)^2 + (y + 3)^2 = 16$ **22.** $(x + 2)^2 + (y + 4)^2 = 25$

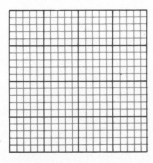   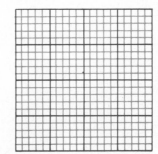

Write the equation of a circle in the form

$(x - h)^2 + (y - k)^2 = R^2$

with radius and center as given.

23. $7, (3, 5)$ _____

24. $2, (4, 1)$ _____

25. $8, (-3, 3)$ _____

26. $6, (5, -2)$ _____

27. $\sqrt{3}, (-4, -1)$ _____

28. $\sqrt{14}, (-7, -5)$ _____

C   *Use method of completing the square to graph each of the following circles. Indicate the center and radius of each.*

29. $x^2 + y^2 - 4x - 6y + 4 = 0$

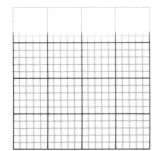

30. $x^2 + y^2 - 6x - 4y + 4 = 0$

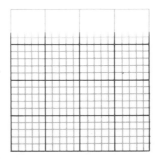

31. $x^2 + y^2 - 6x + 6y + 2 = 0$

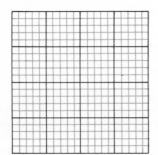

32. $x^2 + y^2 + 6x - 4y - 3 = 0$

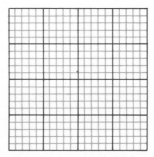

33. $x^2 + y^2 + 6x + 4y + 4 = 0$

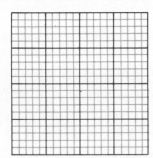

34. $x^2 + y^2 + 4x + 4y - 8 = 0$

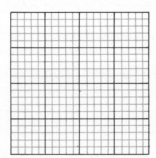

APPLICATION

35. An ancient stone bridge in the form of a circular arc has a span of 80 ft (see Figure 11). If the height of the arch above its ends is 20 ft, find an equation of the circle containing the arch if its center is at the origin as indicated. [HINT: $(40, R - 20)$ must satisfy $x^2 + y^2 = R^2$.]

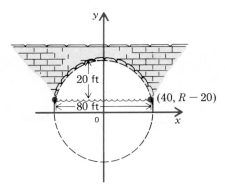

Figure 11

The Check Exercise for this section is on page 443.

8.5 PARABOLAS, ELLIPSES, AND HYPERBOLAS

- •Parabolas
- ••Ellipses
- •••Hyperbolas

This section provides a very brief glimpse into a subject that is treated in detail in a course in analytic geometry. Many results are simply stated without development or motivation. Nevertheless, a brief first exposure to these important curves at this time should help to increase your understanding of a more detailed development in a future course. In addition, you will have gained some concrete experience with graphs other than straight lines and circles.

•Parabolas

We start with a definition of a parabola:

> *Definition of a Parabola*
>
> A **parabola** is the set of all points equidistant from a fixed point and a fixed line. The fixed point is called the **focus,** and the fixed line the **directrix.**

We will begin by finding an equation of a parabola with focus $(a, 0)$, $a > 0$, and directrix $x = -a$ (see Figure 12).

$$d_1 = d_2$$

$$x + a = \sqrt{(x - a)^2 + (y - 0)^2}$$

$$(x + a)^2 = (x - a)^2 + y^2$$

$$x^2 + 2ax + a^2 = x^2 - 2ax + a^2 + y^2$$

$$4ax = y^2$$

$$y^2 = 4ax$$

$$y^2 = kx \qquad \text{where } k = 4a$$

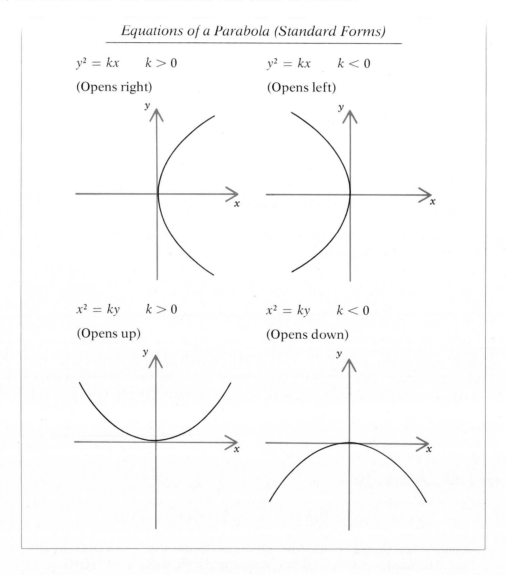

Figure 12

Proceeding in the same way we can obtain similar equations for parabolas opening upward, to the left, and downward. We summarize four cases as follows:

Equations of a Parabola (Standard Forms)

$y^2 = kx \qquad k > 0$

(Opens right)

$y^2 = kx \qquad k < 0$

(Opens left)

$x^2 = ky \qquad k > 0$

(Opens up)

$x^2 = ky \qquad k < 0$

(Opens down)

We now show how particular examples of these four cases can be sketched rather quickly.

Rapid Sketching of Parabolas

Graphing $y^2 = kx$ and $x^2 = ky$.

Step 1. The origin $(0, 0)$ is always part of the graph.
Step 2. Two other points can be found easily by assigning x in $y^2 = kx$ (or y in $x^2 = ky$) a value that will make the right side a positive perfect square.
Step 3. Locate the two points found in step 2; then sketch a parabola through these two points and the origin.

NOTE: If more accuracy is desired, assign x in $y^2 = kx$ (or y in $x^2 = ky$) additional values (making the right side positive) and use a calculator or a square root table.

EXAMPLE 17
Graph $y^2 = -8x$.

Solution
Step 1. The origin $(0, 0)$ is part of the graph.
Step 2. What can we assign x to make the right side a positive perfect square? We let $x = -2$; then

$$y^2 = -8(-2)$$

$$y^2 = 16$$

$$y = \pm \sqrt{16} = \pm 4$$

Thus, the points $(-2, -4)$ and $(-2, 4)$ are also on the graph.
Step 3. We plot $(0, 0)$, $(-2, -4)$, and $(-2, 4)$; then we sketch a parabola through these three points.

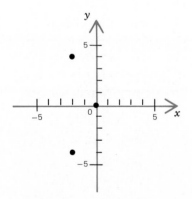 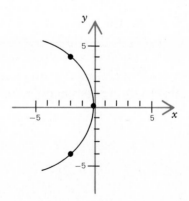

NOTE: With a little practice most of the computation can be done mentally and the graph sketched with little effort. A very common error is to draw the graph opening the wrong way. If you find the three points as directed, you will generally avoid this type of error.

Work Problem 17 and check solution in Solutions to Matched Problems on page 427.

PROBLEM 17   Graph $x^2 = -3y$.

Solution

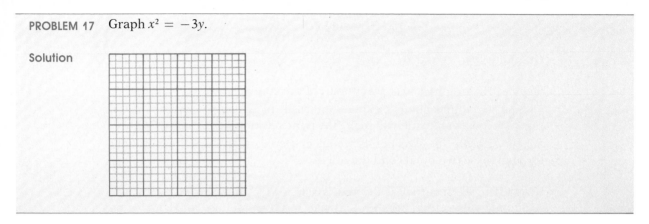

Parabolas are encountered frequently in the physical world. Suspension bridges, arch bridges, reflecting telescopes, radio telescopes, radar equipment, solar furnaces, and searchlights all utilize parabolic forms in their design.

••Ellipses

You are no doubt aware of many uses or occurrences of elliptical forms; orbits of satellites, orbits of planets and comets, gears and cams, and domes in buildings are but a few examples. Formally, we define an ellipse as follows.

Definition of an Ellipse

An **ellipse** is the set of all points such that the sum of the distances of each to two fixed points is constant. The fixed points are called **foci,** and each separately is a **focus.**

An ellipse is easy to draw. Place two pins in a piece of cardboard at the foci and tie a piece of loose string (representing the constant sum) to the pins; then move a pencil within the string, keeping it taut.

With regard to an equation of an ellipse, we will limit ourselves to the cases in which the foci are symmetrically located on either coordinate axis. Thus, if $(-c, 0)$ and $(c, 0)$, $c > 0$, are the foci, and $2a$ is the constant sum of the distances [note from Figure 13 that $2a > 2c$ (why?), hence $a > c$], then

$$d_1 + d_2 = 2a$$
$$\sqrt{(x + c)^2 + y^2} + \sqrt{(x - c)^2 + y^2} = 2a$$

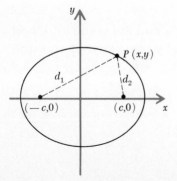

Figure 13

After eliminating radicals and simplifying—a good exercise for the reader—we eventually obtain

$$(a^2 - c^2)(x^2) + a^2y^2 = a^2(a^2 - c^2)$$

or

$$\frac{x^2}{a^2} + \frac{y^2}{a^2 - c^2} = 1$$

Since $a > c$, then $a^2 - c^2 > 0$. To simplify the equation further, we choose to let $b^2 = a^2 - c^2$, $b > 0$. Thus,

$$\frac{x^2}{a^2} + \frac{y^2}{b^2} = 1$$

Proceeding similarly with the foci on the vertical axis, we arrive at

$$\frac{x^2}{b^2} + \frac{y^2}{a^2} = 1$$

Combining these results, we can write

$$\frac{x^2}{m^2} + \frac{y^2}{n^2} = 1$$

as a standard form for an equation of an ellipse located as described above. (We shift over to m and n to simplify our approach and because a and b have special significance in a more advanced treatment of the subject.) It can be shown that if $m > n > 0$, then the foci are on the x axis; and if $n > m > 0$, then the foci are on the y axis. The two cases are summarized as follows:

Equations of Ellipses (Standard Forms)

$$\frac{x^2}{m^2} + \frac{y^2}{n^2} = 1$$

Case 1. $m > n > 0$

Case 2. $n > m > 0$

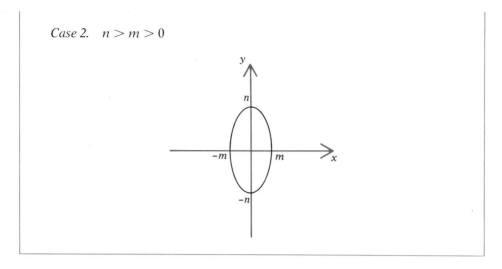

We now show how particular examples of these two cases can be sketched rather quickly.

Rapid Sketching of Ellipses

To graph $\dfrac{x^2}{m^2} + \dfrac{y^2}{n^2} = 1$:

Step 1. Find the x intercepts by letting $y = 0$ and solving for x.
Step 2. Find the y intercepts by letting $x = 0$ and solving for y.
Step 3. Sketch an ellipse passing through these intercepts.

EXAMPLE 18

$$\frac{x^2}{16} + \frac{y^2}{9} = 1.$$

Solution
Step 1. Find x intercepts:

$$\frac{x^2}{16} + \frac{0}{9} = 1$$

$$x^2 = 16$$

$$x = \pm\sqrt{16} = \pm 4$$

Step 2. Find y intercepts:

$$\frac{0}{16} + \frac{y^2}{9} = 1$$

$$y^2 = 9$$

$$y = \pm\sqrt{9} = \pm 3$$

Step 3. Plot the intercepts and draw in the ellipse.

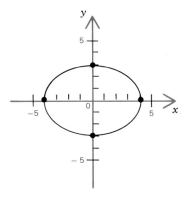

Work Problem 18 and check solution in Solutions to Matched Problems on page 427.

PROBLEM 18 Graph $\dfrac{x^2}{9} + \dfrac{y^2}{16} = 1$.

Solution

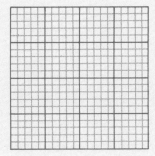

•••Hyperbolas

> *Definition of a Hyperbola*
>
> A **hyperbola** is the set of all points such that the absolute value of the difference of the distances of each to two fixed points is constant. The two fixed points are called **foci.**

As with the ellipse, we will limit our investigation to cases in which the foci are symmetrically located on either coordinate axis. Thus if $(-c, 0)$ and $(c, 0)$ are the foci, and $2a$ is the constant difference (Figure 14), then

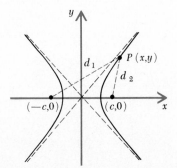

Figure 14

$$|d_1 - d_2| = 2a$$

$$|\sqrt{(x + c)^2 + y^2} - \sqrt{(x - c)^2 + y^2}| = 2a$$

After eliminating radicals and absolute-value signs (by appropriate use of squaring) and simplifying—another good exercise for the reader—we eventually obtain

$$\frac{x^2}{a^2} + \frac{y^2}{a^2 - c^2} = 1$$

which looks exactly like the equation we obtained for the ellipse. However, from Figure 14 we see that $2a < 2c$; hence $a^2 - c^2 < 0$. To simplify the equation further, we let $-b^2 = a^2 - c^2$. Thus

$$\frac{x^2}{a^2} - \frac{y^2}{b^2} = 1$$

Proceeding similarly with the foci on the vertical axis, we obtain

$$\frac{y^2}{a^2} - \frac{x^2}{b^2} = 1$$

Combining these results, we can write

$$\frac{x^2}{m^2} - \frac{y^2}{n^2} = 1 \qquad \frac{y^2}{n^2} - \frac{x^2}{m^2} = 1$$

as standard forms for equations of hyperbolas located as described above. (Again, we shift over to m and n to simplify our approach and because a and b have special significance in a more advanced treatment of the subject.) In summary:

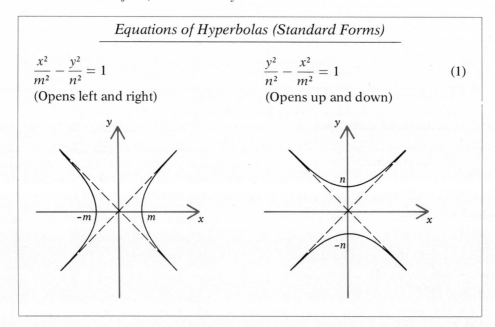

Equations of Hyperbolas (Standard Forms)

$$\frac{x^2}{m^2} - \frac{y^2}{n^2} = 1 \qquad\qquad \frac{y^2}{n^2} - \frac{x^2}{m^2} = 1 \qquad (1)$$

(Opens left and right)   (Opens up and down)

As an aid to graphing equations (1), it can be shown that as x moves away from the origin the graphs get closer and closer to the two straight lines through the origin:

Asymptotes

$$y = \pm \frac{n}{m}x \qquad (2)$$

The two straight lines in (2) are guide lines called **asymptotes.** The graphs of (1) will approach these guide lines, but never touch them, as the graph moves farther and farther away from the origin. Quick sketches of hyperbolas can be made rather easily by following the following steps.

Rapid Sketching of Hyperbolas

To graph $\dfrac{x^2}{m^2} - \dfrac{y^2}{n^2} = 1$ and $\dfrac{y^2}{n^2} - \dfrac{x^2}{m^2} = 1$:

Step 1. Draw a dotted rectangle with intercepts $x = \pm m$ and $y = \pm n$.

Step 2. Draw dotted diagonals of the rectangle and extend to form asymptotes [these are the graphs of equations (2)].

Step 3. Determine the true intercepts of the hyperbola (be particularly careful in this step); then sketch in the hyperbola (both branches).

EXAMPLE 19

Graph $\dfrac{x^2}{25} - \dfrac{y^2}{16} = 1$.

Solution

Step 1. Draw a dotted rectangle with intercepts $x = \pm\sqrt{25} = \pm 5$ and $y = \pm\sqrt{16} = \pm 4$.

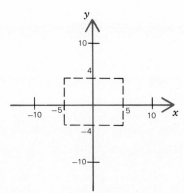

Step 2.   Draw in asymptotes (extended diagonals of rectangle).

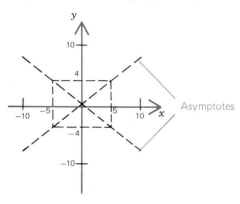

Step 3.   Determine true intercepts for the hyperbola; then sketch it in.

Let $y = 0$; then

$$\frac{x^2}{25} - \frac{0}{16} = 1$$

$$x^2 = 25$$

$$x = \pm\sqrt{25} = \pm 5$$

If we let $x = 0$, then

$$\frac{0}{25} - \frac{y^2}{16} = 1$$

$$y^2 = -16$$

$$y = \pm\sqrt{-16} = \pm 4i$$

These are complex numbers and do not represent real intercepts. We conclude that the only real intercepts are $x = \pm 5$, and the hyperbola opens left and right.

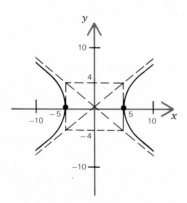

Work Problem 19 and check solution in Solutions to Matched Problems on page 427.

PROBLEM 19   Graph $\dfrac{y^2}{16} - \dfrac{x^2}{25} = 1$.

Solution

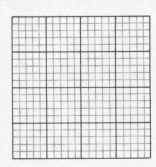

Hyperbolic forms are encountered in the study of comets, the loran system of navigation for ships and aircraft, some modern architectural structures (Figure 15), and optics, to name a few examples of many.

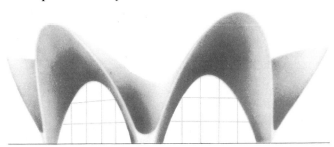

Figure 15
Hyperbolic Paraboloid

SOLUTIONS TO MATCHED PROBLEMS

17. $x^2 = -3y$

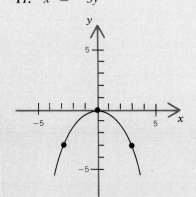

18. $\dfrac{x^2}{9} + \dfrac{y^2}{16} = 1$

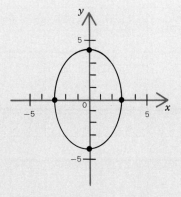

19. $\dfrac{y^2}{16} - \dfrac{x^2}{25} = 1$

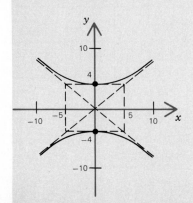

PRACTICE EXERCISE 8.5 _____

Work odd-numbered problems first, check answers, and then work even-numbered problems in areas of weakness. Answers to all problems are in the back of the book. Make every effort to work a problem yourself before you look at an answer.

A *Graph each of the following parabolas.*

1. $y^2 = 4x$

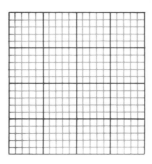

2. $y^2 = x$

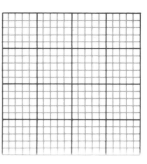

3. $y^2 = -12x$

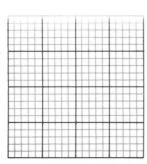

4. $y^2 = -16x$

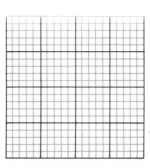

5. $x^2 = y$

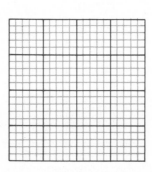

6. $x^2 = 12y$

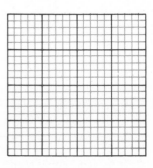

7. $x^2 = -16y$

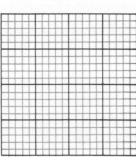

8. $x^2 = -4y$

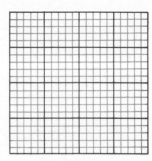

B *Graph each of the following ellipses.*

9. $\dfrac{x^2}{25} + \dfrac{y^2}{4} = 1$

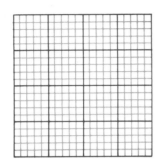

10. $\dfrac{x^2}{9} + \dfrac{y^2}{4} = 1$

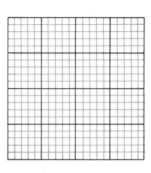

11. $\dfrac{x^2}{4} + \dfrac{y^2}{25} = 1$

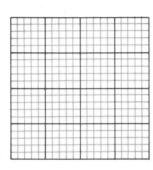

12. $\dfrac{x^2}{4} + \dfrac{y^2}{9} = 1$

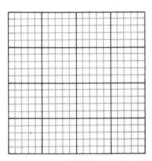

Graph each of the following hyperbolas.

13. $\dfrac{x^2}{4} - \dfrac{y^2}{25} = 1$

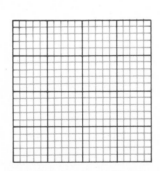

14. $\dfrac{x^2}{4} - \dfrac{y^2}{9} = 1$

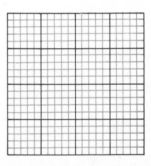

15. $\dfrac{y^2}{25} - \dfrac{x^2}{4} = 1$

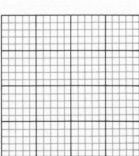

16. $\dfrac{y^2}{9} - \dfrac{x^2}{4} = 1$

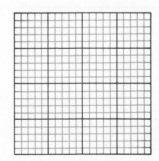

C *Graph each of the following equations after first writing the equation in one of the standard forms discussed in this section. For example,*

$$9x^2 + 4y^2 = 36$$

can be written in the form

$$\frac{x^2}{4} + \frac{y^2}{9} = 1$$

by dividing through by 36.

17. $4y^2 - 8x = 0$

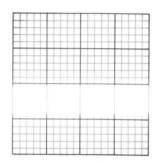

18. $3x^2 + 9y = 0$

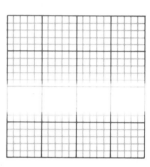

19. $4x^2 + 9y^2 = 36$

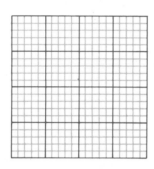

20. $4x^2 + 25y^2 = 100$

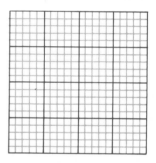

21. $4x^2 - 9y^2 = 36$

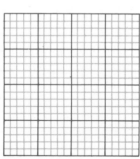

22. $25y^2 - 4x^2 = 100$

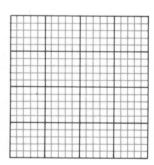

The Check Exercise for this section is on page 445.

8.6 CHAPTER REVIEW

Important terms and symbols

8.1 GRAPHING LINEAR EQUATIONS cartesian coordinate system, coordinate axes, quadrants, coordinates, abscissa, ordinate, origin, graphing a first-degree equation in two variables, x intercept, y intercept, vertical and horizontal lines

8.2 SLOPE AND EQUATIONS OF A LINE slope of a line, slope-intercept form, point-slope form, vertical and horizontal lines, parallel and perpendicular lines

$$m = \frac{y_2 - y_1}{x_2 - x_1}, \qquad x_1 \neq x_2 \qquad y = mx + b \qquad y - y_1 = m(x - x_1)$$

8.3 GRAPHING LINEAR INEQUALITIES half planes, upper half plane, lower half plane

8.4 CONIC SECTIONS, AN INTRODUCTION; CIRCLES circle, ellipse, parabola, hyperbola, distance-between-two-points formula, definition of a circle, equation of a circle

$$d = \sqrt{(x_2 - x_1)^2 + (y_2 - y_1)^2} \qquad (x - h)^2 + (y - k)^2 = R^2$$

$$x^2 + y^2 = R^2$$

8.5 PARABOLAS, ELLIPSES, AND HYPERBOLAS definitions of a parabola, equation of a parabola, definition of an ellipse, equation of an ellipse, definition of a hyperbola, equation of a hyperbola

$$y^2 = kx \qquad x^2 = ky \qquad \frac{x^2}{m^2} + \frac{y^2}{n^2} = 1 \qquad \frac{x^2}{m^2} - \frac{y^2}{n^2} = 1 \qquad \frac{y^2}{n^2} - \frac{x^2}{m^2} = 1$$

DIAGNOSTIC (REVIEW) EXERCISE 8.6

Work through all the problems in this chapter review and check answers in the back of the book. (Answers to all problems are there, and following each answer is a number in italics indicating the section in which that type of problem is discussed.) Where weaknesses show up, review appropriate sections in the text. When you are satisfied that you know the material, take the practice test following this review.

A *Graph each in a rectangular coordinate system.*

1. $y = 2x - 3$

2. $2x + y = 6$

3. $x - y \geq 6$

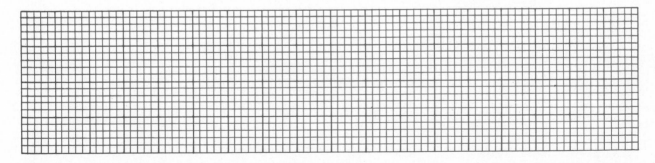

4. $y > x - 1$ **5.** $x^2 + y^2 = 36$

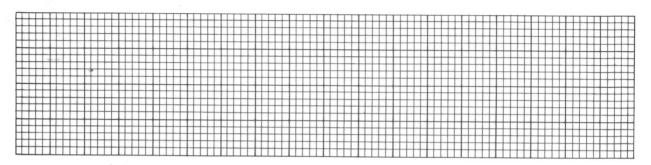

6. What is the slope and y intercept for the graph of $y = -2x - 3$?

_____

7. Write an equation of a line that passes through $(2, 4)$ with slope -2. Write the final answer in the form $y = mx + b$.

_____

8. What is the slope of the line that passes through $(1, 3)$ and $(3, 7)$?

_____

9. What is the equation of a line that passes through $(1, 3)$ and $(3, 7)$? Write the answer in the form $y = mx + b$.

_____

10. What is the distance between the two points $(1, 3)$ and $(3, 7)$?

_____

11. What is an equation of a circle with radius 5 and center at the origin?

_____

B *Graph each in a rectangular coordinate system.*

12. $3x - 2y = 9$ **13.** $y = \frac{1}{3}x - 2$ **14.** $x = -3$

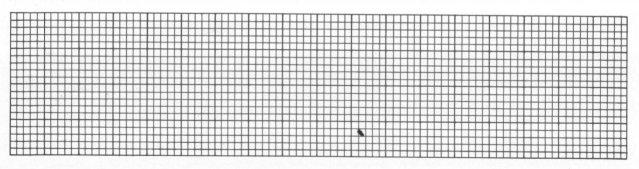

15. $4x - 5y \le 20$ **16.** $y < \dfrac{x}{2} + 1$ **17.** $x \ge -3$

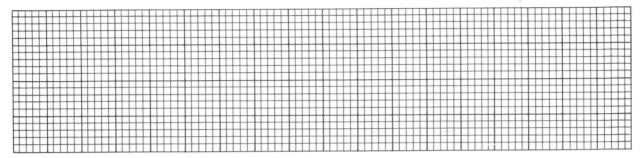

18. $-4 \le y < 3$ **19.** $x^2 + y^2 = 49$ **20.** $(x - 2)^2 + (y + 3)^2 = 16$

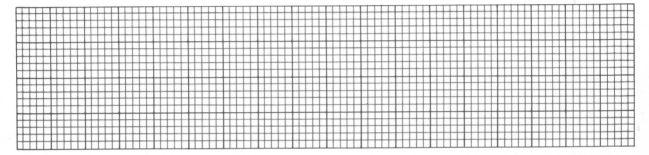

21. $y^2 = -2x$ **22.** $\dfrac{x^2}{9} + \dfrac{y^2}{16} = 1$ **23.** $\dfrac{y^2}{16} - \dfrac{x^2}{9} = 1$

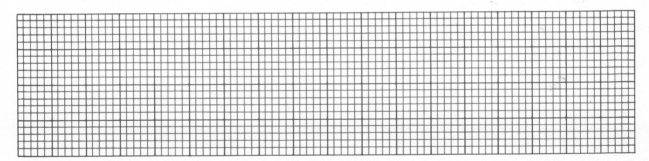

24. What is the slope and y intercept for the graph of $x + 2y = -6$?

_____

25. Write an equation of a line that passes through $(-3, 2)$ with slope $-\frac{1}{3}$. Write the final answer in the form $y = mx + b$.

_____

26. Write the equation of a line that passes through $(-3, 2)$ and $(3, -2)$. Write the final answer in the form $Ax + By = C, A > 0$.

_____

27. Find the equation of a line that passes through $(3, -4)$ and is perpendicular to $x + 2y = -6$. Write the final answer in the form $y = mx + b$.

_____

28. Write the equations of the vertical and horizontal lines that pass through $(5, -2)$.

_____

29. Write an equation of a circle in the form $(x - h)^2 + (y - k)^2 = R^2$ if its center is at $(-3, 4)$ and it has a radius of 7.

_____

C 30. Find an equation of a circle with center at the origin that passes through $(12, -5)$.

_____

31. Write the equation of a line that passes through $(-6, 2)$ and is parallel to $3x - 2y = 5$. Write final answer in the form $y = mx + b$.

_____

32. Transform the equation

$$x^2 + y^2 + 6x - 8y = 0$$

into the form

$$(x - h)^2 + (y - k)^2 = R^2$$

Since the graph is a circle, what is its radius and what are the coordinates of its center?

_____

33. Graph the set

$$\{(x, y) \mid y \geq 0, 1 \leq x \leq 5, 2x + 3y \leq 18\}$$

_____

PRACTICE TEST CHAPTER 8 _____

Take this as if it were a graded test. Allow yourself up to 50 minutes. Work the problems without looking back in the chapter. Correct your work, using the answers (keyed to appropriate sections) in the back of the book.

Graph Problems 1 to 5 in a rectangular coordinate system.

  1. $4x - 2y = 10$ **2.** $4x - 3y \geq 12$ **3.** $(x + 3)^2 + (y - 2)^2 = 9$

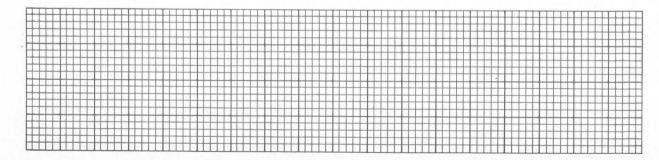

4. $x^2 = -12y$

5. $\dfrac{x^2}{9} - \dfrac{y^2}{16} = 1$

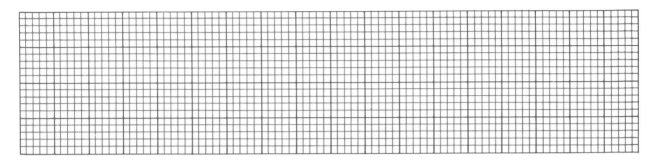

6. Write an equation of a line with y intercept 5 that is parallel to $y = -2x - 3$.

_____

7. What is the equation of the line that passes through $(-2, -3)$ and $(0, -2)$? Write final answer in the form $y = mx + b$.

_____

8. Write the equation of the horizontal line that passes through $(-3, -2)$.

_____

9. Write an equation of a line that passes through $(-3, 4)$ and is perpendicular to $3x - 2y = 4$. Write the final equation in the form $y = mx + b$.

_____

10. Transform $x^2 + y^2 - 8x + 6y + 9 = 0$ into the form $(x - h)^2 + (y - k)^2 = R^2$ and identify the radius and coordinates of the center of the circle.

_____

CHECK EXERCISE 8.1

Work the following problems without looking at any text examples. Show your work in the space provided. Write the letter that best indicates your answer in the answer column.

Graph each equation in the space provided; then out of all the indicated answers select the one (by letter) that matches your graph.

1. $y = \frac{3}{4}x$

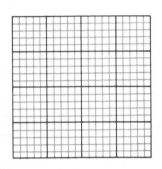 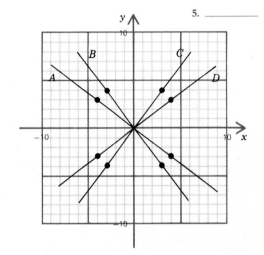

2. $6x - 8y = 24$

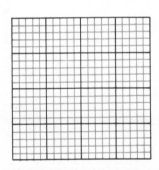 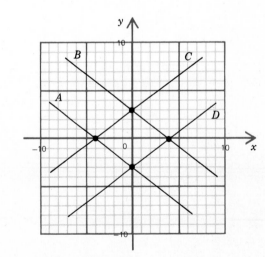

3. $y = -\frac{2}{3}x + 2$

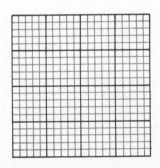 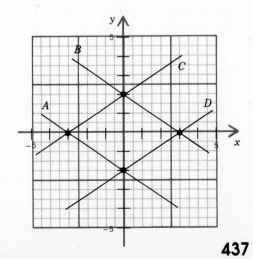

437

4. $3x + 4y = 10$

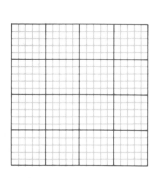

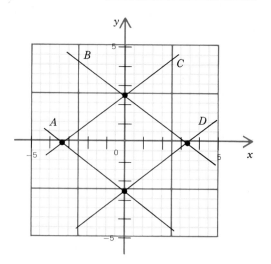

5. $A = 100 + 25t, \; 0 \le t \le 6$

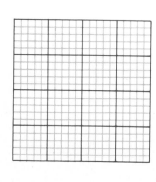

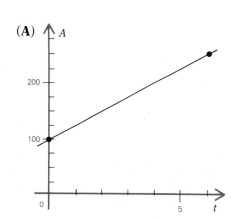

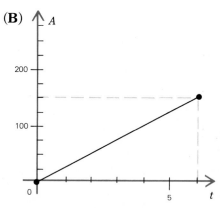

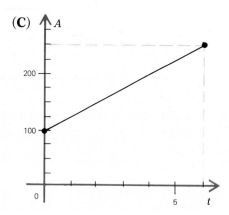

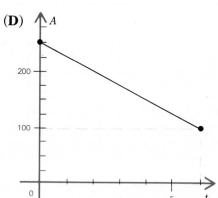

CHECK EXERCISE 8.2

Work the following problems without looking at any text examples. Show your work in the space provided. Write the letter that best indicates your answer in the answer column.

1. What are the slope and y intercept of the graph for $3x - 4y = 8$?

   (A) $m = -\frac{3}{4}$
        y intercept $= 2$

   (B) $m = \frac{3}{4}$
        y intercept $= -2$

   (C) $m = -\frac{3}{4}$
        y intercept $= -2$

   (D)   None of these

2. Write the equation of a line with slope $-\frac{3}{4}$ that passes through $(4, -1)$. Transform the equation into the form $y = mx + b$.

   (A) $y = -\frac{3}{4}x - 2$
   (B) $y = -\frac{3}{4}x - 5$
   (C) $y = -\frac{3}{4}x + 2$
   (D)   None of these

3. Write the equations of the horizontal and vertical lines that pass through $(-5, 3)$.

   (A) $y = -3$ and $x = 5$
   (B) $y = 5$ and $x = -3$
   (C) $y = -5$ and $x = 3$
   (D)   None of these

4. Write the equation of a line that passes through $(-4, 3)$ and $(5, -3)$. Transform the equation into the form $y = mx + b$.

(A) $y = -\frac{3}{2}x - 3$
(B) $y = -\frac{2}{3}x + \frac{17}{3}$
(C) $y = -\frac{2}{3}x + \frac{1}{3}$
(D) None of these

5. Write the equation of the line that is perpendicular to $3x + 4y = 9$ and passes through $(6, -5)$. Transform the equation into the form $y = mx + b$.

(A) $y = \frac{4}{3}x - 13$
(B) $y = \frac{3}{4}x - \frac{19}{2}$
(C) $y = \frac{3}{4}x + \frac{1}{2}$
(D) None of these

CHECK EXERCISE 8.3

Work the following problems without looking at any text examples. Show your work in the space provided. Write the letter that best indicates your answer in the answer column.

Graph each inequality in Problems 1 to 4 in a rectangular coordinate system.

1. $y \geq \dfrac{1}{2}x - 2$

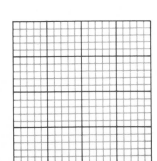

(A)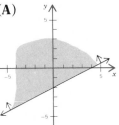

(B)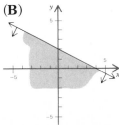

(C)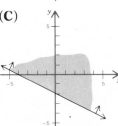

(D) None of these

2. $4x - 3y > 12$

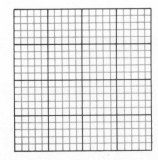

(A)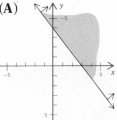

(B)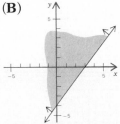

(C)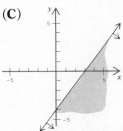

(D) None of these

441

3. $5x + 6y \le 30$

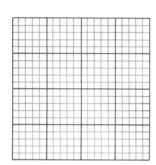

(A)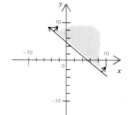

(B)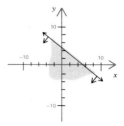

(C)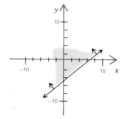

(D)   None of these

4. $-2 \le y < 4$

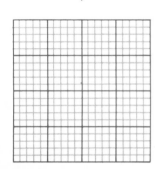

(A)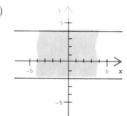

(B)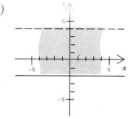

(C)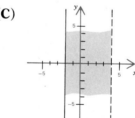

(D)   None of these

5.   Graph the set $\{(x, y) \mid x \ge 0, y \ge 0, 6x + 8y \le 48\}$

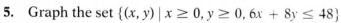

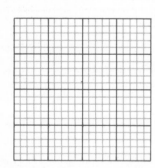

(A)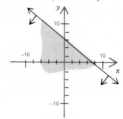

(B)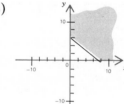

(C)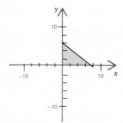

(D)   None of these

CHECK EXERCISE 8.4

Work the following problems without looking at any text examples. Show your work in the space provided. Write the letter that best indicates your answer in the answer column.

1. Find the distance between $(-2, 5)$ and $(3, 2)$. Leave answer in exact radical form.

   (A) $\sqrt{10}$
   (B) $\sqrt{34}$
   (C) $2\sqrt{6}$
   (D) None of these

2. Write the equation of a circle with radius $\sqrt{3}$ and center at the origin.

   (A) $x^2 + y^2 = \sqrt{3}$
   (B) $x^2 + y^2 = 3$
   (C) $x^2 + y^2 = 9$
   (D) None of these

3. Write the equation of a circle with radius 7 and center at $(-2, 3)$.

   (A) $(x - 2)^2 + (y + 3)^2 = 49$
   (B) $(x + 2)^2 + (y - 3)^2 = 7$
   (C) $(x - 2)^2 + (y - 3)^2 = 49$
   (D) None of these

4. Graph $(x - 3)^2 + (y + 2)^2 = 9$

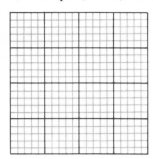

(A)

Center: $(-3, 2)$
Radius: 3

(B)

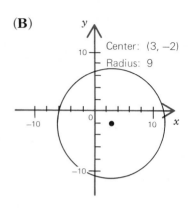

Center: $(3, -2)$
Radius: 9

(C)

Center: $(3, -2)$
Radius: 3

(D) None of these

5. Use method of completing the square as an aid to graphing $x^2 + y^2 + 6x - 4y + 9 = 0$

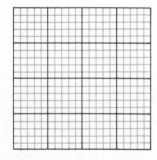

(A)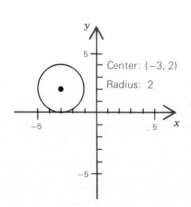

Center: $(-3, 2)$
Radius: 2

(B)

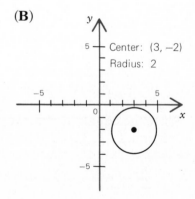

Center: $(3, -2)$
Radius: 2

(C)

Center: $(-3, 2)$
Radius: 4

(D) None of these

CHECK EXERCISE 8.5

Work the following problems without looking at any text examples. Show your work in the space provided. Write the letter that best indicates your answer in the answer column.

1.   Graph $y^2 = 5x$.

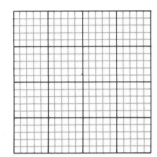

(A)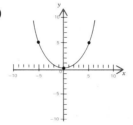

(B)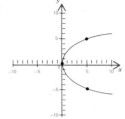

(C)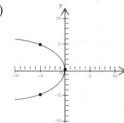

(D)   None of these

2.   Graph $x^2 = -16y$.

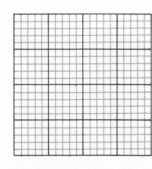

(A)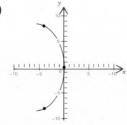

(B)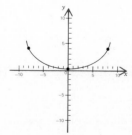

(C)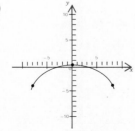

(D)   None of these

445

3. Graph $\dfrac{x^2}{36} + \dfrac{y^2}{9} = 1$.

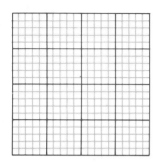

(A)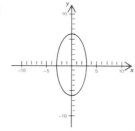

(B)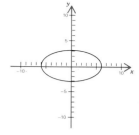

(C)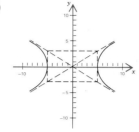

(D) None of these

4. Graph $\dfrac{y^2}{9} - \dfrac{x^2}{25} = 1$.

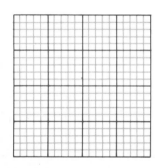

(A)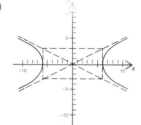

(B)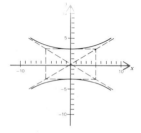

(C)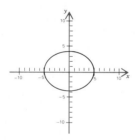

(D) None of these

5. Graph $25x^2 + 4y^2 = 100$.

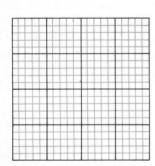

(A)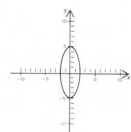

(B)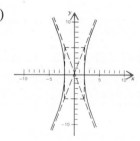

(C)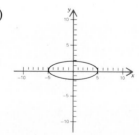

Chapter Nine

SYSTEMS OF EQUATIONS

CONTENTS

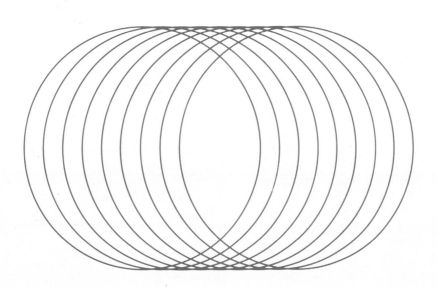

Is the closest opening to the "pipe" on the left or on the right?

INSTRUCTIONS FOR STUDENTS IN A
SELF-PACED CLASS OR LAB

yes — **HAVE YOU HAD INTERMEDIATE ALGEBRA BEFORE THIS COURSE?** — no

1. Work Diagnostic (Review) Exercise 9.5 on page 476. Check answers in back of book; then work through text sections corresponding to problems missed. (Section numbers are in italics following each answer.)
2. When finished with step 1, take Practice Test: Chapter 9 on page 478 as a final check of your understanding of the chapter. Check answers in the back of the book; then review sections where weakness still prevails. (Corresponding section numbers are in italics following each answer.)
3. When you think you are ready, ask your instructor for a graded test for Chapter 9.
4. If your instructor approves, after the test is corrected, go to the next chapter.

1. Work through each section in the chapter as follows:
   (a) Read discussion.
   (b) Read each example and work the corresponding matched problem. Check your solution to the matched problem in Solutions to Matched Problems on the indicated page.
   (c) At the end of a section work the odd-numbered problems in the Practice Exercise and check answers; then work even-numbered problems in areas of weakness. (Answers to *all* Practice Exercise sets are in the back of the book.)
   (d) Work Check Exercise as instructed. Tear out and turn in as directed by your instructor. (Answers are not in the text.)
2. Repeat each step in item 1 for each section in the chapter.
3. After the instructional part of the chapter is completed, proceed with steps 1 to 4 in the box above.

9.1 SYSTEMS OF LINEAR EQUATIONS IN TWO VARIABLES

•Solution by graphing
••Solution by substitution
•••Solution by elimination using addition

Many practical problems can be solved conveniently using two-equation–two-unknown methods. For example, if a 12-ft board is cut in two pieces so that one piece is 2 ft longer than the other piece, how long is each piece? We could solve this problem using one-equation–one unknown methods studied earlier, but we can also proceed as follows, using two variables:

Let x = the length of the longer piece
 y = the length of the shorter piece

then $x + y = 12$

 $x - y = 2$

To **solve** this system is to find all the ordered pairs of real numbers that satisfy both equations at the same time. In general, we are interested in solving linear systems of the type:

Linear System—Standard Form

$ax + by = m$

$cx + dy = n$

$a, b, c, d, m,$ and n are constants
x and y are variables
a, b, c, d not all zero

There are several methods of solving systems of this type. We will consider three that are widely used.

•Solution by graphing

We proceed by graphing both equations on the same coordinate system. Then the coordinates of any points that the graphs have in common must be solutions to the system since they must satisfy both equations.

EXAMPLE 1
Solve by graphing:

$x + y = 12$

$x - y = 2$

Solution
Graph each equation and find coordinates of points
of intersection, if they exist.

CHECK

$x + y = 12$ $x - y = 2$

$7 + 5 \overset{\vee}{=} 12$ $7 - 5 \overset{\vee}{=} 2$

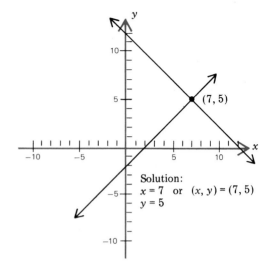

Solution:
$x = 7$ or $(x, y) = (7, 5)$
$y = 5$

Work Problem 1 and check solution in Solutions to Matched Problems on page 455.

PROBLEM 1   Solve by graphing:

$x + y = 10$

$x - y = 6$

It is clear that the system in Example 1 has exactly one solution since the lines have exactly
one point of intersection. In general, two lines in the same rectangular coordinate system must
be related to each other in one of three ways: (1) they intersect at one and only one point, (2) they
are parallel, or (3) they coincide (see Example 2).

EXAMPLE 2
Solve each of the following systems by graphing.

(A) $2x - 3y = 2$ (B) $4x + 6y = 12$ (C) $2x - 3y = -6$
      $x + 2y = 8$ $2x + 3y = -6$ $-x + \frac{3}{2}y = 3$

(A) 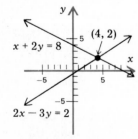   (B) 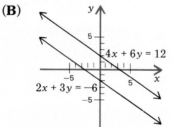   (C)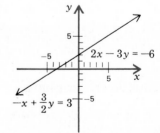

Lines intersect at   Lines are parallel   lines coincide:
one point only: (each has slope $-\frac{2}{3}$):   *infinitely many*
exactly one solution   *no solution*   *solutions*

$x = 4, y = 2$

Now we know exactly what to expect when solving a system of two linear equations in two unknowns:

Possible Solutions to a Linear System

---

1. Exactly one pair of numbers
2. No solution
3. Infinitely many solutions

Work Problem 2 and check solution in Solutions to Matched Problems on page 455.

PROBLEM 2 Solve each of the following systems by graphing:

(A) $2x + 3y = 12$ (B) $x - 3y = -3$ (C) $2x - 3y = 12$
       $x - 3y = -3$ $-2x + 6y = 12$ $-x + \frac{3}{2}y = -6$

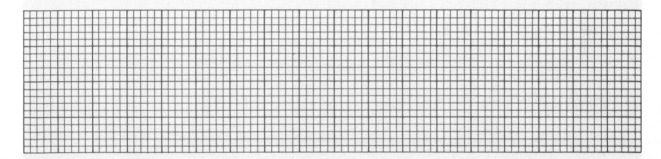

Generally, graphic methods give us only rough approximations of solutions. The methods of substitution and elimination using addition to be considered next will yield results to any decimal accuracy desired—assuming solutions exist.

••Solution by substitution

Choose one of the two equations in a system and solve for one variable in terms of the other (choose an equation that avoids getting involved with fractions, if possible). Then substitute the result into the other equation and solve the resulting linear equation in one variable. Now substitute this result back into either of the original equations to find the second variable. An example should make the process clear.

EXAMPLE 3
Solve by substitution

$$2x - 3y = 7 \tag{1}$$

$$-3x + y = -7 \tag{2}$$

Solution

$$y = \underbrace{3x - 7}$$

Solve (2), the simplest choice, for y in terms of x.

$$2x - 3y = 7$$

$$2x - 3(3x - 7) = 7$$

$$2x - 9x + 21 = 7$$

$$-7x = -14$$

$$x = 2$$

Substitute $x = 2$ into $y = 3x - 7$ and solve for y.

$$y = 3 \cdot 2 - 7$$

$$y = -1$$

Thus, $(2, -1)$ is a solution to the original system, as we can readily check:

CHECK

$(2, -1)$ must satisfy *both* equations.

$$2x - 3y = 7 \qquad\qquad -3x + y = -7$$

$$2(2) - 3(-1) \overset{?}{=} 7 \qquad -3(2) + (-1) \overset{?}{=} -7$$

$$4 + 3 \overset{\checkmark}{=} 7 \qquad\qquad -6 - 1 \overset{\checkmark}{=} -7$$

Work Problem 3 and check solution in Solutions to Matched Problems on page 456.

PROBLEM 3   Solve by substitution and check.

$$3x - 4y = 18$$

$$2x + y = 1$$

•••Solution by elimination using addition

Now we turn to elimination using addition. This is probably the most important method of solution, since it is readily generalized to higher-order systems. The method involves the replacement of systems of equations with simpler *equivalent systems* (by performing appropriate operations) until we obtain a system with an obvious solution. **Equivalent systems** of equations are, as you would expect, systems that have exactly the same solution set. Theorem 1 lists operations that produce equivalent systems.

THEOREM 1

Equivalent systems of equations result if:

(A) Two equations are interchanged

(B) An equation is multiplied by a nonzero constant

(C) A constant multiple of another equation is added to a given equation

Solving systems of equations by use of this theorem is best illustrated by examples.

EXAMPLE 4

Solve by elimination using addition.

$$3x - 2y = 8 \tag{3}$$

$$2x + 5y = -1 \tag{4}$$

Solution

We use Theorem 1 to eliminate one of the variables, and thus obtain a system with an obvious solution.

$$
\begin{array}{r}
15x - 10y = 40 \\
\underline{4x + 10y = -2} \\
19x \qquad\;\; = 38 \\
x = 2
\end{array}
$$

If we multiply (3) by 5 and (4) by 2, and then add, we can eliminate y.

Now substitute $x = 2$ back into either of the original equations, say (4), and solve for y ($x = 2$ paired with either of the two original equations produces an equivalent system):

$$2(2) + 5y = -1$$

$$5y = -5$$

$$y = -1$$

Thus, $(2, -1)$ is a solution to the original system. The check is completed as in Example 3.

Work Problem 4 and check solution in Solutions to Matched Problems on page 456.

PROBLEM 4 Solve by elimination using addition and check. Eliminate the variable x first.

$$2x + 5y = -9$$

$$3x - 4y = -2$$

EXAMPLE 5

Solve by elimination using addition.

$$x + 3y = 2 \tag{5}$$

$$2x + 6y = -3 \tag{6}$$

Solution

$$-2x - 6y = -4 \qquad \text{Multiply (5) by } -2 \text{ and add.}$$
$$\underline{2x + 6y = -3}$$
$$0 = -7 \qquad \text{A contradiction!}$$

Our assumption that there are values for x and y that satisfy (5) and (6) simultaneously must be false (otherwise, we have proved that $0 = -7$). If you check the slope of each line, you will find that they are the same (but the y intercepts are different); hence, the lines are parallel, and the system has no solution. Systems of this type are called **inconsistent**—conditions have been placed on the variables x and y that are impossible to meet.

Work Problem 5 and check solution in Solutions to Matched Problems on page 456.

PROBLEM 5   Solve by elimination using addition.

$$3x - 4y = -2$$
$$-6x + 8y = 1$$

EXAMPLE 6

Solve by elimination using addition.

$$-2x + y = -8 \qquad\qquad (7)$$
$$x - \tfrac{1}{2}y = 4 \qquad\qquad (8)$$

Solution

$$-2x + y = -8 \qquad \text{Multiply (8) by 2 and add.}$$
$$\underline{2x - y = 8}$$
$$0 = 0$$

Both sides have been eliminated. Actually, if we had multiplied (8) by -2, we would have obtained (7). When one equation is a constant multiple of the other, the system is said to be **dependent,** and their graphs will coincide. There are infinitely many solutions to the system—any solution of one equation will be a solution of the other. One way of expressing all solutions is to solve one of the equations for y in terms of x, say the first equation,

$$y = 2x - 8$$

then

$$(x, 2x - 8)$$

is a solution to the system for any real number x, as can easily be checked.

Work Problem 6 and check solution in Solutions to Matched Problems on page 456.

PROBLEM 6   Solve by elimination using addition:

$$6x - 3y = -2$$
$$-2x + y = \tfrac{2}{3}$$

SOLUTIONS TO MATCHED PROBLEMS

1. $x + y = 10$
   $x - y = 6$

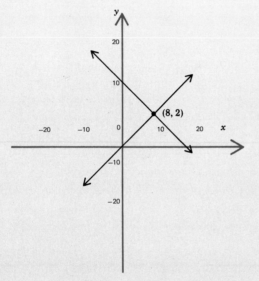

SOLUTION:

$x = 8$
$y = 2$ or $(x, y) = (8, 2)$

2. (A) $2x + 3y = 12$
          $x - 3y = -3$

(B) $x - 3y = -3$
        $-2x + 6y = 12$

(C) $2x - 3y = 12$
        $-x + \tfrac{3}{2}y = -6$

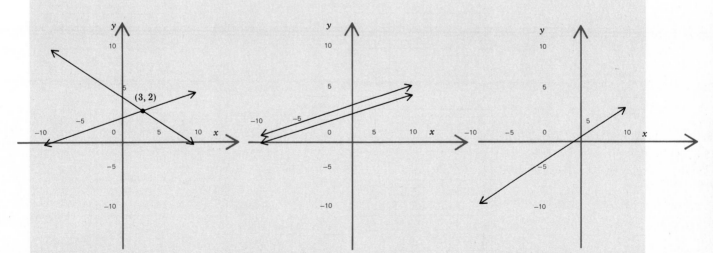

SOLUTION:
$x = 3$
$y = 2$ or $(x, y) = (3, 2)$

SOLUTION:
Lines are parallel;
no solution.

SOLUTION:
Lines coincide;
infinitely many solutions

3. $3x - 4y = 18$ (A)
   $2x + y = 1$ (B)
Solve (B) for y in terms of x; then substitute into (A):
$$y = 1 - 2x$$
$$3x - 4(1 - 2x) = 18$$
$$3x - 4 + 8x = 18$$
$$11x = 22$$
$$x = 2$$
$$y = 1 - 2(2)$$
$$y = -3$$
$(2, -3)$ is the solution

4. $2x + 5y = -9$ (A)
   $3x - 4y = -2$ (B)
Multiply (A) by 3 and (B) by -2; then add to eliminate x:
$$6x + 15y = -27$$
$$\underline{-6x + 8y = 4}$$
$$23y = -23$$
$$y = -1$$
Substitute into (A) to find x:
$$2x + 5(-1) = -9$$
$$2x = -4$$
$$x = -2$$

5. $3x - 4y = -2$ (A)
   $-6x + 8y = 1$ (B)
Multiply (A) by 2 and add:
$$6x - 8y = -4$$
$$\underline{-6x + 8y = 1}$$
$$0 = -3$$
Contradiction! No solution.

6. $6x - 3y = -2$ (A)
   $-2x + y = \frac{2}{3}$ (B)
Multiply (B) by 3 and add:
$$6x - 3y = -2$$
$$\underline{-6x + 3y = 2}$$
$$0 = 0$$
Infinitely many solutions; $(x, 2x + \frac{2}{3})$ is a solution for any real number x.

PRACTICE EXERCISE 9.1

Work odd-numbered problems first, check answers, and then work even-numbered problems in areas of weakness. Answers to all problems are in the back of the book. Make every effort to work a problem yourself before you look at an answer.

A *Solve by graphing.*

1. $3x - 2y = 12$
   $7x + 2y = 8$

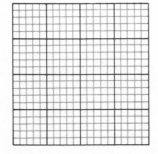

2. $x + 5y = -10$
   $-5x + y = 24$

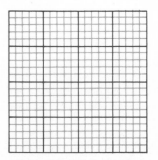

3. $3x + 5y = 15$
   $6x + 10y = -5$

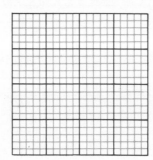

4. $3x - 5y = 15$
   $x - \frac{5}{3}y = 5$

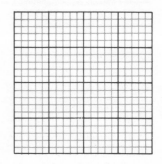

Solve by substitution.

5. $2x + y = 6$
$\qquad y = x + 3$ _____

6. $\qquad m - 2n = 0$
$\qquad -3m + 6n = 8$ _____

7. $3x - y = -3$
$\quad 5x + 3y = -19$ _____

8. $2m - 3n = 9$
$\quad m + 2n = -13$ _____

Solve by elimination using addition.

9. $\quad 3p + 8q = 4$
$\quad 15p + 10q = -10$ _____

10. $3x - y = -3$
$\quad 5x + 3y = -19$ _____

11. $\quad 6x - 2y = 18$
$\quad -3x + y = -9$ _____

12. $4m + 6n = 2$
$\quad 6m - 9n = 15$ _____

B *Solve each system by graphing, by substitution, and by elimination using addition.*

13. $\quad x - 3y = -11$
$\quad 2x + 5y = 11$ _____

14. $5x + y = 4$
$\quad x - 2y = 3$ _____

15. $11x + 2y = 1$
$\quad 9x - 3y = 24$ _____

16. $2x + y = 0$
$\quad 3x + y = 2$ _____

Use any of the methods discussed in this section to solve each system.

17. $\quad y = 3x - 3$
$\quad 6x = 8 + 3y$ _____

18. $3m = 2n$
$\quad n = -7 - 2m$ _____

19. $\quad \frac{1}{2}x - y = -3$
$\quad -x + 2y = 6$ _____

20. $\qquad y = 2x - 1$
$\quad 6x - 3y = -1$ _____

21. $2x + 3y = 2y - 2$
$\quad 3x + 2y = 2x + 2$ _____

22. $2u - 3v = 1 - 3u$
$\qquad 4v = 7u - 2$ _____

C **23.** $0.2x - 0.5y = 0.07$
$\quad 0.8x - 0.3y = 0.79$ _____

24. $0.5m + 0.2n = 0.54$
$\quad 0.3m - 0.6n = 0.18$ _____

25. $\frac{1}{4}x - \frac{2}{3}y = -2$
$\quad \frac{1}{2}x - y = -2$ _____

26. $\frac{2}{3}a + \frac{1}{2}b = 2$
$\quad \frac{1}{2}a + \frac{1}{3}b = 1$ _____

The Check Exercise for this section is on page 479.

9.2 APPLICATIONS

Many of the applications we considered earlier, using one-equation–one-unknown methods, can be set up more naturally using two-equation–two-unknown methods. The following examples will illustrate the process.

EXAMPLE 7

A change machine changes dollar bills into quarters and nickels. If you receive 12 coins after inserting a $1 bill, how many of each type of coin did you receive?

Solution
Let

x = number of quarters
y = number of nickels

$$x + y = 12 \qquad \text{number of coins}$$
$$25x + 5y = 100 \qquad \text{value of coins in cents}$$

$$-5x - 5y = -60$$
$$\underline{25x + 5y = 100}$$
$$20x \qquad = 40$$

$$x = 2 \qquad \text{quarters}$$
$$x + y = 12$$
$$2 + y = 12$$
$$y = 10 \qquad \text{nickels}$$

CHECK
$$2 + 10 = 12 \text{ coins in all}$$
$$25 \cdot 2 + 5 \cdot 10 = 50 + 50 = 100 \text{ cents or } \$1$$

Work Problem 7 and check solution in Solutions to Matched Problems on page 461.

PROBLEM 7   Repeat Example 7 with a receipt of eight coins from a $1 bill.

Solution

EXAMPLE 8

A zoologist wishes to prepare a special diet that contains, among other things, 120 grams of protein and 17 grams of fat. Two available food mixes specify the following percentages of protein and fat:

Mix	Protein, %	Fat, %
A	30	1
B	20	5

How many grams of each mix should be used to prepare the diet mix?

Solution

Let x = number of grams of mix A used

$\quad\;\; y$ = number of grams of mix B used

Set up one equation for the protein requirements and one equation for the fat requirements.

$0.3x + 0.2y = 120 \qquad$ protein requirements

$0.01x + 0.05y = 17 \qquad$ fat requirements

Multiply the top equation by 10 and the bottom equation by 100 to clear decimals (not necessary, but helpful).

$3x + 2y = 1,200 \qquad$ Multiply bottom equation
$\underline{\;x + 5y = 1,700\;} \qquad$ by -3; then add to eliminate x.

$$
\begin{array}{r}
3x + 2y = 1,200 \\
\underline{-3x - 15y = -5,100} \\
-13y = -3,900
\end{array}
$$

$\qquad\qquad y = 300$ grams of mix B

$\qquad x + 5y = 1,700$

$\quad x + 5(300) = 1,700$

$\qquad\qquad x = 200$ grams of mix A

The zoologist should use 200 grams of mix A and 300 grams of mix B to meet the diet requirements.

CHECK

Protein requirement:

(Protein from mix A) + (protein from mix B) $\overset{?}{=}$ 120 grams

$\quad$ (30% of 200 g) $\quad + \quad$ (20% of 300 g) $\quad = 60 + 60 \overset{\swarrow}{=} 120$ grams

Fat requirement:

(Fat from mix A) + (fat from mix B) $\overset{?}{=}$ 17 grams

$\quad$ (1% of 200 g) $\quad + \quad$ (5% of 300 g) $\quad = 2 + 15 \overset{\swarrow}{=} 17$ grams

Work Problem 8 and check solution in Solutions to Matched Problems on page 461.

PROBLEM 8   Repeat Example 8 if the diet mixture is to contain 110 grams of protein and 8 grams of fat.

Solution

EXAMPLE 9

A jeweler has two bars of gold alloy in stock, one 12-carat and the other 18-carat (24-carat gold is pure gold, 12-carat gold is $\frac{12}{24}$ pure, 18-carat gold is $\frac{18}{24}$ pure, and so on). How many grams of each alloy must be mixed to obtain 10 grams of 14-carat gold?

Solution
Let

$x = $ number of grams of 12-carat gold used
$y = $ number of grams of 18-carat gold used

$$x + \quad y = 10 \qquad \text{Amount of new alloy.}$$
$$\tfrac{12}{24}x + \tfrac{18}{24}y = \tfrac{14}{24}(10) \qquad \text{Pure gold present before mixing equals pure gold present after mixing.}$$

$$x + \quad y = 10 \qquad \text{Multiply second equation by } \tfrac{24}{2} \text{ to simplify, and then solve using methods described above. (We use}$$
$$6x + 9y = 70 \qquad \text{elimination here.)}$$

$$\begin{aligned} -6x - 6y &= -60 \\ \underline{6x + 9y} &= \underline{70} \\ 3y &= 10 \end{aligned}$$

$$y = 3\tfrac{1}{3} \text{ grams of 18-carat alloy}$$

$$x + 3\tfrac{1}{3} = 10$$

$$x = 6\tfrac{2}{3} \text{ grams of 12-carat alloy}$$

The checking of solutions is left to the reader.

Work Problem 9 and check solution in Solutions to Matched Problems on page 461.

PROBLEM 9   Repeat Example 9, using the fact that the jeweler has only 10-carat and pure gold in stock.

Solution

SOLUTIONS TO MATCHED PROBLEMS

Checks are left to the reader.

7. Let x = number of quarters
$\quad\quad y$ = number of nickels

$x + y = 8$
$25x + 5y = 100$
$-5x - 5y = -40$
$\underline{25x + 5y = 100}$
$20x \quad\quad = 60$
$\quad\quad x = 3$ quarters

$x + y = 8$
$3 + y = 8$
$\quad\quad y = 5$ nickels

8. Let x = number of grams of mix A used
$\quad\quad y$ = number of grams of mix B used

$0.3x + 0.2y = 110$
$0.01x + 0.05y = 8$
$3x + 2y = 1{,}100$
$x + 5y = 800$

$3x + 2y = 1{,}100$
$\underline{-3x - 15y = -2{,}400}$
$\quad\quad -13y = -1{,}300$
$\quad\quad y = 100$ grams of mix B

$x + 5y = 800$
$x + 5(100) = 800$
$\quad\quad x = 300$ grams of mix A

9. Let x = number of grams of 10-carat gold used
$\quad\quad y$ = number of grams of pure gold used

$x + y = 10$
$\frac{10}{24}x + y = \frac{14}{24}(10)$
$x + y = 10$
$5x + 12y = 70$
$-5x - 5y = -50$
$\underline{5x + 12y = 70}$
$\quad\quad y = \frac{20}{7} \approx 2.86$ grams of pure gold

$x + y = 10$
$x + \frac{20}{7} = 10$
$\quad\quad x = \frac{50}{7} \approx 7.14$ grams of 10-carat gold

PRACTICE EXERCISE 9.2

Work odd-numbered problems first, check answers, and then work even-numbered problems in areas of weakness. Answers to all problems are in the back of the book. Make every effort to work a problem yourself before you look at an answer.

Solve, using two-equation–two-unknown methods.

A 1. PUZZLE   A bank gave you $1.50 in change consisting of only nickels and dimes. If there were 22 coins in all, how many of each type of coin did you receive?

2. PUZZLE   A friend of yours bought 30 stamps and spent $1.32. If they were 4-cent and 5-cent stamps, how many of each type did your friend buy?

3. GEOMETRY   If the sum of two angles in a right triangle is 90° and their difference is 14°, find the two angles.

4. **GEOMETRY**   Find the dimensions of a rectangle with perimeter 72 in, if its length is 1.25 times its width.

_____

B 5. **BUSINESS**   A packing carton contains 144 small packages, some weighing $\frac{1}{4}$ lb each and the others $\frac{1}{2}$ lb each. How many of each type are in the carton if the total contents of the carton weigh 51 lb?

_____

6. **BUSINESS**   Repeat Problem 5 if the box contains 80 packages weighing 28 lb.

_____

7. **BIOLOGY**   A biologist, in a nutrition experiment, wants to prepare a special diet for experimental animals. He requires a food mixture that contains among other things, 20 oz of protein and 6 oz of fat. He is able to purchase food mixes of the following compositions:

Mix	Protein, %	Fat, %
A	20	2
B	10	6

How many ounces of each mix should he use to prepare the diet mix?

_____

8. **BIOLOGY**   Repeat Problem 7 if the diet mixture is to contain 20 oz of protein and 4 oz of fat.

_____

9. **CHEMISTRY**   A chemist has two concentrations of hydrochloric acid in stock, a 50% solution and an 80% solution. How much of each should she mix to obtain 100 ml of a 68% solution?

_____

10. **CHEMISTRY**   Repeat Problem 9 if the chemist wishes to obtain 100 ml of a 62% solution.

_____

C 11. **EARTH SCIENCE**   A ship using sound-sensing devices above and below water recorded a surface explosion 6 sec sooner by its underwater device than its above-water device. Sound travels in air at about 1,100 ft/sec and in seawater at about 5,000 ft/sec.
(**A**)   How long did it take each sound wave to reach the ship?
(**B**)   How far was the explosion from the ship?

(**A**) _____   (**B**) _____

12. **EARTH SCIENCE**   An earthquake emits a primary wave and a secondary wave. Near the surface of the earth the primary wave travels at about 5 mi/sec, and the secondary wave at about 3 mi/sec. From the time lag between the two waves arriving at a given station, it is possible to estimate the distance to the quake. (The *epicenter* can be located by obtaining distance bearings at three or more sta-

tions.) Suppose a station measured a time difference of 16 sec between the arrival of the two waves. How long did each wave travel, and how far was the earthquake from the station?

The Check Exercise for this section is on page 481.

9.3 SYSTEMS OF LINEAR EQUATIONS IN THREE VARIABLES

Having learned how to solve systems of linear equations in two variables, we proceed to higher-order systems. Systems of the form

$$3x - 2y + 4z = 6$$
$$2x + 3y - 5z = -8 \tag{1}$$
$$5x - 4y + 3z = 7$$

as well as higher-order systems are encountered frequently and are worth studying. In fact, systems of equations in many variables are so important in solving real-world problems that there are whole courses on this one topic. A triplet of numbers $x = 0$, $y = -1$, and $z = 1$ [also written as an ordered triplet $(0, -1, 1)$] is a **solution** of system (1) since each equation is satisfied by this triplet. The set of all such ordered triplets of numbers is called the **solution set** of the system. Two systems are said to be **equivalent** if they have the same solution set.

We will use an extension of the method of elimination by addition discussed in Section 9.1 to solve systems in the form of (1). Theorem 1 in Section 9.1 is behind the process.

Steps in Solving Systems of Three Equations in Three Variables

Step 1. Choose two equations from the system and eliminate one of the three variables, using elimination by addition or subtraction. The result is generally one equation in two unknowns.

Step 2. Now eliminate the same variable from the unused equation and one of those used in step 1. We (generally) obtain another equation in two variables.

Step 3. The two equations from steps 1 and 2 form a system of two equations and two unknowns. Solve as in the preceding section.

Step 4. Substitute the solution from step 3 into any of the three original equations and solve for the third variable to complete the solution of the original system.

EXAMPLE 10

Solve: $3x - 2y + 4z = 6$ (2)

$\qquad 2x + 3y - 5z = -8$ (3)

$\qquad 5x - 4y + 3z = 7$ (4)

Solution

Step 1. We look at the coefficients of the various variables and choose to eliminate y from equations (2) and (4) because of the convenient coefficients -2 and -4. Multiply equation (2) by -2 and add to equation (4):

$$\begin{array}{ll} -6x + 4y - 8z = -12 & \quad -2[\text{equation (2)}] \\ \underline{5x - 4y + 3z = 7} & \quad [\text{equation (4)}] \\ -x \quad\quad - 5z = -5 & \end{array}$$

Step 2. Now let us eliminate y (the same variable) from equations (2) and (3). Multiply equation (2) by 3 and equation (3) by 2 and add.

$$\begin{array}{ll} 9x - 6y + 12z = 18 & \quad 3[\text{equation (2)}] \\ \underline{4x + 6y - 10z = -16} & \quad 2[\text{equation (3)}] \\ 13x \quad\quad + 2z = 2 & \end{array}$$

Step 3. From Steps 1 and 2 we obtain the system

$-x - 5z = -5$ (5)

$13x + 2z = 2$ (6)

We solve this system as in the preceding section. Multiply equation (5) by 13 and add to equation (6) to eliminate x.

$$\begin{array}{ll} -13x - 65z = -65 & \quad 13[\text{equation (5)}] \\ \underline{13x + 2z = 2} & \\ \quad\quad - 63z = -63 & \\ \quad\quad\quad z = 1 & \end{array}$$

Substitute $z = 1$ back into either equation (5) or (6) [we choose equation (5)] to find x:

$-x - 5z = -5$

$-x - 5 \cdot 1 = -5$

$\qquad -x = 0$

$\qquad x = 0$

Step 4. Substitute $x = 0$ and $z = 1$ back into any of the three original equations [we choose equation (2)] to find y:

$$3x - 2y + 4z = 6$$
$$3 \cdot 0 - 2y + 4 \cdot 1 = 6$$
$$-2y + 4 = 6$$
$$-2y = 2$$
$$y = -1$$

Thus, the solution to the original system is $(0, -1, 1)$ or $x = 0, y = -1, z = 1$.

CHECK
To check the solution, we must check *each* equation in the original system:

$$3x - 2y + 4z = 6 \qquad\qquad 2x + 3y - 5z = -8$$
$$3 \cdot 0 - 2(-1) + 4 \cdot 1 \overset{?}{=} 6 \qquad 2 \cdot 0 + 3(-1) - 5 \cdot 1 \overset{?}{=} -8$$
$$6 \overset{\checkmark}{=} 6 \qquad\qquad\qquad -8 \overset{\checkmark}{=} -8$$

$$5x - 4y + 3z = 7$$
$$5 \cdot 0 - 4(-1) + 3 \cdot 1 \overset{?}{=} 7$$
$$7 \overset{\checkmark}{=} 7$$

Work Problem 10 and check solution in Solutions to Matched Problems on page 467.

PROBLEM 10 Solve the system:

$$2x + 3y - 5z = -12$$
$$3x - 2y + 2z = 1$$
$$4x - 5y - 4z = -12$$

Solution

If we encounter, in the process described above, an equation that states a contradiction, such as $0 = -2$, then we must conclude that the system has no solution (that is, the system is **inconsistent**). If, on the other hand, one of the equations turns out to be $0 = 0$, the system either has infinitely many solutions or it has none. We must proceed further to determine which. Notice how this last result differs from the two-equation–two-unknown case. There, when we obtained $0 = 0$, we *knew* that there were infinitely many solutions. If a system has infinitely many solutions, then it is said to be **dependent.**

For completeness, let us look at a system that turns out to be dependent to see how the solution set can be represented. Consider the system:

$$x + y - z = 2 \tag{7}$$
$$3x + 2y - z = 5 \tag{8}$$
$$5x + 2y + z = 7 \tag{9}$$

We choose to eliminate z from two equations by adding equation (9) to equation (7) and by adding equation (9) to equation (8). Doing this we obtain the system

$$6x + 3y = 9$$
$$8x + 4y = 12$$

By multiplying the top equation by 1/3 and the bottom equation by 1/4, we obtain the simpler system

$$2x + y = 3 \tag{10}$$
$$2x + y = 3 \tag{11}$$

Since these two equations are the same, the original system must be dependent [if we multiply either equation (10) or (11) by -1 and add the result to the other we will obtain $0 = 0$]. To represent the solution set of the original system, we proceed as follows. Solve (10) for y in terms of x:

$$y = 3 - 2x \tag{12}$$

Now replace y by $3 - 2x$ in any of the original equations and solve for z. We use equation (9):

$$5x + 2y + z = 7$$
$$5x + 2(3 - 2x) + z = 7$$
$$5x + 6 - 4x + z = 7$$
$$z = 1 - x$$

Thus, for *any* real number x, the ordered triplet

$$(x, 3 - 2x, 1 - x)$$

is a solution of the original system. For example:

If $x = 1$, then

$$(1, 3 - 2 \cdot 1, 1 - 1) = (1, 1, 0)$$

is a solution.

If $x = -3$, then

$$(-3, 3 - 2(-3), 1 - (-3)) = (-3, 9, 4)$$

is a solution. And so on.

SOLUTIONS TO MATCHED PROBLEMS

10. $\quad 2x + 3y - 5z = -12 \qquad$ (A)
$\quad 3x - 2y + 2z = 1 \qquad$ (B)
$\quad 4x - 5y - 4z = -12 \qquad$ (C)
We choose to eliminate z from (A) and (B)
$$\begin{array}{ll} 4x + 6y - 10z = -24 & 2\text{(A)} \\ \underline{15x - 10y + 10z = 5} & 5\text{(B)} \\ 19x - 4y \quad\quad = -19 & \end{array}$$
Now we choose to eliminate z from (B) and (C).
$$\begin{array}{ll} 6x - 4y + 4z = 2 & 2\text{(B)} \\ \underline{4x - 5y - 4z = -12} & \text{(C)} \\ 10x - 9y \quad\quad = -10 & \end{array}$$
We now have two equations and two unknowns, which we solve.
$$\begin{array}{ll} 19x - 4y = -19 & \text{(D)} \\ 10x - 9y = -10 & \text{(E)} \end{array}$$
$$\begin{array}{ll} 171x - 36y = -171 & 9\text{(D)} \\ \underline{-40x + 36y = 40} & -4\text{(E)} \\ 131x \quad\quad = -131 & \\ \quad\quad x = -1 & \end{array}$$

Substitute $x = -1$ into either (D) or (E) [we choose (E)] and solve for y.
$$\begin{array}{l} 10x - 9y = -10 \\ 10(-1) - 9y = -10 \\ \quad\quad -9y = 0 \\ \quad\quad y = 0 \end{array}$$
Now substitute $x = -1$ and $y = 0$ into either (A), (B), or (C) [we choose (B)] and solve for z.
$$\begin{array}{l} 3x - 2y + 2z = 1 \\ 3(-1) - 2(0) + 2z = 1 \\ \quad\quad\quad 2z = 4 \\ \quad\quad\quad z = 2 \end{array}$$
Thus, $(-1, 0, 2)$ is the solution to the original system, as the reader can check.

PRACTICE EXERCISE 9.3

Work odd-numbered problems first, check answers, and then work even-numbered problems in areas of weakness. Answers to all problems are in the back of the book. Make every effort to work a problem yourself before you look at an answer.

Solve and check.

A   **1.** $\quad -2x \quad\quad\quad = 2$
$\quad\quad\quad x - 3y \quad\quad = 2$
$\quad\quad -x + 2y + 3z = -7$

2. $\quad\quad 2y + z = -4$
$\quad\quad x - 3y + 2z = 9$
$\quad\quad - y \quad\quad = 3$

3.
$$4y - z = -13$$
$$3y + 2z = 4$$
$$6x - 5y - 2z = 0$$

4.
$$2x + z = -5$$
$$x - 3z = -6$$
$$4x + 2y - z = -9$$

B 5.
$$2x + y - z = 5$$
$$x - 2y - 2z = 4$$
$$3x + 4y + 3z = 3$$

6.
$$x - 3y + z = 4$$
$$-x + 4y - 4z = 1$$
$$2x - y + 5z = -3$$

7.
$$2a + 4b + 3c = 6$$
$$a - 3b + 2c = -7$$
$$-a + 2b - c = 5$$

8.
$$3u - 2v + 3w = 11$$
$$2u + 3v - 2w = -5$$
$$u + 4v - w = -5$$

9.
$$2x - 3y + 3z = -15$$
$$3x + 2y - 5z = 19$$
$$5x - 4y - 2z = -2$$

10.
$$3x - 2y - 4z = -8$$
$$4x + 3y - 5z = -5$$
$$6x - 5y + 2z = -17$$

11.
$$5x - 3y + 2z = 13$$
$$2x + 4y - 3z = -9$$
$$4x - 2y + 5z = 13$$

12.
$$4x - 2y + 3z = 0$$
$$3x - 5y - 2z = -12$$
$$2x + 4y - 3z = -4$$

C 13.
$$x - 8y + 2z = -1$$
$$x - 3y + z = 1$$
$$2x - 11y + 3z = 2$$

14.
$$-x + 2y - z = -4$$
$$4x + y - 2z = 1$$
$$x + y - z = -1$$

15.
$$4w - x = 5$$
$$-3w + 2x - y = -5$$
$$2w - 5x + 4y + 3z = 13$$
$$2w + 2x - 2y - z = -2$$

16.
$$2r - s + 2t - u = 5$$
$$r - 2s + t + u = 1$$
$$-r + s - 3t - u = -1$$
$$-r - 2s + t + 2u = -4$$

APPLICATIONS

17. **GEOMETRY**　The equation of a circle in a rectangular coordinate system can be written in the form $x^2 + y^2 + Dx + Ey + F = 0$. Find D, E, and F so that the circle passes through $(-2, -1)$, $(-1, -2)$, and $(6, -1)$. HINT: The coordinates of each point must satisfy the equation. We thus obtain three equations with three unknowns (D, E, and F) after substitution.

18. **LIFE SCIENCE** A zoologist, in an experiment involving mice, finds she needs a food mix that contains, among other things, 23 grams of protein, 6.2 grams of fat, and 16 grams of moisture. She has on hand mixes of the following compositions:

Mix	Protein, %	Fat, %	Moisture, %
A	20	2	15
B	10	6	10
C	15	5	5

How many grams of each mix should she use to get the desired diet mix?

The Check Exercise for this section is on page 483.

9.4 NONLINEAR SYSTEMS

In this section we will investigate several special types of systems that involve at least one second-degree equation in two variables. The methods used to solve these systems are best illustrated through examples.

EXAMPLE 11

Solve the system:

$$4x^2 + y^2 = 25$$
$$2x + y = 7$$

Solution

In this type of problem the substitution principle is effective. Solve the linear equation for one variable in terms of the other, and then substitute into the nonlinear equation to obtain a quadratic equation in one variable.

$$4x^2 + y^2 = 25 \qquad \text{Solve } 2x + y = 7 \text{ for } y \text{ in terms of } x.$$
$$2x + y = 7$$
$$y = 7 - 2x \qquad \text{Substitute into } 4x^2 + y^2 = 25.$$
$$4x^2 + y^2 = 25$$
$$4x^2 + (7 - 2x)^2 = 25 \qquad \text{Simplify, and write in standard quadratic form } ax^2 + bx + c = 0.$$
$$4x^2 + 49 - 28x + 4x^2 = 25$$

$$8x^2 - 28x + 24 = 0 \qquad \text{Multiply both sides by 1/4.}$$

$$2x^2 - 7x + 6 = 0 \qquad \text{Solve—we use factoring method.}$$

$$(2x - 3)(x - 2) = 0$$

$$2x - 3 = 0 \quad \text{or} \quad x - 2 = 0$$

$$x = \tfrac{3}{2} \qquad\qquad x = 2$$

These values are substituted back into the linear equation $y = 7 - 2x$ to find the corresponding values for y. (Note that if we substitute these values back into the second-degree equation, we may obtain "extraneous" roots; try it and see why. Recall that a solution of a system is an ordered pair of numbers that satisfies *both* equations.)

For $x = \tfrac{3}{2}$,

$$y = 7 - 2(\tfrac{3}{2})$$

$$y = 4$$

For $x = 2$,

$$y = 7 - 2 \cdot 2$$

$$y = 3$$

Thus $(\tfrac{3}{2}, 4)$ and $(2, 3)$ are the solutions to the system, as can easily be checked.

Solve Problem 11 and check solution in solutions to Matched Problems on page 473.

PROBLEM 11   Solve the system:

$$2x^2 - y^2 = 1$$

$$3x + y = 2$$

Solution

EXAMPLE 12

Solve:

$$x^2 - y^2 = 5$$

$$x^2 + 2y^2 = 17$$

Solution

$$x^2 - y^2 = 5$$
$$\underline{x^2 + 2y^2 = 17}$$

Proceed as with linear equations— subtract to eliminate x.

$$-3y^2 = -12$$
$$y^2 = 4$$
$$y = \pm 2$$

For $y = 2$,

Substitute into either original equations; then solve for x.

$$x^2 - (2)^2 = 5$$
$$x^2 = 9$$
$$x = \pm 3$$

For $y = -2$,

Substitute into either original equation; then solve for x.

$$x^2 - (-2)^2 = 5$$
$$x^2 = 9$$
$$x = \pm 3$$

Thus $(3, -2)$, $(3, 2)$, $(-3, -2)$, and $(-3, 2)$ are the four solutions to the system. The checking of the solutions is left to the reader.

Solve Problem 12 and check solution in Solutions to Matched Problems on page 473.

PROBLEM 12 Solve the system:

$$2x^2 - 3y^2 = 5$$
$$3x^2 + 4y^2 = 16$$

Solution

EXAMPLE 13
Solve:

$$x^2 + 3xy + y^2 = 20 \tag{1}$$
$$xy - y^2 = 0 \tag{2}$$

Solution

$$y(x - y) = 0 \qquad \text{Factor equation (2).}$$

$$y = 0 \quad \text{or} \quad x - y = 0 \qquad \text{Substitute each of these in turn into}$$
$$\phantom{y = 0 \quad \text{or} \quad} y = x \qquad \text{equation (1), and proceed as before.}$$

For $y = 0$ [replace y with 0 in equation (1) and solve for x],

$$x^2 + 3x(0) + (0)^2 = 20$$
$$x^2 = 20$$
$$x = \pm 2\sqrt{5}$$

For $y = x$ [replace y with x in equation (1) and solve for x],

$$x^2 + 3xx + x^2 = 20$$
$$5x^2 = 20$$
$$x^2 = 4$$
$$x = \pm 2 \qquad \text{Substitute these values back into } y = x$$
$$ \text{to find corresponding values of } y.$$

For $x = 2, y = 2$; for $x = -2, y = -2$.

Thus $(2\sqrt{5}, 0), (-2\sqrt{5}, 0), (2, 2),$ and $(-2, -2)$ are the four solutions to the system. The checking of the solutions is left to the reader.

COMMON ERROR

If we had started by solving $xy - y^2 = 0$ for x as follows:

$$xy - y^2 = 0$$
$$xy = y^2$$
$$x = y$$

we would have lost two solutions, $(2\sqrt{5}, 0)$ and $(-2\sqrt{5}, 0)$. Dividing both sides of $xy = y^2$ by the variable y is behind this loss. **Do not divide both sides of an equation by an expression involving a variable for which you are solving.** The operation often results in the loss of solutions (and the loss of points on a test).

Solve Problem 13 and check solution in Solutions to Matched Problems on page 474.

PROBLEM 13   Solve the system:

$$x^2 + xy - y^2 = 4$$
$$2x^2 - xy = 0$$

Example 13 is somewhat specialized. However, it suggests a procedure that, when used alone or in combination with other procedures, is effective for some problems.

To obtain an idea of how many real solutions one might expect from a system of second-degree equations in two variables, one has only to look at the number of ways two conics can intersect. In general, it turns out that a system of one linear and one quadratic equation can have at most two solutions, and a system of two quadratic equations can have at most four solutions. Of course, some of the solutions may be complex. (See Figure 1 for several examples.)

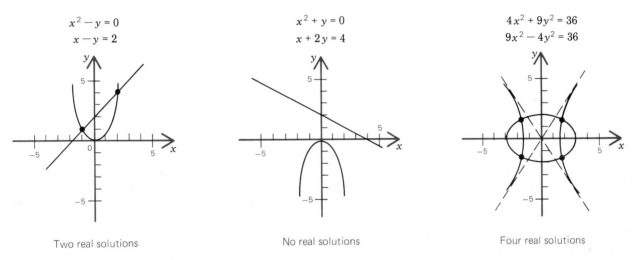

$$x^2 - y = 0$$
$$x - y = 2$$

$$x^2 + y = 0$$
$$x + 2y = 4$$

$$4x^2 + 9y^2 = 36$$
$$9x^2 - 4y^2 = 36$$

Two real solutions   No real solutions   Four real solutions

Figure 1

SOLUTIONS TO MATCHED PROBLEMS

Checks are left to the reader.

11.
$$2x^2 - y^2 = 1$$
$$3x + y = 2$$
$$y = 2 - 3x$$
$$2x^2 - (2 - 3x)^2 = 1$$
$$2x^2 - (4 - 12x + 9x^2) = 1$$
$$2x^2 - 4 + 12x - 9x^2 = 1$$
$$-7x^2 + 12x - 5 = 0$$
$$7x^2 - 12x + 5 = 0$$
$$(7x - 5)(x - 1) = 0$$
$$7x - 5 = 0 \quad \text{or} \quad x - 1 = 0$$
$$x = \tfrac{5}{7} \qquad\qquad x = 1$$

For x = 1:
$$y = 2 - 3x$$
$$y = 2 - 3(1) = -1$$

For x = \tfrac{5}{7}:
$$y = 2 - 3x$$
$$y = 2 - 3(\tfrac{5}{7}) = -\tfrac{1}{7}$$

Thus, $(1, -1)$ and $(\tfrac{5}{7}, -\tfrac{1}{7})$ are the solutions to the system.

12.
$$2x^2 - 3y^2 = 5 \qquad \text{(A)}$$
$$3x^2 + 4y^2 = 16 \qquad \text{(B)}$$
$$8x^2 - 12y^2 = 20 \qquad 4\text{(A)}$$
$$\underline{9x^2 + 12y^2 = 48} \qquad 3\text{(B)}$$
$$17x^2 \qquad\quad = 68$$
$$x^2 = 4$$
$$x = \pm 2$$

For x = 2:
$$2x^2 - 3y^2 = 5$$
$$2(2)^2 - 3y^2 = 5$$
$$-3y^2 = -3$$
$$y^2 = 1$$
$$y = \pm 1$$

For x = -2:
$$2x^2 - 3y^2 = 5$$
$$2(-2)^2 - 3y^2 = 5$$
$$-3y^2 = -3$$
$$y^2 = 1$$
$$y = \pm 1$$

Thus, $(2, 1), (2, -1), (-2, 1),$ and $(-2, -1)$ are the solutions to the system.

13. $x^2 + xy - y^2 = 4$ *For $y = 2x$:*

$\qquad 2x^2 - xy = 0 \qquad\qquad\qquad x^2 + xy - y^2 = 4$

$\qquad\quad x(2x - y) = 0 \qquad\qquad x^2 + x(2x) - (2x)^2 = 4$

$\qquad x = 0 \quad$ or $\quad 2x - y = 0 \qquad\qquad -x^2 = 4$

$\qquad\qquad\qquad\qquad\quad y = 2x \qquad\qquad\qquad x^2 = -4$

$\qquad\qquad\qquad\qquad\qquad\qquad\qquad\qquad x = \pm\sqrt{-4} = \pm 2i$

For $x = 0$:   *For $x = 2i$:*   *For $x = -2i$:*

$x^2 + xy - y^2 = 4 \qquad\qquad\quad y = 2x \qquad\quad y = 2x$

$0^2 + 0y - y^2 = 4 \qquad\qquad\quad y = 2(2i) = 4i \quad y = 2(-2i) = -4i$

$\qquad\qquad y^2 = -4$

$\qquad\qquad\quad y = \pm\sqrt{-4} = \pm 2i$

Thus, $(0, 2i), (0, -2i), (2i, 4i),$ and $(-2i, -4i)$ are the solutions to the system.

PRACTICE EXERCISE 9.4

Work odd-numbered problems first, check answers, and then work even-numbered problems in areas of weakness. Answers to all problems are in the back of the book. Make every effort to work a problem yourself before you look at an answer.

Solve each system.

A **1.** $x^2 + y^2 = 25$
$\qquad\quad y = -4$

2. $x^2 + y^2 = 169$
$\qquad\quad\quad x = -12$

3. $y^2 = 2x$
$\quad\; x = y - \frac{1}{2}$

4. $8x^2 - y^2 = 16$
$\qquad\qquad y = 2x$

5. $x^2 + 4y^2 = 32$
$\quad\; x + 2y = 0$

6. $2x^2 - 3y^2 = 25$
$\qquad x + y = 0$

7. $x^2 = 2y$
$\;\; 3x = y + 5$

8. $\qquad y^2 = -x$
$\quad\; x - 2y = 5$

9. $x^2 - y^2 = 3$
$\quad\; x^2 + y^2 = 5$

10. $2x^2 + y^2 = 24$
$\qquad x^2 - y^2 = -12$

11. $x^2 - 2y^2 = 1$
$\quad\;\; x^2 + 4y^2 = 25$

12. $\qquad x^2 + y^2 = 10$
$\quad\; 16x^2 + y^2 = 25$

B **13.** $xy - 6 = 0$
  $x - y = 4$

14. $xy = -4$
  $y - x = 2$

_____

_____

15. $x^2 + xy - y^2 = -5$
  $y - x = 3$

16. $x^2 - 2xy + y^2 = 1$
  $x - 2y = 2$

_____

_____

17. $2x^2 - 3y^2 = 10$
  $x^2 + 4y^2 = -17$

18. $2x^2 + 3y^2 = -4$
  $4x^2 + 2y^2 = 8$

_____

_____

19. $x^2 + y^2 = 20$
  $x^2 = y$

20. $x^2 - y^2 = 2$
  $y^2 = x$

_____

_____

21. $x^2 + y^2 = 16$
  $y^2 = 4 - x$

22. $x^2 + y^2 = 5$
  $x^2 = 4(2 - y)$

_____

_____

23. Find the dimensions of a rectangle with area 32 ft^2 and perimeter 36 ft.

_____

24. Find two numbers such that their sum is 1 and their product is 1.

_____

C **25.** $2x^2 + y^2 = 18$
  $xy = 4$

26. $x^2 - y^2 = 3$
  $xy = 2$

_____

_____

27. $x^2 + 2xy + y^2 = 36$
  $x^2 - xy = 0$

28. $2x^2 - xy + y^2 = 8$
  $(x - y)(x + y) = 0$

_____

_____

29. $x^2 - 2xy + 2y^2 = 16$
  $x^2 - y^2 = 0$

30. $x^2 + xy - 3y^2 = 3$
  $x^2 + 4xy + 3y^2 = 0$

_____

_____

The Check Exercise for this section is on page 485.

9.5 CHAPTER REVIEW

Important terms and symbols

9.1 SYSTEMS OF LINEAR EQUATIONS IN TWO VARIABLES   solution by graphing, solution by substitution, solution by elimination using addition, equivalent systems, inconsistent systems, dependent systems

9.2 APPLICATIONS

9.3 SYSTEMS OF LINEAR EQUATIONS IN THREE VARIABLES   solution, solution set, equivalent systems

9.4 NONLINEAR SYSTEMS

DIAGNOSTIC (REVIEW) EXERCISE 9.5

Work through all the problems in this chapter review and check answers in the back of the book. (Answers to all problems are there, and following each answer is a number in italics indicating the section in which that type of problem is discussed.) Where weaknesses show up, review appropriate sections in the text. When you are satisfied that you know the material, take the practice test following this review.

A   1.   Solve graphically: $x - y = 5$
$\qquad\qquad\qquad x + y = 7$

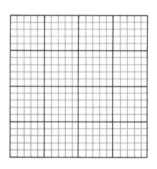

_____

2.   Solve by substitution: $2x + 3y = 7$
$\qquad\qquad\qquad\qquad 3x - \ y = 5$

_____

3.   Solve by elimination using addition: $2x + 3y = 7$
$\qquad\qquad\qquad\qquad\qquad\qquad\qquad 3x - \ y = 5$

_____

Solve each system in Problems 4 to 6.

4. $\quad\ y + 2z = 4$
$\ \ x \qquad\ - z = -2$
$\ \ x + y \qquad = 1$

5. $\quad x^2 + y^2 = 2$
$\qquad 2x - y\ = 3$

_____

_____

6. $x^2 - y^2 = 7$
$\ \ x^2 + y^2 = 25$

_____

B 7. Solve graphically: $3x - 2y = 6$
$x + 4y = 16$

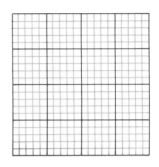

_____

8. Solve by substitution: $3x - 2y = 6$
$x + 4y = 16$

9. Solve: $5m - 3n = 4$
$-2m + 4n = -10$

_____   _____

Solve each system.

10. $3x - 2y = -1$
$-6x + 4y = 3$

11. $3x - 2y - 7z = -6$
$-x + 3y + 2z = -1$
$x + 5y + 3z = 3$

_____   _____

12. $3x^2 - y^2 = -6$
$2x^2 + 3y^2 = 29$

13. $x^2 = y$
$y = 2x - 2$

_____   _____

C 14. $2x - 6y = -3$
$-\frac{2}{3}x + 2y = 1$

15. $x^2 + 2xy - y^2 = -4$
$x^2 - xy = 0$

_____   _____

APPLICATIONS
Solve, using two-equation–two-unknown methods.

16. If you have 30 nickels and dimes in your pocket, worth \$2.30, how many of each do you have?

_____

17. If \$6,000 is to be invested, part at 10 percent and the rest at 6 percent, how much should be invested at each rate so that the total annual return from both investments is \$440?

_____

18. The perimeter of a rectangle is 22 cm. If its area is 30 cm², find the length of each side.

_____

19. A chemist has one 40% and one 70% solution of acid in stock. How much of each should she take to get 100 grams of a 49% solution?

_____

20. A container contains 120 packages. Some of the packages weigh $\frac{1}{2}$ lb each and the rest weigh $\frac{1}{3}$ lb each. If the total contents of the container weigh 48 lb, how many are there of each type of package?

_____

PRACTICE TEST CHAPTER 9 _____

Take this as if it were a graded test. Allow yourself up to 50 minutes. Work the problems without looking back in the chapter. Correct your work, using the answers (keyed to appropriate sections) in the back of the book.

1. Solve graphically: $2x + 3y = 18$
$\qquad\qquad\qquad x - 3y = -9$

_____

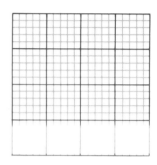

2. Solve by substitution: $2x + 3y = 18$
$\qquad\qquad\qquad\quad x - 3y = -9$

_____

Solve each system.

3. $4x - 3y = -11$
$\quad 3x + 2y = -4$

_____

4. $\quad 4x - 6y = -12$
$\quad -6x + 9y = 9$

_____

5. $\quad x - y + z = 4$
$\quad 2x + y - 3z = -5$
$\quad -x - 3y + 2z = 6$

_____

6. $2x^2 - y^2 = 7$
$\quad 3x^2 + 2y^2 = -14$

_____

7. $2m^2 - n^2 = 1$
$\quad 3m + n = 2$

_____

8. $x^2 + xy - y^2 = -5$
$\quad\quad\; 2xy + y^2 = 0$

_____

Solve Problems 9 and 10, using two-equation–two-unknown methods.

9. If \$8,000 is to be invested, part at 9 percent and the rest at 5 percent, how much should be invested at each rate so that the total annual return from both investments is \$500?

_____

10. A chemist has two concentrations of alcohol in stock; one is a 30% solution and the other is an 80% solution. How much of each should be used to obtain 50 grams of a 60% solution?

_____

CHECK EXERCISE 9.1

Work the following problems without looking at any text examples. Show your work in the space provided. Write the letter that best indicates your answer in the answer column.

1. Solve by graphing:
   $3x + 2y = 13$
   $2x - y = 4$

   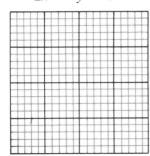

   (A)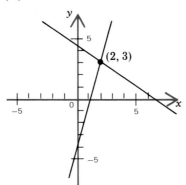

   (B)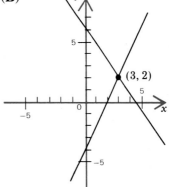

   (C) $(x, 2x - 6)$ for any real number x

   (D) None of these

Solve Problems 2 to 5 by substitution or by elimination using addition.

2. $2x - 4y = 4$
   $x = 10 - 2y$

   (A) $(2, 6)$
   (B) $(10 - 2y, y)$ for any real number y
   (C) $(6, 2)$
   (D) None of these

3. $4x - 6y = -2$
   $-6x + 9y = 6$

   (A) $(1, 1)$
   (B) $(2, 2)$
   (C) No solution
   (D) None of these

479

4. $2x - 3y = 2$
  $3x - 7y = -2$

    (**A**) $(4, 2)$
    (**B**) $(1, 0)$
    (**C**)   No solution
    (**D**)   None of these

5. $8x - 4y = 12$
  $-2x + \ y = -3$

    (**A**) $(x, 2x - 3)$ for any real number x
    (**B**) $(1, -1)$
    (**C**)   No solution
    (**D**)   None of these

CHECK EXERCISE 9.2

NAME _____

CLASS _____

SCORE _____

Work the following problems without looking at any text examples. Show your work in the space provided. Write the letter that best indicates your answer in the answer column.

ANSWER
COLUMN

1. _____

2. _____

3. _____

4. _____

5. _____

Solve, using two-equation–two-unknown methods.

1. If a parking meter contains 56 coins in nickels and dimes, worth $4.05, how many of each type of coin is in the meter?

(A)   40 nickels, 16 dimes
(B)   31 nickels, 25 dimes
(C)   11 nickels, 35 dimes
(D)   None of these

2. Find the dimensions of a rectangle with perimeter 320 meters, if its length is 3 times its width.

(A)   30 by 120 meters
(B)   35 by 105 meters
(C)   80 by 240 meters
(D)   None of these

3. A jeweler has two bars of silver alloy in stock, one 50% silver and the other 100% silver. How much of each should be used to obtain 100 grams of an alloy that is 60% silver?

(A)   80 grams of 50%, 20 grams of 100%
(B)   20 grams of 50%, 80 grams of 100%
(C)   70 grams of 50%, 30 grams of 100%
(D)   None of these

481

4. A shopping carton contains 86 small packages, some weighing $\frac{1}{3}$ lb each and the rest $\frac{3}{4}$ lb each. How many of each type of package are in the carton if the carbon weighs 57 lb total?

    (A) $42\frac{1}{3}$-lb packages, $44\frac{3}{4}$-lb packages
    (B) $45\frac{1}{3}$-lb packages, $41\frac{3}{4}$-lb packages
    (C) $18\frac{1}{3}$-lb packages, $68\frac{3}{4}$-lb packages
    (D) None of these

5. A farmer placed an order with a chemical company for a chemical fertilizer that would contain, among other things, 120 lb of nitrogen and 90 lb of phosphoric acid. The company had two mixtures on hand with the following compositions:

Mix	Nitrogen, %	Phosphoric acid, %
A	20	10
B	6	6

How many pounds of each mixture should the chemist mix to fill the order?

    (A) 400 lb mix A; 667 lb mix B
    (B) 360 lb mix A; 900 lb mix B
    (C) 300 lb mix A; 1,000 lb mix B
    (D) None of these

CHECK EXERCISE 9.3

Work the following problems without looking at any text examples. Show your work in the space provided. Write the letter that best indicates your answer in the answer column.

Solve and check.

1. $3x - y \quad\;\; = 5$
$\quad x + 2y - z = 5$
$\quad 2x - 3y + z = 0$

    (A) $x = 3, y = 4, z = 6$
    (B) $x = 1, y = -2, z = -8$
    (C) $x = 2, y = 1, z = -1$
    (D) None of these

2. $2x + y + \;\; z = 3$
$\quad\;\; x - 2y - \;\; z = 8$
$\quad 3x - \;\; y + 2z = 9$

    (A) $x = -1, y = 2, z = 3$
    (B) $x = 3, y = -2, z = -1$
    (C) No solution
    (D) None of these

3. $\quad 2x - 3y + 4z = 4$
$\quad -4x + 2y - 2z = -6$
$\quad\;\; 3x - 5y + 3z = 10$

    (A) $x = 4, y = 0, z = -1$
    (B) $x = 1, y = -2, z = -1$
    (C) No solution
    (D) None of these

4. $4x + y - 2z = 2$
  $3x - y + z = -2$
  $x + 2y - 3z = 1$

(A) $x = 1, y = 2, z = 2$
(B) $x = -2, y = 0, z = -5$
(C) No solution
(D) None of these

5. The equation of a circle in a rectangular coordinate system can be written in the form $x^2 + y^2 + Dx + Ey + F = 0$. Find $D, E,$ and F so that the circle passes through $(-2, 5)$, $(5, 6)$, and $(7, 2)$.

(A) $D = 4, E = 4, F = 17$
(B) $D = -4, E = -4, F = -17$
(C) No solution
(D) None of these

CHECK EXERCISE 9.4

Work the following problems without looking at any text examples. Show your work in the space provided. Write the letter that best indicates your answer in the answer column.

Solve each system.

1. $2x^2 - y^2 = 9$
  $y - x = 0$

    (A) $(3, 3), (3, -3), (-3, 3), (-3, -3)$
    (B) $(3, 3), (-3, -3)$
    (C) $(3, 3)$
    (D) None of these

2. $x^2 - 3y^2 = -3$
  $x - 2y = 1$

    (A) $(-3, -2)$
    (B) $(3, -2), (-3, -2)$
    (C) $(3, 2), (3, -2), (-3, 2), (-3, -2)$
    (D) None of these

3. $2x^2 - 3y^2 = 15$
  $3x^2 + 4y^2 = 31$

    (A) $(3, 1), (-3, -1)$
    (B) $(3, 1)$
    (C) $(3, 1), (3, -1), (-3, 1), (-3, -1)$
    (D) None of these

4. $x^2 - xy + y^2 = 7$
    $\quad\quad x - y = 3$

    (A) $(2, -1)$
    (B) $(2, 1), (2, -1), (-2, 1), (-2, -1)$
    (C) $(1, -2), (2, -1)$
    (D) None of these

5. $x^2 + xy - y^2 = -9$
    $\quad\quad 2xy - y^2 = 0$

    (A) $(0, 3i), (0, -3i)$
    (B) $(3i, 0), (-3i, 0)$
    (C) $(3i, 0), (-3i, 0), (\sqrt{3}, 2\sqrt{3}), (-\sqrt{3}, -2\sqrt{3})$
    (D) None of these

Chapter Ten

RELATIONS AND FUNCTIONS

CONTENTS

Are you looking down from above or up from below at the "box" in the figure?

INSTRUCTIONS FOR STUDENTS IN A
SELF-PACED CLASS OR LAB

yes — **HAVE YOU HAD INTERMEDIATE ALGEBRA BEFORE THIS COURSE?** — no

1. Work Diagnostic (Review) Exercise 10.6 on page 533. Check answers in back of book; then work through text sections corresponding to problems missed. (Section numbers are in italics following each answer.)
2. When finished with step 1, take Practice Test Chapter 10 on page 537 as a final check of your understanding of the chapter. Check answers in the back of the book; then review sections where weakness still prevails. (Corresponding section numbers are in italics following each answer.)
3. When you think you are ready, ask your instructor for a graded test for Chapter 10.
4. If your instructor approves, after the test is corrected, go to the next chapter.

1. Work through each section in the chapter as follows:
   (a) Read discussion.
   (b) Read each example and work the corresponding matched problem. Check your solution to the matched problem in Solutions to Matched Problems on the indicated page.
   (c) At the end of a section work the odd-numbered problems in the Practice Exercise and check answers; then work even-numbered problems in areas of weakness. (Answers to *all* Practice Exercise sets are in the back of the book.)
   (d) Work Check Exercise as instructed. Tear out and turn in as directed by your instructor. (Answers are not in the text.)
2. Repeat each step in item 1 for each section in the chapter.
3. After the instructional part of the chapter is completed, proceed with steps 1 to 4 in the box above.

10.1 RELATIONS AND FUNCTIONS

•Introduction
••Relations and functions
•••Common ways of specifying relations and functions

•Introduction

Relations and functions are two of the most important concepts in mathematics. Efforts made to understand and use these concepts will be rewarded many times.

You have already encountered relations in everyday life. For example:

To each item on the shelf in a grocery store there corresponds a price

To each name in a telephone book there corresponds one or more telephone numbers

To each square there corresponds an area

To each number there corresponds its cube

To each student there corresponds a grade point average

and so on.

One of the most important aspects of science is establishing relations between various types of phenomena. Once a relation is known, predictions can be made. An engineer can use a formula to predict pressures on a bridge for various wind speeds; an economist would like to predict unemployment rates given various levels of government spending; a chemist can use a gas law to predict the pressure of an enclosed gas given its temperature; and so on.

Establishing and working with relations is so fundamental to both pure and applied science that people have found it desirable to describe them in the precise language of mathematics.

••Relations and functions

What do all of the above examples of relations have in common? Each deals with the matching of elements from a first set, called the **domain** of the relation, with elements in a second set, called the **range** of the relation. Let us consider two of the above examples in more detail. Suppose from a student record office we select five names with their corresponding grade point averages (GPAs) and telephone numbers. The information is summarized in the following two tables:

Relation I

Domain (name)	Range (GPA)
Jones, Robert ⟶	2.4
Jones, Ruth ⟶	2.9
Jones, Sally ⟶	3.8
Jones, Samuel ⟶	3.4
Jones, Sandra ⟋	

Relation II

Domain (name)	Range (telephone number)
Jones, Robert ⟶	841-2315
	841-2403
Jones, Ruth ⟶	838-5106
Jones, Sally ⟋	
Jones, Samuel ⟶	715-0176
Jones, Sandra ⟶	732-1934

The first relation is an example of a *function*. These two very important terms, relation and function, are now defined.

Definition of a Relation and of a Function

A **relation** is a rule (process or method) that produces a correspondence between a first set called the **domain** and a second set called the **range** such that to each element in the domain there corresponds *one or more* elements in the range. A **function** is a relation with the added restriction that to each domain element there corresponds *one and only one* range element. (All functions are relations, but some relations are not functions.)

In the GPA and telephone-number examples above, both are relations. Relation I is also a function since to each domain element (name) there corresponds one and only one range element (GPA). On the other hand, Relation II is not a function since in one case two range elements (telephone numbers) correspond to one domain element (name)—Robert Jones has two telephone numbers.

In most of the material that follows domains and ranges of relations and functions will be sets of numbers. In such cases we will refer to the domain and range elements as *values.*

EXAMPLE 1

Out of the following three relations, two are functions:

(A) Domain   Range   (B) Domain   Range   (C) Domain   Range
        1 ⟶ 5   −2 ⟶ −1   3 ⟶ 1
        2 ⟶ 7   0 ⟶ 0   ⟶ 3
        3 ⟶ 9   2 ⟶ 0   7 ⟶ 8
                                       4 ⟶ 1   9 ⟶ 9

      Function   *Function*   *Not a function*

(Exactly one range   (Exactly one range   (Two range values
value corresponds   value corresponds   correspond to the
to each domain   to each domain   domain value 3.)
value.)   value.)

Work Problem 1 and check solution in Solutions to Matched Problems on page 495.

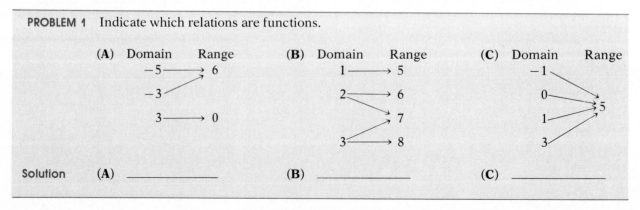

PROBLEM 1   Indicate which relations are functions.

     (A) Domain   Range   (B) Domain   Range   (C) Domain   Range
        −5 ⟶ 6   1 ⟶ 5   −1
        −3   2 ⟶ 6   0 ⟶
        3 ⟶ 0   ⟶ 7   1   5
                        3 ⟶ 8   3

Solution   (A) _____   (B) _____   (C) _____

One of the main objectives of this section is to expose you to the more common ways relations and functions are specified and to provide tests and experience in determining whether a given relation is or is not a function.

•••Common ways of specifying relations and functions

The arrow method of specifying relations and functions illustrated above is convenient for an introduction to the subject—using it we can easily identify which relations are functions. However, in actual practice, relations and functions are more generally specified by a rule (such as an equation), by a table, by a set of ordered pairs of elements,* or by a graph. We will often use equations to specify relations and functions.

Common ways of specifying relations and functions

Method	Illustration	For example,
•Equations	$y = x^2$	$x = 3$ corresponds to $y = 9$
•Tables	$\begin{array}{c\|c} u & v \\ \hline 3 & 2 \\ 1 & 1 \\ 5 & 2 \end{array}$	$u = 3$ corresponds to $v = 2$
•Sets of ordered pairs of elements	$\{(3, 2), (1, 1), (5, 2)\}$ $\{(x, y) \mid y = 2x - 1, x \in R\}$	5 corresponds to 2 $x = 5$ corresponds to $y = 9$
•Graphs		4 corresponds to ± 2

For a given function or relation it is often convenient to be able to shift from one representation to another or to use more than one representation. Consider the following example.

EXAMPLE 2
A relation specified by a table and a graph:

A laboratory experiment

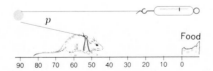

---

*More precisely (but less understandable to many), a **relation** is often defined by mathematicians to be any set of ordered pairs of elements, and a **function** is a relation with the added restriction that no two distinct pairs can have the same first component. The set of first components in a relation (or function) is called the **domain,** and the set of second components is called the **range.** All the methods of specifying relations and functions illustrated yield ordered pairs of elements.

yields a table   or a graph

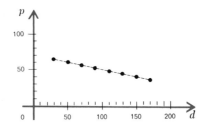

d (distance in centimeters)	p (pull toward food in grams)
30	64
50	60
70	56
90	52
110	48
130	44
150	40
170	36

Both the table and the graph in Example 2 establish the same correspondence between domain values d and range values p; hence both specify the same relation.

> NOTE: It is the usual practice to associate domain values with the horizontal axis and range values with the vertical axis. Thus, the first coordinate (abscissa) of the coordinates of a point on the graph is a domain value and the second coordinate (ordinate) is a range value.

Work Problem 2 and check solution in Solutions to Matched Problems on page 495.

PROBLEM 2   Is the relation in Example 2 a function? Explain.

If a relation is specified by a set of ordered pairs of elements, then the set of first components form the domain and the set of second components form the range.

EXAMPLE 3
A relation specified by a set of ordered pairs:

Given the set

$$F = \{(0, 0), (1, -1), (1, 1), (4, -2), (4, 2)\}$$

(A)   Write this relation, using arrows as in Example 1. Indicate domain and range.
(B)   Graph the set in a rectangular coordinate system.
(C)   Is the relation a function? Explain.

Solution
(A)   Domain   Range   **(B)**

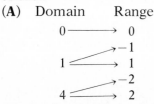

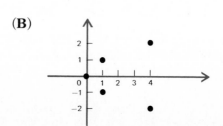

(C) The relation is not a function, since more than one range value corresponds to a given domain value. (For F to be a function no two ordered pairs in F can have the same first coordinate.)

Work Problem 3 and check solution in Solutions to Matched Problems on page 495.

PROBLEM 3 Given the set $F = \{(-2, 4), (-1, 1), (0, 0), (1, 1), (2, 4)\}$

(A) Write the relation, using arrows as in Example 1. Indicate domain and range.

(B) Graph the relation.

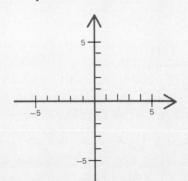

(C) Is the relation a function? Explain.

It is very easy to determine whether a relation is a function if one has its graph:

> *Vertical-Line Test for a Function*
>
> A relation is a function if each vertical line in the coordinate system passes through *at most* one point on the graph of the relation.

If a vertical line passes through more than one point on the graph of a relation, then all points of intersection will have the same first coordinate (same domain value) and different second coordinates (different range values). Thus, since more than one range value is associated with a given domain value, the relation is not a function.

The graph in Example 2 passes the vertical-line test, but the graph in Example 3 does not:

Graph from Example 2

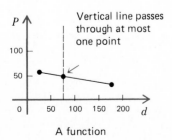

A function

Graph from Example 3

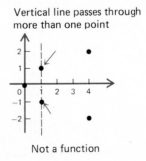

Not a function

We now turn to **relations specified by equations in two variables.** Consider the equation

$$y = 2x - 3$$

For each *input x* we obtain one *output y*. For example,

If $x = 2$, then $y = 2(2) - 3 = 1$

If $x = -1$, then $y = 2(-1) - 3 = -5$

The **input** values are **domain** values and the output values are range values. The equation (a rule) assigns each domain value x a range value y. The variable x is called an *independent variable* (since values are independently assigned to x from the domain) and y is called a *dependent variable* (since y's value depends on the value assigned to x). In general,

---

Definition of Independent and Dependent Variables
_____

Any variable used as a placeholder for domain values is called an **independent variable;** any variable that is used as a placeholder for range values is called a **dependent variable.**

Unless stated to the contrary, we shall adhere to the following convention regarding domains and ranges for relations and functions specified by equations.

Agreement on Domains and Ranges
_____

If a relation or function is specified by an equation and the domain is not indicated, then we shall assume that the domain is the set of all real number replacements of the independent variable (inputs) that produce real values for the dependent variable (outputs). The range is the set of all outputs corresponding to input values.

Most equations in two variables specify relations, but when does an equation specify a function?

Equations and Functions
_____

If in an equation in two variables, there corresponds exactly one value of the dependent variable (output) for each value of the independent variable (input), then the equation specifies a function. If there is more than one output for at least one input, then the equation does not specify a function.

EXAMPLE 4

Relations specified by equations. Given the relations with independent variables x and dependent variables y:

$R: x^2 + y^2 = 4$

$G: x^2 + y = 4$

Which is a function? Explain.

Solution

Test relation R. Is there a value of x (input) that will produce more than one value of y (output)? Yes. For example, if $x = 0$, then

$$0^2 + y^2 = 4$$
$$y^2 = 4$$
$$y = \pm 2$$

Thus, two outputs result from one input [both $(0, -2)$ and $(0, 2)$ belong to the relation R], therefore, the relation R is not a function.

Test relation G. Is there a value of x (input) that will produce more than one value of y (output)? Write G in the form

$$y = 4 - x^2$$

We see that for each value of x (input), we square it and subtract the result from 4 to obtain a single y (output). For example, if $x = 3$, then

$$y = 4 - 3^2$$
$$= 4 - 9 = -5 \qquad \text{and no other number}$$

Therefore, the relation G is a function.

Work Problem 4 and check solution in Solutions to Matched Problems on page 495.

PROBLEM 4   Repeat Example 4 for

$$R: y = x^2 - 3$$

$$G: y^2 = x - 3$$

SOLUTIONS TO MATCHED PROBLEMS

1. **(A)** Function   **(B)** Not a function   **(C)** Function
2. Yes, each domain value corresponds to exactly one range value.
3. **(A)** Domain   Range   **(B)**

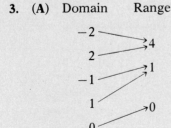

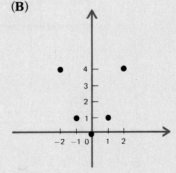

(C) The relation is a function, since each domain value corresponds to exactly one range value.

4. R is a function: each replacement of x (input) produces exactly one y (output). G is not a function: for example, if $x = 4$, then $y = \pm 1$ (two outputs for one input).

PRACTICE EXERCISE 10.1

Work odd-numbered problems first, check answers, and then work even-numbered problems in areas of weakness. Answers to all problems are in the back of the book. Make every effort to work a problem yourself before you look at an answer.

A *Indicate whether each relation in Problems 1 to 6 is or is not a function.*

1. Domain   Range

    3 ⟶ 0
    5 ⟶ 1
    7 ⟶ 2

    _____

2. Domain   Range

    −1 ⟶ 5
    −2 ⟶ 7
    −3 ⟶ 9

    _____

3. Domain   Range

    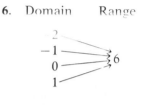

    _____

4. Domain   Range

    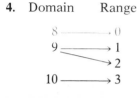

    _____

5. Domain   Range

    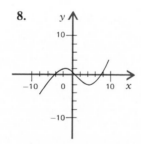

    _____

6. Domain   Range

    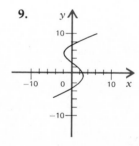

    _____

Each relation in Problems 7 to 12 is specified by a graph. Indicate whether the relation is a function.

7.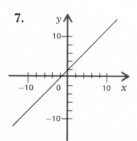

    _____

8.

    _____

9.

    _____

10.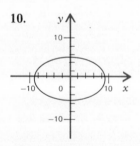

    _____

11.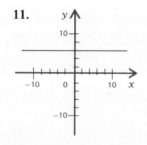

    _____

12.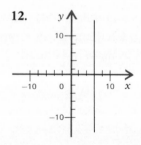

    _____

B *Each equation specifies a relation. Which specify functions given that x is an independent variable?*

13. $y = 3x - 1$ _____

14. $y = \dfrac{x}{2} - 1$ _____

15. $y = x^2 - 3x + 1$ _____

16. $y = x^3$ _____

17. $y^2 = x$ _____

18. $x^2 + y^2 = 25$ _____

19. $x = y^2 - y$ _____

20. $x = (y - 1)(y + 2)$ _____

21. $y = x^4 - 3x^2$ _____

22. $2x - 3y = 5$ _____

23. $y = \dfrac{x + 1}{x - 1}$ _____

24. $y = \dfrac{x^2}{1 - x}$ _____

Graph each relation in Problems 25 to 34. State its domain and range and indicate which are functions. The variable x is independent.

25. $F = \{(1, 1), (2, 1), (3, 2), (3, 3)\}$

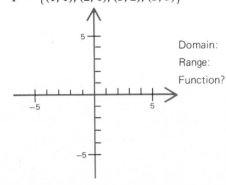

Domain:

Range:

Function?

26. $f = \{(2, 4), (4, 2), (2, 0), (4, -2)\}$

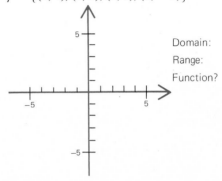

Domain:

Range:

Function?

27. $G = \{(-1, -2), (0, -1), (1, 0), (2, 1), (3, 2), (4, 1)\}$

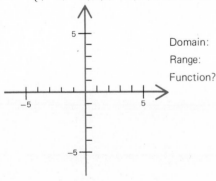

Domain:

Range:

Function?

28. $g = \{(-2, 0), (0, 2), (2, 0)\}$

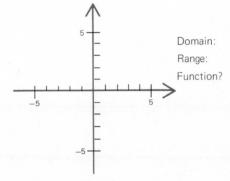

Domain:

Range:

Function?

C 29. $y = 5 - 2x, x \in \{0, 1, 2, 3, 4\}$

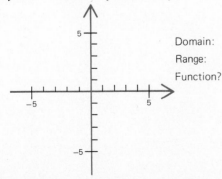

Domain:

Range:

Function?

30. $y = \dfrac{x}{2} - 4, x \in \{0, 2, 4\}$

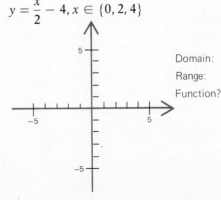

Domain:

Range:

Function?

31. $y^2 = x, x \in \{0, 1, 4\}$

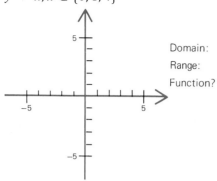

Domain:

Range:

Function?

32. $y = x^2, x \in \{-2, 0, 2\}$

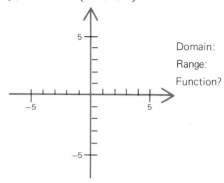

Domain:

Range:

Function?

33. $x^2 + y^2 = 4, x \in \{-2, 0, 2\}$

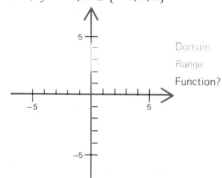

Domain:

Range:

Function?

34. $x^2 + y^2 = 9, x \in \{-3, 0, 3\}$

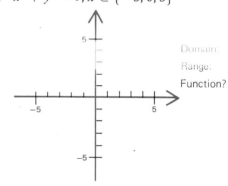

Domain:

Range:

Function?

The Check Exercise for this section is on page 539.

10.2 FUNCTION NOTATION

•The function symbol f(x)
••Use of the function symbol f(x)

•The function symbol f(x)

We have just seen that a function involves two sets of elements, a domain and a range, and a rule of correspondence that enables one to assign each element in the domain to exactly one element in the range. We use different letters to denote names for numbers; in essentially the same way, we will now use different letters to denote names for functions. For example, f and g may be used to name the two functions

$$f: \quad y = 2x + 1$$

$$g: \quad y = x^2 + 2x - 3$$

If x represents an element in the domain of a function f, then we will often use the symbol

$$f(x)$$

in place of y to designate the number in the range of the function f to which x is paired

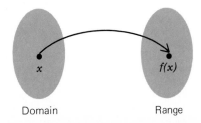

Domain   Range

It is important not to confuse this new function symbol and think of it as the product of f and x. The symbol $f(x)$ is read "f of x" or "the value of f at x." The variable x is an independent variable; both y and $f(x)$ are dependent variables.

••Use of the function symbol f(x)

This new function notation is extremely important, and its correct use should be mastered as early as possible. For example, in place of the more formal representation of the functions f and g above, we can now write

$$f(x) = 2x + 1 \quad \text{and} \quad g(x) = x^2 + 2x - 3$$

The function symbols $f(x)$ and $g(x)$ have certain advantages over the variable y in certain situations. For example, if we write $f(3)$ and $g(5)$, then each symbol indicates in a concise way that these are range values of particular functions associated with particular domain values. Let us find $f(3)$ and $g(5)$.

To find $f(3)$, we replace x by 3 wherever x occurs in

$$f(x) = 2x + 1$$

and evaluate the right side:

$$f(3) = 2 \cdot 3 + 1$$
$$= 6 + 1$$
$$= 7$$

Thus,

$$f(3) = 7$$ The function f assigns the range value 7 to the domain value 3; the ordered pair (3, 7) belongs to f.

To find $g(5)$, we replace x by 5 wherever x occurs in

$$g(x) = x^2 + 2x - 3$$

and evaluate the right side:

$$g(5) = 5^2 + 2 \cdot 5 - 3$$
$$= 25 + 10 - 3$$
$$= 32$$

Thus,

$$g(5) = 32$$ The function g assigns the range value 32 to the domain value 5; the ordered pair (5, 32) belongs to g.

It is very important to understand and remember the definition of $f(x)$:

The Function Symbol $f(x)$

For any element x in the domain of the function f, the function symbol

$$f(x)$$

represents the element in the range of f corresponding to x in the domain of f. [If x is an input value, then $f(x)$ is an output value; or, symbolically, $f: \quad x \rightarrow f(x)$.] The ordered pair $(x, f(x))$ belongs to the function f.

Figure 1, illustrating a "function machine," may give you additional insight into the nature of function and the new function symbol $f(x)$. We can think of a "function machine" as a device that produces exactly one output (range) value for each input (domain) value. (If more than one output value is produced for an input value, then the machine would be a "relation machine" and not a "function machine.")

For the function $f(x) = 2x + 1$, the machine takes each domain value (input), multiplies it by 2, then adds 1 to the result to produce the range value (output). Different rules inside the machine result in different functions.

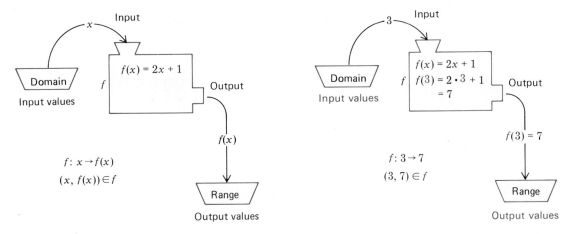

Figure 1
"Function Machine"—Exactly One Output for Each Input

EXAMPLE 5

Let $f(x) = \dfrac{x}{2} + 1$ and $g(x) = 1 - x^2$. Then,

(A) $f(6) = \frac{6}{2} + 1 = 3 + 1 = 4$

(B) $g(-2) = 1 - (-2)^2 = 1 - 4 = -3$

(C) $f(4) + g(0) = \overset{f(4)}{(\frac{4}{2} + 1)} + \overset{g(0)}{(1 - 0^2)} = 3 + 1 = 4$

(D) $\dfrac{2g(-3) + 6}{f(8)} = \dfrac{2[1 - \overset{g(-3)}{(-3)^2}] + 6}{\underset{f(8)}{\frac{8}{2} + 1}} = \dfrac{2(-8) + 6}{5} = \dfrac{-10}{5} = -2$

(E) $\dfrac{f(2)}{g(-1)} = \dfrac{\frac{2}{2} + 1}{1 - (-1)^2} = \dfrac{2}{0}$ Not defined

Work Problem 5 and check solution in Solutions to Matched Problems on page 503.

PROBLEM 5 If $f(x) = \dfrac{x}{3} - 2$ and $g(x) = 4 - x^2$, find:

(A) $f(9) =$

(B) $g(-2) =$

(C) $f(0) + g(2) =$

(D) $\dfrac{4g(-1) - 4}{f(12)} =$

(E) $\dfrac{g(2)}{f(6)} =$

EXAMPLE 6

For $g(x) = 1 - x^2$:

(A) $g(2 + h) = 1 - (2 + h)^2 = 1 - (4 + 4h + h^2)$

$= -3 - 4h - h^2$

Replace x in $g(x) = 1 - x^2$ with $(2 + h)$.

(B) $\dfrac{g(2 + h) - g(2)}{h} = \dfrac{[1 - (2 + h)^2] - (1 - 2^2)}{h}$

Be careful here! The brackets are important.

$= \dfrac{-3 - 4h - h^2 + 3}{h}$

$= \dfrac{-4h - h^2}{h} = -4 - h$

(C) $\dfrac{g(x + h) - g(x)}{h} = \dfrac{\overset{g(x+h)}{\overbrace{[1 - (x + h)^2]}} - \overset{g(x)}{\overbrace{(1 - x^2)}}}{h}$

Replace x in $g(x) = 1 - x^2$ with $(x + h)$ to find $g(x + h)$.

$= \dfrac{1 - x^2 - 2hx - h^2 - 1 + x^2}{h}$

$= \dfrac{-2hx - h^2}{h} = \dfrac{\overset{1}{\cancel{h}}(-2x - h)}{\underset{1}{\cancel{h}}} = -2x - h$

Work Problem 6 and check solution in Solutions to Matched Problems on page 503.

PROBLEM 6   For $g(x) = 4 - x^2$, find:

(A) $g(3 + h) =$

(B) $\dfrac{g(3 + h) - g(3)}{h} =$

(C) $\dfrac{g(x + h) - g(x)}{h} =$

NOTE: Expressions of the form $\dfrac{f(x + h) - f(x)}{h}$ are fundamental to certain concepts in calculus—a course some of you will be taking before too long. Problems 47 and 48 in Practice Exercise 10.2 suggest a physical interpretation.

EXAMPLE 7

For $f(x) = \dfrac{x}{2} + 1$ and $g(x) = 1 - x^2$:

(A) $f[g(3)] = f(1 - 3^2)$ Evaluate $g(3)$ first; then evaluate f for this value.

$$= f(-8) = \dfrac{-8}{2} + 1 = -3$$

(B) $g[f(-4)] = g\left(\dfrac{-4}{2} + 1\right)$

$$= g(-1) = 1 - (-1)^2 = 0$$

Work Problem 7 and check solution in Solutions to Matched Problems on page 503.

PROBLEM 7   For $f(x) = \dfrac{x}{3} - 2$ and $g(x) = 3 - x^2$, find:

(A) $f[g(3)] =$

(B) $g[f(-3)] =$

SOLUTIONS TO MATCHED PROBLEMS

5. **(A)** $f(9) = \frac{9}{3} - 2 = 3 - 2 = 1$ **(B)** $g(-2) = 4 - (-2)^2 = 4 - 4 = 0$

    (C) $f(0) + g(2) = (\frac{0}{3} - 2) + (4 - 2^2) = (-2) + (0) = -2$

    (D) $\dfrac{4g(-1) - 4}{f(12)} = \dfrac{4[4 - (-1)^2] - 4}{\frac{12}{3} - 2} = \dfrac{4(3) - 4}{2} = \dfrac{8}{2} = 4$

    (E) $\dfrac{g(2)}{f(6)} = \dfrac{4 - 2^2}{\frac{6}{3} - 2} = \dfrac{0}{0}$ Not defined

6. **(A)** $g(3 + h) = 4 - (3 + h)^2 = 4 - (9 + 6h + h^2) = 4 - 9 - 6h - h^2 = -5 - 6h - h^2$

    (B) $\dfrac{g(3 + h) - g(3)}{h} = \dfrac{[4 - (3 + h)^2] - (4 - 3^2)}{h} = \dfrac{-5 - 6h - h^2 + 5}{h} = \dfrac{-6h - h^2}{h} = -6 - h$

    (C) $\dfrac{g(x + h) - g(x)}{h} = \dfrac{[4 - (x + h)^2] - (4 - x^2)}{h} = \dfrac{4 - x^2 - 2hx - h^2 - 4 + x^2}{h}$

$$= \dfrac{-2hx - h^2}{h} = -2x - h$$

7. **(A)** $f[g(3)] = f(3 - 3^2) = f(-6) = \dfrac{-6}{3} - 2 = -2 - 2 = -4$

    (B) $g[f(-3)] = g\left(\dfrac{-3}{3} - 2\right) = g(-3) = 3 - (-3)^2 = -6$

PRACTICE EXERCISE 10.2 ────────────────────────────

Work odd-numbered problems first, check answers, and then work even-numbered problems in areas of weakness. Answers to all problems are in the back of the book. Make every effort to work a problem yourself before you look at an answer.

A *If $f(x) = 3x - 2$, find:*

1. $f(2)$ ─────────────   2. $f(1)$ ─────────────   3. $f(-2)$ ─────────────

4. $f(-1)$ ─────────────   5. $f(0)$ ─────────────   6. $f(4)$ ─────────────

If $g(x) = x - x^2$, find:

7. $g(2)$ ─────────────   8. $g(1)$ ─────────────   9. $g(4)$ ─────────────

10. $g(5)$ ─────────────   11. $g(-2)$ ─────────────   12. $g(-1)$ ─────────────

B *Problems 13 to 36 refer to the functions*

$$f(x) = 10x - 7 \qquad g(t) = 6 - 2t \qquad F(u) = 3u^2 \qquad G(v) = v - v^2$$

Evaluate as indicated:

13. $f(-2)$ ─────────────   14. $F(-1)$ ─────────────   15. $g(2)$ ─────────────

16. $G(-3)$ ─────────────   17. $g(0)$ ─────────────   18. $G(0)$ ─────────────

19. $f(3) + g(2)$ ─────────────   20. $F(2) + G(3)$ ─────────────   21. $2g(-1) - 3G(-1)$ ─────────────

22. $4G(-2) - g(-3)$ ─────────────   23. $\dfrac{f(2) \cdot g(-4)}{G(-1)}$ ─────────────   24. $\dfrac{F(-1) \cdot G(2)}{g(-1)}$ ─────────────

25. $g(2 + h)$ ─────────────   26. $F(2 + h)$ ─────────────   27. $\dfrac{g(2 + h) - g(2)}{h}$ ─────────────

28. $\dfrac{F(2 + h) - F(2)}{h}$ ─────────────   29. $\dfrac{f(3 + h) - f(3)}{h}$ ─────────────   30. $\dfrac{G(2 + h) - G(2)}{h}$ ─────────────

31. $F[g(1)]$ ─────────────   32. $G[F(1)]$ ─────────────   33. $g[f(1)]$ ─────────────

34. $g[G(0)]$ ─────────────   35. $f[G(1)]$ ─────────────   36. $G[g(2)]$ ─────────────

37. If $A(w) = \dfrac{w - 3}{w + 5}$, find $A(5)$, $A(0)$, and $A(-5)$.   38. If $h(s) = \dfrac{s}{s - 2}$, find $h(3)$, $h(0)$, and $h(2)$.

─────────────────────────────   ─────────────────────────────

For $f(x) = 10x - 7$ and $g(t) = 6 - 2t$, find:

39. $\dfrac{f(x + h) - f(x)}{h}$ ─────────────────────────   40. $\dfrac{g(t + h) - g(t)}{h}$ ─────────────────────────

C *For F(u) = u − u² and G(t) = t² − t + 1, find:*

41. $\dfrac{F(u + h) - F(u)}{h}$ _____

42. $\dfrac{G(t + h) - G(t)}{h}$ _____

APPLICATIONS

Each of the statements in Problems 43 to 46 can be described by a function. Write an equation that specifies the function.

43. COST FUNCTION   The cost $C(x)$ of x records at $5 per record. (The cost depends on the number of records purchased.)

_____

44. COST FUNCTION   The total daily cost $C(x)$ of manufacturing x pairs of skis if fixed costs are $800 per day and the variable costs are $60 per pair of skis. (The total daily cost per day depends on the number of skis manufactured per day.)

_____

45. TEMPERATURE CONVERSION   The temperature in Celsius degrees C can be found from the temperature in Fahrenheit degrees F by subtracting 32 from the Fahrenheit temperature and multiplying the difference by $\frac{5}{9}$.

_____

46. EARTH SCIENCE   The pressure $P(d)$ in the ocean in pounds per square inch depends on the depth d. To find the pressure, divide the depth by 33, add 1 to the quotient, and then multiply the result by 15.

_____

47. DISTANCE-RATE-TIME   Let the distance that a car travels at 30 mi/hr in t hr be given by $d(t) = 30t$. Find

(A) $d(1), d(10)$

(B) $\dfrac{d(2 + h) - d(2)}{h}$

_____   _____

48. PHYSICS   The distance in feet that an object falls in t sec in a vacuum is given by $s(t) = 16t^2$. Find

(A) $s(0), s(1), s(2),$ and $s(3)$

(B) $\dfrac{s(2 + h) - s(2)}{h}$

_____   _____

What happens as h tends to 0? Interpret physically.

_____

The Check Exercise for this section is on page 541.

10.3 GRAPHING LINEAR AND QUADRATIC FUNCTIONS

> •Linear functions
> ••Quadratic functions
> •••Graphing by completing the square

In Chapters 7 and 8 we studied linear and quadratic forms in some detail. These forms, with certain restrictions, define linear and quadratic functions.

•Linear functions

Any nonvertical line in a rectangular coordinate system defines a linear function. [A vertical line does not define a function. (Why?)] Thus, any function defined by an equation of the form

$$f(x) = ax + b \qquad \text{Linear function (standard form)}$$

where a and b are constants, is called a **linear function.** We know from Section 8.2 that the graph of this equation is a straight line (nonvertical) with slope a and y intercept b.

EXAMPLE 8

Graph the linear function defined by $f(x) = x/3 + 1$, and indicate its slope and y intercept.

Solution

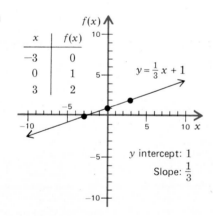

x	$f(x)$
-3	0
0	1
3	2

$y = \frac{1}{3}x + 1$

y intercept: 1

Slope: $\frac{1}{3}$

Work Problem 8 and check solution in Solutions to Matched Problems on page 512.

PROBLEM 8   Graph the linear function defined by $f(x) = -x/2 + 3$, and indicate its slope and y intercept.

Solution

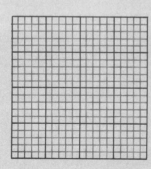

••Quadratic functions

Any function defined by an equation of the form

$$f(x) = ax^2 + bx + c \qquad a \neq 0 \qquad \text{Quadratic function (standard form)}$$

where a, b, and c are constants and x is a variable, is called a **quadratic function.** *

Let us start by graphing two simple quadratic functions:

$$f(x) = x^2 \qquad \text{and} \qquad g(x) = -x^2$$

We evaluate these functions for integer values from their domains, find corresponding range values, and then plot the resulting ordered pairs and join these points with a smooth curve. The work in the dotted boxes in the following calculations is usually done mentally or on scratch paper.

Graphing $f(x) = x^2$:

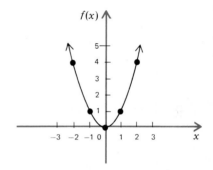

x	y = f(x)	(x, f(x))
-2	$y = f(-2) = (-2)^2 = 4$	$(-2, 4)$
-1	$y = f(-1) = (-1)^2 = 1$	$(-1, 1)$
0	$y = f(0) = 0^2 = 0$	$(0, 0)$
1	$y = f(1) = 1^2 = 1$	$(1, 1)$
2	$y = f(2) = 2^2 = 4$	$(2, 4)$
Domain values	Range values	Elements of f

Graphing $g(x) = -x^2$:

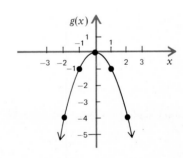

x	y = g(x)	(x, g(x))
-2	$y = g(-2) = -(-2)^2 = -4$	$(-2, -4)$
-1	$y = g(-1) = -(-1)^2 = -1$	$(-1, -1)$
0	$y = g(0) = -0^2 = 0$	$(0, 0)$
1	$y = g(1) = -1^2 = -1$	$(1, -1)$
2	$y = g(2) = -2^2 = -4$	$(2, -4)$
Domain values	Range values	Elements of g

Both these curves are called **parabolas.** A more detailed discussion of parabolas can be found in Section 8.5.

It is shown in a course in analytic geometry that the graph of any quadratic function is also a parabola. In general,

---

*More generally, we have **polynomial functions** that are defined by equations of the form

$$f(x) = a_n x^n + a_{n-1} x^{n-1} + \cdots + a_1 x + a_0$$

where n is a nonnegative integer and each a_i is a constant. For example,

$$f(x) = 2x^4 - 3x^3 + 2x - 5$$

is a polynomial function of degree 4. Linear and quadratic functions are polynomial functions of degrees 1 and 2, respectively.

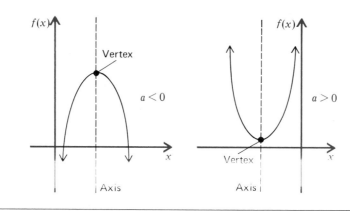

$$Graph\ of\ f(x) = ax^2 + bx + c,\ a \neq 0$$

The graph of a quadratic function f is a parabola that has its **axis** (line of symmetry*) parallel to the vertical axis. It opens upward if $a > 0$ and downward if $a < 0$. The intersection point of the axis and parabola is called the **vertex.**

From the examples above, we observe:

$$f(x) = x^2 \qquad a = 1 > 0$$

Parabola opens upward.

$$g(x) = -x^2 \qquad a = -1 < 0$$

Parabola opens downward.

In addition to the point-by-point method of graphing quadratic functions described above, let us consider another approach that will give added insight into these functions. (A brief review of Section 7.3 on completing the square may prove useful first.)

•••Graphing by completing the square

We illustrate the method through an example and then generalize the results. Our problem is to graph

$$f(x) = 2x^2 - 8x + 5$$

*Intuitively, a line is a **line of symmetry** for a graph if, when the paper on which the graph is drawn is folded along the line, the parts of the graph on either side of the line match exactly.

We start by transforming this equation into the form

$$f(x) = a(x - h)^2 + k \qquad a, h, k \text{ constants}$$

by completing the square.

$f(x) = 2x^2 - 8x + 5$ Factor the coefficient of x^2 out of the first two terms.

$f(x) = 2(x^2 - 4x) + 5$

$\quad = 2(x^2 - 4x + ?) + 5$ Complete the square within the parentheses.

$\quad = 2(x^2 - 4x + 4) + 5 - 8$ We added 4 to complete the square inside the parentheses. But because of the 2 on the outside we have actually added 8, so we must subtract 8.

$\quad = 2(x - 2)^2 - 3$ The transformation is complete.

Thus,

$$f(x) = \underbrace{2(x - 2)^2}_{\substack{\text{Never negative} \\ \text{(why?)}}} - 3$$

When $x = 2$, the first term on the right vanishes, and we add 0 to -3. For *any* other value of x we will add a positive number to -3, thus making $f(x)$ larger. Therefore,

$$f(2) = -3$$

is the minimum value of $f(x)$ for *all* x. A very important result!

    The point $(2, -3)$ is the lowest point on the parabola and is also the vertex. The vertical line $x = 2$ is the axis of the parabola. We plot the vertex and the axis and a couple of points on either side of the axis to complete the graph.

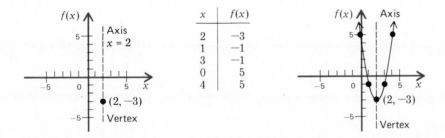

x	$f(x)$
2	-3
1	-1
3	-1
0	5
4	5

    Note the important results we have obtained with this approach. We have found:

1. The axis of the parabola
2. Its vertex
3. The minimum value of $f(x)$
4. The graph of $y = f(x)$

    Proceeding in the same way for the general quadratic function $f(x) = ax^2 + bx + c, a \neq 0$, one can obtain the following general results:

Additional Properties of $f(x) = ax^2 + bx + c, a \neq 0$

---

1. Maximum or minimum value of $f(x)$:

$f\left(-\dfrac{b}{2a}\right)$ Minimum if $a > 0$
                Maximum if $a < 0$

2. Axis (of symmetry) of the parabola:

$x = -\dfrac{b}{2a}$

3. Vertex of the parabola:

$\left(-\dfrac{b}{2a}, f\left(-\dfrac{b}{2a}\right)\right)$

To graph a quadratic function using the method of completing the square, we can either actually complete the square as in the earlier example or use the derived properties in the box—some people can more readily remember a formula, others a process. We will use the derived properties in the next example.

EXAMPLE 9

Graph, finding axis of symmetry, maximum or minimum of $f(x)$, and vertex:

$f(x) = 12x - 2x^2$

Solution

$f(x) = -2x^2 + 12x$ Write $f(x) = 12x - 2x^2$ in standard form $f(x) = ax^2 + bx + c$ and note that $a = -2$ and $b = 12$.

Axis of symmetry:

$x = -\dfrac{b}{2a} = -\dfrac{12}{2(-2)}$

$x = 3$ Axis of symmetry.

Maximum or minimum value of f(x):

Since $a = -2 < 0$, $f(-b/2a)$ is a maximum.

Maximum $f(x) = f\left(-\dfrac{b}{2a}\right)$

$\qquad = f(3) = -2(3)^2 + 12(3) = 18$

Vertex:

$\left(-\dfrac{b}{2a}, f\left(-\dfrac{b}{2a}\right)\right) = (3, 18)$

Graph:

To graph the function f, locate the axis of symmetry and the vertex; then find a few points on either side of $x = 3$.

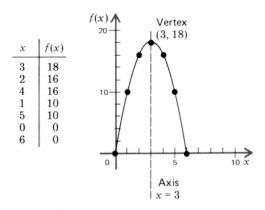

x	$f(x)$
3	18
2	16
4	16
1	10
5	10
0	0
6	0

Work Problem 9 and check solution in Solutions to Matched Problems on page 512.

PROBLEM 9   Graph as in Example 9:

$$f(x) = x^2 - 2x - 3$$

Solution

EXAMPLE 10

Without graphing, find the maximum or minimum value for $f(x) = 3x^2 - 12x + 1$.

Solution

Since $a = 3 > 0$, $f(-b/2a)$ is a minimum.

$$\text{Minimum } f(x) = f\left(-\frac{b}{2a}\right) = f\left(-\frac{-12}{2(3)}\right)$$

$$= f(2) = 3(2)^2 - 12(2) + 1 = -11$$

Work Problem 10 and check solution in Solutions to Matched Problems on page 512.

PROBLEM 10   Without graphing, find the maximum or minimum value of $f(x) = 7 - 16x - 2x^2$.

Solution

SOLUTIONS TO MATCHED PROBLEMS _____

8.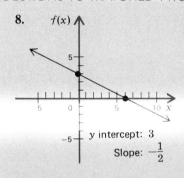

y intercept: 3

Slope: $-\dfrac{1}{2}$

9.   *Axis of symmetry:*

$$x = -\frac{b}{2a} = -\frac{-2}{2(1)} = 1$$

Minimum (since a = 1 > 0):
$f(1) = 1^2 - 2(1) - 3 = -4$
Vertex: $(1, -4)$

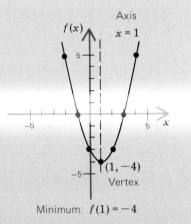

10. $f(x) = -2x^2 - 16x + 7$ (in standard form)
Since $a = -2 < 0$, $f(-b/2a)$ is a maximum.

$$\text{Maximum } f(x) = \left(-\frac{b}{2a}\right) = f\left(-\frac{-16}{2(-2)}\right)$$
$$= f(-4) = -2(-4)^2 - 16(-4) + 7 = 39$$

PRACTICE EXERCISE 10.3 _____

Work odd-numbered problems first, check answers, and then work even-numbered problems in areas of weakness. Answers to all problems are in the back of the book. Make every effort to work a problem yourself before you look at an answer.

A *Graph the following linear functions. Indicate the slope and y intercepts for each.*

1. $f(x) = 2x - 4$ **2.** $g(x) = \dfrac{x}{2}$ **3.** $h(x) = 4 - 2x$

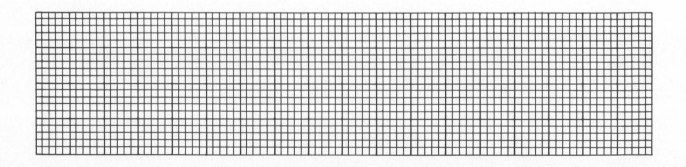

4. $f(x) = -\dfrac{x}{2} + 3$ **5.** $g(x) = -\dfrac{2}{3}x + 4$ **6.** $f(x) = 3$

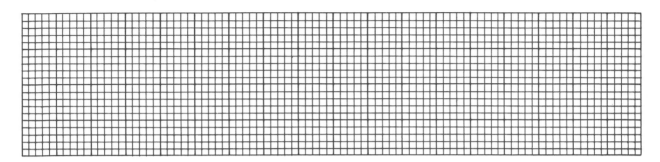

B *Graph each of the following quadratic functions. Include axis of symmetry, vertex, and maximum or minimum value for each.*

7. $f(x) = x^2 + 8x + 16$ **8.** $h(x) = x^2 - 2x - 3$ **9.** $f(u) = u^2 - 2u + 4$

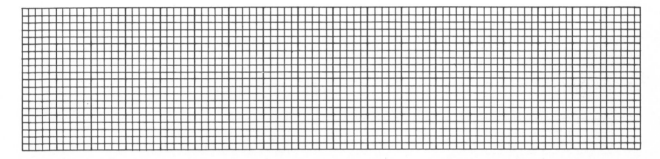

10. $f(x) = x^2 - 10x + 25$ **11.** $h(x) = 2 + 4x - x^2$ **12.** $g(x) = -x^2 - 6x - 4$

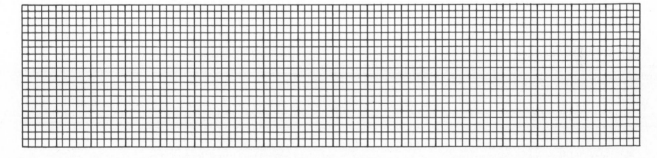

13. $f(x) = 6x - x^2$ **14.** $G(x) = 16x - 2x^2$ **15.** $F(s) = s^2 - 4$

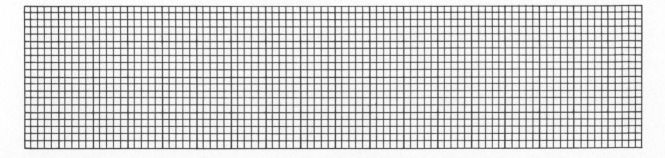

16. $g(t) = t^2 + 4$ **17.** $F(x) = 4 - x^2$ **18.** $G(x) = 9 - x^2$

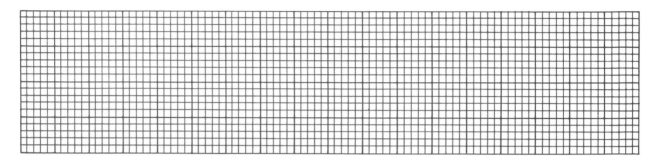

C *Find the maximum or minimum value of each function. Do not graph.*

19. $f(x) = 4x^2 - 16x + 9$ **20.** $h(x) = 3 + 18x - 3x^2$

21. $f(t) = 2t(t - 24)$ **22.** $g(x) = 3x(x + 12)$

23. $A(x) = x(50 - x)$ **24.** $A(x) = x(100 - 2x)$

Graph each of the following quadratic functions. Include axis of symmetry, vertex, and maximum or minimum value for each.

25. $f(x) = x^2 - 7x + 10$ **26.** $g(t) = t^2 - 5t + 2$

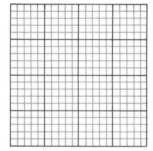   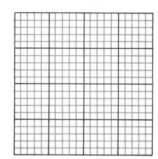

27. $g(t) = 4 + 3t - t^2$ **28.** $h(x) = 2 - 5x - x^2$

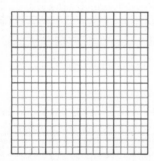   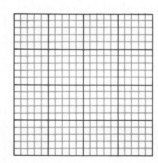

APPLICATIONS

29. COST EQUATION   The cost equation for a particular company to produce stereos is found to be

$$C = g(n) = 96{,}000 + 80n$$

where \$96,000 is fixed costs (tooling and overhead) and \$80 is the variable cost per unit (material, labor, etc.). Graph this function for $0 \le n \le 1{,}000$.

_____

30. DEMAND EQUATION   After extensive surveys the research department in a stereo company produced the demand equation

$$n = f(p) = 8{,}000 - 40p \qquad 100 \le p \le 200$$

where n is the number of units that retailers are likely to purchase per week at a price of p dollars per unit. Graph the function for the indicated domain.

_____

The Check Exercise for this section is on page 543.

10.4 INVERSE RELATIONS AND FUNCTIONS

•**Inverses**
••**One-to-one correspondence and inverses**
•••**Geometric interpretation**

In this section we are going to discuss an important method for obtaining new relations and functions from old relations and functions. We will use this method in Chapter 11 to obtain the logarithmic functions from the exponential functions.

•Inverses

Given a relation R, if we interchange the order of the components in each ordered pair belonging to R, we obtain a new relation R^{-1}, called the **inverse of R.** (NOTE: R^{-1} is a relation-function symbol; it does not mean $1/R$.) For example, if

$$R = \{(3, 5), (5, -1), (7, 0)\}$$

then

$$R^{-1} = \{(5, 3), (-1, 5), (0, 7)\} \qquad \text{R^{-1} is the inverse of R.}$$

It follows from the definition (and is evident from the example) that the domain of R is the **range of R^{-1}** and the range of R is the **domain of R^{-1}**.

Symbolically,

***

Inverse of R

***

If R is a relation, the inverse of R, denoted by R^{-1}, is given by

$$R^{-1} = \{(b, a) \,|\, (a, b) \in R\}$$

***

If a relation R is specified by an equation, say

$$R: y = 2x - 1 \tag{1}$$

then how do we find R^{-1}? The answer is easy: We interchange the variables in (1). Thus,

$$R^{-1}: x = 2y - 1 \tag{2}$$

or, solving for y,

$$R^{-1}: y = \frac{x + 1}{2} \tag{3}$$

Any ordered pair of numbers that satisfies equation (1), when reversed in order, will satisfy equations (2) and (3). For example, (3, 5) satisfies equation (1) and (5, 3) satisfies (2) and (3), as can easily be checked. The graphs of R and R^{-1} are given in Figure 2.

It is useful to sketch in the line $y = x$ and observe that if we fold the paper along this line, then R and R^{-1} will match. Actually, we can graph R^{-1} by drawing R with wet ink and folding the paper along $y = x$ before the ink dries; R will then print R^{-1}. [To prove this, one has to show that the line $y = x$ is the perpendicular bisector of the line segment joining (a, b) to (b, a).] Knowing that the graph of R and R^{-1} are symmetric relative to the line $y = x$ makes it easy to graph R^{-1} if R is known and vice versa.

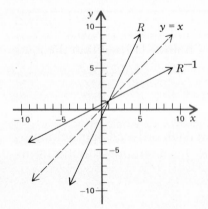

Figure 2

In Figure 2 observe that R and R^{-1} are both functions. This is not always the case, however. *Inverses of some functions may not be functions and inverses of some relations that are not functions may be functions.* We must check each case.

EXAMPLE 11

If R is given by $y = x^2$,

(A) Find R^{-1}.

(B) Graph R, R^{-1}, and $y = x$.

(C) Indicate which (R or R^{-1}) is a function.

Solution

(A) $R: y = x^2$

$R^{-1}: x = y^2$ or $y = \pm \sqrt{x}$

(B)

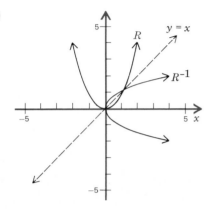

(C) R is a function; R^{-1} is not a function.

Work Problem 11 and check solution in Solutions to Matched Problems on page 521.

PROBLEM 11   Given the relation $R: y = x^2 + 1$,

(A)   Find R^{-1}.

(B)   Graph R, R^{-1}, and $y = x$ all in the same coordinate system.

(C)   Indicate which (R or R^{-1}) is a function.

EXAMPLE 12

Given $f(x) = 2x - 1$, find

(A) $f^{-1}(x)$

Solution
First write f in the form

$f: y = 2x - 1$ Replace $f(x)$ with y.

Interchange variables to obtain f^{-1}:

$f^{-1}: x = 2y - 1$

Solve for y to obtain $f^{-1}(x)$:

$y = \dfrac{x + 1}{2}$ or $f^{-1}(x) = \dfrac{x + 1}{2}$ Replace y with $f^{-1}(x)$.

(B) $f^{-1}(2)$

Solution

$f^{-1}(2) = \dfrac{2 + 1}{2} = \dfrac{3}{2}$

(C) $f[f^{-1}(3)]$

Solution

$f[f^{-1}(3)] = f\left(\dfrac{3 + 1}{2}\right) = f(2) = 2 \cdot 2 - 1 = 3$

Work Problem 12 and check solution in Solutions to Matched Problems on page 521.

PROBLEM 12   Given $g(x) = \dfrac{x}{2} + 1$, find:

   (A) $g^{-1}(x)$

   (B) $g^{-1}(3)$

   (C) $g^{-1}[g(4)]$

••One-to-one correspondence and inverses

Consider the following functions:

R			H		
Domain	Range		Domain	Range	

R

Domain   Range

1 ⟶ 5
2 ⟶ 7
3 ⟶ 9

H

Domain   Range

1 ⟶ 3
4 ⟶
6 ⟶ 7

Function R is an example of a one-to-one correspondence; function H is not a one-to-one correspondence. In general, a **one-to-one correspondence** exists between two sets if each element in the first set corresponds to exactly one element in the second set, and each element in the second set corresponds to exactly one element in the first set. For the function H, the range value 3 corresponds to two domain values; hence, the function is not a one-to-one correspondence.

Now notice what happens when we form the inverses of each function:

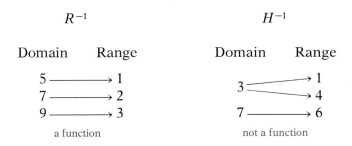

Only the function that is a one-to-one correspondence has an inverse that is a function. In addition, note that

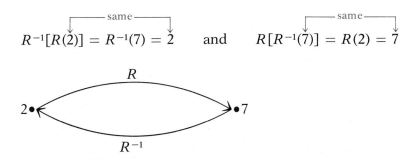

$$R^{-1}[R(2)] = R^{-1}(7) = 2 \quad \text{and} \quad R[R^{-1}(7)] = R(2) = 7$$

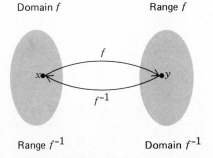

R "sends" 2 into 7 and R^{-1} "sends" 7 right back into 2. Similarly, R^{-1} "sends" 7 into 2 and R "sends" 2 right back into 7. The inverse function R^{-1} reverses what the function R does and vice versa. The same result does not hold for the function H, since it is not a one-to-one correspondence. (Try it to see why.) The observations for the function R are generalized completely in the following important theorem.

THEOREM 1

A function f has an inverse that is a function if and only if there exists a one-to-one correspondence between domain and range values. In this case,

$$f^{-1}[f(x)] = x \quad \text{and} \quad f[f^{-1}(x)] = x$$

Theorem 1 is interpreted schematically as follows:

Domain f Range f

x f y

            f^{-1}

Range f^{-1} Domain f^{-1}

EXAMPLE 13
From Example 12,

$$f(x) = 2x - 1 \quad \text{and} \quad f^{-1}(x) = \frac{x+1}{2}$$

Compute $f^{-1}[f(2)]$.

Solution

$$f^{-1}[f(2)] = f^{-1}(2 \cdot 2 - 1)$$
$$= f^{-1}(3)$$
$$= \frac{3+1}{2} = 2$$

Note that f "sends"
2 into 3 and f^{-1}
"sends" 3 right back to 2.

$$\begin{array}{cc} f & f^{-1} \\ 2 \to 3 & 3 \to 2 \end{array}$$

Thus,

$$f^{-1}[f(2)] = 2 \qquad \text{Verifying Theorem 1.}$$

   Work Problem 13 and check solution in Solutions to Matched Problems on page 521.

PROBLEM 13   For f and f^{-1} in Example 13, find $f[f^{-1}(5)]$.

Solution

•••Geometric interpretation

Functions that are one-to-one* (each has an inverse that is a function):

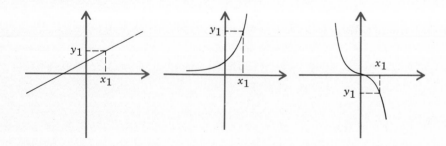

*When we refer to a function as being one-to-one, we mean that there exists a one-to-one correspondence between its domain and range values.

Functions that are not one-to-one (neither has an inverse that is a function):

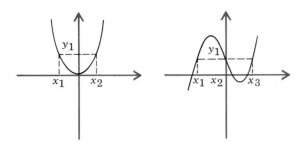

SOLUTIONS TO MATCHED PROBLEMS

11. (A) $R: y = x^2 + 1$
$R^{-1}: x = y^2 + 1$
    or $y = \pm\sqrt{x - 1}$

(C) R is a function.
R^{-1} is not a function.

(B)

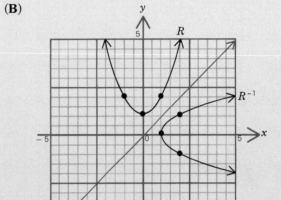

12. (A) $g: y = \dfrac{x}{2} + 1$

$g^{-1}: x = \dfrac{y}{2} + 1$

$\dfrac{y}{2} = x - 1$

$y = 2x - 2$
or $g^{-1}(x) = 2x - 2$

(B) $g^{-1}(x) = 2x - 2$
$g^{-1}(3) = 2(3) - 2 = 4$

(C) $g^{-1}[g(4)] = g^{-1}(\frac{4}{2} + 1)$
$= g^{-1}(3) = 2(3) - 2 = 4$

13. $f[f^{-1}(5)] = f\left(\dfrac{5 + 1}{2}\right) = f(3) = 2(3) - 1 = 5$ As it should, since f is one-to-one.

PRACTICE EXERCISE 10.4

Work odd-numbered problems first, check answers, and then work even-numbered problems in areas of weakness. Answers to all problems are in the back of the book. Make every effort to work a problem yourself before you look at an answer.

A *Find the inverse for each of the following relations.*

1. $R = \{(-2, 1), (0, 3), (2, 2)\}$ _____

2. $F = \{(-3, -1), (0, 1), (3, 2)\}$ _____

3. $G = \{(-2, 4), (-1, 1), (0, 0), (1, 1), (2, 4)\}$ _____

4. $H = \{(-5, 0), (-2, 1), (0, 0), (2, 1), (5, 0)\}$ _____

Graph on the same coordinate system along with $y = x$.

5. R and R^{-1} in Problem 1   6. F and F^{-1} in Problem 2

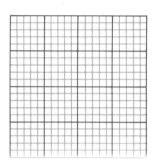 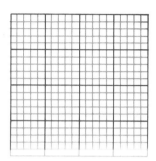

7. G and G^{-1} in Problem 3   8. H and H^{-1} in Problem 4

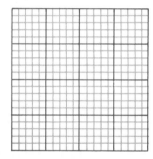 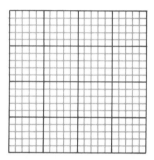

Indicate which are functions.

9. R or R^{-1} in Problem 1 _____   10. F or F^{-1} in Problem 2 _____

11. G or G^{-1} in Problem 3 _____   12. H or H^{-1} in Problem 4 _____

B *For each of the following relations find the inverse in the form of an equation.*

13. $f\colon y = 3x - 2$ _____   14. $g\colon y = 2x + 3$ _____

15. $F\colon y = \dfrac{x}{3} - 2$ _____   16. $G\colon y = \dfrac{x}{2} + 5$ _____

17. $h\colon y = \dfrac{x^2}{2}$ _____   18. $H\colon y = |2x|$ _____

                                                        (For H^{-1} don't solve for y in terms of x.)

In Problems 19 to 22 graph each pair of relations on the same coordinate system along with $y = x$. Indicate which relations are functions and which are one-to-one.

19. f and f^{-1} in Problem 13

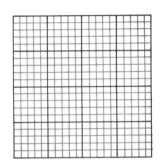

20. g and g^{-1} in Problem 14

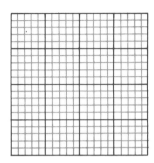

21. h and h^{-1} in Problem 17

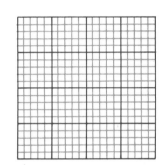

22. H and H^{-1} in Problem 18

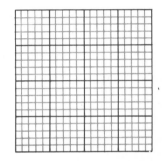

23. For $f(x) = 3x - 2$, find:

    (A) $f^{-1}(x) =$ _____   **(B)** $f^{-1}(2) =$ _____   **(C)** $f[f^{-1}(3)] =$ _____

24. For $g(x) = 2x + 3$, find:

    (A) $g^{-1}(x) =$ _____   **(B)** $g^{-1}(5) =$ _____   **(C)** $g[g^{-1}(4)] =$ _____

25. For $F(x) = \dfrac{x}{3} - 2$, find:

    (A) $F^{-1}(x) =$ _____   **(B)** $F^{-1}(-1) =$ _____   **(C)** $F^{-1}[F(4)] =$ _____

26. For $G(x) = \dfrac{x}{2} + 5$, find:

    (A) $G^{-1}(x) =$ _____   **(B)** $G^{-1}(8) =$ _____   **(C)** $G^{-1}[G(-4)] =$ _____

C 27. For $f(x) = \dfrac{x}{3} + 2$, find:

    (A) $f^{-1}(x) =$ _____   **(B)** $f[f^{-1}(a)] =$ _____

28. For $g(x) = 4x + 2$, find:

(A) $g^{-1}(x) = $ _____   **(B)** $g^{-1}[g(a)] = $ _____

The Check Exercise for this section is on page 545.

10.5 VARIATION PROBLEMS

•Direct variation
••Inverse variation
•••Joint variation
••••Combined variation

In reading scientific material, one is likely to come across statements such as "The pressure of an enclosed gas varies directly as the absolute temperature," or "The frequency of vibration of air in an organ pipe varies inversely as the length of the pipe," or even more complicated statements, such as, "The force of attraction between two bodies varies jointly as their masses and inversely as the square of the distance between the two bodies." These statements have precise mathematical meaning in that they represent particular types of functions. The purpose of this section is to investigate these special functions.

•**Direct variation**

The statement y **varies directly as** x means

$$y = kx \qquad k \neq 0$$

where k is a constant called the **constant of variation.** Similarly, the statement "y varies directly as the square of x" means

$y = kx^2 \qquad k \neq 0$

and so on. The first equation defines a linear function, and the second a quadratic function.
    Direct variation is illustrated by the familiar formulas

$C = \pi D \qquad$ and $\qquad A = \pi r^2$

where the first formula asserts that the circumference of a circle varies directly as the diameter, and the second that the area of a circle varies directly as the square of the radius. In both cases, π is the constant of variation.

EXAMPLE 14
Translate each statement into an appropriate equation, and find the constant of variation if $y = 16$ when $x = 4$.

(A) y varies directly as x

Solution
$y = kx$ Don't forget to put in k.

To find the constant of variation k, substitute $x = 4$ and $y = 16$ and solve for k.

$16 = k \cdot 4$

$k = \dfrac{16}{4} = 4$

Thus, $k = 4$, and the equation of variation is

$y = 4x$

(B) y varies directly as the cube of x

Solution
$y = kx^3$ Don't forget to put in k.

To find k, substitute $x = 4$ and $y = 16$.

$16 = k \cdot 4^3$

$k = \dfrac{16}{64} = \dfrac{1}{4}$

Thus, the equation of variation is

$y = \dfrac{1}{4}x^3$

Work Problem 14 and check solution in Solutions to Matched Problems on page 529.

PROBLEM 14   If $y = 4$ when $x = 8$, find the equation of variation for each statement:

(A) y varies directly as x

Solution

(B) y varies directly as the cube root of x

Solution

••Inverse variation

The statement y **varies inversely as** x means

$$y = \frac{k}{x} \qquad k \neq 0$$

where k is a constant (the constant of variation). As in the case of direct variation, we also discuss y varying inversely as the square of x, and so on.

An illustration of inverse variation is given in the distance-rate-time formula, $d = rt$, in the form $t = d/r$ for a fixed distance d. In driving a fixed distance, say $d = 40$ mi, time varies inversely as the rate. That is,

$$t = \frac{40}{r}$$

where 40 is the constant of variation—as the rate increases, the time decreases, and vice versa.

EXAMPLE 15
Translate each statement into an appropriate equation, and find the constant of variation if $y = 16$ when $x = 4$.

(A) y varies inversely as x

Solution
$$y = \frac{k}{x} \qquad \text{Don't forget to put in } k.$$

To find k, substitute $x = 4$ and $y = 16$.

$$16 = \frac{k}{4}$$

$$k = 64$$

Thus, the equation of variation is

$$y = \frac{64}{x}$$

(B) y varies inversely as the square root of x.

Solution

$$y = \frac{k}{\sqrt{x}}$$ Dont' forget to put in k.

To find k, substitute $x = 4$ and $y = 16$.

$$16 = \frac{k}{\sqrt{4}}$$

$$k = 32$$

Thus, the equation of variation is

$$y = \frac{32}{\sqrt{x}}$$

Work Problem 15 and check solution in Solutions to Matched Problems on page 529.

PROBLEM 15   If $y = 4$ when $x = 8$, find the equation of variation for each statement:

(A) y varies inversely as x

Solution

(B) y varies inversely as the square of x

Solution

•••Joint variation

The statement w **varies jointly as** x **and** y means

$$w = kxy \qquad k \neq 0$$

where k is a constant (the constant of variation). Similarly, if

$$w = kxyz^2 \qquad k \neq 0$$

we would say that "w varies jointly as x, y, and the square of z," and so on. For example, the area of a rectangle varies jointly as its length and width (recall $A = lw$), and the volume of a right circular cylinder varies jointly as the square of its radius and its height (recall $V = \pi r^2 h$). What is the constant of variation in each case?

••••Combined variation

The above types of variation are often combined. For example, the statement "w varies jointly as x and y, and inversely as the square of z" means

$$w = k\frac{xy}{z^2} \qquad k \neq 0 \qquad \text{We do not write } w = \frac{kxy}{kz^2}.$$

Thus the statement, "The force of attraction F between two bodies varies jointly as their masses m_1 and m_2, and inversely as the square of the distance d between the two bodies," means

$$F = k\frac{m_1 m_2}{d^2} \qquad k \neq 0$$

If (assuming k is positive) either of the two masses is increased, the force of attraction increases; on the other hand, if the distance is increased, the force of attraction decreases.

EXAMPLE 16

The pressure P of an enclosed gas varies directly as the absolute temperature T, and inversely as the volume V. If 500 ft³ of gas yields a pressure of 10 lb/ft² at a temperature of 300 K (absolute temperature*), what will be the pressure of the same gas if the volume is decreased to 300 ft³ and the temperature increased to 360 K?

Solution

Method 1   Write the equation of variation $P = k\dfrac{T}{V}$, and find k, using the first set of values:

$$10 = k\tfrac{300}{500}$$

$$k = \tfrac{50}{3}$$

Hence, the equation of variation for this particular gas is $P = \tfrac{50}{3}\dfrac{T}{V}$.

Now find the new pressure P, using the second set of values:

$$P = \tfrac{50}{3} \cdot \tfrac{360}{300} = 20 \text{ lb/ft}^2$$

Method 2 (generally faster than Method 1)   Write the equation of variation $P = k\dfrac{T}{V}$; then convert it to the equivalent form:

$$\frac{PV}{T} = k$$

If P_1, V_1, and T_1 are the first set of values for the gas, and P_2, V_2, and T_2 are the second set, then

$$\frac{P_1 V_1}{T_1} = k \qquad \text{and} \qquad \frac{P_2 V_2}{T_2} = k$$

*A kelvin (absolute) and a Celsius degree are the same size, but 0 on the Kelvin scale is $-273°$ on the Celsius scale. This is the point at which molecular action is supposed to stop and is called absolute zero.

Hence

$$\frac{P_1 V_1}{T_1} = \frac{P_2 V_2}{T_2}$$

Since all values are known except P_2, substitute and solve. Thus

$$\frac{(10)(500)}{300} = \frac{P_2(300)}{360}$$

$$P_2 = 20 \text{ lb/ft}^2$$

Work Problem 16 and check solution in Solutions to Matched Problems on page 530.

PROBLEM 16 The length L of skid marks of a car's tires (when brakes are applied) varies directly as the square of the speed v of the car. If skid marks of 20 ft are produced at 30 mi/hr, how fast would the same car be going if it produced skid marks of 80 ft? Solve two ways (see Example 16).

Solution   *Method 1:*

Method 2:

SOLUTIONS TO MATCHED PROBLEMS

14. **(A)** $y = kx$
$4 = k \cdot 8$
$k = \frac{1}{2}$
$y = \frac{1}{2}x$

(B) $y = k\sqrt[3]{x}$
$4 = k\sqrt[3]{8}$
$k = 2$
$y = 2\sqrt[3]{x}$

15. **(A)** $y = \frac{k}{x}$
$4 = \frac{k}{8}$
$k = 32$

$y = \frac{32}{x}$

(B) $y = \frac{k}{x^2}$
$4 = \frac{k}{8^2}$
$k = 256$

$y = \frac{256}{x^2}$

16. *Method 1*

$$L = kv^2$$
$$20 = k(30)^2$$
$$k = \frac{2}{90}$$
$$L = \frac{2}{90}v^2$$

$$80 = \frac{2}{90}v^2$$
$$v^2 = \frac{(80)(90)}{2}$$
$$= 3{,}600$$
$$v = \sqrt{3{,}600} = 60 \text{ mi/hr}$$

Method II

$$L = kv^2$$
$$\frac{L}{v^2} = k$$
$$\frac{L_1}{v_1^2} = \frac{L_2}{v_2^2}$$

Now find v_2, given $L_1 = 20$,

$$\frac{20}{30^2} = \frac{80}{v_2^2}$$
$$v_2^2 = \frac{(900)(80)}{20}$$
$$v_2 = \sqrt{3{,}600} = 60 \text{ mi/hr}$$

PRACTICE EXERCISE 10.5

Work odd-numbered problems first, check answers, and then work even-numbered problems in areas of weakness. Answers to all problems are in the back of the book. Make every effort to work a problem yourself before you look at an answer.

A *Translate each statement into an equation, using k as the constant of variation.*

1. F varies directly as the square of r. _____

2. u varies directly as v. _____

3. The pitch or frequency of a guitar string f of a given length varies directly as the square root of the tension T of the string.

    _____

4. Geologists have found in studies of earth erosion that the erosive force (sediment-carrying power) P of a swiftly flowing stream varies directly as the sixth power of the velocity v of the water.

    _____

5. y varies inversely as the square root of x. _____

6. I varies inversely as t. _____

7. The biologist Reaumur suggested in 1735 that the length of time t that it takes fruit to ripen during the growing season varies inversely as the sum of the average daily temperatures T during the growing season.

    _____

8. In a study on urban concentration, F. Auerback discovered an interesting law. After arranging all the cities of a given country according to their population size, starting with the largest, it was found that the population p of a city varied inversely as the number n indicating its position in the ordering.

    _____

9. R varies jointly as S, T, and V. _____

10. g varies jointly as x and the square of y. _____

11. The volume of a cone varies jointly as its height h and the square of the radius r of its base.

    _____

12. The amount of heat Q put out by an electrical appliance (in calories) varies jointly as time t, resistance R in the circuit, and the square of the current I.

———————————————

Solve, using either of the two methods illustrated in Example 16.

13. u varies directly as the square root of v. If $u = 2$ when $v = 2$, find u when $v = 8$.

———————————————

14. y varies directly as the square of x. If $y = 20$ when $x = 2$, find y when $x = 5$.

———————————————

15. L varies inversely as the square root of M. If $L = 9$ when $M = 9$, find L when $M = 3$.

———————————————

16. I varies inversely as the cube of t. If $I = 4$ when $t = 2$, find I when $t = 4$.

———————————————

B *Translate each statement into an equation, using k as the constant of variation.*

17. U varies jointly as a and b, and inversely as the cube of c. ————————————————

18. w varies directly as the square of x and inversely as the square root of y. ————————————

19. The maximum safe load L for a horizontal beam varies jointly as its width w and the square of its height h, and inversely as its length l.

———————————————

20. Joseph Cavanaugh, a sociologist, found that the number of long-distance phone calls n between pairs of cities in a given time period varied (approximately) jointly as the populations P_1 and P_2 of the two cities, and inversely as the distance d between the two cities.

———————————————

Solve, using either of the two methods illustrated in Example 16.

21. Q varies jointly as m and the square of n, and inversely as P. If $Q = -4$ when $m = 6, n = 2$, and $P = 12$, find Q when $m = 4, n = 3$, and $P = 6$.

———————————————

22. w varies jointly as x, y, and z, and inversely as the square of t. If $w = 2$ when $x = 2, y = 3, z = 6$, and $t = 3$, find w when $x = 3, y = 4, z = 2$, and $t = 2$.

———————————————

23. The weight w of an object on or above the surface of the earth varies inversely as the square of the distance d between the object and the center of the earth. If a person weighs 100 lb on the surface of the earth, how much would that person weigh (to the nearest pound) 400 mi above the earth's surface? (Assume the radius of the earth is 4,000 mi.)

24. A child was struck by a car in a crosswalk. The driver of the car had slammed on the brakes and left skid marks 160 ft long. The driver told the police that the car was traveling 30 mi/hr. The police know that the length of skid marks L (when brakes are applied) varies directly as the square of the speed of the car v, and that at 30 mi/hr (under ideal condition) skid marks would be 40 ft long. How fast was the driver actually going before the brakes were applied?

C 25. Ohm's law states that the current I in a wire varies directly as the electromotive force E and inversely as the resistance R. If $I = 22$ amperes when $E = 110$ volts and $R = 5$ ohms, find I if $E = 220$ volts and $R = 11$ ohms.

26. Anthropologists, in their study of human genetic groupings, often use an index called the cephalic index. The cephalic index C varies directly as the width w of the head and inversely as the length l of the head (both looking down from the top). If an Indian in Baja California (Mexico) has $C = 75$, $w = 6$ in, and $l = 8$ in, then what would C be for an Indian in northern California with $w = 8.1$ in and $l = 9$ in?

The Check Exercise for this section is on page 547.

10.6 CHAPTER REVIEW

Important terms and symbols

10.1 RELATIONS AND FUNCTIONS   relation, function, domain, range, input, output, independent variable, dependent variable

10.2 FUNCTION NOTATION $f(x)$

10.3 GRAPHING LINEAR AND QUADRATIC FUNCTIONS   linear functions, quadratic functions, $f(x) = ax + b, f(x) = ax^2 + bx + c, a \neq 0$, axis of symmetry, vertex, graphing by completing the square, maximum and minimum value of a quadratic function

10.4 INVERSE RELATIONS AND FUNCTIONS   inverse relation R^{-1}, domain of R^{-1}, range of R^{-1}, graph of R and R^{-1}, one-to-one correspondence

10.5 VARIATION PROBLEMS   constant of variation

y varies directly as x: $y = kx$

y varies inversely as x: $y = \dfrac{k}{x}$

w varies jointly as x and y: $w = kxy$

w varies directly as x and inversely as y: $y = k\dfrac{x}{y}$

DIAGNOSTIC (REVIEW) EXERCISE 10.6 ——————————————————

Work through all the problems in this chapter review and check answers in the back of the book. (Answers to all problems are there, and following each answer is a number in italics indicating the section in which that type of problem is discussed.) Where weaknesses show up, review appropriate sections in the text. When you are satisfied that you know the material, take the practice test following this review.

A *Which of the following relations are functions? The variable x is independent.*

1. Domain   Range

2. Domain   Range

3. Domain   Range

3 ——————→ 2
5 ——————→ 4
7 ——————→ 6

————————————

————————————

————————————

4.

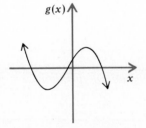

5.

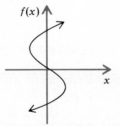

6.

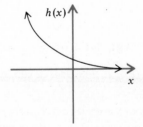

————————————

————————————

————————————

7. $y = x^3 - 2x$

8. $y^2 = x$

9. $x^2 + y^2 = 25$

————————————

————————————

————————————

10. $\{(1, 2), (1, -2), (0, 3)\}$

11. $\{(-1, 2), (1, 3), (2, 4)\}$

12. $\{(-2, 3), (0, 3), (2, 3)\}$

————————————

————————————

————————————

For Problems 13 to 15 write the domain and range of the relation in the indicated problem.

13. Problem 10

14. Problem 11

15. Problem 12

————————————

————————————

————————————

16. If $f(x) = 6 - x$, find:

    (A) $f(6)$ _____   **(B)** $f(0)$ _____   **(C)** $f(-3)$ _____   **(D)** $f(m)$ _____

17. If $G(z) = z - 2z^2$, find:

    (A) $G(2)$ _____   **(B)** $G(0)$ _____   **(C)** $G(-1)$ _____   **(D)** $G(c)$ _____

18. Graph $f(x) = 2x - 4$. Indicate its slope and y intercept.

19. Graph $g(x) = \dfrac{x^2}{2}$.

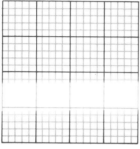

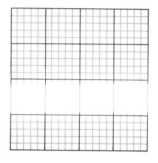

20. Graph the relation $M = \{(0, 5), (2, 7), (2, 3)\}$, its inverse M^{-1}, and $y = x$, all on the same coordinate system. Indicate whether M or M^{-1} is a function.

_____

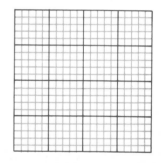

21. What is the domain and range of M^{-1} in the preceding problem? _____

In Problems 22 to 25 translate each statement into an equation, using k as the constant of variation.

22. m varies directly as the square of n. _____

23. P varies inversely as the cube of Q. _____

24. A varies jointly as a and b. _____

25. y varies directly as the cube of x and inversely as the square root of z. _____

B 26. Is every relation a function? Explain. _____

27. If $f(t) = 4 - t^2$ and $g(t) = t - 3$, find:

    (A) $f(0) - g(0)$ _____   **(B)** $\dfrac{g(6)}{f(-1)}$ _____

   (C) $g(x) - f(x)$ _____   **(D)** $f[g(2)]$ _____

28.   If $f(x) = 2x - 3$, find:

   (A) $f(3 + h)$ _____   **(B)** $\dfrac{f(3 + h) - f(3)}{h}$ _____

29.   Graph $g(t) = -\frac{3}{2}t + 6$ and indicate its slope   **30.**   Graph $f(x) = x^2 - 4x + 5$. Indicate its vertex,
         and y intercept.   axis, and maximum or minimum.

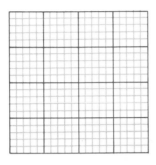   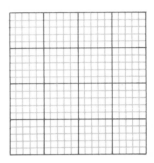

31.   Find the maximum or minimum value of $g(x) = 3 - 8x - 2x^2$ without graphing g.

         _____

In Problems 32 to 35 indicate which relations are one-to-one correspondences.

32.   Problems 1 to 3 _____   **33.**   Problems 4 to 6 _____

34.   Problems 7 to 9 _____   **35.**   Problems 10 to 12 _____

36.   Which of the functions in Problems 4 and 6 have inverses that are functions?

         _____

37.   Which of the functions in Problems 2 and 3 have inverses that are functions?

         _____

38.   For $f: y = 1 - x/2$ find f^{-1} in the form of an equation in x and y. Solve for y in terms of x.

         _____

39.   Let $M(x) = \dfrac{x + 3}{2}$.

   (A)   Find $M^{-1}(x)$. _____   **(B)**   Are both M and M^{-1} functions? _____

   (C)   Find $M^{-1}(2)$. _____   **(D)**   Find $M^{-1}[M(3)]$. _____

40.   If y varies directly as x and inversely as z:

   (A)   Write the equation of variation. _____

   (B)   If $y = 4$ when $x = 6$ and $z = 2$, find y when $x = 4$ and $z = 4$. _____

41. If $g(t) = 1 - t^2$, find:

   (A) $g(2 + h)$ _____

   (B) $\dfrac{g(2 + h) - g(2)}{h}$ _____

42. Graph $g(t) = 96t - 16t^2$. Indicate its vertex, axis, and the maximum or minimum.

   _____

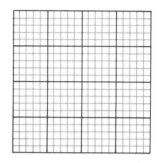

APPLICATIONS

43. COST FUNCTION   The cost $C(x)$ for renting a business copying machine is $200 per month plus 5 cents a copy for x copies. Express this functional relationship in terms of an equation and graph it for $0 \le x \le 3,000$.

   _____

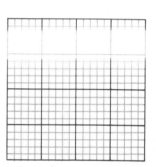

44. REVENUE FUNCTION   The revenue function for a company producing stereo radios (for a particular model) is

   $$R = f(p) = 6,000p - 30p^2 \qquad 0 \le p \le 200$$

   where p is the price per unit.
   (A) Graph f.
   (B) At what price p will the revenue be maximum? What is the maximum?

   _____

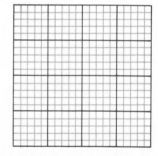

45. VARIATION   The time t required for an elevator to lift a weight varies jointly as the weight w and the distance d through which it is lifted, and inversely as the power P of the motor. Write the equation of variation, using k as the constant of variation. If it takes a 400-horsepower motor 4 sec to lift an 800-kg elevator 8 meters, how long will it take the same motor to lift a 1,600-kg elevator 24 meters?

   _____

PRACTICE TEST CHAPTER 10 ──────────────────────────────────

Take this as if it were a graded test. Allow yourself up to 50 minutes. Work the problems without looking back . in the chapter. Correct your work, using the answers (keyed to appropriate sections) in the back of the book.

1. Which of the following relations are functions (*x* is independent):

   (A) Domain   Range   **(B)** $f(x)$

   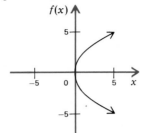

   (C) $y = x^2 - 3x + 1$ ──────────────   **(D)** $\{(-1, 2), (0, 0), (-1, 3)\}$ ──────────

2. For $f(x) = x^2 - 2x + 1$, find $f(3)$ and $f(0)$. ─────────────────

3. For $f(x) = x^2 - 2x + 1$, find $f(-2)$ and $f(a)$. ─────────────────

4. For $f(t) = t - 2$ and $g(t) = 6 - t^2$, find $f(4) + g(2)$ and $3g(0)$. ─────────────────

5. For $f(t) = t - 2$ and $g(t) = 6 - t^2$, find $f(m) + g(m)$ and $f[g(2)]$.

   ──────────────────────

6. Write the relation in Problem **1D** in the form of **1A** (that is, identify the domain and range and use arrows to indicate the correspondence). Is the relation a one-to-one correspondence?

   ──────────────────────

7. Given $f(x) = \dfrac{x - 2}{4}$, find $f^{-1}(x)$ and $f^{-1}(1)$. ──────────────────

8. For f and f^{-1} in Problem 7, find $f^{-1}[f(2)]$ and $f[f^{-1}(a)]$. ─────────────────────

9. For $f(x) = 3^x$, $x \in \{-1, 0, 1\}$, graph f, f^{-1}, and $y = x$, all on the same coordinate system.

   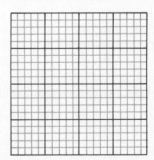

10. Which of f and f^{-1} in Problem 9 are functions?

    ──────────────────────

11. For $g(x) = 2x - 1$, find $\dfrac{g(a + h) - g(a)}{h}$.

_____

12. Graph $f(x) = -\dfrac{x}{2} + 4$.

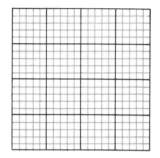

_____

13. Indicate the slope and y intercept of the graph in Problem 12.

_____

14. Graph $f(x) = x^2 - 4x + 3$.

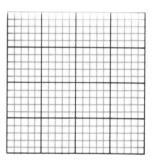

15. What are the equation of the axis and the coordinates of the vertex for the graph of f in Problem 14?

_____

16. What is the maximum or minimum of $f(x)$ in Problem 14?

_____

17. Find the maximum or minimum value of $G(t) = 240t - 40t^2$ without graphing G.

_____

18. Write the equation of variation equivalent to: w varies jointly as x and the square root of y.

_____

19. Write the equation of variation equivalent to: m varies inversely as n to the two-thirds' power.

_____

20. If y varies directly as the square of x and inversely as z, and if $y = 8$ when $x = 2$ and $z = 3$, find y when $x = 6$ and $z = 4$.

_____

CHECK EXERCISE 10.1

Work the following problems without looking at any text examples. Show your work in the space provided. Write the letter that best indicates your answer in the answer column.

1. One of the relations below is not a function. Indicate which one.

   (A) $1 \longrightarrow 0$
         $3 \longrightarrow 7$
         5

   (B) $-2 \longrightarrow 4$
         2
         $-1 \longrightarrow 1$
         1

   (C) $4 \longrightarrow 2$
         $ \longrightarrow -2$
         $9 \longrightarrow 3$
         $ \longrightarrow -3$

   (D) -2
         $0 \longrightarrow 0$
         2

2. One of the relations below is not a function. Indicate which one.

   (A)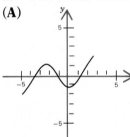

   (B)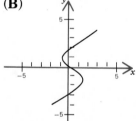

   (C)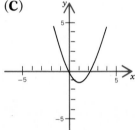

   (D)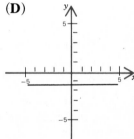

3. One of the equations below does not specify a function. Indicate which one, given that x is the independent variable.

   (A) $y = 2x^2 + 1$ (B) $y^2 = 4x$ (C) $y = x^4$ (D) $y - x = 1$

539

4. Given the relation $F = \{(-2, 1), (0, 1), (-2, 3)\}$; write down the domain and indicate whether F is a function.

    (A) $\{-2, 0\}$; a function
    (B) $\{1, 3\}$; not a function
    (C) $\{1, 3\}$; a function
    (D) $\{-2, 0\}$; not a function

5. Given the relation $y = x^2 - x, x \in \{1, 2, 3\}$; write down the range and indicate whether the relation is a function.

    (A) $\{0, 2, 6\}$; a function
    (B) $\{1, 2, 3\}$; a function
    (C) $\{1, 2, 3\}$; not a function
    (D) $\{0, 2, 6\}$; not a function

CHECK EXERCISE 10.2

Work the following problems without looking at any text examples. Show your work in the space provided. Write the letter that best indicates your answer in the answer column.

Problems 1 to 9 refer to the functions

$$f(x) = 2x - 4 \qquad g(t) = 3 - t \qquad F(u) = 2u^2 \qquad G(v) = 2v - v^2$$

Evaluate as indicated.

1. $g(-3) =$

    (A)   0
    (B)   6
    (C) -6
    (D)   None of these

2. $G(-2) =$

    (A)   0
    (B)   8
    (C) -6
    (D)   None of these

3. $f(3) - F(2) =$

    (A) -6
    (B) -14
    (C) -2
    (D)   None of these

4. $3f(0) - g(0) =$

    (A) -15
    (B)   1
    (C)   0
    (D)   None of these

5. $\dfrac{G(-1)}{2F(2) - 13} =$

    (A) $-\frac{1}{3}$
    (B) -1
    (C)   1
    (D)   None of these

6. $\dfrac{g(2 + h) - g(2)}{h} =$

    (A) 1

    (B) $\dfrac{2 + h}{h}$

    (C) -1

    (D) None of these

7. $f[g(4)] =$

    (A) -1

    (B) -6

    (C) -20

    (D) None of these

8. $F[G(-1)] =$

    (A) -18

    (B) 2

    (C) 18

    (D) None of these

9. $\dfrac{G(v + h) - G(v)}{h} =$

    (A) $2 - 2v - h$

    (B) $2 + 2v + h$

    (C) $\dfrac{2h + 2hv + h^2 + v^2}{h}$

    (D) None of these

10. Write an equation that specifies the function C: The total cost per week $C(x)$ of manufacturing x electronic games if fixed costs are \$4,000 per week and variable costs are \$9 per game.

    (A) $C(x) = 9x$

    (B) $C(x) = 4{,}000 + 9x$

    (C) $C(x) = 4{,}009x$

    (D) None of these

CHECK EXERCISE 10.3

Work the following problems without looking at any text examples. Show
your work in the space provided. Write the letter that best indicates your
answer in the answer column.

In Problems 1 to 3 graph each function in the space provided; then out of all
the indicated answers select the one (by letter) that matches your answer.

1. $f(x) = 3 - \dfrac{x}{2}$

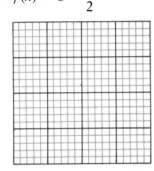

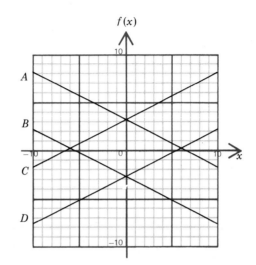

2. $g(t) = t^2 - 4t$

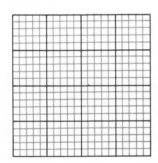

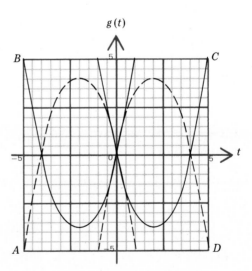

543

3. $F(m) = -8 - 8m - m^2$

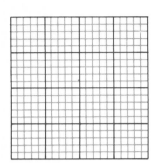

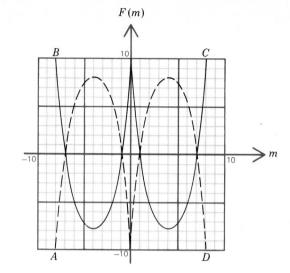

4. Without graphing, find the equation of the axis of symmetry and the coordinates of the vertex for the parabola given by $f(x) = 3x^2 + 12x - 7$.

   (**A**) $x = -2, (-2, -19)$
   (**B**) $x = 2, (2, 29)$
   (**C**) $y = -2, (-2, -19)$
   (**D**) None of these

5. Without graphing, find the maximum or minimum (indicate which) value of $f(x) = 4 - 16x - 4x^2$.

   (**A**) Min $f(x) = f(-2) = 20$
   (**B**) Min $f(x) = f(2) = -44$
   (**C**) Max $f(x) = f(-2) = 20$
   (**D**) None of these

CHECK EXERCISE 10.4

Work the following problems without looking at any text examples. Show
your work in the space provided. Write the letter that best indicates your
answer in the answer column.

1. For $f(x) = 3 - \dfrac{x}{2}$, find $f^{-1}(x)$

    and $f^{-1}(2)$.

    (A) $f^{-1}(x) = \dfrac{1}{3 - x/2}; f^{-1}(2) = \frac{1}{2}$

    (B) $f^{-1}(x) = 6 - 2x; f^{-1}(2) = 2$

    (C) $f^{-1}(x) = 3 - \dfrac{y}{2}; f^{-1}(2) = 2$

    (D) None of these

2. For f and f^{-1} in Problem 1, find $f^{-1}[f(6)]$ and $f^{-1}[f(a)]$.

    (A) $6; a$
    (B) $0; -a$
    (C) $6; -a$
    (D) None of these

3. For $F: y = |x| + 1$, find F^{-1} in the form of an equation.

    (A) $F^{-1}: y = \dfrac{1}{|x| + 1}$
    (B) $F^{-1}: |y| = x + 1$
    (C) $F^{-1}: x = |y| + 1$
    (D) None of these

4. Graph F and F^{-1} from Problem 3 and $y = x$, all on the same coordinate system.

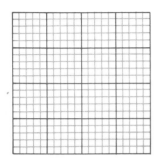

(A)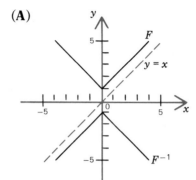

(B)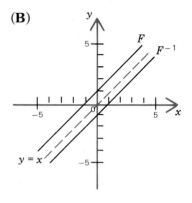

(C)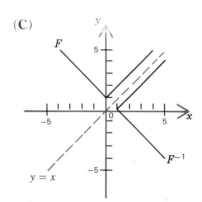

(D) None of these

5. Indicate which of F and F^{-1} from Problems 3 and 4 are functions.

     (A) Both are functions
     (B) Neither are functions
     (C) F is a function, F^{-1} is not a function
     (D) F is not a function, F^{-1} is a function

CHECK EXERCISE 10.5

Work the following problems without looking at any text examples. Show your work in the space provided. Write the letter that best indicates your answer in the answer column.

In Problems 1 to 6 translate each statement into an equation, using k as the constant of variation.

ANSWER COLUMN

1. _____
2. _____
3. _____
4. _____
5. _____
6. _____
7. _____
8. _____
9. _____
10. _____

1. *M* varies directly as the square root of *N*.

   (A) $M = \sqrt{N}$

   (B) $M = \dfrac{k}{\sqrt{N}}$

   (C) $M = k\sqrt{N}$

   (D) None of these

2. *x* varies inversely as the cube of *t*.

   (A) $x = \dfrac{1}{t^3}$

   (B) $x = \dfrac{k}{t^3}$

   (C) $x = kt^3$

   (D) None of these

3. *P* varies jointly as *u* and *v*.

   (A) $P = uv$

   (B) $P = k\dfrac{u}{v}$

   (C) $P = k\dfrac{v}{u}$

   (D) None of these

4. *Q* varies jointly as *m*, *n*, and the cube of *s*.

   (A) $Q = kmns^3$

   (B) $Q = (kmns)^3$

   (C) $Q = mns^3$

   (D) None of these

5. *F* varies directly as the square of *c* and inversely as the cube root of *d*.

   (A) $F = \dfrac{kc^2}{k\sqrt[3]{d}}$

   (B) $F = k\dfrac{c^2}{\sqrt[3]{d}}$

   (C) $F = \dfrac{c^2}{\sqrt[3]{d}}$

   (D) None of these

547

6. G varies jointly as u and the cube of v, and inversely as the square of w.

(A) $\quad G = k\dfrac{uv^3}{w^2}$

(B) $\quad G = \dfrac{kuv^2}{kw^2}$

(C) $\quad G = k\dfrac{w^2}{uv^3}$

(D) None of these

Solve Problems 7 to 10.

7. If y varies directly as x^2, and $y = 60$ when $x = 6$, find y when $x = 3$.

(A) 15
(B) 240
(C) 60
(D) None of these

8. If w varies jointly as x and y, and inversely as z, and if $w = 60$ when $x = 3$, $y = 50$, and $z = 2$, find w when $x = 2$, $y = 30$, and $z = 3$.

(A) 16
(B) 225
(C) 160
(D) None of these

9. The horsepower P required to drive a speedboat through water varies directly as the cube of the speed v of the boat. If 10 horsepower is required to drive the boat at 20 mi/hr, how many horsepower is required to drive it at 30 mi/hr?

(A) 10 horsepower
(B) 33.75 horsepower
(C) 15 horsepower
(D) None of these

10. The frequency of vibration f of a musical string varies directly with the square root of the tension T and inversely as the length L of the string. If the frequency is 440 Hz when $T = 36$ kg and $L = 60$ cm, find the frequency when $T = 49$ kg and $L = 50$ cm.

(A) 314 Hz
(B) 520 Hz
(C) 616 Hz
(D) None of these

Chapter Eleven

EXPONENTIAL AND LOGARITHMIC FUNCTIONS

CONTENTS

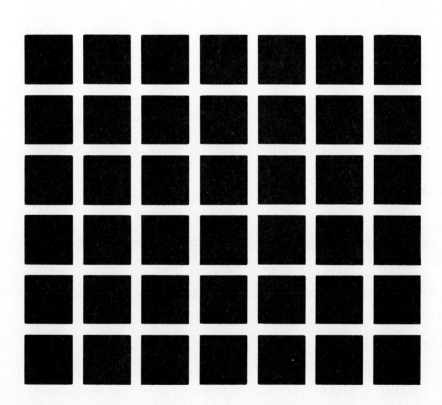

There are no gray areas where the white lines intersect.

INSTRUCTIONS FOR STUDENTS IN A
SELF-PACED CLASS OR LAB

(yes) — **HAVE YOU HAD INTERMEDIATE ALGEBRA BEFORE THIS COURSE?** — **(no)**

1. Work Diagnostic (Review) Exercise 11.6 on page 586. Check answers in back of book; then work through text sections corresponding to problems missed. (Section numbers are in italics following each answer.)
2. When finished with step 1, take Practice Test Chapter 11 on page 587 as a final check of your understanding of the chapter. Check answers in the back of the book; then review sections where weakness still prevails. (Corresponding section numbers are in italics following each answer.)
3. When you think you are ready, ask your instructor for a graded test for Chapter 11.
4. If your instructor approves, after the test is corrected, review for a final examination.

1. Work through each section in the chapter as follows:
   (a) Read discussion.
   (b) Read each example and work the corresponding matched problem. Check your solution to the matched problem in Solutions to Matched Problems on the indicated page.
   (c) At the end of a section work the odd-numbered problems in the Practice Exercise and check answers; then work even-numbered problems in areas of weakness. (Answers to *all* Practice Exercise sets are in the back of the book.)
   (d) Work Check Exercise as instructed. Tear out and turn in as directed by your instructor. (Answers are not in the text.)
2. Repeat each step in item 1 for each section in the chapter.
3. After the instructional part of the chapter is completed, proceed with steps 1 to 4 in the box above.

11.1 EXPONENTIAL FUNCTIONS

•**Exponential functions**
••**Graphing an exponential function**
•••**Typical types of exponential graphs**
••••**Base e**
•••••**Basic exponential properties**

•Exponential functions

In this and the next section we will consider two new kinds of functions that use variable exponents in their definitions. To start, note that

$$f(x) = 2^x \qquad \text{and} \qquad g(x) = x^2$$

are not the same function. The function g is a quadratic function, which we have just discussed (Section 10.3), and the function f is a new type of function called an exponential function. An **exponential function** is a function defined by an equation of the form

Exponential Function
$f(x) = b^x \qquad b > 0, b \neq 1$

where b is a constant, called the **base,** and the exponent x is a variable. The replacement set for the exponent, the **domain of f,** is the set of real numbers R. The **range of f** is the set of positive real numbers.

••Graphing an exponential function

Many students, if asked to graph an exponential function such as $f(x) = 2^x$, would not hesitate at all. They would likely make up a table by assigning integers to x, plot the resulting points, and then join these points with a smooth curve (see Figure 1). The only catch is that 2^x has not been defined for all real numbers. We know what 2^5, 2^{-3}, $2^{2/3}$, $2^{-3/5}$, $2^{1.4}$, and $2^{-3.15}$ all mean (that is, 2^p, where p is a rational number—see Chapter 6), but what does

$$2^{\sqrt{2}}$$

mean? The question is not easy to answer at this time. In fact, a precise definition of $2^{\sqrt{2}}$ must

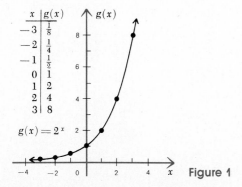

Figure 1

wait for more advanced courses, where we can show that

$$b^x$$

names a real number for b a positive real number and x any real number, and that the graph of $g(x) = 2^x$ is as indicated in Figure 1.

•••Typical types of exponential graphs

It is useful to compare the graphs of $y = 2^x$ and $y = (\frac{1}{2})^x = 2^{-x}$ by plotting both on the same coordinate system (Figure 2a). The graph of

$$f(x) = b^x \qquad b > 1 \text{ (Figure 2b)}$$

will look very much like the graph of $y = 2^x$, and the graph of

$$f(x) = b^x \qquad 0 < b < 1 \text{ (Figure 2b)}$$

will look very much like the graph of $y = (\frac{1}{2})^x$.

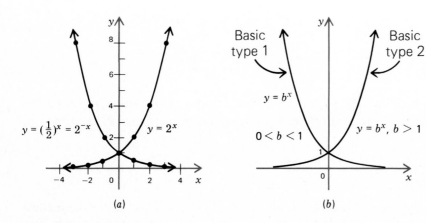

(a)   (b)   **Figure 2**

EXAMPLE 1

Graph $y = \frac{1}{2}(4^x)$ for $-3 \le x \le 3$.

Solution

x	y
-3	0.01
-2	0.03
-1	0.13
0	0.50
1	2.00
2	8.00
3	32.00

Work Problem 1 and check solution in Solutions to Matched Problems on page 555.

PROBLEM 1   Graph $y = \frac{1}{2}(4^{-x})$ for $-3 \leq x \leq 3$.

Solution

x	y

Exponential functions are often referred to as growth functions because of their widespread use in describing different kinds of growth phenomena. These functions are used to describe population growth of people, animals, and bacteria; radioactive decay (negative growth); growth of a new chemical substance in a chemical reaction; increase or decline in the temperature of a substance being heated or cooled; growth of money at compound interest; light absorption (negative growth) as it passes through air, water, or glass; decline of atmospheric pressure as altitude is increased; and growth of learning a skill such as swimming or typing, relative to practice.

••••Base e

For introductory purposes, the bases 2 and $\frac{1}{2}$ were convenient choices; however, a certain irrational number, denoted by e, is by far the most frequently used exponential base for both theoretical and practical purposes. In fact,

$$f(x) = e^x$$

is often referred to as *the* exponential function because of its widespread use. The reasons for the preference for e as a base are made clear in more advanced courses. And at that time, it is shown that e is approximated by $(1 + 1/n)^n$ to any decimal accuracy desired by making n (an integer) sufficiently large. The irrational number e to five decimal places is

$$e \approx 2.71828$$

Similarly, e^x can be approximated by using $(1 + 1/n)^{nx}$ for sufficiently large n. Because of the importance of e^x and e^{-x}, tables for their evaluation are readily available. In fact, many hand calculators can evaluate these functions directly. A short table (Table 2) for e^x and e^{-x} can be found in Appendix D for those not using a calculator.

EXAMPLE 2
Graph $y = 10e^{-0.5x}$, $-3 \leq x \leq 3$, using a hand calculator or Table 2.

Solution

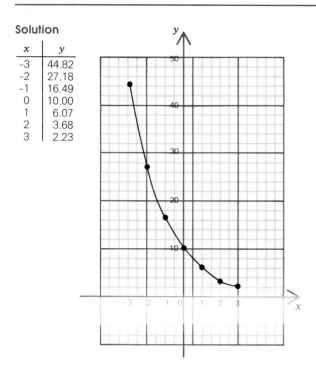

x	y
-3	44.82
-2	27.18
-1	16.49
0	10.00
1	6.07
2	3.68
3	2.23

Work Problem 2 and check solution in Solutions to Matched Problems on page 555.

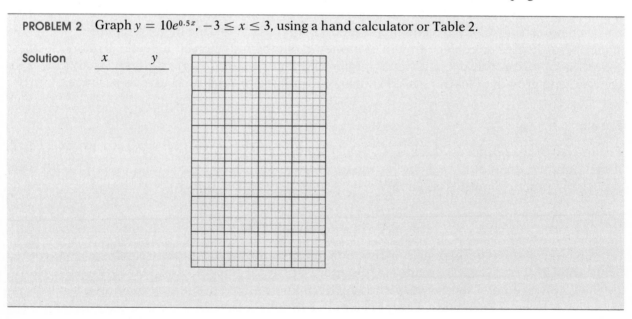

PROBLEM 2   Graph $y = 10e^{0.5x}$, $-3 \leq x \leq 3$, using a hand calculator or Table 2.

Solution   x   y

••••Basic exponential properties

Earlier (Section 6.3) we discussed five laws of exponents for rational exponents. It can be shown that these same laws hold for irrational exponents. Thus, we now assume that all five laws of exponents hold for any real exponents so long as all bases involved are positive.

Finally, we state without proof that for $b > 0$, $b \neq 1$,

$$b^m = b^n \quad \text{if and only if} \quad m = n$$

Thus, if $2^{15} = 2^{3x}$, then $3x = 15$ and $x = 5$.

SOLUTIONS TO MATCHED PROBLEMS

1. $y = \frac{1}{2}(4^{-x})$

x	y
-3	32.00
-2	8.00
-1	2.00
0	0.50
1	0.13
2	0.03
3	0.01

2. $y = 10e^{0.5x}$

x	y
-3	2.23
-2	3.68
-1	6.07
0	10.00
1	16.49
2	27.18
3	44.82

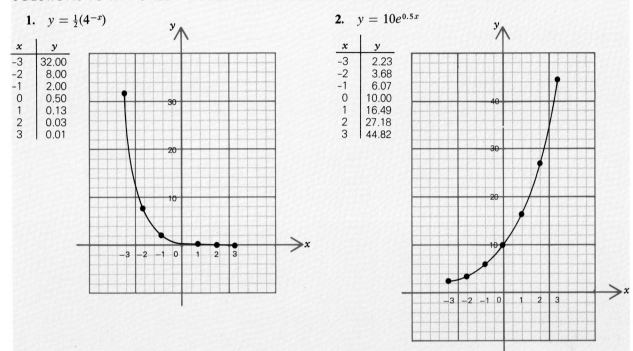

PRACTICE EXERCISE 11.1

Work odd-numbered problems first, check answers, and then work even-numbered problems in areas of weakness. Answers to all problems are in the back of the book. Make every effort to work a problem yourself before you look at an answer.

A *Graph Problems 1 to 10 for* $-3 \le x \le 3$ *by using integers for x, and then join the points with a smooth curve.*

1. $y = 3^x$

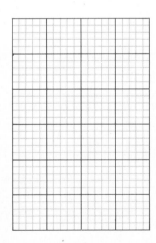

2. $y = 2^x$

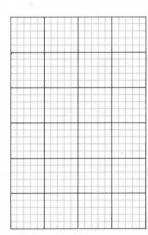

3. $y = \left(\frac{1}{3}\right)^x = 3^{-x}$

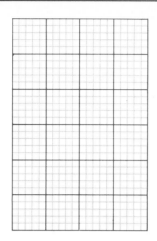

4. $y = \left(\frac{1}{2}\right)^x = 2^{-x}$

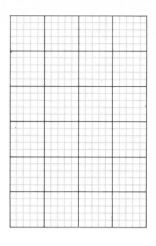

5. $y = 4 \cdot 3^x$
(NOTE: $4 \cdot 3^x \neq 12^x$)

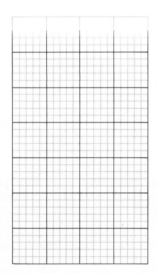

6. $y = 5 \cdot 2^x$

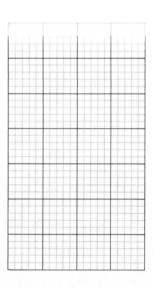

B   **7.** $y = 2^{x+3}$

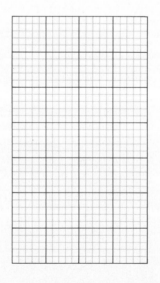

8. $y = 3^{x+1}$

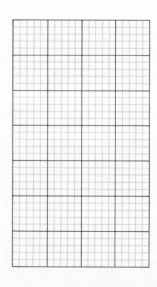

9. $y = 7(\frac{1}{2})^{2x} = 7 \cdot 2^{-2x}$

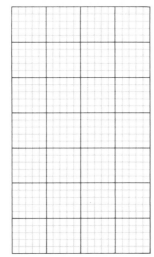

10. $y = 11 \cdot 2^{-2x}$

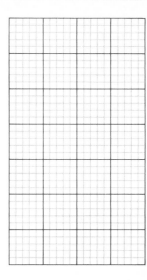

Graph Problems 11 to 14 for $-3 \le x \le 3$, using a calculator or Table 2.

11. $y = e^x$

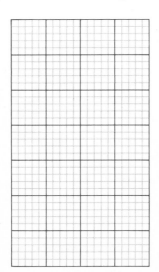

12. $y = e^{-x}$

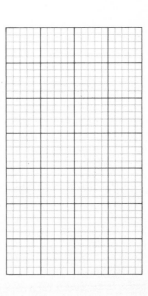

C 13. $y = 10e^{-0.12x}$

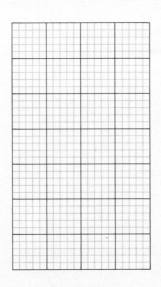

14. $y = 100e^{0.25x}$

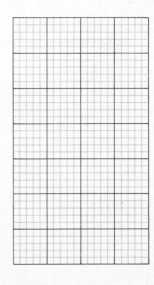

APPLICATIONS

15. If we start with 2 cents on the first day and double the amount each succeeding day, at the end of n days we will have 2^n cents. Graph $f(n) = 2^n$ for $1 \le n \le 10$.

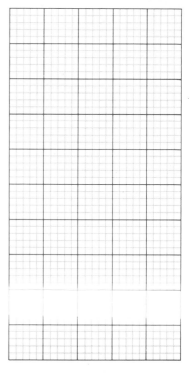

16. COMPOUND INTEREST   If a certain amount of money P, called principal, is invested at $100r$ percent interest compounded annually, the amount of money A after t years is given by

$$A = P(1 + r)^t$$

Graph this equation for $P = \$10$, $r = 0.10$, and $0 \le t \le 10$.

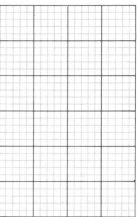

*17. EARTH SCIENCE   The atmospheric pressure P, in pounds per square inch, can be calculated approximately by the formula

$$P = 14.7e^{-0.21x}$$

where x is altitude relative to sea level in miles. Graph the equation for $-1 \le x \le 5$.

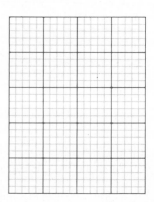

The Check Exercise for this section is on page 589.

11.2 LOGARITHMIC FUNCTIONS

•Logarithmic functions
••From logarithmic to exponential form and vice versa
•••Finding x, b, or y in $y = \log_b x$
••••Inverse relationships

•Logarithmic functions

Now we will define a new class of functions, called **logarithmic functions,** as inverses of exponential functions. Here you will see why we placed special emphasis on the general concept of inverse function in Section 10.4. If we start with the exponential function

$$f: y = 2^x$$

and interchange the variables x and y, we obtain the inverse of f

$$f^{-1}: x = 2^y$$

Graphing f and f^{-1} on the same coordinate system (Figure 3), we see

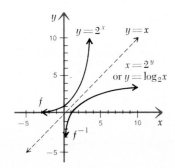

f		f^{-1}	
x	$y = 2^x$	$x = 2^y$	y
-3	$\frac{1}{8}$	$\frac{1}{8}$	-3
-2	$\frac{1}{4}$	$\frac{1}{4}$	-2
-1	$\frac{1}{2}$	$\frac{1}{2}$	-1
0	1	1	0
1	2	2	1
2	4	4	2

$\left.\begin{array}{c}\text{Ordered}\\\text{pairs}\\\text{reversed}\end{array}\right)$

Figure 3

that f^{-1} is also a function. This new function is given the name **logarithmic function with base 2,** and is symbolized as follows:

$$y = \log_2 x$$

Thus,

$$y = \log_2 x \qquad \text{is equivalent to} \qquad x = 2^y$$

that is, $\log_2 x$ is the power to which 2 must be raised to obtain x. Note that the domain of $y = \log_2 x$ is the set of positive real numbers since the range of $y = 2^x$ is the set of positive real numbers.

In general, we define the logarithmic function with base b to be the inverse of the exponential function with base b $(b > 0, b \neq 1)$. Symbolically,

---

Definition of the Logarithmic Function

For $b > 0$ and $b \neq 1$

$y = \log_b x$ is equivalent to $x = b^y$

(The log to the base b of x is the power y to which b must be raised to obtain x.)

$y = \log_3 x$ is equivalent to $x = 3^y$
$y = \log_{10} x$ is equivalent to $x = 10^y$

It is very important to remember that $y = \log_b x$ and $x = b^y$ **define the same function,** and as such can be used interchangeably.

Since the domain of an exponential function includes all real numbers and its range is the set of positive real numbers, the **domain of a logarithmic function** is the set of all positive real numbers and its **range** is the set of all real numbers. Thus, $\log_{10} 3$ is defined, but $\log_{10} 0$ and $\log_{10} (-5)$ are not defined (3 is a logarithmic domain value, but 0 and -5 are not). Typical logarithmic curves are shown in Figure 4.

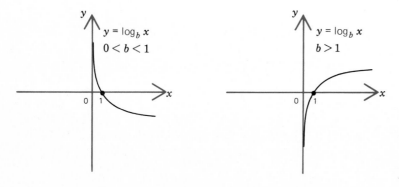

Figure 4
Typical Logarithmic Graphs

••From logarithmic to exponential form and vice versa

Let us now look into the matter of converting logarithmic forms into equivalent exponential forms and vice versa.

EXAMPLE 3
From logarithmic form to exponential form:

(A) $\log_2 8 = 3$ is equivalent to $8 = 2^3$
(B) $\log_{25} 5 = \frac{1}{2}$ is equivalent to $5 = 25^{1/2}$
(C) $\log_2 \frac{1}{4} = -2$ is equivalent to $\frac{1}{4} = 2^{-2}$

Work Problem 3 and check solution in Solutions to Matched Problems on page 563.

PROBLEM 3   Change to an equivalent exponential form

(A) $\log_3 27 = 3$ is equivalent to

(B) $\log_{36} 6 = \frac{1}{2}$ is equivalent to

(C) $\log_3 \frac{1}{9} = -2$ is equivalent to

EXAMPLE 4
From exponential form to logarithmic form:

(A) $49 = 7^2$ is equivalent to $\log_7 49 = 2$
(B) $3 = \sqrt{9}$ is equivalent to $\log_9 3 = \frac{1}{2}$
(C) $\frac{1}{5} = 5^{-1}$ is equivalent to $\log_5 \frac{1}{5} = -1$

Work Problem 4 and check solution in Solutions to Matched Problems on page 563.

PROBLEM 4   Change to an equivalent logarithmic form.

(A) $64 = 4^3$ is equivalent to

(B) $2 = \sqrt[3]{8}$ is equivalent to

(C) $\frac{1}{16} = 4^{-2}$ is equivalent to

•••Finding x, b, or y in y = log$_b$ x

To gain a deeper understanding of logarithmic functions and their relationship to the exponential functions we will look at a few problems where one is to find one of x, b, or y in $y = \log_b x$, given the other two values. All values are chosen so that the problems can be solved without tables or a calculator.

EXAMPLE 5
Find x, b, or y as indicated.

(A) Find y: $y = \log_4 8$

Solution
Write in equivalent exponential form

$8 = 4^y$ Write each member to the same base 2.

$2^3 = 2^{2y}$

Thus, since $b^m = b^n$ if and only if $m = n$,

$2y = 3$ and $y = \frac{3}{2}$

(B) Find x: $\log_3 x = -2$

Solution
Write in equivalent exponential form

$x = 3^{-2}$

Thus,

$x = \frac{1}{9}$

(C)   Find b: $\log_b 1{,}000 = 3$

Solution
Write in equivalent exponential form

$1{,}000 = b^3$ Write the left member as a third power.

$10^3 = b^3$

Thus,

$b = 10$

Work Problem 5 and check solution in Solutions to Matched Problems on page 563.

PROBLEM 5   Find x, b, or y as indicated.

   (A)   Find y: $y = \log_9 27$ **(B)**   Find x: $\log_2 x = -3$

   (C)   Find b: $\log_b 100 = 2$

••••Inverse relationships

Recall from the preceding chapter that, if f and f^{-1} are both functions (that is, if f is one-to-one), then

$$f[f^{-1}(x)] = x \quad \text{and} \quad f^{-1}[f(x)] = x$$

Since the logarithmic and exponential functions are inverses of each other for the same base, it follows that

$$\boxed{\begin{aligned} \log_b b^x &= x \\ b^{\log_b x} &= x \qquad x > 0 \end{aligned}}$$

Let $f(x) = b^x$; then $f^{-1}(x) = \log_b x$.
$f[f^{-1}(x)] = x \qquad f^{-1}[f(x)] = x$
$f[\log_b x] = x \qquad f^{-1}[b^x] = x$
$b^{\log_b x} = x \qquad \log_b b^x = x$

Even though both of these properties are very useful and are encountered frequently in more advanced courses, we will be mainly concerned with the first property.

EXAMPLE 6

(A) $\log_{10} 10^5 = 5$

(B) $\log_{10} 0.01 = \log_{10} 10^{-2} = -2$

(C) $\log_2 8 = \log_2 2^3 = 3$

(D) $\log_4 1 = \log_4 4^0 = 0$

(E) $10^{\log_{10} 7} = 7$

Work Problem 6 and check solution in Solutions to Matched Problems on page 563.

PROBLEM 6 Find each of the following:

(A) $\log_{10} 10^{-5} =$ (B) $\log_5 25 =$

(C) $\log_{10} 100 =$ (D) $\log_8 1 =$

(E) $2^{\log_2 9} =$

SOLUTIONS TO MATCHED PROBLEMS

3. (A) $\log_3 27$ is equivalent to $27 = 3^3$ (B) $\log_{36} 6 = \frac{1}{2}$ is equivalent to $6 = 36^{1/2}$
   (C) $\log_3 \frac{1}{9} = -2$ is equivalent to $\frac{1}{9} = 3^{-2}$

4. (A) $64 = 4^3$ is equivalent to $\log_4 64 = 3$ (B) $2 = \sqrt[3]{8}$ is equivalent to $\log_8 2 = \frac{1}{3}$
   (C) $\frac{1}{16} = 4^{-2}$ is equivalent to $\log_4 \frac{1}{16} = -2$

5. (A) $y = \log_9 27$ (B) $\log_2 x = -3$ (C) $\log_b 100 = 2$
   $27 = 9^y$ $x = 2^{-3}$ $100 = b^2$
   $3^3 = 3^{2y}$ $x = \dfrac{1}{2^3} = \dfrac{1}{8}$ $10^2 = b^2$
   $2y = 3$ $b = 10$
   $y = \frac{3}{2}$

6. (A) $\log_{10} 10^{-5} = -5$ (B) $\log_5 25 = \log_5 5^2 = 2$ (C) $\log_{10} 100 = \log_{10} 10^2 = 2$
   (D) $\log_8 1 = \log_8 8^0 = 0$ (E) $2^{\log_2 9} = 9$

PRACTICE EXERCISE 11.2

Work odd-numbered problems first, check answers, and then work even-numbered problems in areas of weakness. Answers to all problems are in the back of the book. Make every effort to work a problem yourself before you look at an answer.

A *Rewrite in exponential form.*

1. $\log_3 9 = 2$ _____   2. $\log_2 4 = 2$ _____

3. $\log_3 81 = 4$ _____   4. $\log_5 125 = 3$ _____

5. $\log_{10} 1,000 = 3$ _____   6. $\log_{10} 100 = 2$ _____

7. $\log_e 1 = 0$ _____   8. $\log_8 1 = 0$ _____

Rewrite in logarithmic form.

9. $64 = 8^2$ _____

10. $25 = 5^2$ _____

11. $10,000 = 10^4$ _____

12. $1,000 = 10^3$ _____

13. $u = v^x$ _____

14. $a = b^c$ _____

15. $9 = 27^{2/3}$ _____

16. $8 = 4^{3/2}$ _____

Find each of the following:

17. $\log_{10} 10^5$ _____

18. $\log_5 5^3$ _____

19. $\log_2 2^{-4}$ _____

20. $\log_{10} 10^{-7}$ _____

21. $\log_6 36$ _____

22. $\log_3 9$ _____

23. $\log_{10} 1,000$ _____

24. $\log_{10} 0.001$

Find x, y, or b as indicated.

25. $\log_2 x = 2$ _____

26. $\log_3 x = 2$ _____

27. $\log_4 16 = y$ _____

28. $\log_8 64 = y$ _____

29. $\log_b 16 = 2$ _____

30. $\log_b 10^{-3} = -3$ _____

B *Rewrite in exponential form.*

31. $\log_{10} 0.001 = -3$ _____

32. $\log_{10} 0.01 = -2$ _____

33. $\log_{81} 3 = \frac{1}{4}$ _____

34. $\log_4 2 = \frac{1}{2}$ _____

35. $\log_{1/2} 16 = -4$ _____

36. $\log_{1/3} 27 = -3$ _____

37. $\log_a N = e$ _____

38. $\log_k u = v$ _____

Rewrite in logarithmic form.

39. $0.01 = 10^{-2}$ _____

40. $0.001 = 10^{-3}$ _____

41. $1 = e^0$ _____

42. $1 = (\frac{1}{2})^0$ _____

43. $\frac{1}{8} = 2^{-3}$ _____

44. $\frac{1}{8} = (\frac{1}{2})^3$ _____

45. $\frac{1}{3} = 81^{-1/4}$ _____

46. $\frac{1}{2} = 32^{-1/5}$ _____

47. $7 = \sqrt{49}$ _____

48. $11 = \sqrt{121}$ _____

Find each of the following:

49. $\log_b b^u$ _____

50. $\log_b b^{uv}$ _____

51. $\log_e e^{1/2}$ _____   **52.** $\log_e e^{-3}$ _____

53. $\log_2 \sqrt{8}$ _____   **54.** $\log_5 \sqrt[3]{5}$ _____

55. $\log_{23} 1$ _____   **56.** $\log_{17} 1$ _____

57. $\log_4 8$ _____   **58.** $\log_4 \frac{1}{4}$ _____

Find x, y, or b as indicated.

59. $\log_4 x = \frac{1}{2}$ _____   **60.** $\log_{25} x = \frac{1}{2}$ _____

61. $\log_{1/3} 9 = y$ _____   **62.** $\log_{49} \frac{1}{7} = y$ _____

63. $\log_b 1{,}000 = \frac{3}{2}$ _____   **64.** $\log_b 4 = \frac{2}{3}$ _____

C 65. $\log_b 1 = 0$ _____   **66.** $\log_b b = 1$ _____

67. $2^{\log_2 3} = ?$ _____   **68.** $5^{\log_5 8} = ?$ _____

69. Given $f: y = 10^x$
   (A) Graph f and f^{-1}, using the same coordinate axes.

   (B) Discuss the domain and range of f and f^{-1}. _____

   (C) What other name could you use for the inverse of f? _____

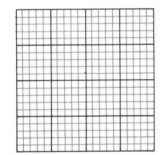

The Check Exercise for this section is on page 591.

11.3 PROPERTIES OF LOGARITHMIC FUNCTIONS

•Basic logarithmic properties
••Use of the logarithmic properties

•Basic logarithmic properties

Logarithmic functions have several very useful properties that follow directly from the fact that they are inverses of exponential functions. These properties will enable us to convert multiplication problems into addition problems, division problems into subtraction problems, and power and root problems into multiplication problems. In addition, we will be able to solve exponential equations such as $2 = 10^x$.

THEOREM 1

Properties of Logarithmic Functions

If b, M, and N are positive·real numbers, $b \neq 1$, and p is a real number, then

1. $\log_b b^u = u$
2. $\log_b MN = \log_b M + \log_b N$
3. $\log_b \dfrac{M}{N} = \log_b M - \log_b N$
4. $\log_b M^p = p \log_b M$
5. $\log_b 1 = 0$

The first property follows directly from the definition of a logarithmic function. The proof of the second property is based on the laws of exponents. To bring exponents into the proof, we let

$$u = \log_b M \quad \text{and} \quad v = \log_b N$$

and convert these to the equivalent exponential forms

$$M = b^u \quad \text{and} \quad N = b^v$$

Now, see if you can provide the reasons for each of the following steps:

$$\log_b MN = \log_b b^u b^v = \log_b b^{u+v} = u + v = \log_b M + \log_b N$$

The other properties are established in a similar manner.

••Use of the logarithmic properties

We will now see how logarithmic properties can be used to convert multiplication problems into addition problems, division problems into subtraction problems, and power and root problems into multiplication problems.

EXAMPLE 7

(A) $\log_{10} 10^5 = 5$ $\log_b b^u = u$

(B) $\log_b 3x = \log_b 3 + \log_b x$ $\log_b MN = \log_b M + \log_b N$

(C) $\log_b \dfrac{x}{5} = \log_b x - \log_b 5$ $\log_b \dfrac{M}{N} = \log_b M - \log_b N$

(D) $\log_b x^7 = 7 \log_b x$ $\log_b M^p = p \log_b M$

(E) $\log_b \dfrac{mn}{pq} = \log_b mn - \log_b pq$ $\log_b \dfrac{M}{N} = \log_b M - \log_b N$

$\qquad\qquad = \log_b m + \log_b n - (\log_b p + \log_b q)$ $\log_b MN = \log_b M + \log_b N$

$\qquad\qquad = \log_b m + \log_b n - \log_b p - \log_b q$

(F) $\log_b (mn)^{2/3} = \tfrac{2}{3} \log_b mn$ $\log_b M^p = p \log_b M$

$\qquad\qquad\quad = \tfrac{2}{3}(\log_b m + \log_b n)$ $\log_b MN = \log_b M + \log_b N$

(G) $\log_b \dfrac{x^8}{y^{1/5}} = \log_b x^8 - \log_b y^{1/5}$ $\log_b \dfrac{M}{N} = \log_b M - \log_g N$

$\qquad\qquad\quad = 8 \log_b x - \tfrac{1}{5} \log_b y$ $\log_b M^p = p \log_b M$

Work Problem 7 and check solution in Solutions to Matched Problems on page 569.

PROBLEM 7   Write in terms of simpler logarithmic forms, as in Example 7.

(A) $\log_b \dfrac{r}{uv} =$

(B) $\log_b \left(\dfrac{m}{n}\right)^{3/5} =$

(C) $\log_b \dfrac{u^{1/3}}{v^5} =$

EXAMPLE 8

If $\log_e 3 = 1.10$ and $\log_e 7 = 1.95$, find

(A) $\log_e \tfrac{7}{3}$

Solution

$\log_e \tfrac{7}{3} = \log_e 7 - \log_e 3 = 1.95 - 1.10 = 0.85$

(B) $\log_e \sqrt[3]{21}$

Solution

$\log_e \sqrt[3]{21} = \log_e (21)^{1/3} = \frac{1}{3} \log_e (3 \cdot 7) = \frac{1}{3}(\log_e 3 + \log_e 7)$

$\qquad = \frac{1}{3}(1.10 + 1.95) = 1.02$

Work Problem 8 and check solution in Solutions to Matched Problems on page 569.

PROBLEM 8   If $\log_e 5 = 1.609$ and $\log_e 8 = 2.079$, find:

(A) $\log_e \dfrac{5^{10}}{8} =$

(B) $\log_e \sqrt[4]{\dfrac{8}{5}} =$

Finally, we state without proof that for n and m any positive real numbers

$$\log_b m = \log_b n \qquad \text{if and only if} \qquad m = n$$

Thus, if $\log_{10} x = \log_{10} 32.15$, then $x = 32.15$.

The following example and problem, though somewhat artificial will give you additional practice in using the properties in Theorem 1.

EXAMPLE 9

Find x so that $\log_b x = \frac{2}{3} \log_b 27 + 2 \log_b 2 - \log_b 3$. Do not use a table or a calculator.

Solution

$\log_b x = \frac{2}{3} \log_b 27 + 2 \log_b 2 - \log_b 3$ Express right side in terms of a single log.

$\qquad = \log_b 27^{2/3} + \log_b 2^2 - \log_b 3$ Property 4.

$\qquad = \log_b 9 + \log_b 4 - \log_b 3$ $27^{2/3} = 9$, $2^2 = 4$

$\qquad = \log_b \dfrac{9 \cdot 4}{3} = \log_b 12$ Properties 2 and 3.

Thus,

$\log_b x = \log_b 12$

Hence,

$x = 12$

Work Problem 9 and check solution in Solutions to Matched Problems on page 569.

PROBLEM 9   Find x so that $\log_b x = \frac{2}{3} \log_b 8 + \frac{1}{2} \log_b 9 - \log_b 6$.

Solution

SOLUTIONS TO MATCHED PROBLEMS

7. **(A)** $\log_b \dfrac{r}{uv} = \log_b r - \log_b (uv)$

$\qquad\qquad = \log_b r - (\log_b u + \log_b v)$
$\qquad\qquad = \log_b r - \log_b u - \log_b v$

(B) $\log_b \left(\dfrac{m}{n}\right)^{3/5} = \dfrac{3}{5}\log_b \dfrac{m}{n}$

$\qquad\qquad\qquad = \dfrac{3}{5}(\log_b m - \log_b n)$

(C) $\log_b \dfrac{u^{1/3}}{v^5} = \log_b u^{1/3} - \log_b v^5$

$\qquad\qquad = \tfrac{1}{3}\log_b u - 5\log_b v$

8. **(A)** $\log_e \dfrac{5^{10}}{8} = \log_e 5^{10} - \log_e 8$

$\qquad\qquad = 10\log_e 5 - \log_e 8$
$\qquad\qquad = 10(1.609) - (2.079)$
$\qquad\qquad = 14.01 \qquad$ (to four significant figures)

(B) $\log_e \sqrt[4]{\dfrac{8}{5}} = \log_e \left(\dfrac{8}{5}\right)^{1/4}$

$\qquad\qquad = \tfrac{1}{4}\log_e \tfrac{8}{5}$
$\qquad\qquad = \tfrac{1}{4}(\log_e 8 - \log_e 5)$
$\qquad\qquad = \tfrac{1}{4}(2.079 - 1.609)$
$\qquad\qquad = 0.1175 \qquad$ (to four significant figures)

9. $\log_b x = \tfrac{2}{3}\log_b 8 + \tfrac{1}{2}\log_b 9 - \log_b 6 = \log_b 8^{2/3} + \log_b 9^{1/2} - \log_b 6$

$\qquad = \log_b 4 + \log_b 3 - \log_b 6 = \log_b \dfrac{4 \cdot 3}{6} = \log_b 2$

$\log_b x = \log_b 2$
$\qquad x = 2$

PRACTICE EXERCISE 11.3

Work odd-numbered problems first, check answers, and then work even-numbered problems in areas of weakness. Answers to all problems are in the back of the book. Make every effort to work a problem yourself before you look at an answer.

A *Write in terms of simpler logarithmic forms (going as far as you can with logarithmic properties—see Example 7).*

1. $\log_b uv$ _____

2. $\log_b rt$ _____

3. $\log_b \dfrac{A}{B}$ _____

4. $\log_b \dfrac{p}{q}$ _____

5. $\log_b u^5$ _____

6. $\log_b w^{25}$ _____

7. $\log_b N^{3/5}$ _____

8. $\log_b u^{-2/3}$ _____

9. $\log_b \sqrt{Q}$ _____

10. $\log_b \sqrt[5]{M}$ _____

11. $\log_b uvw$ _____

12. $\log_b \dfrac{u}{vw}$ _____

Write each expression in terms of a single logarithm with a coefficient of 1. Example: $\log_b u^2 - \log_b v = \log_b (u^2/v)$.

13. $\log_b A + \log_b B$ _____

14. $\log_b P + \log_b Q + \log_b R$ _____

15. $\log_b X - \log_b Y$ _____

16. $\log_b x^2 - \log_b y^3$ _____

17. $\log_b w + \log_b x - \log_b y$ _____

18. $\log_b w - \log_b x - \log_b y$ _____

If $\log_e 2 = 0.69$, $\log_e 3 = 1.10$, and $\log_e 5 = 1.61$, find the logarithm to the base e of each of the following numbers.

19. $\log_e 30$ _____

20. $\log_e 6$ _____

21. $\log_e \frac{2}{5}$ _____

22. $\log_e \frac{5}{3}$ _____

23. $\log_e 27$ _____

24. $\log_e 16$ _____

B *Write in terms of simpler logarithmic forms (going as far as you can with logarithmic properties—see Example 7).*

25. $\log_b u^2 v^7$ _____

26. $\log_b u^{1/2} v^{1/3}$ _____

27. $\log_b \dfrac{1}{a}$ _____

28. $\log_b \dfrac{1}{M^3}$ _____

29. $\log_b \dfrac{\sqrt[3]{N}}{p^2 q^3}$ _____

30. $\log_b \dfrac{m^5 n^3}{\sqrt{p}}$ _____

31. $\log_b \sqrt[4]{\dfrac{x^2 y^3}{\sqrt{z}}}$ _____

32. $\log_b \sqrt[5]{\left(\dfrac{x}{y^4 z^9}\right)^3}$ _____

Write each expression in terms of a single logarithm with a coefficient of 1.

33. $2 \log_b x - \log_b y$ _____

34. $\log_b m - \frac{1}{2} \log_b n$ _____

35. $3 \log_b x + 2 \log_b y - 4 \log_b z$ _____

36. $\frac{1}{3} \log_b w - 3 \log_b x - 5 \log_b y$ _____

37. $\frac{1}{5}(2 \log_b x + 3 \log_b y)$ _____

38. $\frac{1}{3}(\log_b x - \log_b y)$ _____

If $\log_b 2 = 0.69$, $\log_b 3 = 1.10$, and $\log_b 5 = 1.61$, find the logarithm to the base b of the following numbers.

39. $\log_b 7.5$ _____

40. $\log_b 1.5$ _____

41. $\log_b \sqrt[3]{2}$ _____

42. $\log_b \sqrt{3}$ _____

43. $\log_b \sqrt{0.9}$ _____

44. $\log_b \sqrt{\frac{3}{2}}$ _____

C 45. Find x so that $\frac{3}{2} \log_b 4 - \frac{2}{3} \log_b 8 + 2 \log_b 2 = \log_b x$. _____

46. Find x so that $3 \log_b 2 + \frac{1}{2} \log_b 25 - \log_b 20 = \log_b x$. _____

The Check Exercise for this section is on page 593.

11.4 COMMON AND NATURAL LOGARITHMS

•**Common logarithms (base 10)**
••**Natural logarithms (base e)**

John Napier (1550–1617) is credited with the invention of logarithms, which grew out of a desire to reduce tedious computations involved in astronomical research. This new computational tool was immediately accepted by the scientific world. Now, with the availability of powerful, inexpensive hand calculators, logarithms have lost most of their importance as a computational device. However, the logarithmic concept has been greatly generalized since its introduction, and logarithmic functions are now used widely in both science and mathematics. For example, even with a very good scientific hand calculator, we still need logarithmic functions to solve the simple-looking exponential equation from population growth studies and mathematics of finance: $2 = 1.08^x$.

Of all possible logarithmic bases, the bases e and 10 are used most widely. Finding base 10 and base e logarithms for *any* positive number is the subject matter of this section.

•Common logarithms (base 10)

Logarithms with base 10 are called **common logarithms.** To find the common logarithm of any power of 10 is easy because of the logarithmic property:

$$\log_b b^x = x$$

Thus,

$$\log_{10} 10^k = k \qquad \log_{10} 10^{-5} = -5 \qquad \log_{10} 10^3 = 3$$

But how do we find common logarithms of numbers such as 33,800 or 0.003 51? If you have a hand calculator with a common log button (LOG), then you simply enter either number and push LOG. If you do not have such a calculator but do have a table such as Table 3 in Appendix D, then you proceed as follows: Recalling that any decimal fraction can be written in scientific notation (see Section 6.2), we see that

$$\log_{10} 33{,}800 = \log_{10} (3.38 \times 10^4)$$
$$= \log_{10} 3.38 + \log_{10} 10^4$$
$$= (\log_{10} 3.38) + 4$$

and that

$$\log_{10} 0.003\,51 = \log_{10} (3.51 \times 10^{-3})$$
$$= \log_{10} 3.51 + \log_{10} 10^{-3}$$
$$= (\log_{10} 3.51) - 3$$

Thus, if common logarithms of numbers r, $1 \leq r < 10$, are known, then we can find the common logarithm of any positive decimal numeral.

Using methods of advanced mathematics, a table of common logarithms of numbers from

1 to 10 can be computed to any decimal accuracy desired. Table 3 in Appendix D is such a table to four-decimal-place accuracy.

> NOTE: If x is between 1 and 10, then $\log_{10} x$ is between 0 and 1. (See Figure 5).

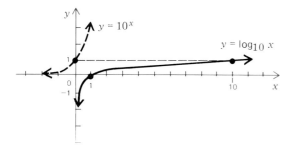

Figure 5

To illustrate the use of Table 3, a small portion of it is reproduced in Figure 6. To find $\log_{10} 3.47$, for example, we first locate 3.4 under the x heading; then we move across to the column headed 7, where we find 0.5403. Thus $\log_{10} 3.47 = 0.5403$.

Now let us finish finding the common logarithms of 33,800 and 0.003 51.

Figure 6

x	0	1	2	3	4	5	6	7	8	9
3.2	0.5051	0.5065	0.5079	0.5092	0.5105	0.5119	0.5132	0.5145	0.5159	0.5172
3.3	0.5185	0.5198	0.5211	0.5224	0.5237	0.5250	0.5263	0.5276	0.5289	0.5302
3.4	0.5315	0.5328	0.5340	0.5353	0.5366	0.5378	0.5391	0.5403	0.5416	0.5428
3.5	0.5441	0.5453	0.5465	0.5478	0.5490	0.5502	0.5514	0.5527	0.5539	0.5551

$\log_{10} 3.47 = 0.5403$

NOTE: Throughout the rest of this chapter we will use $=$ in place of $\approx$ in many places, realizing values are only approximately equal. Occasionally, we will use $\approx$ when a special emphasis is desired.

EXAMPLE 10

(A) $\log_{10} 33{,}800 = \log_{10} (3.38 \times 10^4)$

$= \log_{10} 3.38 + \log_{10} 10^4$

$= 0.5289 + 4$ Use Figure 6.

$= 4.5289$

(B) $\log_{10} 0.003\ 51 = \log_{10} (3.51 \times 10^{-3})$

$= \log_{10} 3.51 + \log_{10} 10^{-3}$

$= 0.5453 - 3$ Use Figure 6.

$= -2.4547$

Work Problem 10 and check solution in Solutions to Matched Problems on page 576.

PROBLEM 10   Use the table in Figure 6 to find each logarithm.

       (A) $\log_{10} 328{,}000 =$

       (B) $\log_{10} 0.000\,342 =$

We see that the common logarithm of a positive decimal fraction has two main parts: a non-negative decimal fraction between 0 and 1, called the **mantissa,** and an integer part, called the **characteristic.** (See the second line in each dotted box in Example 10). In general,

> If a number N is written in scientific notation
>
> $$N = r \times 10^{k} \qquad 1 \le r < 10, \, k \text{ an integer}$$
>
> then
>
> $$\log_{10} N = \log_{10} (r \times 10^{k})$$
> $$= \log_{10} r + \log_{10} 10^{k}$$
>
> $$= (\underset{\text{mantissa}}{\log_{10} r}) + \underset{\text{characteristic}}{k}$$

Now let us reverse the problem; that is, given the log of a number, find the number. To find the number, we first write the log of the number in the form

$$m + c$$

where m (the mantissa) is a nonnegative number between 0 and 1, and c (the characteristic) is an integer; then reverse the process illustrated in Example 10.

EXAMPLE 11
(A)   If $\log_{10} x = 2.5224$, find x.

Solution

$$\log_{10} x = 2.5224$$ Write 2.5224 in the form $m + c$, $0 \le m < 1$ and c an integer.

$$= 0.5224 + 2$$ Look for 0.5224 in the body of Figure 6 or Table 3. Thus, we see that $0.5224 = \log_{10} 3.33$.

$$= \log_{10} 3.33 + \log_{10} 10^{2}$$

$$= \log_{10} (3.33 \times 10^{2})$$

Thus,

$$x = 3.33 \times 10^{2} \qquad \text{or} \qquad 333$$

(B)   If $\log_{10} x = 0.5172 - 4$, find x.

Solution

$\log_{10} x = 0.5172 - 4$ Use Table 3 or Figure 6.

$\phantom{\log_{10} x} = \log_{10} 3.29 + \log_{10} 10^{-4}$

$\phantom{\log_{10} x} = \log_{10} (3.29 \times 10^{-4})$

Thus,

$x = 3.29 \times 10^{-4}$ or $0.000\,329$

(C)   If $\log_{10} x = -4.4685$, find x.

Solution

$\log_{10} x = -4.4685$ Convert -4.4685 to $m + c$ form with $0 \le m < 1$ by adding and subtracting 5:

$\phantom{\log_{10} x} = 0.5315 - 5$

$\phantom{\log_{10} x} = \log_{10} 3.40 + \log_{10} 10^{-5}$ $\begin{aligned} 5.0000 &- 5 \\ -4.4685 & \\ \hline 0.5315 &- 5 \end{aligned}$

$\phantom{\log_{10} x} = \log_{10} (3.40 \times 10^{-5})$

Thus,

$x = 3.40 \times 10^{-5}$ or $0.000\,034\,0$

Work Problem 11 and check solution in Solutions to Matched Problems on page 576.

PROBLEM 11   Use Figure 6 or Table 3 to find x if:

       (A) $\log_{10} x = 5.5378$

       (B) $\log_{10} x = 0.5289 - 3$

       (C) $\log_{10} x = -2.4921$

What if a number does not appear in a table? We then use the nearest table value, or, if more accuracy is desired, we use a hand calculator. (Interpolation is discussed in Appendix A for those desiring to use this technique to improve accuracy in table use.)

EXAMPLE 12

(A) $\log_{10} 32{,}683 \approx \log_{10} (3.27 \times 10^4)$ Round 3.2683 to nearest table value, that is, to 3.27.

$$= \log_{10} 3.27 + \log_{10} 10^4$$

$$= 0.5145 + 4$$

$$= 4.5145$$

(B) To find x if $\log_{10} x = 0.5241 - 3$, we observe (in Figure 6) that 0.5241 is between 0.5237 and 0.5250, but is closer to 0.5237. Thus, we write

$\log_{10} x = 0.5241 - 3$ Since 0.5241 is not in the body of the table, select value that is closest, that is, 0.5237.

$$\approx 0.5237 - 3$$

$$= \log_{10} 3.34 + \log_{10} 10^{-3}$$

$$= \log_{10} (3.34 \times 10^{-3})$$

Hence,

$$x \approx 3.34 \times 10^{-3} = 0.003\ 34$$

Work Problem 12 and check solution in Solutions to Matched Problems on page 577.

PROBLEM 12   Find:

(A) $\log_{10} 0.034\ 319 =$

(B) x if $\log_{10} x = 6.5473$

(C) x if $\log_{10} x = -4.4942$

••Natural logarithms (base e)

Logarithms of numbers to the base e are called **natural logarithms;** these are used almost exclusively from calculus onward. The notation "$\log_e$" is often shortened to "ln"; that is,

$$\ln x = \log_e x$$

The "ln" symbol is also used in many hand calculators to represent natural logarithms. (Recall "log" is often used to represent common logarithms; that is, base 10 logarithms.)

If you do not have a hand calculator with an LN button, then you can use a natural logarithmic table (Table 4 in Appendix D) much in the same way we use a common logarithmic table (Table 3), but the arithmetic is often a little more involved.

EXAMPLE 13

(A) $\ln 52{,}400 = \ln (5.24 \times 10^4)$

$= \ln 5.24 + 4 \ln 10$

[ln 5.24 is read out of the main table (Table 4), and 4 ln 10 is obtained from the list at the top of the table.]

$= 1.6563 + 9.2103$

$= 10.8666$

(B) $\ln 0.00278 = \ln (2.78 \times 10^{-3})$

$= \ln 2.78 - 3 \ln 10$

$= 1.0225 - 6.9078$

$= -5.8853$

Work Problem 13 and check solution in Solutions to Matched Problems on page 577.

PROBLEM 13   Use Table 4 to find:

(A) $\ln 0.000\ 683 =$

(B) $\ln 328{,}000 =$

(C) $\ln 23{,}582 =$

SOLUTIONS TO MATCHED PROBLEMS _____

10. **(A)** $\log_{10} 328{,}000 = \log_{10} (3.28 \times 10^5)$

$= \log_{10} 3.28 + \log_{10} 10^5$

$= 0.5159 + 5 = 5.5159$

(B) $\log_{10} 0.000\ 342 = \log_{10} (3.42 \times 10^{-4})$

$= \log_{10} 3.42 + \log_{10} 10^{-4}$

$= 0.5340 - 4 = -3.4660$

11. **(A)** $\log_{10} x = 5.5378$

$= 0.5378 + 5$

$= \log_{10} 3.45 + \log_{10} 10^5$

$= \log_{10} (3.45 \times 10^5)$

$x = 3.45 \times 10^5 \text{ or } 345{,}000$

(B) $\log_{10} x = 0.5289 - 3$

$= \log_{10} 3.38 + \log_{10} 10^{-3}$

$= \log_{10} (3.38 \times 10^{-3})$

$x = 3.38 \times 10^{-3} \text{ or } 0.003\ 38$

(C) $\log_{10} x = -2.4921$ Add and subtract 3: $\begin{array}{r} 3.0000 - 3 \\ -2.4921 \\ \hline 0.5079 - 3 \end{array}$

$\quad\quad\quad = 0.5079 - 3$

$\quad\quad\quad\quad = \log_{10} 3.22 + \log_{10} 10^{-3}$
$\quad\quad\quad\quad = \log_{10} (3.22 \times 10^{-3})$

$\quad\quad\quad x = 3.22 \times 10^{-3}$ or $0.003\,22$

12. (A) $\log_{10} 0.034\,319 \approx \log_{10} (3.43 \times 10^{-2})$ Nearest table value

$\quad\quad\quad = \log_{10} 3.43 + \log_{10} 10^{-2}$

$\quad\quad\quad = 0.5353 - 2 = 1.4647$

(B) $\log_{10} x = 6.5473$

$\quad\quad\quad = 0.5473 + 6$

$\quad\quad\quad \approx 0.5478 + 6$ Nearest table value
$\quad\quad\quad = \log_{10} 3.53 + \log_{10} 10^6$
$\quad\quad\quad = \log_{10} (3.53 \times 10^6)$

$\quad\quad\quad x = 3.53 \times 10^6$ or $3{,}530{,}000$

$\begin{array}{r} 5.0000 - 5 \end{array}$ Add and subtract 5.

(C) $\log_{10} x = \underline{-4.4942}$

$\quad\quad\quad = \quad 0.5058 - 5$

$\quad\quad\quad \approx 0.5051 - 5$ Nearest table value
$\quad\quad\quad = \log_{10} 3.20 + \log_{10} 10^{-5}$
$\quad\quad\quad = \log_{10} (3.20 \times 10^{-5})$

$\quad\quad\quad x = 3.20 \times 10^{-5}$ or $0.000\,032\,0$

13. (A) $\ln 0.000\,683 = \ln (6.83 \times 10^{-4})$

$\quad\quad\quad = \ln 6.83 - 4 \ln 10$

$\quad\quad\quad = 1.9213 - 9.2103$

$\quad\quad\quad = -7.2890$

(B) $\ln 328{,}000 = \ln (3.28 \times 10^5)$

$\quad\quad\quad = \ln 3.28 + 5 \ln 10$

$\quad\quad\quad = 1.1878 + 11.5130$

$\quad\quad\quad = 12.7008$

(C) $\ln 23{,}582 \approx \ln (2.36 \times 10^4)$ Nearest table value

$\quad\quad\quad = \ln 2.36 + 4 \ln 10$

$\quad\quad\quad = 0.8587 + 9.2103$

$\quad\quad\quad = 10.0690$

PRACTICE EXERCISE 11.4

Work odd-numbered problems first, check answers, and then work even-numbered problems in areas of weakness. Answers to all problems are in the back of the book. Make every effort to work a problem yourself before you look at an answer.

NOTE: *Optional exercises are included in which a calculator may be used.*

A *Use Table 3 in Appendix D to find the following:*

1. $\log_{10} 6.37$ _____

2. $\log_{10} 7.29$ _____

3. $\log_{10} 327$ _____

4. $\log_{10} 2{,}120$ _____

5. $\log_{10} 0.0218$ _____

6. $\log_{10} 0.003\,72$ _____

Find x, using nearest table values in Table 3.

7. $\log_{10} x = 0.3712$ _____

8. $\log_{10} x = 0.8174$ _____

9. $\log_{10} x = 4.5023$ _____

10. $\log_{10} x = 7.1032$ _____

11. $\log_{10} x = 0.3177 - 4$ _____

12. $\log_{10} x = 0.7819 - 5$ _____

B *Use nearest values in Table 3 to approximate each of the following.*

13. $\log_{10} 82{,}734$ _____

14. $\log_{10} 304{,}918$ _____

15. $\log_{10} 0.061\,94$ _____

16. $\log_{10} 0.004\,769$ _____

Find x, using nearest values in Table 3.

17. $\log_{10} x = 9.113\,64$ _____

18. $\log_{10} x = 7.437\,15$ _____

19. $\log_{10} x = -3.4128$ _____

20. $\log_{10} x = -4.8013$ _____

(Optional) Use a calculator to find each of the following to four decimal places.

21. $\log_{10} 82{,}734$ _____

22. $\log_{10} 304{,}918$ _____

23. $\log_{10} 0.061\,94$ _____

24. $\log_{10} 0.004\,769$ _____

(Optional) Switch to exponential form and use a calculator to find x to four significant figures. (For example, to find x in $\log_{10} x = 3.1723$, we write $x = 10^{3.1723}$ and use a calculator that can compute powers to arrive at $x = 1{,}487$ to four significant figures.)

25. $\log_{10} x = 9.11364$ _____

26. $\log_{10} x = 5.371922$ _____

27. $\log_{10} x = -2.0138$ _____

28. $\log_{10} x = -1.3884$ _____

Find each of the following: (A) Using nearest values in Table 4; (B) (Optional) using a calculator and rounding to four decimal places.

29. $\ln 2.35$ _____

30. $\ln 7.02$ _____

31. $\ln 5{,}233$ _____

32. $\ln 603{,}517$ _____

33. $\ln 0.071\,33$ _____

34. $\ln 0.003\,168\,7$ _____

C *(Optional) Swtich to exponential form and find x, using a calculator. Round answers to four significant figures.*

35. $\ln x = 3.103$ _____

36. $\ln x = 5.2174$ _____

37. $\ln x = -2.521$ _____

38. $\ln x = -3.0118$ _____

The Check Exercise for this section is on page 595.

11.5 EXPONENTIAL AND LOGARITHMIC EQUATIONS

Often, when dealing with exponential functions, one must solve exponential equations in which a variable appears in an exponent. Logarithmic functions and their properties play an important role in the solution of such equations.

EXAMPLE 14

Solve $2^{3x-2} = 5$ for x.

Solution

$$2^{3x-2} = 5 \qquad \text{How can we get } x \text{ out of the exponent?}$$

$$\log_{10} 2^{3x-2} = \log_{10} 5 \qquad \text{Use logs! If two positive quantities are equal, their logs are equal.}$$

$$(3x - 2) \log_{10} 2 = \log_{10} 5 \qquad \log_b N^p = p \log_b N$$

$$3x - 2 = \frac{\log_{10} 5}{\log_{10} 2} \qquad \frac{\log_{10} 5}{\log_{10} 2} \neq \log_{10} 5 - \log_{10} 2$$

$$x = \frac{1}{3}\left(2 + \frac{\log_{10} 5}{\log_{10} 2}\right)$$

$$\approx \frac{1}{3}\left(2 + \frac{0.6990}{0.3010}\right)$$

$$= 1.4408 \qquad \text{To four decimal places.}$$

Work Problem 14 and check solution in Solutions to Matched Problems on page 583.

PROBLEM 14 Solve $35^{1-2x} = 7$ for x.

Solution

The next example illustrates an approach to solving some types of logarithmic equations.

EXAMPLE 15

Solve $\log_{10}(x + 3) + \log_{10} x = 1$ exactly without use of a calculator or table.

Solution

$\log_{10} (x + 3) + \log_{10} x = 1$

$\log_{10} [x(x + 3)] = 1$ Change to equivalent exponential form.

$x(x + 3) = 10^1$ Write in $ax^2 + bx + c$ form.

$x^2 + 3x - 10 = 0$ Solve (in this case by factoring).

$(x + 5)(x - 2) = 0$

$x = -5, 2$ $\log_{10} (-5 + 3) = \log_{10} (-2)$ is not defined.

Thus 2 is the only solution.

REMEMBER: Answers must be checked in the original equation to see if any must be discarded.

Work Problem 15 and check solution in Solutions to Matched Problems on page 583.

PROBLEM 15   Solve $\log_{10} (x - 15) = 2 - \log_{10} x$ exactly without use of a calculator or table.

Solution

EXAMPLE 16

If a certain amount of money P (principal) is invested at $100r$ percent interest compounded annually, then the amount of money A in the account after n years, assuming no withdrawals, is given by

$A = P(1 + r)^n$

(A) How much will $1,000 be worth in 4 years at 6 percent compounded annually?
(B) How long will it take money to double if it is invested at 6 percent compounded annually?

Solution

(A) $A = 1,000(1 + 0.06)^4$ $100r = 6; r = 0.06; P = 1,000; n = 4$

$= 1,000(1.06)^4$

$\log_{10} A = \log_{10} 1,000 + 4 \log_{10} 1.06$ Use common logs.

$\approx 3 + 4(0.0253)$

$= 3.1012$ Now find A.

$= 0.1012 + 3$

$= \log_{10} 1.26 + \log_{10} 10^3$ Nearest table value.

$= \log_{10} (1.26 \times 10^3)$

Thus,

$A = 1.26 \times 10^3 = \$1,260$ To three significant figures.

(NOTE: This problem can be solved directly with some hand calculators without using logarithms. Try it if you have one. Use the y^x button.)

(B)   To find the doubling time, we find n in $A = P(1.06)^n$ where A is replaced with $2P$; that is, we solve $2P = P(1.06)^n$ for n.

$2P = P(1.06)^n$ Divide both sides by P.

$2 = 1.06^n$ Take the log of both sides.

$\log_{10} 2 = \log_{10} 1.06^n$

$\log_{10} 2 = n \log_{10} 1.06$ Note how log properties are used.

$n = \dfrac{\log_{10} 2}{\log_{10} 1.06}$ Note that $\dfrac{\log_{10} 2}{\log_{10} 1.06} \neq \log_{10} 2 - \log_{10} 1.06$.

$\approx \dfrac{0.3010}{0.0253}$

$= 11.9$ years

≈ 12 years   To nearest year.

It would take 12 years for the money to double, since interest is paid at the end of each year. (We note that properties of logarithms played an essential role in the solution of this problem. Even though a hand calculator is helpful in the latter part of the problem, it can't solve the problem directly at the beginning.)

Work Problem 16 and check solution in Solutions to Matched Problems on page 583.

PROBLEM 16   Repeat Example 16, changing the interest rate from 6 percent compounded annually to 9 percent compounded annually.

Solution   **(A)**   **(B)**

EXAMPLE 17

The atmospheric pressure P (in pounds per square inch) at x mi above sea level is given approximately by

$$P = 14.7e^{-0.21x}$$

At what height will the atmospheric pressure be half of sea-level pressure?

Solution

Sea-level pressure is the pressure at $x = 0$. Thus,

$$P = 14.7e^0 = 14.7$$

One-half of sea-level pressure is $14.7/2 = 7.35$. Now our problem is to find x so that $P = 7.35$; that is, we solve $7.35 = 14.7e^{-0.21x}$ for x.

$7.35 = 14.7e^{-0.21x}$	Divide both sides by 14.7 to simplify.
$0.5 = e^{-0.21x}$	Take natural logs of both sides.
$\ln 0.5 = \ln e^{-0.21x}$	Why use natural logs? Compare with common log to see why.
$\ln 0.5 = -0.21x$	
$x = \dfrac{\ln 0.5}{-0.21}$	Use Table 4 or a hand calculator.
$\approx \dfrac{-0.6931}{-0.21} = 3.3 \text{ mi}$	To two significant figures.

Work Problem 17 and check solution in Solutions to Matched Problems on page 583.

PROBLEM 17 Using the formula in Example 17, find the altitude in miles so that the atmospheric pressure will be one-eighth that at sea level.

Solution

SOLUTIONS TO MATCHED PROBLEMS

14.

$$35^{1-2x} = 7$$
$$\log_{10} 35^{1-2x} = \log_{10} 7$$
$$(1 - 2x)(\log_{10} 35) = \log_{10} 7$$
$$1 - 2x = \frac{\log_{10} 7}{\log_{10} 35}$$
$$-2x = \frac{\log_{10} 7}{\log_{10} 35} - 1$$
$$x = -\frac{1}{2}\left(\frac{\log_{10} 7}{\log_{10} 35} - 1\right)$$
$$\approx -\frac{1}{2}\left(\frac{0.8451}{1.5441} - 1\right)$$
$$= 0.2263$$

15.

$$\log_{10} (x - 15) = 2 - \log_{10} x$$
$$\log_{10} (x - 15) + \log_{10} x = 2$$
$$\log_{10} x(x - 15) = 2$$
$$x(x - 15) = 10^2$$
$$x^2 - 15x - 100 = 0$$
$$(x - 20)(x + 5) = 0$$
$$x = -5, 20$$

CHECK:

$\underline{x = -5:}$

Neither side of the equation is defined for $x = -5$.

$\underline{x = 20:}$

$$\log_x (20 - 15) \overset{?}{=} 2 - \log_{10} 20$$
$$0.6990 \overset{\checkmark}{=} 0.6990$$

Thus, $x = 20$ is the only solution.

16. **(A)**

$$A = P(1 + r)^n$$
$$A = 1,000(1.09)^4$$
$$\log_{10} A = \log_{10} 1,000 + 4 \log_{10} 1.09$$
$$= 3 + 4(0.0374)$$
$$= 3.1496$$
$$= 0.1496 + 3 \quad \text{\small Choose nearest table value.}$$
$$= \log_{10} 1.41 + \log_{10} 10^3$$
$$= \log_{10} (1.41 \times 10^3)$$
$$A = 1.41 \times 10^3 \quad \text{or} \quad \$1,410$$

(B)

$$A = P(1 + r)^n$$
$$2P = P(1.09)^n$$
$$2 = 1.09^n$$
$$\log_{10} 2 = \log_{10} 1.09^n$$
$$\log_{10} 2 = n \log_{10} 1.09$$
$$n = \frac{\log_{10} 2}{\log_{10} 1.09}$$
$$\approx \frac{0.3010}{0.0374} = 8.05$$

It will more than double in 9 years, but not quite double in 8 years.

17.

$$P = 14.7e^{-0.21x}$$
$$\frac{14.7}{8} = 14.7e^{-0.21x}$$
$$0.125 = e^{-0.21x}$$
$$\ln 0.125 = \ln e^{-0.21x}$$
$$\ln 0.125 = -0.21x$$
$$x = \frac{\ln 0.125}{-0.21} \approx \frac{-2.0794}{-0.21} = 9.9 \text{ mi}$$

PRACTICE EXERCISE 11.5

Work odd-numbered problems first, check answers, and then work even-numbered problems in areas of weakness. Answers to all problems are in the back of the book. Make every effort to work a problem yourself before you look at an answer.

Use logarithmic properties and appropriate tables or a hand calculator to solve to three significant figures.

A **1.** $3^x = 23$ _____

2. $2^x = 5$ _____

3. $10^{5x-2} = 348$ _____

4. $10^{3x+1} = 92$ _____

   5. $3^{-x} = 0.074$ _____   6. $2^{-x} = 0.238$ _____

   7. $\log_{10} x - \log_{10} 8 = 1$ _____   8. $\log_{10} 5 + \log_{10} x = 2$ _____

B 9. $3 = 1.06^x$ _____   10. $2 = 1.05^x$ _____

   11. $12 = e^{-x}$ _____   12. $42.1 = e^x$ _____

   13. $e^{0.32x} = 632$ _____   14. $e^{-1.4x} = 13$ _____

   15. $\log_{10} (x - 9) + \log_{10} 100x = 3$ _____   16. $\log_{10} x + \log_{10} (x - 3) = 1$ _____

C 17. $438 = 100e^{0.25x}$ _____   18. $123 = 500e^{-0.12x}$ _____

APPLICATIONS

19. **COMPOUND INTEREST** If a certain amount of money P (called the principal) is invested at r percent interest compounded annually, the amount of money A after t years is given by the equation

   $$A = P(1 + r)^t$$

   How much will $1,000 be worth in 5 years at 6 percent interest compounded annually?

   _____

20. **COMPOUND INTEREST** How long (to the nearest year) will it take for a sum of money to double if invested at 5 percent interest compounded annually? See Problem 19.

   _____

21. **BACTERIA GROWTH** A single cholera bacterium divides every $\frac{1}{2}$ hr to produce two complete cholera bacteria. If we start with a colony of 5,000 bacteria, then after t hr we will have

   $$A = 5,000 \cdot 2^{2t}$$

   bacteria. How long will it take for A to equal 1,000,000?

   _____

22. **ASTRONOMY** An optical instrument is required to observe stars beyond the sixth magnitude, the limit of ordinary vision. However, even optical instruments have their limitations. The limiting magnitude L of any optical telescope with lens diameter D in inches is given by

   $$L = 8.8 + 5.1 \log_{10} D$$

   (A) Find the limiting magnitude for a homemade 6-in-diameter reflecting telescope.

   _____

   (B) Find the diameter of a lens that would have a limiting magnitude of 20.6.

   _____

*23. WORLD POPULATION A mathematical model for world population growth over short periods of time is given by

$$P = P_0 e^{rt}$$

where P_0 = population at $t = 0$
r = rate compounded continuously
t = time in years
P = population at time t

How long will it take the earth's population to double if it continues to grow at its current rate of 2 percent per year (compounded continuously)? (HINT: Given $r = 0.02$, find t so that $P = 2P_0$.)

**24. WORLD POPULATION If the world population is now 4 billion people and if it continues to grow at 2 percent per year (compounded continuously), how long will it be before there is only 1yd² of land per person? Use the formula in the preceding problem and the fact that there is 1.7×10^{14} yd² of land on earth.

---

*25. NUCLEAR REACTORS—STRONTIUM 90 Radioactive strontium 90 is used in nuclear reactors and decays according to

$$A = Pe^{-0.0248t}$$

where P is the amount present at $t = 0$, and A is the amount remaining after t years. Find the half-life of strontium 90; that is, find t so that $A = 0.5P$.

---

*26. ARCHEOLOGY—CARBON-14 DATING Cosmic-ray bombardment of the atmosphere produces neutrons, which in turn react with nitrogen to produce radioactive carbon 14. Radioactive carbon 14 enters all living tissues through carbon dioxide which is first absorbed by plants. As long as a plant or animal is alive, carbon 14 is maintained at a constant level in its tissues. Once dead, however, it ceases taking in carbon and, to the slow beat of time, the carbon 14 diminishes by radioactive decay according to the equation

$$A = A_0 e^{-0.000\ 124t}$$

where t is time in years. Estimate the age of a skull uncovered in an archeological site if 10 percent of the original amount of carbon 14 was still present. (HINT: Find t such that $A = 0.1A_0$.)

The Check Exercise for this section is on page 597.

11.6 CHAPTER REVIEW

Important terms and symbols

11.1 EXPONENTIAL FUNCTIONS   exponential functions; graphing for real exponents; general types of exponential graphs; base e; $f(x) = b^x$, $b > 0$, $b \neq 1$

11.2 LOGARITHMIC FUNCTIONS   logarithmic functions; from logarithmic to exponential and vice versa; finding x, b, or y in $y = \log_b x$; inverse relationships; $\log_b b^x = x$

11.3 PROPERTIES OF LOGARITHMIC FUNCTIONS   basic logarithmic properties; use of logarithmic properties; $\log_b b^x = x$, $\log_b MN = \log_b M + \log_b N$; $\log_b M/N = \log_b M - \log_b N$; $\log_b M^p = p \log_b M$; $\log_b 1 = 0$

11.4 COMMON AND NATURAL LOGARITHMS   common logarithms (base 10), natural logarithms (base e)

11.5 EXPONENTIAL AND LOGARITHMIC EQUATIONS

DIAGNOSTIC (REVIEW) EXERCISE 11.6

Work through all the problems in this chapter review and check answers in the back of the book. (Answers to all problems are there, and following each answer is a number in italics indicating the section in which that type of problem is discussed.) Where weaknesses show up, review appropriate sections in the text. When you are satisfied that you know the material, take the practice test following this review.

A **1.** Write $\log_{10} x = y$ in exponential form. _____

2. Write $m = 10^n$ in logarithmic form with base 10. _____

Solve Problems 3 to 5 for x, without the use of a table or calculator.

3. $\log_2 x = 2$ _____   **4.** $\log_x 25 = 2$ _____   **5.** $\log_2 8 = x$ _____

6. Solve for x to three significant figures: $10^x = 17.5$. _____

7. Solve for x: $\log_{10} 4 + \log_{10} x = 2$. _____

B **8.** Write $\ln y = x$ in exponential form. _____

9. Write $y = e^x$ in logarithmic form with base e. _____

Solve Problems 10 to 15 for x, without the use of any table or calculator.

10. $\log_{1/4} 16 = x$ _____   **11.** $\log_x 9 = -2$ _____   **12.** $\log_{16} x = \frac{3}{2}$ _____

13. $\log_x e^5 = 5$ _____   **14.** $\log_{10} 10^x = 33$ _____   **15.** $\ln x = 0$ _____

Solve Problems 16 and 17 for x to three significant figures. Use tables or a calculator as necessary.

16. $25 = 5(2)^x$ _____   **17.** $0.01 = e^{-0.05x}$ _____

Solve Problems 18 and 19 exactly without a table or calculator.

18. $\log_{10} 3x^2 - \log_{10} 9x = 1$ _____   **19.** $\log_{10} 10^{x^2} = 4$ _____

C 20.   For $f: y = 2^x$, graph f and f^{-1}, using the same coordinate axes. What are the domains and ranges of f and f^{-1}?

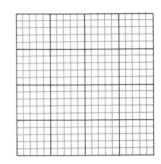

_____

APPLICATIONS

21.   POPULATION GROWTH   Many countries in the world have a population growth rate of 3 percent (or more) per year. At this rate how long, to the nearest year, will it take a population to double? Use the population growth model $P = P_0(1.03)^t$.

_____

22.   CARBON-14 DATING   How long, to three significant figures, will it take for the carbon 14 to diminish to 1 percent of the original amount after the death of a plant or animal? ($A = A_0 e^{-0.000124t}$, where t is time in years.)

_____

PRACTICE TEST CHAPTER 11 _____

Take this as if it were a graded test. Allow yourself up to 50 minutes. Work the problems without looking back in the chapter. Correct your work, using the answers (keyed to appropriate sections) in the back of the book.

1.   Write $\log_{16} 8 = \frac{3}{4}$ in exponential form. _____

2.   Write $\log_{10} y = 0.2x$ in exponential form. _____

3.   Write $A = e^{0.08t}$ in logarithmic form using base e. _____

In Problems 4 to 8 solve for x, without using a calculator or table.

4. $\log_2 x = 4$ _____   **5.** $\log_{16} 4 = x$ _____

6. $\log_x 16 = -2$ _____   **7.** $\log_{10} 2x + \log_{10} 5x = 3$ _____

8. $\log_e e^{-x} = -2$ _____

For Problems 9 to 15 use the nearest table values in Table 3 or Table 4 to find the indicated numbers.

9. $\log_{10} 432{,}000$ _____   **10.** $\log_{10} 0.000\ 020\ 3$ _____

11. x, if $\log_{10} x = 4.8311$ _____   **12.** x, if $\log_{10} x = 0.7318 - 1$ _____

13. x, if $\log_{10} x = -4.1689$ _____   **14.** $\ln 213$ _____

15. $\ln 0.000\ 803$ _____

Solve Problems 16 to 18 to three significant figures.

16. $10^{2x+1} = 252$ _____   **17.** $24 = 4(3)^x$ _____

18. $0.1 = e^{-0.4x}$ _____

19. If a country has a population growth rate of 4 percent per year, how long to the nearest year will it take the population to double? [Use $P = P_0(1.04)^t$ and solve for t when $P = 2P_0$.]

_____

20. Given $f: y = 3^x$, graph f and f^{-1} on the same coordinate system.

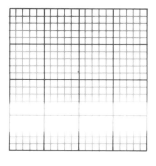

CHECK EXERCISE 11.1

Work the following problems without looking at any text examples. Show your work in the space provided. Write the letter that best indicates your answer in the answer column.

Graph Problems 1 to 4 in the space provided; then from the following graphs write the letters in the answer column that correspond to your graphs.

(A)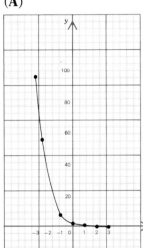

(B)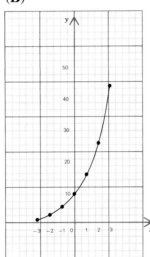

(C)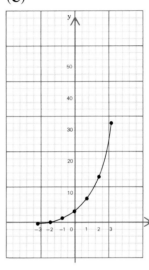

(D)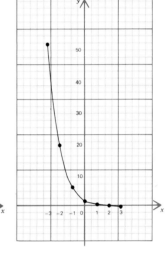

1. $y = 4 \cdot 2^x$, $-3 \le x \le 3$

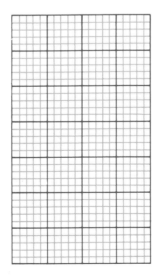

2. $y = 2 \cdot 3^{-x}$, $-3 \le x \le 3$

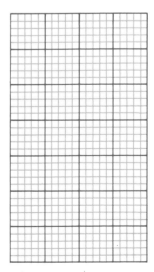

3. $y = 1.5 \cdot 2^{-2x}, \; -3 \le x \le 3$

4. $y = 10e^{0.5x}, \; -3 \le x \le 3$. Use a calculator or Table 2.

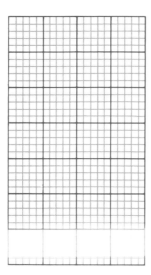

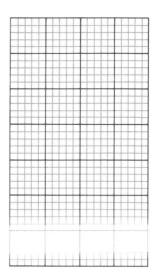

5. The atmospheric pressure P, in pounds per square inch, can be calculated approximately using $P = 14.7e^{-0.21x}$, where x is the altitude in miles relative to sea level. What is the atmospheric pressure in a mine shaft 0.5 mi below sea level? Use a calculator or Table 2.

(A)   13.23 lb/in²

(B)   16.33 lb/in²

(C)   14.7 lb/in²

(D)   None of these

CHECK EXERCISE 11.2

*Work the following problems without looking at any text examples. Show
your work in the space provided. Write the letter that best indicates your
answer in the answer column.*

1. Rewrite in exponential form: $\log_3 81 = 4$

    (A) $4^3 = 81$
    (B) $81^{1/4} = 3$
    (C) $81 = 3^4$
    (D) None of these

2. Rewrite in exponential form: $\log_{10} 0.0001 = -4$

    (A) $0.0001^{-4} = 10$
    (B) $(-4)^{10} = 0.0001$
    (C) $0.0001 = 10^{-4}$
    (D) None of these

3. Rewrite in exponential form: $y = \log_b x,\ b > 0,\ b \neq 1$

    (A) $y = b^x$
    (B) $x = b^y$
    (C) $x = y^b$
    (D) None of these

4. Rewrite in logarithmic form: $\frac{1}{16} = 2^{-4}$

    (A) $\log_2 \frac{1}{16} = -4$
    (B) $\log_{-4} \frac{1}{16} = 2$
    (C) $\log_{1/16} 2 = -4$
    (D) None of these

5. Rewrite in logarithmic form: $1 = 5^0$

    (A) $\log_5 0 = 1$
    (B) $\log_1 0 = 5$
    (C) $\log_1 5 = 0$
    (D) None of these

6. Rewrite in logarithmic form: $x = b^y$, $b > 0$, $b \neq 1$

(A) $x = \log_b y$
(B) $y = \log_b x$
(C) $y = \log_x b$
(D) None of these

7. Find: $\log_3 \sqrt{27}$

(A) $\frac{3}{2}$
(B) $\frac{2}{3}$
(C) $\frac{1}{2}$
(D) None of these

8. Find x: $\log_2 x = 3$

(A) 8
(B) 9
(C) 6
(D) None of these

9. Find y: $y = \log_{1/2} 4$

(A) 2
(B) -2
(C) 4
(D) None of these

10. Find b: $\log_b \frac{1}{8} = -3$

(A) -2
(B) 2
(C) 8
(D) None of these

CHECK EXERCISE 11.3

Work the following problems without looking at any text examples. Show your work in the space provided. Write the letter that best indicates your answer in the answer column.

Write Problems 1 to 4 in terms of simpler logarithmic forms (going as far as you can using logarithmic properties).

1. $\log_b \dfrac{w}{xy} =$

(A) $\log_b w - \log_b x + \log_b y$
(B) $\log_b w - \log_b xy$
(C) $\log_b w - \log_b x - \log_b y$
(D) None of these

2. $\log_b \sqrt{MN} =$

(A) $\frac{1}{2} \log_b M + \log_b N$
(B) $\log_b \sqrt{M} + \log_b \sqrt{N}$
(C) $\sqrt{\log_b M + \log_b N}$
(D) None of these

3. $\log_b \sqrt{\dfrac{x^3}{y^5}} =$

(A) $\frac{1}{2} \log_b x^3 - \log_b y^5$
(B) $\frac{3}{2} \log_b x - \frac{5}{2} \log_b y$
(C) $\frac{3}{2} \log_b x - 5 \log_b y$
(D) None of these

4. $\log_b \dfrac{\sqrt[4]{A}}{2B^3} =$

(A) $\frac{1}{4} \log_b A - \log_b (2B^3)$
(B) $\frac{1}{4} \log_b A - \log_b 2 - 3 \log_b B$
(C) $\frac{1}{4} \log_b A - \log_b 2 + 3 \log_b B$
(D) None of these

Write Problems 5 to 7 in terms of a single logarithm with a coefficient of 1.

5. $2 \log_b u - 3 \log_b v - 4 \log_b w =$

(A) $\log_b \dfrac{u}{6vw}$

(B) $-5 \log_b \dfrac{u}{vw}$

(C) $\log_b \dfrac{u^2}{v^3 w^4}$

(D) None of these

6. $\frac{1}{3}(\log_b A - \log_b B) =$

     (A) $\sqrt[3]{\log_b A - \log_b B}$

     (B) $\sqrt[3]{\dfrac{\log_b A}{\log_b B}}$

     (C) $\sqrt[3]{\log_b \dfrac{A}{B}}$

     (D)   None of these

7. $-3 \log_b M - 2 \log_b N =$

     (A) $\log_b \dfrac{1}{M^3 N^2}$

     (B) $\log_b \dfrac{M^3}{N^2}$

     (C) $\log_b \dfrac{N^2}{M^3}$

     (D)   None of these

If $\log_e 2 = 0.69$, $\log_e 3 = 1.10$, and $\log_e 5 = 1.61$, find the indicated logarithms in Problems 8 and 9 to two decimal places.

8. $\log_e 40 =$

     (A)   3.68
     (B)   6.61
     (C)   1.60
     (D)   None of these

9. $\log_e \sqrt{0.3} =$

     (A) -0.26
     (B)   1.01
     (C) -0.60
     (D)   None of these

10.   Find x so that $\frac{1}{3} \log_b 8 - 2 \log_b 2 = \log_b x$.

     (A) $x = 2$
     (B) $x = \frac{1}{2}$
     (C) $x = -\frac{1}{2}$
     (D)   None of these

CHECK EXERCISE 11.4

Work the following problems without looking at any text examples. Show your work in the space provided. Write the letter that best indicates your answer in the answer column.

Use nearest values in Table 3 (Appendix D) to approximate Problems 1 to 7.

1. $\log_{10} 0.00677 =$

    (A) 0.8306
    (B) -2.1694
    (C) -3.8306
    (D) None of these

2. $\log_{10} 43,662 =$

    (A) 0.6405
    (B) 4.6395
    (C) 4.6405
    (D) None of these

3. $\log_{10} 0.13748 =$

    (A) -0.8633
    (B) 0.1367
    (C) -0.8665
    (D) None of these

4. x if $\log_{10} x = 0.9250$

    (A) 8.42
    (B) 8.41
    (C) 0.9661
    (D) None of these

5. x if $\log_{10} x = 4.2723$

    (A) 1,870
    (B) 18,800
    (C) 187,000
    (D) None of these

6. x if $\log_{10} x = 0.9565 - 1$

(A)   0.905
(B)   0.0905
(C)   90.5
(D)   None of these

7. x if $\log_{10} x = -2.8021$

(A)   0.0634
(B)   0.001 58
(C)   0.000 158
(D)   None of these

Use nearest table values in Table 4 to approximate Problems 8 to 10.

8. $\ln 6.45 =$

(A)   1.8641
(B)   1.8485
(C)   1.8795
(D)   None of these

9. $\ln 4{,}079 =$

(A)   1.4061
(B)   6.0113
(C)   8.3139
(D)   None of these

10. $\ln 0.006\ 13 =$

(A)   8.7210
(B)   1.8132
(C)   -5.0946
(D)   None of these

CHECK EXERCISE 11.5

NAME _____

CLASS _____

SCORE _____

Work the following problems without looking at any text examples. Show
your work in the space provided. Write the letter that best indicates your
answer in the answer column.

Use logarithmic properties and appropriate tables (or a hand calculator)
to solve Problems 1 to 3 to three significant figures.

ANSWER
COLUMN

1. _____

2. _____

3. _____

4. _____

5. _____

1. $5^{2x-3} = 3$

      (**A**)   1.84
      (**B**)   1.39
      (**C**)   3.65
      (**D**)   None of these

2. $\log_{10} x + \log_{10} (x - 9) = 1$

      (**A**) $x = -1, 10$
      (**B**) $x = -10, 1$
      (**C**) $x = 10$
      (**D**)   None of these

3. $57 = 10e^{-2.3t}$

      (**A**) $t = 0.757$
      (**B**) $t = -0.757$
      (**C**) $t = 2.09$
      (**D**)   None of these

4. How long to the nearest year will it take money to triple if invested at 12 percent compounded annually? $[A = P(1 + r)^n]$

   (A) 8 years
   (B) 9 years
   (C) 10 years
   (D) None of these

5. The atmospheric pressure P (in pounds per square inch) at x miles above sea level is given approximately by

   $$P = 14.7e^{-0.21x}$$

   At what height (to two decimal places) will the pressure be one-tenth that at sea level?

   (A) 10.96 miles
   (B) 1.56 miles
   (C) 0.68 miles

   (D) None of these

APPENDIXES

APPENDIX A

LINEAR INTERPOLATION FOR LOGARITHMS

NOTE: In this section we will use $\log x$ in place of $\log_{10} x$. Thus,

$$\boxed{\log x = \log_{10} x}$$

Any printed table is necessarily limited to a finite number of entries. The logarithmic function

$$y = \log x \qquad 1 \leq x < 10$$

is defined for an infinite number of values—all real numbers between 1 and 10. What, then, do we do about values of x not in a table? How do we find, for example, $\log 3.276$? If a certain amount of accuracy can be sacrificed, we can round 3.276 to the closest entry in the table and proceed as before. Thus,

$$\log 3.276 \approx \log 3.28 = 0.5159$$

We can do better than this, however, without too much additional work, by using a process called **linear interpolation** (see Figure 1).

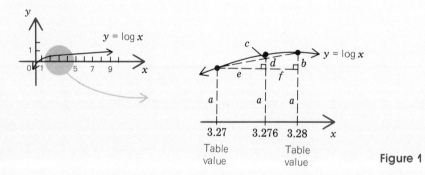

Figure 1

In Figure 1, we would like to find $a + d + c$. Using proportional parts of similar triangles, we will be able to find d without difficulty, since a and $a + b$ can be determined from the table. Hence, we will settle for $a + d$ as an approximation for $a + d + c$. This approximation is better than what appears in the figure, since the curve is distorted in the drawing for clarity. The logarithmic curve is actually much flatter.

Now to the process of linear interpolation. We find d, using the proportion

$$\frac{d}{b} = \frac{e}{e + f}$$

600

We organize the work as follows for convenient computation.

$$
0.010 \left\{ \begin{array}{cc} x & \log x \\ 3.280 & 0.5159 \\ 0.006\left\{ \begin{array}{cc} 3.276 & n \\ 3.270 & 0.5145 \end{array} \right\}d \end{array} \right\} 0.0014
$$

NOTE: Larger values are placed on top to make the mental subtraction easier.

Here $b = 0.0014$, $e = 0.006$, and $e + f = 0.010$, thus

$$\frac{d}{0.0014} = \frac{0.006}{0.010}$$

NOTE: We will use $=$, understanding that only approximately equal ($\approx$) is correct in many places.

$$d = 0.0008$$

Hence,

$$\log 3.276 = 0.5145 + 0.0008 = 0.5153$$

In practice, the linear interpolation process is carried out by means of a few key operational steps (often done mentally), as indicated in the next example. Notice how decimal points have been dropped to simplify the arithmetic. This is convenient, and no harm is done as long as the decimal points are properly reintroduced at the end of the calculation. If in doubt, proceed with all the decimal points, as in the example above.

EXAMPLE 1
Use linear interpolation to find log 3,514.

Solution
$$\log 3{,}514 = \log (3.514 \times 10^3) = (\log 3.514) + 3$$

$$
10 \left\{ \begin{array}{cc} x & \log x \\ 3.520 & 0.5465 \\ 4\left\{ \begin{array}{cc} 3.514 & n \\ 3.510 & 0.5453 \end{array} \right\}d \end{array} \right\} 12
$$

$$\frac{4}{10} = \frac{d}{12}$$

$$d = 5$$

Thus,

$$\log 3.514 = 0.5453 + 0.0005 = 0.5458$$

and

$$\log 3{,}514 = 0.5458 + 3 = 3.5458$$

Work Problem 1 and check solution in Solutions to Matched Problems on page 603.

PROBLEM 1   Use linear interpolation to find log 326.6.

Solution

EXAMPLE 2

Use linear interpolation to find x, given $\log x = 2.5333$.

$\log x = 2.5333$

$\qquad = 0.5333 + 2$

$$10\left\{ \begin{array}{l} \begin{array}{cc} x & \log x \\ 3.420 & 0.5340 \end{array} \\ d\left\{\begin{array}{cc} n & 0.5333 \\ 3.410 & 0.5328 \end{array}\right\}5 \end{array}\right\}12$$

$\dfrac{d}{10} = \dfrac{5}{12}$

$d = 4$

$n = 3.410 + 0.004 = 3.414$

Thus,

$x = 3.414 \times 10^2$ or   341.4

Work Problem 2 and check solution in Solutions to Matched Problems on page 603.

PROBLEM 2   Use linear interpolation to find x, given $\log x = 7.5230$.

Solution

SOLUTIONS TO MATCHED PROBLEMS

1. $\log 326.6 = \log (3.266 \times 10^2)$

$= \log 3.266 + \log 10^2$
$= 0.5140 + 2$

$= 2.5140$

Linear Interpolation

	x	$\log x$	
	3.270	0.5145	
$10\begin{cases} 6\begin{cases} \\ \end{cases} \end{cases}$	3.266	n	$\begin{cases} d \\ \end{cases} 13$
	3.260	0.5132	

$$\frac{d}{13} = \frac{6}{10}$$

$$d \approx 8$$

$$n = 0.5132 + 0.0008 = 0.5140$$

2. $\log x = 7.5230$
$= 0.5230 + 7$

$= \log 3.335 + \log 10^7$
$= \log (3.335 \times 10^7)$

$x = 3.335 \times 10^7$

Linear Interpolation

	x	$\log x$	
	3.340	0.5237	
$10\begin{cases} d\begin{cases} \\ \end{cases} \end{cases}$	n	0.5230	$\begin{cases} \\ 6 \end{cases} 13$
	3.330	0.5224	

$$\frac{d}{10} = \frac{6}{13}$$

$$d \approx 5$$

$$n = 3.330 + 0.005 = 3.335$$

PRACTICE EXERCISE APPENDIX A

Work odd-numbered problems first, check answers, and then work even-numbered problems in areas of weakness. Answers to all problems are in the back of the book. Make every effort to work a problem yourself before you look at an answer.

Use linear interpolation to find each logarithm.

1. $\log 2.317$ _____

2. $\log 5.143$ _____

3. $\log 703,400$ _____

4. $\log 28,430$ _____

5. $\log 65.03$ _____

6. $\log 20.35$ _____

7. $\log 0.004006$ _____

8. $\log 0.03713$ _____

9. $\log 0.9008$ _____

10. $\log 0.6413$ _____

11. $\log 692,300$ _____

12. $\log 84,660$ _____

Use linear interpolation to find x.

13. $\log x = 0.7163$ _____

14. $\log x = 0.4085$ _____

15. $\log x = 5.5458$ _____

16. $\log x = 2.4735$ _____

17. $\log x = 3.4303$ _____

18. $\log x = 1.9141$ _____

19. $\log x = 0.6038 - 3$ _____

20. $\log x = 0.2177 - 1$ _____

21. $\log x = 0.8392 - 1$ _____

22. $\log x = 0.8509 - 4$ _____

23. $\log x = -0.8315$ _____

24. $\log x = -2.6651$ _____

APPENDIX B

DETERMINANTS

•Second-order determinants

••Third-order determinants

Determinants arise quite naturally in many areas in mathematics, including the solving of linear systems, vector analysis, calculus, and so on. We consider a few of their uses in this appendix and in Appendix C.

•Second-order determinants

A square array of four real numbers, such as

$$\begin{vmatrix} 2 & -3 \\ 5 & 1 \end{vmatrix}$$

is called a determinant of order 2. (It is important to note that the array of numbers is between parallel lines and not square brackets. If square brackets are used, then the symbol has another meaning.) The above determinant has two **rows** and two **columns**—rows are across and columns are up and down. Each number in the determinant is called an **element** of the determinant.

In general, we can symbolize a **second-order determinant** as follows:

$$\begin{vmatrix} a_{11} & a_{12} \\ a_{21} & a_{22} \end{vmatrix}$$

where we use a single letter with a **double subscript** to facilitate generalization to higher-order determinants. The first subscript number indicates the row in which the element lies, and the second subscript number indicates the column. Thus a_{21} is the element in the second row and first column, and a_{12} is the element in the first row and second column. Each second-order determinant represents a real number given by the formula:

$$\begin{vmatrix} a_{11} & a_{12} \\ a_{21} & a_{22} \end{vmatrix} = a_{11}a_{22} - a_{21}a_{12}$$

EXAMPLE 1

$$\begin{vmatrix} -1 & 2 \\ -3 & -4 \end{vmatrix} = (-1)(-4) - (-3)(2) = 4 - (-6) = 10$$

Work Problem 1 and check solution in Solutions to Matched Problems on page 608.

PROBLEM 1 Find:

$$\begin{vmatrix} 3 & -5 \\ 4 & -2 \end{vmatrix} =$$

••Third-order determinants

A determinant of order 3 is a square array of nine elements, and represents a real number given by the formula

$$\begin{vmatrix} a_{11} & a_{12} & a_{13} \\ a_{21} & a_{22} & a_{23} \\ a_{31} & a_{32} & a_{33} \end{vmatrix} = \begin{aligned} & a_{11}a_{22}a_{33} - a_{11}a_{32}a_{23} + a_{21}a_{32}a_{13} - a_{21}a_{12}a_{33} \\ & + a_{31}a_{12}a_{23} - a_{31}a_{22}a_{13} \end{aligned} \tag{1}$$

Note that each term in the expansion on the right of (1) contains exactly one element from each row and each column. Don't panic! You do not need to memorize formula (1). After we introduce the ideas of "minor" and "cofactor," we will state a theorem that can be used to obtain the same result with much less memory strain.

The **minor of an element** in a third-order determinant is a second-order determinant obtained by deleting the row and column that contains the element. For example, in the determinant in formula (1)

$$\text{The minor of } a_{23} = \begin{vmatrix} a_{11} & a_{12} & a_{13} \\ a_{21} & a_{22} & a_{23} \\ a_{31} & a_{32} & a_{33} \end{vmatrix} = \begin{vmatrix} a_{11} & a_{12} \\ a_{31} & a_{32} \end{vmatrix}$$

$$\text{The minor of } a_{32} = \begin{vmatrix} a_{11} & a_{12} & a_{13} \\ a_{21} & a_{22} & a_{23} \\ a_{31} & a_{32} & a_{33} \end{vmatrix} = \begin{vmatrix} a_{11} & a_{13} \\ a_{21} & a_{23} \end{vmatrix}$$

A quantity closely associated with the minor of an element is the cofactor of an element. The **cofactor of an element** a_{ij} (from the ith row and jth column) is the product of the minor of a_{ij} and $(-1)^{i+j}$. That is,

$$\boxed{\text{Cofactor of } a_{ij} = (-1)^{i+j}(\text{minor of } a_{ij})}$$

Thus, a cofactor of an element is nothing more than a signed minor. The sign is determined by raising -1 to a power that is the sum of the numbers indicating the row and column in which the element lies. Note that $(-1)^{i+j}$ is -1 if $i + j$ is odd and 1 if $i + j$ is even. Referring again to the determinant in formula (1),

$$\text{The cofactor of } a_{23} = (-1)^{2+3}\begin{vmatrix} a_{11} & a_{12} \\ a_{31} & a_{32} \end{vmatrix} = -\begin{vmatrix} a_{11} & a_{12} \\ a_{31} & a_{32} \end{vmatrix}$$

$$\text{The cofactor of } a_{11} = (-1)^{1+1}\begin{vmatrix} a_{22} & a_{23} \\ a_{32} & a_{33} \end{vmatrix} = \begin{vmatrix} a_{22} & a_{23} \\ a_{32} & a_{33} \end{vmatrix}$$

EXAMPLE 2
Find the cofactor of -2 and 5 in the determinant

$$\begin{vmatrix} -2 & 0 & 3 \\ 1 & -6 & 5 \\ -1 & 2 & 0 \end{vmatrix}$$

Solution

The cofactor of $-2 = (-1)^{1+1} \begin{vmatrix} -6 & 5 \\ 2 & 0 \end{vmatrix} = \begin{vmatrix} -6 & 5 \\ 2 & 0 \end{vmatrix}$

$$= (-6)(0) - (2)(5) = -10$$

The cofactor of $5 = (-1)^{2+3} \begin{vmatrix} -2 & 0 \\ -1 & 2 \end{vmatrix} = - \begin{vmatrix} -2 & 0 \\ -1 & 2 \end{vmatrix}$

$$= -[(-2)(2) - (-1)(0)] = 4$$

Work Problem 2 and check solution in Solutions to Matched Problems on page 608.

PROBLEM 2   Find the cofactors of 2 and 3 in the determinant in Example 2.

The cofactor of $2 =$

The cofactor of $3 =$

NOTE: The sign in front of the minor, $(-1)^{i+j}$, can be determined rather mechanically by using a checkerboard pattern of $+$ and $-$ signs over the determinant, starting with $+$ in the upper left-hand corner:

$$\begin{array}{ccc} + & - & + \\ - & + & - \\ + & - & + \end{array}$$

Use either the checkerboard or the exponent method, whichever is easier for you, to determine the sign in front of the minor.

Now we are ready for the central theorem of this appendix. It will provide us with an efficient means of evaluating third-order determinants. In addition, it is worth noting that the theorem generalizes completely to include determinants of arbitrary order.

THEOREM 1
The value of a determinant of order 3 is the sum of three products obtained by multiplying each element of any one row (or each element of any one column) by its cofactor.

To prove this theorem one must show that the expansions indicated by the theorem for any row or any column (six cases) produce the expression on the right of formula (1) above. We omit the details, and accept the theorem as stated.

EXAMPLE 3
Evaluate by expanding by (**A**) the first row and (**B**) the second column:

$$\begin{vmatrix} 2 & -2 & 0 \\ -3 & 1 & 2 \\ 1 & -3 & -1 \end{vmatrix}$$

Solution

(**A**) $\begin{vmatrix} 2 & -2 & 0 \\ -3 & 1 & 2 \\ 1 & -3 & -1 \end{vmatrix}$ Expand by first row.

$$= a_{11}\begin{pmatrix} \text{cofactor} \\ \text{of } a_{11} \end{pmatrix} + a_{12}\begin{pmatrix} \text{cofactor} \\ \text{of } a_{12} \end{pmatrix} + a_{13}\begin{pmatrix} \text{cofactor} \\ \text{of } a_{13} \end{pmatrix}$$

$$= 2\left((-1)^{1+1}\begin{vmatrix} 1 & 2 \\ -3 & -1 \end{vmatrix}\right) + (-2)\left((-1)^{1+2}\begin{vmatrix} -3 & 2 \\ 1 & -1 \end{vmatrix}\right) + 0$$

$$= (2)(1)[(1)(-1) - (-3)(2)] + (-2)(-1)[(-3)(-1) - (1)(2)]$$

$$= (2)(5) + (2)(1) = 12$$

(**B**) $\begin{vmatrix} 2 & -2 & 0 \\ -3 & 1 & 2 \\ 1 & -3 & -1 \end{vmatrix}$ Expand by second column.

$$= a_{12}\begin{pmatrix} \text{cofactor} \\ \text{of } a_{12} \end{pmatrix} + a_{22}\begin{pmatrix} \text{cofactor} \\ \text{of } a_{22} \end{pmatrix} + a_{32}\begin{pmatrix} \text{cofactor} \\ \text{of } a_{32} \end{pmatrix}$$

$$= (-2)\left((-1)^{1+2}\begin{vmatrix} -3 & 2 \\ 1 & -1 \end{vmatrix}\right) + (1)\left((-1)^{2+2}\begin{vmatrix} 2 & 0 \\ 1 & -1 \end{vmatrix}\right)$$

$$+ (-3)\left((-1)^{3+2}\begin{vmatrix} 2 & 0 \\ -3 & 2 \end{vmatrix}\right)$$

$$= (-2)(-1)[(-3)(-1) - (1)(2)] + (1)(1)[(2)(-1) - (1)(0)]$$

$$+ (-3)(-1)[(2)(2) - (-3)(0)]$$

$$= (2)(1) + (1)(-2) + (3)(4)$$

$$= 12$$

Work Problem 3 and check solutions in Solutions to Matched Problems on page 608.

PROBLEM 3   (**A**)   Evaluate by expanding by the first row:

$$\begin{vmatrix} 2 & 1 & -1 \\ -2 & -3 & 0 \\ -1 & 2 & 1 \end{vmatrix} =$$

(B) Evaluate by expanding by the third column:

$$\begin{vmatrix} 2 & 1 & -1 \\ -2 & -3 & 0 \\ -1 & 2 & 1 \end{vmatrix} =$$

It should now be clear that we can greatly reduce the work involved in evaluating a determinant by choosing to expand by a row or column with the greatest number of zeros.

Where are determinants used? Many equations and formulas have particularly simple and compact representations in determinant form that are easily remembered. See, for example, Problems 35 and 36 in Practice Exercise B, and Cramer's rule in Appendix C.

SOLUTIONS TO MATCHED PROBLEMS

1. $\begin{vmatrix} 3 & -5 \\ 4 & -2 \end{vmatrix} = (3)(-2) - (4)(-5) = (-6) - (-20) = 14$

2. Cofactor of 2 $= (-1)^{3+2}\begin{vmatrix} -2 & 3 \\ 1 & 5 \end{vmatrix} = (-1)[(-2)(5) - (1)(3)] = (-1)[(-10) - (3)] = 13$

Cofactor of 3 $= (-1)^{1+3}\begin{vmatrix} 1 & -6 \\ -1 & 2 \end{vmatrix} = (+1)[(1)(2) - (-1)(-6)] = (2) - (6) = -4$

3. **(A)** $\begin{vmatrix} 2 & 1 & -1 \\ -2 & -3 & 0 \\ -1 & 2 & 1 \end{vmatrix} = (2)\left((-1)^{1+1}\begin{vmatrix} -3 & 0 \\ 2 & 1 \end{vmatrix}\right) + (1)\left((-1)^{1+2}\begin{vmatrix} -2 & 0 \\ -1 & 1 \end{vmatrix}\right) + (-1)\left((-1)^{1+3}\begin{vmatrix} -2 & -3 \\ -1 & 2 \end{vmatrix}\right)$

$= (2)(1)[(-3)(1) - 0] + (1)(-1)[(-2)(1) - 0] + (-1)(1)[(-2)(2) - (-1)(-3)]$

$= 2[(-3) - 0] + (-1)[(-2) - 0] + (-1)[(-4) - (3)]$

$= (-6) + (2) + (7) = 3$

(B) $\begin{vmatrix} 2 & 1 & -1 \\ -2 & -3 & 0 \\ -1 & 2 & 1 \end{vmatrix} = (-1)\left((-1)^{1+3}\begin{vmatrix} -2 & -3 \\ -1 & 2 \end{vmatrix}\right) + 0 + (1)\left((-1)^{3+3}\begin{vmatrix} 2 & 1 \\ -2 & -3 \end{vmatrix}\right)$

$= (-1)(1)[(-2)(2) - (-1)(-3)] + (1)(1)[(2)(-3) - (-2)(1)]$

$= (-1)[(-4) - (3)] + (1)[(-6) - (-2)]$

$= 7 + (-4) = 3$

PRACTICE EXERCISE APPENDIX B

Work odd-numbered problems first, check answers, and then work even-numbered problems in areas of weakness. Answers to all problems are in the back of the book. Make every effort to work a problem yourself before you look at an answer.

A *Evaluate each second-order determinant.*

1. $\begin{vmatrix} 2 & 4 \\ 3 & -1 \end{vmatrix}$ _____

2. $\begin{vmatrix} 2 & 2 \\ -3 & 1 \end{vmatrix}$ _____

3. $\begin{vmatrix} 5 & -4 \\ -2 & 2 \end{vmatrix}$ _____

4. $\begin{vmatrix} 6 & -2 \\ -1 & -3 \end{vmatrix}$ _____

5. $\begin{vmatrix} 3 & -3.1 \\ -2 & 1.2 \end{vmatrix}$ _____

6. $\begin{vmatrix} -1.4 & 3 \\ -0.5 & -2 \end{vmatrix}$ _____

Given the determinant

$$\begin{vmatrix} a_{11} & a_{12} & a_{13} \\ a_{21} & a_{22} & a_{23} \\ a_{31} & a_{32} & a_{33} \end{vmatrix}$$

write the minor of each of the following elements.

7. a_{11}

8. a_{33}

_____

_____

9. a_{23}

10. a_{22}

_____

_____

Write the cofactor of each of the following elements.

11. a_{11}

12. a_{33}

_____

_____

13. a_{23}

14. a_{22}

_____

_____

Given the determinant

$$\begin{vmatrix} -2 & 3 & 0 \\ 5 & 1 & -2 \\ 7 & -4 & 8 \end{vmatrix}$$

write the minor of each of the following elements. (Leave answer in determinant form.)

15. a_{11}

16. a_{22}

_____

_____

17. a_{32}

18. a_{21}

_____

_____

Write the cofactor of each of the following elements and evaluate each.

19. a_{11}

20. a_{22}

_____

_____

21. a_{32}

22. a_{21}

_____

_____

Evaluate each of the following determinants, using cofactors.

B 23. $\begin{vmatrix} 1 & 0 & 0 \\ -2 & 4 & 3 \\ 5 & -2 & 1 \end{vmatrix}$ _____

24. $\begin{vmatrix} 2 & -3 & 5 \\ 0 & -3 & 1 \\ 0 & 6 & 2 \end{vmatrix}$ _____

25. $\begin{vmatrix} 0 & 1 & 5 \\ 3 & -7 & 6 \\ 0 & -2 & -3 \end{vmatrix}$ _____

26. $\begin{vmatrix} 4 & -2 & 0 \\ 9 & 5 & 4 \\ 1 & 2 & 0 \end{vmatrix}$ _____

27. $\begin{vmatrix} 4 & -4 & 6 \\ 2 & 8 & -3 \\ 0 & -5 & 0 \end{vmatrix}$ _____

28. $\begin{vmatrix} 3 & -2 & -8 \\ -2 & 0 & -3 \\ 1 & 0 & -4 \end{vmatrix}$ _____

29. $\begin{vmatrix} -1 & 2 & -3 \\ -2 & 0 & -6 \\ 4 & -3 & 2 \end{vmatrix}$ _____

30. $\begin{vmatrix} 0 & 2 & -1 \\ -6 & 3 & 1 \\ 7 & -9 & -2 \end{vmatrix}$ _____

31. $\begin{vmatrix} 1 & 4 & 1 \\ 1 & 1 & -2 \\ 2 & 1 & -1 \end{vmatrix}$ _____

32. $\begin{vmatrix} 3 & 2 & 1 \\ -1 & 5 & 1 \\ 2 & 3 & 1 \end{vmatrix}$ _____

33. $\begin{vmatrix} 1 & 4 & 3 \\ 2 & 1 & 6 \\ 3 & -2 & 9 \end{vmatrix}$ _____

34. $\begin{vmatrix} 4 & -6 & 3 \\ -1 & 4 & 1 \\ 5 & -6 & 3 \end{vmatrix}$ _____

C 35. It can be shown that the equation of a line that passes through (x_1, y_1) and (x_2, y_2) is given by

$$\begin{vmatrix} x & y & 1 \\ x_1 & y_1 & 1 \\ x_2 & y_2 & 1 \end{vmatrix} = 0$$

(A) Write the equation of a line that passes through $(2, 3)$ and $(-1, 2)$ in determinant form.

_____

(B) Expand the determinant in part **A**, using the first row, and write the final equation in the form $Ax + By = C$.

_____

36. It can be shown that the area of a triangle with vertices (x_1, y_1), (x_2, y_2), and (x_3, y_3) is the absolute value of

$$\frac{1}{2}\begin{vmatrix} x_1 & y_1 & 1 \\ x_2 & y_2 & 1 \\ x_3 & y_3 & 1 \end{vmatrix}$$

Use this result to find the area of a triangle with vertices $(-1, 4)$, $(4, 8)$, and $(1, 1)$.

_____

APPENDIX C

CRAMER'S RULE

Now let us see how determinants arise rather naturally in the process of solving systems of linear equations. We will start by investigating two equations and two unknowns, and then extend any results to three equations and three unknowns.

Instead of thinking of each system of linear equations in two unknowns as a different problem, let us see what happens when we attempt to solve the general system

$$a_{11}x + a_{12}y = k_1 \tag{1A}$$

$$a_{21}x + a_{22}y = k_2 \tag{1B}$$

once and for all in terms of the unspecified real constants a_{11}, a_{12}, a_{21}, a_{22}, k_1, and k_2.

We proceed by multiplying equations (1A) and (1B) by suitable constants so that when the resulting equations are added, left side to left side and right side to right side, one of the variables drops out. Suppose we choose to eliminate y; what constants should we use to make the coefficients of y the same except for the signs? Multiply (1A) by a_{22} and equation (1B) by $-a_{12}$; then add.

$$\begin{array}{ll} a_{11}a_{22}x + a_{12}a_{22}y = k_1a_{22} & a_{22}(1A) \\ -a_{21}a_{12}x - a_{12}a_{22}y = -k_2a_{12} & -a_{12}(1B) \\ \hline \end{array}$$

$$a_{11}a_{22}x - a_{21}a_{12}x + 0y = k_1a_{22} - k_2a_{12}$$

$$(a_{11}a_{22} - a_{21}a_{12})x = k_1a_{22} - k_2a_{12}$$

$$x = \frac{k_1a_{22} - k_2a_{12}}{a_{11}a_{22} - a_{21}a_{12}} \qquad a_{11}a_{22} - a_{21}a_{12} \neq 0$$

What do the numerator and denominator remind you of? From your experience with determinants in the preceding section you should recognize these expressions as

$$x = \frac{\begin{vmatrix} k_1 & a_{12} \\ k_2 & a_{22} \end{vmatrix}}{\begin{vmatrix} a_{11} & a_{12} \\ a_{21} & a_{22} \end{vmatrix}}$$

Similarly, starting with system (1) and eliminating x (this is left as an exercise), we obtain

$$y = \frac{\begin{vmatrix} a_{11} & k_1 \\ a_{21} & k_2 \end{vmatrix}}{\begin{vmatrix} a_{11} & a_{12} \\ a_{21} & a_{22} \end{vmatrix}}$$

These results are summarized in the following theorem, which is named after the Swiss mathematician G. Cramer (1704–1752):

THEOREM 1

Cramer's Rule for Two Equations and Two Unknowns

Given the system

$$a_{11}x + a_{12}y = k_1$$
$$a_{21}x + a_{22}y = k_2$$

with

$$D = \begin{vmatrix} a_{11} & a_{12} \\ a_{21} & a_{22} \end{vmatrix} \neq 0$$

then

$$x = \frac{\begin{vmatrix} k_1 & a_{12} \\ k_2 & a_{22} \end{vmatrix}}{D} \quad \text{and} \quad y = \frac{\begin{vmatrix} a_{11} & k_1 \\ a_{21} & k_2 \end{vmatrix}}{D}$$

The determinant D is called the **coefficient determinant.** If $D \neq 0$, then the system has exactly one solution, which is given by Cramer's rule. If, on the other hand, $D = 0$, then it can be shown that the system is either inconsistent or dependent; that is, the system either has no solutions or has an infinite number of solutions.

EXAMPLE 1
Solve, using Cramer's rule:

$$2x - 3y = 7$$
$$-3x + y = -7$$

Solution

$$D = \begin{vmatrix} 2 & -3 \\ -3 & 1 \end{vmatrix} = -7$$

$$x = \frac{\begin{vmatrix} 7 & -3 \\ -7 & 1 \end{vmatrix}}{-7} = \frac{-14}{-7} = 2 \qquad y = \frac{\begin{vmatrix} 2 & 7 \\ -3 & -7 \end{vmatrix}}{-7} = \frac{7}{-7} = -1$$

Work Problem 1 and check solution in Solutions to Matched Problems on page 616.

PROBLEM 1   Solve, using Cramer's rule:

$$3x + 2y = -3$$
$$-4x + 3y = -13$$

Solution

Cramer's rule generalizes completely for any size linear system that has the same number of unknowns as equations. We state without proof the rule for three equations and three unknowns.

THEOREM 2

Cramer's Rule for Three Equations and Three Unknowns

Given the system

$$a_{11}x + a_{12}y + a_{13}z = k_1$$
$$a_{21}x + a_{22}y + a_{23}z = k_2$$
$$a_{31}x + a_{32}y + a_{33}z = k_3$$

with

$$D = \begin{vmatrix} a_{11} & a_{12} & a_{13} \\ a_{21} & a_{22} & a_{23} \\ a_{31} & a_{32} & a_{33} \end{vmatrix} \neq 0$$

then

$$x = \frac{\begin{vmatrix} k_1 & a_{12} & a_{13} \\ k_2 & a_{22} & a_{23} \\ k_3 & a_{32} & a_{33} \end{vmatrix}}{D} \qquad y = \frac{\begin{vmatrix} a_{11} & k_1 & a_{13} \\ a_{21} & k_2 & a_{23} \\ a_{31} & k_3 & a_{33} \end{vmatrix}}{D} \qquad z = \frac{\begin{vmatrix} a_{11} & a_{12} & k_1 \\ a_{21} & a_{22} & k_2 \\ a_{31} & a_{32} & k_3 \end{vmatrix}}{D}$$

It is easy to remember these determinant formulas for x, y, and z if one observes the following:

1. Determinant D is formed from the coefficients of x, y, and z keeping the same relative position in the determinant as found in the system.
2. Determinant D appears in the denominators for x, y, and z.
3. The numerator for x can be obtained from D by replacing the coefficients of x—a_{11}, a_{21}, and

a_{31}—with the constants k_1, k_2, and k_3, respectively. Similar statements can be made for the numerators for y and z.

EXAMPLE 2
Use Cramer's rule to solve:

$$
\begin{aligned}
x + y \quad &= 1 \\
3y - z &= -4 \\
x \quad + z &= 3
\end{aligned}
$$

Solution

$$
D = \begin{vmatrix} 1 & 1 & 0 \\ 0 & 3 & -1 \\ 1 & 0 & 1 \end{vmatrix} = 2 \qquad \text{Missing variables have 0 coefficients.}
$$

$$
x = \frac{\begin{vmatrix} 1 & 1 & 0 \\ -4 & 3 & -1 \\ 3 & 0 & 1 \end{vmatrix}}{2} = \frac{4}{2} = 2 \qquad
y = \frac{\begin{vmatrix} 1 & 1 & 0 \\ 0 & -4 & -1 \\ 1 & 3 & 1 \end{vmatrix}}{2} = \frac{-2}{2} = -1
$$

$$
z = \frac{\begin{vmatrix} 1 & 1 & 1 \\ 0 & 3 & -4 \\ 1 & 0 & 3 \end{vmatrix}}{2} = \frac{2}{2} = 1
$$

Work Problem 2 and check solution in Solutions to Matched Problems on page 616.

PROBLEM 2 Use Cramer's rule to solve:

$$
\begin{aligned}
3x \quad - z &= 5 \\
x - y + z &= 0 \\
x + y \quad &= 0
\end{aligned}
$$

Solution

SOLUTIONS TO MATCHED PROBLEMS

1. $\begin{aligned} 3x + 2y &= -3 \\ -4x + 3y &= -13 \end{aligned}$ $D = \begin{vmatrix} 3 & 2 \\ -4 & 3 \end{vmatrix} = 17$ $x = \dfrac{\begin{vmatrix} -3 & 2 \\ -13 & 3 \end{vmatrix}}{17} = \dfrac{17}{17} = 1$

$$y = \frac{\begin{vmatrix} 3 & -3 \\ -4 & -13 \end{vmatrix}}{17} = \frac{-51}{17} = -3$$

2. $\begin{aligned} 3x \quad\;\; - z &= 5 \\ x - y + z &= 0 \\ x + y \quad\;\; &= 0 \end{aligned}$ $D = \begin{vmatrix} 3 & 0 & -1 \\ 1 & -1 & 1 \\ 1 & 1 & 0 \end{vmatrix} = -5$ $x = \dfrac{\begin{vmatrix} 5 & 0 & -1 \\ 0 & -1 & 1 \\ 0 & 1 & 0 \end{vmatrix}}{-5} = \dfrac{-5}{-5} = 1$

$$y = \frac{\begin{vmatrix} 3 & 5 & -1 \\ 1 & 0 & 1 \\ 1 & 0 & 0 \end{vmatrix}}{-5} = \frac{5}{-5} = -1 \qquad z = \frac{\begin{vmatrix} 3 & 0 & 5 \\ 1 & -1 & 0 \\ 1 & 1 & 0 \end{vmatrix}}{-5} = \frac{10}{-5} = -2$$

PRACTICE EXERCISE APPENDIX C

Work odd-numbered problems first, check answers, then work even-numbered problems in areas of weakness. Answers to all problems are in the back of the book. Make every effort to work a problem yourself before you look at an answer.

Solve, using Cramer's rule:

A
  1. $\begin{aligned} x + 2y &= 1 \\ x + 3y &= -1 \end{aligned}$ _____

  2. $\begin{aligned} x + 2y &= 3 \\ x + 3y &= 5 \end{aligned}$ _____

  3. $\begin{aligned} 2x + y &= 1 \\ 5x + 3y &= 2 \end{aligned}$ _____

  4. $\begin{aligned} x + 3y &= 1 \\ 2x + 8y &= 0 \end{aligned}$ _____

  5. $\begin{aligned} 2x - y &= -3 \\ -x + 3y &= 4 \end{aligned}$ _____

  6. $\begin{aligned} 2x + y &= 1 \\ 5x + 3y &= 2 \end{aligned}$ _____

B
  7. $\begin{aligned} x + y \quad\;\; &= 0 \\ 2y + z &= -5 \\ -x + \quad\;\; z &= -3 \end{aligned}$ _____

  8. $\begin{aligned} x + y \quad\;\; &= -4 \\ 2y + z &= 0 \\ -x + \quad\;\; z &= 5 \end{aligned}$ _____

  9. $\begin{aligned} x + y \quad\;\; &= 1 \\ 2y + z &= 0 \\ -x + \quad\;\; z &= 0 \end{aligned}$ _____

  10. $\begin{aligned} x + y \quad\;\; &= -4 \\ 2y + z &= 3 \\ -x + \quad\;\; z &= 7 \end{aligned}$ _____

  11. $\begin{aligned} y + z &= -4 \\ x + \quad 2z &= 0 \\ x - y \quad\;\; &= 5 \end{aligned}$ _____

  12. $\begin{aligned} x \quad\;\; - z &= 2 \\ 2x - y \quad\;\; &= 8 \\ x + y + z &= 2 \end{aligned}$ _____

  13. $\begin{aligned} 2y + z &= -4 \\ x - y - z &= 0 \\ x - y + 2z &= 6 \end{aligned}$ _____

  14. $\begin{aligned} 2x + y \quad\;\; &= 2 \\ x - y + z &= -1 \\ x + y + z &= -1 \end{aligned}$ _____

  15. $\begin{aligned} 2a + 4b + 3c &= 6 \\ a - 3b + 2c &= -7 \\ -a + 2b - c &= 5 \end{aligned}$ _____

  16. $\begin{aligned} 3u - 2v + 3w &= 11 \\ 2u + 3v - 2w &= -5 \\ u + 4v - w &= -5 \end{aligned}$ _____

APPENDIX D

TABLES

TABLE 1
Squares and square roots (0 to 199)

n	n²	√n	n	n²	√n	n	n²	√n	n	n²	√n
0	0	0.000	50	2,500	7.071	100	10,000	10.000	150	22,500	12.247
1	1	1.000	51	2,601	7.141	101	10,201	10.050	151	22,801	12.288
2	4	1.414	52	2,704	7.211	102	10,404	10.100	152	23,104	12.329
3	9	1.732	53	2,809	7.280	103	10,609	10.149	153	23,409	12.369
4	16	2.000	54	2,916	7.348	104	10,816	10.198	154	23,716	12.410
5	25	2.236	55	3,025	7.416	105	11,025	10.247	155	24,025	12.450
6	36	2.449	56	3,136	7.483	106	11,236	10.296	156	24,336	12.490
7	49	2.646	57	3,249	7.550	107	11,449	10.344	157	24,649	12.530
8	64	2.828	58	3,364	7.616	108	11,664	10.392	158	24,964	12.570
9	81	3.000	59	3,481	7.681	109	11,881	10.440	159	25,281	12.610
10	100	3.162	60	3,600	7.746	110	12,100	10.488	160	25,600	12.649
11	121	3.317	61	3,721	7.810	111	12,321	10.536	161	25,921	12.689
12	144	3.464	62	3,844	7.874	112	12,544	10.583	162	26,244	12.728
13	169	3.606	63	3,969	7.937	113	12,769	10.630	163	26,569	12.767
14	196	3.742	64	4,096	8.000	114	12,996	10.677	164	26,896	12.806
15	225	3.873	65	4,225	8.062	115	13,225	10.724	165	27,225	12.845
16	256	4.000	66	4,356	8.124	116	13,456	10.770	166	27,556	12.884
17	289	4.123	67	4,489	8.185	117	13,689	10.817	167	27,889	12.923
18	324	4.243	68	4,624	8.246	118	13,924	10.863	168	28,224	12.961
19	361	4.359	69	4,761	8.307	119	14,161	10.909	169	28,561	13.000
20	400	4.472	70	4,900	8.367	120	14,400	10.954	170	28,900	13.038
21	441	4.583	71	5,041	8.426	121	14,641	11.000	171	29,241	13.077
22	484	4.690	72	5,184	8.485	122	14,884	11.045	172	29,584	13.115
23	529	4.796	73	5,329	8.544	123	15,129	11.091	173	29,929	13.153
24	576	4.899	74	5,476	8.602	124	15,376	11.136	174	30,276	13.191
25	625	5.000	75	5,625	8.660	125	15,625	11.180	175	30,625	13.229
26	676	5.099	76	5,776	8.718	126	15,876	11.225	176	30,976	13.266
27	729	5.196	77	5,929	8.775	127	16,129	11.269	177	31,329	13.304
28	784	5.292	78	6,084	8.832	128	16,384	11.314	178	31,684	13.342
29	841	5.385	79	6,241	8.888	129	16,641	11.358	179	32,041	13.379
30	900	5.477	80	6,400	8.994	130	16,900	11.402	180	32,400	13.416
31	961	5.568	81	6,561	9.000	131	17,161	11.446	181	32,761	13.454
32	1,024	5.657	82	6,724	9.055	132	17,424	11.489	182	33,124	13.491
33	1,089	5.745	83	6,889	9.110	133	17,689	11.533	183	33,489	13.528
34	1,156	5.831	84	7,056	9.165	134	17,956	11.576	184	33,856	13.565
35	1,225	5.916	85	7,225	9.220	135	18,225	11.619	185	34,225	13.601
36	1,296	6.000	86	7,396	9.274	136	18,496	11.662	186	34,596	13.638
37	1,369	6.083	87	7,569	9.327	137	18,769	11.705	187	34,969	13.675
38	1,444	6.164	88	7,744	9.381	138	19,044	11.747	188	35,344	13.711
39	1,521	6.245	89	7,921	9.434	139	19,321	11.790	189	35,721	13.748
40	1,600	6.325	90	8,100	9.487	140	19,600	11.832	190	36,100	13.784
41	1,681	6.403	91	8,281	9.539	141	19,881	11.874	191	36,481	13.820
42	1,764	6.481	92	8,464	9.592	142	20,164	11.916	192	36,864	13.856
43	1,849	6.557	93	8,649	9.644	143	20,449	11.958	193	37,249	13.892
44	1,936	6.633	94	8,836	9.659	144	20,736	12.000	194	37,636	13.928
45	2,025	6.708	95	9,025	9.747	145	21,025	12.042	195	38,025	13.964
46	2,116	6.782	96	9,216	9.798	146	21,316	12.083	196	38,416	14.000
47	2,209	6.856	97	9,409	9.849	147	21,609	12.124	197	38,809	14.036
48	2,304	6.928	98	9,604	9.899	148	21,904	12.166	198	39,204	14.071
49	2,401	7.000	99	9,801	9.950	149	22,201	12.207	199	39,601	14.107
n	n²	√n	n	n²	√n	n	n²	√n	n	n²	√n

TABLE 2
Values of e^x and e^{-x} (0.00 to 3.00)

x	e^x	e^{-x}	x	e^x	e^{-x}	x	e^x	e^{-x}
0.00	1.000	1.000	0.50	1.649	0.607	1.00	2.718	0.358
0.01	1.010	0.990	0.51	1.665	0.600	1.01	2.746	0.364
0.02	1.020	0.980	0.52	1.682	0.595	1.02	2.773	0.361
0.03	1.031	0.970	0.53	1.699	0.589	1.03	2.801	0.357
0.04	1.041	0.961	0.54	1.716	0.583	1.04	2.829	0.353
0.05	1.051	0.951	0.55	1.733	0.577	1.05	2.858	0.350
0.06	1.062	0.942	0.56	1.751	0.571	1.06	2.886	0.346
0.07	1.073	0.932	0.57	1.768	0.566	1.07	2.915	0.343
0.08	1.083	0.923	0.58	1.786	0.560	1.08	2.945	0.340
0.09	1.094	0.914	0.59	1.804	0.554	1.09	2.974	0.336
0.10	1.105	0.905	0.60	1.822	0.549	1.10	3.004	0.333
0.11	1.116	0.896	0.61	1.840	0.543	1.11	3.034	0.330
0.12	1.127	0.887	0.62	1.859	0.538	1.12	3.065	0.326
0.13	1.139	0.878	0.63	1.878	0.533	1.13	3.096	0.323
0.14	1.150	0.869	0.64	1.896	0.527	1.14	3.127	0.320
0.15	1.162	0.861	0.65	1.916	0.522	1.15	3.158	0.317
0.16	1.174	0.852	0.66	1.935	0.517	1.16	3.190	0.313
0.17	1.185	0.844	0.67	1.954	0.512	1.17	3.222	0.310
0.18	1.197	0.835	0.68	1.974	0.507	1.18	3.254	0.307
0.19	1.209	0.827	0.69	1.994	0.502	1.19	3.287	0.304
0.20	1.221	0.819	0.70	2.014	0.497	1.20	3.320	0.301
0.21	1.234	0.811	0.71	2.034	0.492	1.21	3.353	0.298
0.22	1.246	0.803	0.72	2.054	0.487	1.22	3.387	0.295
0.23	1.259	0.795	0.73	2.075	0.482	1.23	3.421	0.292
0.24	1.271	0.787	0.74	2.096	0.477	1.24	3.456	0.289
0.25	1.284	0.779	0.75	2.117	0.472	1.25	3.490	0.287
0.26	1.297	0.771	0.76	2.138	0.468	1.26	3.525	0.284
0.27	1.310	0.763	0.77	2.160	0.463	1.27	3.561	0.281
0.28	1.323	0.756	0.78	2.182	0.458	1.28	3.597	0.278
0.29	1.336	0.748	0.79	2.203	0.454	1.29	3.633	0.275
0.30	1.350	0.741	0.80	2.226	0.449	1.30	3.669	0.273
0.31	1.363	0.733	0.81	2.248	0.445	1.31	3.706	0.270
0.32	1.377	0.726	0.82	2.270	0.440	1.32	3.743	0.267
0.33	1.391	0.719	0.83	2.293	0.436	1.33	3.781	0.264
0.34	1.405	0.712	0.84	2.316	0.432	1.34	3.819	0.262
0.35	1.419	0.705	0.85	2.340	0.427	1.35	3.857	0.259
0.36	1.433	0.698	0.86	2.363	0.423	1.36	3.896	0.257
0.37	1.448	0.691	0.87	2.387	0.419	1.37	3.935	0.254
0.38	1.462	0.684	0.88	2.411	0.415	1.38	3.975	0.252
0.39	1.477	0.677	0.89	2.435	0.411	1.39	4.015	0.249
0.40	1.492	0.670	0.90	2.460	0.407	1.40	4.055	0.247
0.41	1.507	0.664	0.91	2.484	0.403	1.41	4.096	0.244
0.42	1.522	0.657	0.92	2.509	0.399	1.42	4.137	0.242
0.43	1.537	0.651	0.93	2.535	0.395	1.43	4.179	0.239
0.44	1.553	0.644	0.94	2.560	0.391	1.44	4.221	0.237
0.45	1.568	0.638	0.95	2.586	0.387	1.45	4.263	0.235
0.46	1.584	0.631	0.96	2.612	0.383	1.46	4.306	0.232
0.47	1.600	0.625	0.97	2.638	0.379	1.47	4.349	0.230
0.48	1.616	0.619	0.98	2.664	0.375	1.48	4.393	0.228
0.49	1.632	0.613	0.99	2.691	0.372	1.49	4.437	0.225
x	e^x	e^{-x}	x	e^x	e^{-x}	x	e^x	e^{-x}

TABLE 2 (*continued*)

x	e^x	e^{-x}	x	e^x	e^{-x}	x	e^x	e^{-x}
1.50	4.482	0.223	2.00	7.389	0.135	2.50	12.182	0.082
1.51	4.527	0.221	2.01	7.463	0.134	2.51	12.305	0.081
1.52	4.572	0.219	2.02	7.538	0.133	2.52	12.429	0.080
1.53	4.618	0.217	2.03	7.614	0.131	2.53	12.554	0.080
1.54	4.665	0.214	2.04	7.691	0.130	2.54	12.680	0.079
1.55	4.712	0.212	2.05	7.768	0.129	2.55	12.807	0.078
1.56	4.759	0.210	2.06	7.846	0.127	2.56	12.936	0.077
1.57	4.807	0.208	2.07	7.925	0.126	2.57	13.066	0.077
1.58	4.855	0.206	2.08	8.004	0.125	2.58	13.197	0.076
1.59	4.904	0.204	2.09	8.085	0.124	2.59	13.330	0.075
1.60	4.953	0.202	2.10	8.166	0.122	2.60	13.464	0.074
1.61	5.003	0.200	2.11	8.248	0.121	2.61	13.599	0.074
1.62	5.053	0.198	2.12	8.331	0.120	2.62	13.736	0.073
1.63	5.104	0.196	2.13	8.415	0.119	2.63	13.874	0.072
1.64	5.155	0.194	2.14	8.499	0.118	2.64	14.013	0.071
1.65	5.207	0.192	2.15	8.585	0.116	2.65	14.154	0.071
1.66	5.259	0.190	2.16	8.671	0.115	2.66	14.296	0.070
1.67	5.312	0.188	2.17	8.758	0.114	2.67	14.440	0.069
1.68	5.366	0.186	2.18	8.846	0.113	2.68	14.585	0.069
1.69	5.420	0.185	2.19	8.935	0.112	2.69	14.732	0.068
1.70	5.474	0.183	2.20	9.025	0.111	2.70	14.880	0.067
1.71	5.529	0.181	2.21	9.116	0.110	2.71	15.029	0.067
1.72	5.585	0.179	2.22	9.207	0.109	2.72	15.180	0.066
1.73	5.641	0.177	2.23	9.300	0.108	2.73	15.333	0.065
1.74	5.697	0.176	2.24	9.393	0.106	2.74	15.487	0.065
1.75	5.755	0.174	2.25	9.488	0.105	2.75	15.643	0.064
1.76	5.812	0.172	2.26	9.583	0.104	2.76	15.800	0.063
1.77	5.871	0.170	2.27	9.679	0.103	2.77	15.959	0.063
1.78	5.930	0.169	2.28	9.777	0.102	2.78	16.119	0.062
1.79	5.989	0.167	2.29	9.875	0.101	2.79	16.281	0.061
1.80	6.050	0.165	2.30	9.974	0.100	2.80	16.445	0.061
1.81	6.110	0.164	2.31	10.074	0.099	2.81	16.610	0.060
1.82	6.172	0.162	2.32	10.176	0.098	2.82	16.777	0.060
1.83	6.234	0.160	2.33	10.278	0.097	2.83	16.945	0.059
1.84	6.297	0.159	2.34	10.381	0.096	2.84	17.116	0.058
1.85	6.360	0.157	2.35	10.486	0.095	2.85	17.288	0.058
1.86	6.424	0.156	2.36	10.591	0.094	2.86	17.462	0.057
1.87	6.488	0.154	2.37	10.697	0.093	2.87	17.637	0.057
1.88	6.553	0.153	2.38	10.805	0.093	2.88	17.814	0.056
1.89	6.619	0.151	2.39	10.913	0.092	2.89	17.993	0.056
1.90	6.686	0.150	2.40	11.023	0.091	2.90	18.174	0.055
1.91	6.753	0.148	2.41	11.134	0.090	2.91	18.357	0.054
1.92	6.821	0.147	2.42	11.246	0.089	2.92	18.541	0.054
1.93	6.890	0.145	2.43	11.359	0.088	2.93	18.728	0.053
1.94	6.959	0.144	2.44	11.473	0.087	2.94	18.916	0.053
1.95	7.029	0.142	2.45	11.588	0.086	2.95	19.106	0.052
1.96	7.099	0.141	2.46	11.705	0.085	2.96	19.298	0.052
1.97	7.171	0.139	2.47	11.822	0.085	2.97	19.492	0.051
1.98	7.243	0.138	2.48	11.941	0.084	2.98	19.688	0.051
1.99	7.316	0.137	2.49	12.061	0.083	2.99	19.886	0.050
						3.00	20.086	0.050
x	e^x	e^{-x}	x	e^x	e^{-x}	x	e^x	e^{-x}

TABLE 3
Common logarithms

N	0	1	2	3	4	5	6	7	8	9
1.0	0.0000	0.004321	0.008600	0.01284	0.01703	0.02119	0.02531	0.02938	0.03342	0.03743
1.1	0.04139	0.04532	0.04922	0.05308	0.05690	0.06070	0.06446	0.06819	0.07188	0.07555
1.2	0.07918	0.08279	0.08636	0.08991	0.09342	0.09691	0.1004	0.1038	0.1072	0.1106
1.3	0.1139	0.1173	0.1206	0.1239	0.1271	0.1303	0.1335	0.1367	0.1399	0.1430
1.4	0.1461	0.1492	0.1523	0.1553	0.1584	0.1614	0.1644	0.1673	0.1703	0.1732
1.5	0.1761	0.1790	0.1818	0.1847	0.1875	0.1903	0.1931	0.1959	0.1987	0.2014
1.6	0.2041	0.2068	0.2095	0.2122	0.2148	0.2175	0.2201	0.2227	0.2253	0.2279
1.7	0.2304	0.2330	0.2355	0.2380	0.2405	0.2430	0.2455	0.2480	0.2504	0.2529
1.8	0.2553	0.2577	0.2601	0.2625	0.2648	0.2673	0.2695	0.2718	0.2742	0.2765
1.9	0.2788	0.2810	0.2833	0.2856	0.2878	0.2900	0.2923	0.2945	0.2967	0.2989
2.0	0.3010	0.3032	0.3054	0.3075	0.3096	0.3118	0.3139	0.3160	0.3181	0.3201
2.1	0.3222	0.3243	0.3263	0.3284	0.3304	0.3324	0.3345	0.3365	0.3385	0.3404
2.2	0.3224	0.3444	0.3464	0.3483	0.3502	0.3522	0.3541	0.3560	0.3579	0.3598
2.3	0.3617	0.3636	0.3655	0.3674	0.3692	0.3711	0.3729	0.3747	0.3766	0.3784
2.4	0.3802	0.3820	0.3838	0.3856	0.3874	0.3892	0.3909	0.3927	0.3945	0.3962
2.5	0.3979	0.3997	0.4014	0.4031	0.4048	0.4065	0.4082	0.4099	0.4116	0.4133
2.6	0.4150	0.4166	0.4183	0.4200	0.4216	0.4232	0.4249	0.4265	0.4281	0.4298
2.7	0.4314	0.4330	0.4346	0.4362	0.4378	0.4393	0.4409	0.4425	0.4440	0.4456
2.8	0.4472	0.4487	0.4502	0.4518	0.4533	0.4548	0.4564	0.4579	0.4594	0.4609
2.9	0.4624	0.4639	0.4654	0.4669	0.4683	0.4698	0.4713	0.4728	0.4742	0.4757
3.0	0.4771	0.4786	0.4800	0.4814	0.4829	0.4843	0.4857	0.4871	0.4886	0.4900
3.1	0.4914	0.4928	0.4942	0.4955	0.4969	0.4983	0.4997	0.5011	0.5024	0.5038
3.2	0.5051	0.5065	0.5079	0.5092	0.5105	0.5119	0.5132	0.5145	0.5159	0.5172
3.3	0.5185	0.5198	0.5211	0.5224	0.5237	0.5250	0.5263	0.5276	0.5289	0.5302
3.4	0.5315	0.5328	0.5340	0.5353	0.5366	0.5378	0.5391	0.5403	0.5416	0.5428
3.5	0.5441	0.5453	0.5465	0.5478	0.5490	0.5502	0.5514	0.5527	0.5539	0.5551
3.6	0.5563	0.5575	0.5587	0.5599	0.5611	0.5623	0.5635	0.5647	0.5658	0.5670
3.7	0.5682	0.5694	0.5705	0.5717	0.5729	0.5740	0.5752	0.5763	0.5775	0.5786
3.8	0.5798	0.5809	0.5821	0.5832	0.5843	0.5855	0.5866	0.5877	0.5888	0.5899
3.9	0.5911	0.5922	0.5933	0.5944	0.5955	0.5966	0.5977	0.5988	0.5999	0.6010
4.0	0.6021	0.6031	0.6042	0.6053	0.6064	0.6075	0.6085	0.6096	0.6107	0.6117
4.1	0.6128	0.6138	0.6149	0.6160	0.6170	0.6180	0.6191	0.6201	0.6212	0.6222
4.2	0.6232	0.6243	0.6253	0.6263	0.6274	0.6284	0.6294	0.6304	0.6314	0.6325
4.3	0.6335	0.6345	0.6355	0.6365	0.6375	0.6385	0.6395	0.6405	0.6415	0.6425
4.4	0.6435	0.6444	0.6454	0.6464	0.6474	0.6484	0.6493	0.6503	0.6513	0.6522
4.5	0.6532	0.6542	0.6551	0.6561	0.6571	0.6580	0.6590	0.6599	0.6609	0.6618
4.6	0.6628	0.6637	0.6646	0.6656	0.6665	0.6675	0.6684	0.6693	0.6702	0.6712
4.7	0.6721	0.6730	0.6739	0.6749	0.6758	0.6767	0.6776	0.6785	0.6794	0.6803
4.8	0.6812	0.6821	0.6830	0.6839	0.6848	0.6857	0.6866	0.6875	0.6884	0.6893
4.9	0.6902	0.6911	0.6920	0.6928	0.6937	0.6946	0.6955	0.6964	0.6972	0.6981
5.0	0.6990	0.6998	0.7007	0.7016	0.7024	0.7033	0.7042	0.7050	0.7059	0.7067
5.1	0.7076	0.7084	0.7093	0.7101	0.7110	0.7118	0.7126	0.7135	0.7143	0.7152
5.2	0.7160	0.7168	0.7177	0.7185	0.7193	0.7202	0.7210	0.7218	0.7226	0.7235
5.3	0.7243	0.7251	0.7259	0.7267	0.7275	0.7284	0.7292	0.7300	0.7308	0.7316
5.4	0.7324	0.7332	0.7340	0.7348	0.7356	0.7364	0.7372	0.7380	0.7388	0.7396
5.5	0.7404	0.7412	0.7419	0.7427	0.7435	0.7443	0.7451	0.7459	0.7466	0.7474
5.6	0.7482	0.7490	0.7497	0.7505	0.7513	0.7520	0.7528	0.7536	0.7543	0.7551
5.7	0.7559	0.7566	0.7574	0.7582	0.7589	0.7597	0.7604	0.7612	0.7619	0.7627
5.8	0.7634	0.7642	0.7649	0.7657	0.7664	0.7672	0.7679	0.7686	0.7694	0.7701
5.9	0.7709	0.7716	0.7723	0.7731	0.7738	0.7745	0.7752	0.7760	0.7767	0.7774
N	0	1	2	3	4	5	6	7	8	9

TABLE 3 (continued)

N	0	1	2	3	4	5	6	7	8	9
6.0	0.7782	0.7789	0.7796	0.7803	0.7810	0.7818	0.7825	0.7832	0.7839	0.7846
6.1	0.7853	0.7860	0.7868	0.7875	0.7882	0.7889	0.7896	0.7903	0.7910	0.7917
6.2	0.7924	0.7931	0.7938	0.7945	0.7952	0.7959	0.7966	0.7973	0.7980	0.7987
6.3	0.7993	0.8000	0.8007	0.8014	0.8021	0.8028	0.8035	0.8041	0.8048	0.8055
6.4	0.8062	0.8069	0.8075	0.8082	0.8089	0.8096	0.8102	0.8109	0.8116	0.8122
6.5	0.8129	0.8136	0.8142	0.8149	0.8156	0.8162	0.8169	0.8176	0.8182	0.8189
6.6	0.8195	0.8202	0.8209	0.8215	0.8222	0.8228	0.8235	0.8241	0.8248	0.8254
6.7	0.8261	0.8267	0.8274	0.8280	0.8287	0.8293	0.8299	0.8306	0.8312	0.8319
6.8	0.8325	0.8331	0.8338	0.8344	0.8351	0.8357	0.8363	0.8370	0.8376	0.8382
6.9	0.8388	0.8395	0.8401	0.8407	0.8414	0.8420	0.8426	0.8432	0.8439	0.8445
7.0	0.8451	0.8457	0.8463	0.8470	0.8476	0.8482	0.8488	0.8494	0.8500	0.8506
7.1	0.8513	0.8519	0.8525	0.8531	0.8537	0.8543	0.8549	0.8555	0.8561	0.8567
7.2	0.8573	0.8579	0.8585	0.8591	0.8597	0.8603	0.8609	0.8615	0.8621	0.8627
7.3	0.8633	0.8639	0.8645	0.8651	0.8657	0.8663	0.8669	0.8675	0.8681	0.8686
7.4	0.8692	0.8698	0.8704	0.8710	0.8716	0.8722	0.8727	0.8733	0.8739	0.8745
7.5	0.8751	0.8756	0.8762	0.8768	0.8774	0.8779	0.8785	0.8791	0.8797	0.8802
7.6	0.8808	0.8814	0.8820	0.8825	0.8831	0.8837	0.8842	0.8848	0.8854	0.8859
7.7	0.8865	0.8871	0.8876	0.8882	0.8887	0.8893	0.8899	0.8904	0.8910	0.8915
7.8	0.8921	0.8927	0.8932	0.8938	0.8943	0.8949	0.8954	0.8960	0.8965	0.8971
7.9	0.8976	0.8982	0.8987	0.8993	0.8998	0.9004	0.9009	0.9015	0.9020	0.9025
8.0	0.9031	0.9036	0.9042	0.9047	0.9053	0.9058	0.9063	0.9069	0.9074	0.9079
8.1	0.9085	0.9090	0.9096	0.9101	0.9106	0.9112	0.9117	0.9122	0.9128	0.9133
8.2	0.9138	0.9143	0.9149	0.9154	0.9159	0.9165	0.9170	0.9175	0.9180	0.9186
8.3	0.9191	0.9196	0.9201	0.9206	0.9212	0.9217	0.9222	0.9227	0.9232	0.9238
8.4	0.9243	0.9248	0.9253	0.9258	0.9263	0.9269	0.9274	0.9279	0.9284	0.9289
8.5	0.9294	0.9299	0.9304	0.9309	0.9315	0.9320	0.9325	0.9330	0.9335	0.9340
8.6	0.9345	0.9350	0.9355	0.9360	0.9365	0.9370	0.9375	0.9380	0.9385	0.9390
8.7	0.9395	0.9400	0.9405	0.9410	0.9415	0.9420	0.9425	0.9430	0.9435	0.9440
8.8	0.9445	0.9450	0.9455	0.9460	0.9465	0.9469	0.9474	0.9479	0.9484	0.9489
8.9	0.9494	0.9499	0.9504	0.9509	0.9513	0.9518	0.9523	0.9528	0.9533	0.9538
9.0	0.9542	0.9547	0.9552	0.9557	0.9562	0.9566	0.9571	0.9576	0.9581	0.9586
9.1	0.9590	0.9595	0.9600	0.9605	0.9609	0.9614	0.9619	0.9624	0.9628	0.9633
9.2	0.9638	0.9643	0.9647	0.9652	0.9657	0.9661	0.9666	0.9671	0.9675	0.9680
9.3	0.9685	0.9689	0.9694	0.9699	0.9703	0.9708	0.9713	0.9717	0.9722	0.9727
9.4	0.9731	0.9736	0.9741	0.9745	0.9750	0.9754	0.9759	0.9763	0.9768	0.9773
9.5	0.9777	0.9782	0.9786	0.9791	0.9795	0.9800	0.9805	0.9809	0.9814	0.9818
9.6	0.9823	0.9827	0.9832	0.9836	0.9841	0.9845	0.9850	0.9854	0.9859	0.9863
9.7	0.9868	0.9872	0.9877	0.9881	0.9886	0.9890	0.9894	0.9899	0.9903	0.9908
9.8	0.9912	0.9917	0.9921	0.9926	0.9930	0.9934	0.9939	0.9943	0.9948	0.9952
9.9	0.9956	0.9961	0.9965	0.9969	0.9974	0.9978	0.9983	0.9987	0.9991	0.9996
N	0	1	2	3	4	5	6	7	8	9

TABLE 4
Natural logarithms (ln N = log$_e$ N)

ln 10 = 2.3026   5 ln 10 = 11.5130   9 ln 10 = 20.7233
2 ln 10 = 4.6052   6 ln 10 = 13.8155   10 ln 10 = 23.0259
3 ln 10 = 6.9078   7 ln 10 = 16.1181   Note: ln 35,200 = ln (3.52×10^4) = ln 3.52 + 4 ln 10
4 ln 10 = 9.2103   8 ln 10 = 18.4207   ln 0.00864 = ln (8.64×10^{-3}) = ln 8.64 − 3 ln 10

N	0.00	0.01	0.02	0.03	0.04	0.05	0.06	0.07	0.08	0.09
1.0	0.0000	0.0100	0.0198	0.0296	0.0392	0.0488	0.0583	0.0677	0.0770	0.0862
1.1	0.0953	0.1044	0.1133	0.1222	0.1310	0.1398	0.1484	0.1570	0.1655	0.1740
1.2	0.1823	0.1906	0.1989	0.2070	0.2151	0.2231	0.2311	0.2390	0.2469	0.2546
1.3	0.2624	0.2700	0.2776	0.2852	0.2927	0.3001	0.3075	0.3148	0.3221	0.3293
1.4	0.3365	0.3436	0.3507	0.3577	0.3646	0.3716	0.3784	0.3853	0.3920	0.3988
1.5	0.4055	0.4121	0.4187	0.4253	0.4318	0.4383	0.4447	0.4511	0.4574	0.4637
1.6	0.4700	0.4762	0.4824	0.4886	0.4947	0.5008	0.5068	0.5128	0.5188	0.5247
1.7	0.5306	0.5365	0.5423	0.5481	0.5534	0.5596	0.5653	0.5710	0.5766	0.5822
1.8	0.5878	0.5933	0.5988	0.6043	0.6098	0.6152	0.6206	0.6259	0.6313	0.6366
1.9	0.6419	0.6471	0.6523	0.6575	0.6627	0.6678	0.6729	0.6780	0.6831	0.6881
2.0	0.6931	0.6981	0.7031	0.7080	0.7129	0.7178	0.7227	0.7275	0.7324	0.7372
2.1	0.7419	0.7467	0.7514	0.7561	0.7608	0.7655	0.7701	0.7747	0.7793	0.7839
2.2	0.7885	0.7930	0.7975	0.8020	0.8065	0.8109	0.8154	0.8198	0.8242	0.8286
2.3	0.8329	0.8372	0.8416	0.8459	0.8502	0.8544	0.8587	0.8629	0.8671	0.8713
2.4	0.8755	0.8796	0.8838	0.8879	0.8920	0.8961	0.9002	0.9042	0.9083	0.9123
2.5	0.9163	0.9203	0.9243	0.9282	0.9322	0.9361	0.9400	0.9439	0.9478	0.9517
2.6	0.9555	0.9594	0.9632	0.9670	0.9708	0.9746	0.9783	0.9821	0.9858	0.9895
2.7	0.9933	0.9969	1.0006	1.0043	1.0080	1.0116	1.0152	1.0188	1.0225	1.0260
2.8	1.0296	1.0332	1.0367	1.0403	1.0438	1.0473	1.0508	1.0543	1.0578	1.0613
2.9	1.0647	1.0682	1.0716	1.0750	1.0784	1.0818	1.0852	1.0886	1.0919	1.0953
3.0	1.0986	1.1019	1.1053	1.1086	1.1119	1.1151	1.1184	1.1217	1.1249	1.1282
3.1	1.1314	1.1346	1.1378	1.1410	1.1442	1.1474	1.1506	1.1537	1.1569	1.1600
3.2	1.1632	1.1663	1.1694	1.1725	1.1756	1.1787	1.1817	1.1848	1.1878	1.1909
3.3	1.1939	1.1969	1.2000	1.2030	1.2060	1.2090	1.2119	1.2149	1.2179	1.2208
3.4	1.2238	1.2267	1.2296	1.2326	1.2355	1.2384	1.2413	1.2442	1.2470	1.2499
3.5	1.2528	1.2556	1.2585	1.2613	1.2641	1.2669	1.2698	1.2726	1.2754	1.2782
3.6	1.2809	1.2837	1.2865	1.2892	1.2920	1.2947	1.2975	1.3002	1.3029	1.3056
3.7	1.3083	1.3110	1.3137	1.3164	1.3191	1.3218	1.3244	1.3271	1.3297	1.3324
3.8	1.3350	1.3376	1.3403	1.3429	1.3455	1.3481	1.3507	1.3533	1.3558	1.3584
3.9	1.3610	1.3635	1.3661	1.3686	1.3712	1.3737	1.3762	1.3788	1.3813	1.3838
4.0	1.3863	1.3888	1.3913	1.3938	1.3962	1.3987	1.4012	1.4036	1.4061	1.4085
4.1	1.4110	1.4134	1.4159	1.4183	1.4207	1.4231	1.4255	1.4279	1.4303	1.4327
4.2	1.4351	1.4375	1.4398	1.4422	1.4446	1.4469	1.4493	1.4516	1.4540	1.4563
4.3	1.4586	1.4609	1.4633	1.4656	1.4679	1.4702	1.4725	1.4748	1.4770	1.4793
4.4	1.4816	1.4839	1.4861	1.4884	1.4907	1.4929	1.4951	1.4974	1.4996	1.5019
4.5	1.5041	1.5063	1.5085	1.5107	1.5129	1.5151	1.5173	1.5195	1.5217	1.5239
4.6	1.5261	1.5282	1.5304	1.5326	1.5347	1.5369	1.5390	1.5412	1.5433	1.5454
4.7	1.5476	1.5497	1.5518	1.5539	1.5560	1.5581	1.5602	1.5623	1.5644	1.5665
4.8	1.5686	1.5707	1.5728	1.5748	1.5769	1.5790	1.5810	1.5831	1.5851	1.5872
4.9	1.5892	1.5913	1.5933	1.5953	1.5974	1.5994	1.6014	1.6034	1.6054	1.6074
5.0	1.6094	1.6114	1.6134	1.6154	1.6174	1.6194	1.6214	1.6233	1.6253	1.6273
5.1	1.6292	1.6312	1.6332	1.6351	1.6371	1.6390	1.6409	1.6429	1.6448	1.6467
5.2	1.6487	1.6506	1.6525	1.6544	1.6563	1.6582	1.6601	1.6620	1.6639	1.6658
5.3	1.6677	1.6696	1.6715	1.6734	1.6752	1.6771	1.6790	1.6808	1.6827	1.6845
5.4	1.6864	1.6882	1.6901	1.6919	1.6938	1.6956	1.6974	1.6993	1.7011	1.7029
N	0.00	0.01	0.02	0.03	0.04	0.05	0.06	0.07	0.08	0.09

TABLE 4 (*continued*)

N	0.00	0.01	0.02	0.03	0.04	0.05	0.06	0.07	0.08	0.09
5.5	1.7047	1.7066	1.7084	1.7102	1.7120	1.7138	1.7156	1.7174	1.7192	1.7210
5.6	1.7228	1.7246	1.7263	1.7281	1.7299	1.7317	1.7334	1.7352	1.7370	1.7387
5.7	1.7405	1.7422	1.7440	1.7457	1.7475	1.7492	1.7509	1.7527	1.7544	1.7561
5.8	1.7579	1.7596	1.7613	1.7630	1.7647	1.7664	1.7681	1.7699	1.7716	1.7733
5.9	1.7750	1.7766	1.7783	1.7800	1.7817	1.7834	1.7851	1.7867	1.7884	1.7901
6.0	1.7918	1.7934	1.7951	1.7967	1.7984	1.8001	1.8017	1.8034	1.8050	1.8066
6.1	1.8083	1.8099	1.8116	1.8132	1.8148	1.8165	1.8181	1.8197	1.8213	1.8229
6.2	1.8245	1.8262	1.8278	1.8294	1.8310	1.8326	1.8342	1.8358	1.8374	1.8390
6.3	1.8405	1.8421	1.8437	1.8453	1.8469	1.8485	1.8500	1.8516	1.8532	1.8547
6.4	1.8563	1.8579	1.8594	1.8610	1.8625	1.8641	1.8656	1.8672	1.8687	1.8703
6.5	1.8718	1.8733	1.8749	1.8764	1.8779	1.8795	1.8810	1.8825	1.8840	1.8856
6.6	1.8871	1.8886	1.8901	1.8916	1.8931	1.8946	1.8961	1.8976	1.8991	1.9006
6.7	1.9021	1.9036	1.9051	1.9066	1.9081	1.9095	1.9110	1.9125	1.9140	1.9155
6.8	1.9169	1.9184	1.9199	1.9213	1.9228	1.9242	1.9257	1.9272	1.9286	1.9301
6.9	1.9315	1.9330	1.9344	1.9359	1.9373	1.9387	1.9402	1.9416	1.9430	1.9445
7.0	1.9459	1.9473	1.9488	1.9502	1.9516	1.9530	1.9544	1.9559	1.9573	1.9587
7.1	1.9601	1.9615	1.9629	1.9643	1.9657	1.9671	1.9685	1.9699	1.9713	1.9727
7.2	1.9741	1.9755	1.9769	1.9782	1.9796	1.9810	1.9824	1.9838	1.9851	1.9865
7.3	1.9879	1.9892	1.9906	1.9920	1.9933	1.9947	1.9961	1.9974	1.9988	2.0001
7.4	2.0015	2.0028	2.0042	2.0055	2.0069	2.0082	2.0096	2.0109	2.0122	2.0136
7.5	2.0149	2.0162	2.0176	2.0189	2.0202	2.0215	2.0229	2.0242	2.0255	2.0268
7.6	2.0281	2.0295	2.0308	2.0321	2.0334	2.0347	2.0360	2.0373	2.0386	2.0399
7.7	2.0412	2.0425	2.0438	2.0451	2.0464	2.0477	2.0490	2.0503	2.0516	2.0528
7.8	2.0541	2.0554	2.0567	2.0580	2.0592	2.0605	2.0618	2.0631	2.0643	2.0656
7.9	2.0669	2.0681	2.0694	2.0707	2.0719	2.0732	2.0744	2.0757	2.0769	2.0782
8.0	2.0794	2.0807	2.0819	2.0832	2.0844	2.0857	2.0869	2.0882	2.0894	2.0906
8.1	2.0919	2.0931	2.0943	2.0956	2.0968	2.0980	2.0992	2.1005	2.1017	2.1029
8.2	2.1041	2.1054	2.1066	2.1078	2.1090	2.1102	2.1114	2.1126	2.1138	2.1150
8.3	2.1163	2.1175	2.1187	2.1199	2.1211	2.1223	2.1235	2.1247	2.1258	2.1270
8.4	2.1282	2.1294	2.1306	2.1318	2.1330	2.1342	2.1353	2.1365	2.1377	2.1389
8.5	2.1401	2.1412	2.1424	2.1436	2.1448	2.1459	2.1471	2.1483	2.1494	2.1506
8.6	2.1518	2.1529	2.1541	2.1552	2.1564	2.1576	2.1587	2.1599	2.1610	2.1622
8.7	2.1633	2.1645	2.1656	2.1668	2.1679	2.1691	2.1702	2.1713	2.1725	2.1736
8.8	2.1748	2.1759	2.1770	2.1782	2.1793	2.1804	2.1815	2.1827	2.1838	2.1849
8.9	2.1861	2.1872	2.1883	2.1894	2.1905	2.1917	2.1928	2.1939	2.1950	2.1961
9.0	2.1972	2.1983	2.1994	2.2006	2.2017	2.2028	2.2039	2.2050	2.2061	2.2072
9.1	2.2083	2.2094	2.2105	2.2116	2.2127	2.2138	2.2148	2.2159	2.2170	2.2181
9.2	2.2192	2.2203	2.2214	2.2225	2.2235	2.2246	2.2257	2.2268	2.2279	2.2289
9.3	2.2300	2.2311	2.2322	2.2332	2.2343	2.2354	2.2364	2.2375	2.2386	2.2396
9.4	2.2407	2.2418	2.2428	2.2439	2.2450	2.2460	2.2471	2.2481	2.2492	2.2502
9.5	2.2513	2.2523	2.2534	2.2544	2.2555	2.2565	2.2576	2.2586	2.2597	2.2607
9.6	2.2618	2.2628	2.2638	2.2649	2.2659	2.2670	2.2680	2.2690	2.2701	2.2711
9.7	2.2721	2.2732	2.2742	2.2752	2.2762	2.2773	2.2783	2.2793	2.2803	2.2814
9.8	2.2824	2.2834	2.2844	2.2854	2.2865	2.2875	2.2885	2.2895	2.2905	2.2915
9.9	2.2925	2.2935	2.2946	2.2956	2.2966	2.2976	2.2986	2.2996	2.3006	2.3016
N	0.00	0.01	0.02	0.03	0.04	0.05	0.06	0.07	0.08	0.09

APPENDIX E

CHAPTER 1

Practice Exercise 1.1

ODD **1.** T   **3.** T   **5.** T   **7.** T   **9.** T   **11.** $\{1, 2, 3, 4, 5\}$ **13.** $\{3\}$ **15.** $\varnothing$ **17.** $\{5\}$
19. $\varnothing$ **21.** $\{-2, 2\}$ **23.** (A) F   (B) T   (C) T   (D) T   (E) T   (F) F   **25.** (A) $B = \{-10, 10\}$
(B) $B = \{x \mid x^2 = 100\}$ **27.** $\{1, 2, 3, 4, 6\}$ **29.** 6

EVEN **2.** F   **4.** F   **6.** F   **8.** T   **10.** T   **12.** $\{3, 4, 5, 6, 7\}$ **14.** $\{3, 4\}$ **16.** $\varnothing$
18. $\{-3\}$ **20.** $\varnothing$ **22.** $\{-3, 3\}$ **24.** (A) F   (B) T   (C) T   (D) T   (E) F   (F) T
26. (A) $M = \{-8, 8\}$ (B) $M = \{x \mid x^2 = 64\}$ **28.** $\{2, 4\}$ **30.** 4

Practice Exercise 1.2

ODD **1.** 1   **3.** 10   **5.** 1   **7.** 14   **9.** 2   **11.** 8   **13.** 5   **15.** 14   **17.** 0   **19.** 18
21. 10   **23.** 3   **25.** 8   **27.** 8   **29.** 46   **31.** 110   **33.** 10   **35.** 9   **37.** 16   **39.** 38
41. $18 = 3x$ **43.** $26 = x - 12$ **45.** $43 = 4x - 7$ **47.** $6x = 3x + 4$ **49.** $x - 6 = 5(x + 7)$ **51.** 42
53. 6   **55.** Wrong use of equal sign; $4 \neq$ even number   **57.** $2x + 2(3x - 10) = 210$

EVEN **2.** 4   **4.** 14   **6.** 28   **8.** 100   **10.** 0   **12.** 3   **14.** 15   **16.** 24   **18.** 8   **20.** 12
22. 180   **24.** 8   **26.** 14   **28.** 18   **30.** 105   **32.** 16   **34.** 8   **36.** 28   **38.** 6   **40.** 129
42. $80 = 2x + 3$ **44.** $32 = x - 5$ **46.** $62 = 5x - 9$ **48.** $7x = 4x + 12$ **50.** $x + 5 = 3(x - 4)$ **52.** 3
54. 10   **56.** $x(2x - 3) = 90$

Practice Exercise 1.3

ODD **1.** Commutative add.   **3.** Associative add.   **5.** Commutative mult.   **7.** Associative mult.
9. Distributive   **11.** Distributive   **13.** $x + 9$ **15.** $20y$ **17.** $u + 15$ **19.** $21x$
21. Commutative add.   **23.** Commutative mult.   **25.** Distributive   **27.** Commutative add.   **29.** Distributive
31. $x + y + z + 12$ **33.** $3x + 4y + 11$ **35.** $36mnp$ **37.** (A) T   (B) F; for example, $5 - 1 \neq 1 - 5$ (C) T
(D) F, for example, $6 \div 3 \neq 3 \div 6$

EVEN **2.** Commutative add.   **4.** Associative add.   **6.** Commutative mult.   **8.** Associative mult.
10. Distributive   **12.** Distributive   **14.** $8 + m$ **16.** $48n$ **18.** $x + 17$ **20.** $12y$
22. Commutative mult.   **24.** Commutative add.   **26.** Distributive   **28.** Associative add.   **30.** Distributive
32. $18 + m + n + p$ **34.** $3a + 5b + 9$ **36.** $64xyz$ **38.** (A) T   (B) F; for example, $(8 - 4) - 1 \neq 8 - (4 - 1)$
(C) T   (D) F; for example, $(16 \div 8) \div 4 \neq 16 \div (8 \div 4)$

Practice Exercise 1.4

ODD **1.** -7 **3.** 6   **5.** 2   **7.** 27   **9.** 0   **11.** -10 **13.** 4   **15.** -6 **17.** 12
19. Sometimes   **21.** -3 **23.** -2 **25.** -6 **27.** -5 **29.** -5 **31.** -5 **33.** -1
35. -6 **37.** 28   **39.** -5 **41.** 7 or -7 **43.** -5 **45.** 5   **47.** -5 **49.** -3 **51.** 6
53. \$23

EVEN   **2.** -12 **4.** 8   **6.** 9   **8.** 32   **10.** 0   **12.** -4 **14.** -4 **16.** 12   **18.** -6
20. Sometimes, since $|0| = 0$ **22.** 6   **24.** -3 **26.** -6 **28.** -5 **30.** 5   **32.** -3 **34.** -2
36. -7 **38.** -5 **40.** 8   **42.** No real number   **44.** -11 **46.** -6 **48.** -10 **50.** -3
52. 4   **54.** 20,550 ft

Practice Exercise 1.5

ODD   **1.** 15   **3.** 3   **5.** -18 **7.** -3 **9.** 0   **11.** 0   **13.** Not defined   **15.** Not defined
17. -7 **19.** -1 **21.** 11   **23.** 9   **25.** Both are 8   **27.** -10 **29.** 3   **31.** -14 **33.** -5
35. -70 **37.** 0   **39.** 10   **41.** -12 **43.** Not defined   **45.** 56   **47.** 0   **49.** -6 **51.** 8
53. Never   **55.** Always   **57.** -50 **59.** 0

EVEN   **2.** 28   **4.** 5   **6.** -18 **8.** -3 **10.** 0   **12.** 0   **14.** Not defined   **16.** Not defined
18. 9   **20.** -13 **22.** -19 **24.** -1 **26.** Both are -3 **28.** -9 **30.** -5 **32.** -8
34. -6 **36.** 72   **38.** 0   **40.** -22 **42.** 8   **44.** 0   **46.** 6   **48.** 0   **50.** 72   **52.** -12
54. Sometimes; zero cannot be divided by zero   **56.** Never   **58.** -3 **60.** -9

Practice Exercise 1.6

ODD   **1.** 9   **3.** 5   **5.** 12   **7.** 2   **9.** $u^7 v^7$ **11.** 4   **13.** $\dfrac{a^8}{b^8}$ **15.** 3   **17.** 6   **19.** 2

21. 7   **23.** 12   **25.** $10x^{11}$ **27.** $3x^2$ **29.** $\dfrac{3}{4m^2}$ **31.** $x^{10}y^{10}$ **33.** $\dfrac{m^5}{n^5}$ **35.** $12y^{10}$

37. 35×10^{17} **39.** 10^{14} **41.** x^6 **43.** $m^6 n^{15}$ **45.** $\dfrac{c^6}{d^{15}}$ **47.** $\dfrac{3u^4}{v^2}$ **49.** $2^4 s^8 t^{16}$ or $16 s^8 t^{16}$

51. $6x^5 y^{15}$ **53.** $\dfrac{m^4 n^{12}}{p^8 q^4}$ **55.** $\dfrac{u^3}{v^9}$ **57.** $9x^4$ **59.** -1 **61.** $\dfrac{(x-y)^2}{2(x+y)^2}$

EVEN   **2.** 12   **4.** 4   **6.** v^6 **8.** 2   **10.** 5   **12.** 3   **14.** 4   **16.** 7   **18.** 4   **20.** 7

22. 4   **24.** 7   **26.** $6x^{10}$ **28.** $2x^2$ **30.** $\dfrac{2}{u^4}$ **32.** $c^{12}d^{12}$ **34.** $\dfrac{x^6}{y^6}$ **36.** $6x^9$ **38.** 6×10^{15}

40. 10^{20} **42.** y^{20} **44.** $x^8 y^{12}$ **46.** $\dfrac{a^{12}}{b^8}$ **48.** $\dfrac{y^6}{3x^4}$ **50.** $3^3 a^9 b^6$ **52.** $2x^8 y^4$ **54.** $\dfrac{x^6 y^3}{8w^6}$

56. $\dfrac{y^3}{16x^4}$ **58.** $\dfrac{-x^2}{32}$ **60.** -1 **62.** $\dfrac{2}{(u-v+w)^3}$

Diagnostic (Review) Exercise 1.7

1. **(A)** F   **(B)** T   **(C)** T   **(D)** F (1.1) **2.** **(A)** $\{1, 3, 4, 5\}$ **(B)** $\{3, 5\}$ **(C)** $\varnothing$ **(D)** $\{2, 4, 6\}$ (1.1)
3. 17 (1.2) **4.** 13 (1.2) **5.** -5 (1.4) **6.** -13 (1.4) **7.** 6 (1.4) **8.** -3 (1.4) **9.** 3 (1.4)
10. -12 (1.4) **11.** 28 (1.5) **12.** -18 (1.5) **13.** -4 (1.5) **14.** 6 (1.5) **15.** Not defined (1.5)
16. 0 (1.5) **17.** 4 (1.5) **18.** -14 (1.5) **19.** -5 (1.5) **20.** -12 (1.5) **21.** 8 (1.4) **22.** 5 (1.4)
23. -3 (1.4) **24.** -2 (1.4) **25.** -5 (1.4) **26.** -5 (1.4) **27.** $x + 10$ (1.3) **28.** $15x$ (1.3)
29. $8xy$ (1.3) **30.** $x + y + z + 12$ (1.3) **31.** 5^{14} (1.6) **32.** x^{12} (1.6) **33.** x^{12} (1.6)
34. $2^3 x^3 y^3$ or $8x^3 y^3$ (1.6) **35.** $\dfrac{c^4}{d^4}$ (1.6) **36.** $\dfrac{x^4}{y^6}$ (1.6) **37.** u^5 (1.6) **38.** $\dfrac{1}{y^4}$ (1.6) **39.** 6 (1.2)
40. 2 (1.4) **41.** 4 (1.4) **42.** 1 (1.4) **43.** 4 (1.5) **44.** -26 (1.5) **45.** 10 (1.5) **46.** 35 (1.5)
47. 6 (1.2) **48.** -2 (1.4) **49.** 8 (1.4) **50.** -6 (1.4) **51.** -9 (1.5) **52.** -7 (1.5)
53. Commutative add. (1.3) **54.** Associative add. (1.3) **55.** Commutative mult. (1.3) **56.** Associative mult. (1.3)
57. Distributive (1.3) **58.** Distributive (1.3) **59.** $9x^6 y^8$ **60.** $12u^6 v^4$ (1.6) **61.** $\dfrac{2x}{y^4}$ (1.6) **62.** $\dfrac{x^4}{2y^5}$ (1.6)
63. $\dfrac{8x^6}{y^{15}}$ (1.6) **64.** $6x^{11}$ (1.6) **65.** -4 (1.6) **66.** 12×10^7 (1.6) **67.** $3x = 2x - 8$ (1.2)
68. $x + 4 = 2(x - 3)$ (1.2) **69.** $2(x + 5) + 2x = 43$ (1.2) **70.** $\{-10, 10\}$ (1.1) **71.** $\varnothing$ (1.1)
72. $\{-2, -1, 0, 1, 2\}$ $(1.1, 1.3)$ **73.** -42 (1.5) **74.** 6 (1.5) **75.** $4(x + y - z)^3$ (1.6) **76.** $\dfrac{x+y}{(x-y)^4}$ (1.6)
77. $2(x - 3) = 3x - 5$ (1.2)

Practice Test Chapter 1

1. -10 *(1.4)*   **2.** 3 *(1.4, 1.5)*   **3.** 36 *(1.4, 1.5)*   **4.** 39 *(1.4, 1.5)*   **5.** 5 *(1.4)*   **6.** -15 *(1.4, 1.5)*
7. 9 *(1.4, 1.5)*   **8.** B *(1.5)*   **9.** C *(1.1)*   **10.** A *(1.3)*   **11.** $(5 + 1 + 4)x$ *(1.3)*   **12.** $\{-3, -2, -1, 0, 1\}$ *(1.1, 1.3)*
13. $\{x \mid x^2 = 9\}$ *(1.1)*   **14.** $x + 5 = 2x - 6$ *(1.2)*   **15.** $74 = 2x + 2(2x - 2)$ *(1.2)*   **16.** $12m^7n^5$ *(1.6)*
17. $\dfrac{4x^8}{y^4}$ *(1.6)*   **18.** $\dfrac{2}{u^5v^4}$ *(1.6)*   **19.** -4 *(1.6)*   **20.** $\dfrac{(u + 2v)^2}{(u - v)^3}$ *(1.6)*

CHAPTER 2

Practice Exercise 2.2

ODD   **1.** -3 **3.** 3 **5.** 1 **7.** $17x$ **9.** x **11.** $8x$ **13.** $-13t$ **15.** $3x + 5y$
17. $4m - 6n$ **19.** $9u - 4v$ **21.** $-2m - 24n$ **23.** $5u - 6v$ **25.** $-u + 4v$ **27.** $9x - 3$
29. $-2x - 4$ **31.** $7x^2 - x - 12$ **33.** $-x + 1$ **35.** $-y^2 - 2$ **37.** $-3x^2y$ **39.** $3y^3 + 4y^2 - y - 3$
41. $3a^2 - b^2$ **43.** $-7x + 9y$ **45.** $-5x + 3y$ **47.** $4x - 6$ **49.** $-8x + 12$ **51.** $10t - 18$
53. $m - 2n$ **55.** $-m + 2n$ **57.** $-x + y - z$ **59.** $2x^4 + 3x^3 + 7x^2 - x - 8$ **61.** $-3x^3 + x^2 + 3x - 2$
63. $-2m^3 - 5$ **65.** $-t + 27$ **67.** $2x - w$ **69.** $P = 2x + 2(x - 5) = 4x - 10$

EVEN   **2.** -1 **4.** 1 **6.** 7 **8.** $10x$ **10.** $4x$ **12.** $8x$ **14.** $-2x$ **16.** $7x + 3y$
18. $-5x + 3y$ **20.** $6m - 2n$ **22.** $-7x + 4y$ **24.** $3x - 2y$ **26.** $-x + 8y$ **28.** $5x - 2$
30. $5x - 3$ **32.** $6x^2 - x + 3$ **34.** $-2x + 12$ **36.** $x^2 - 3x$ **38.** $-9r^3t^3$ **40.** $2x^2 + 2x - 3$
42. $-2x^2y - 6xy + 4xy^2$ **44.** $-a + 3b$ **46.** $-2x - y$ **48.** $2t - 20$ **50.** $-3y + 4$ **52.** $x - 14$
54. $3x - y$ **56.** $-3x + y$ **58.** $x - y + z$ **60.** $3x^3 + x^2 + x + 6$ **62.** $-x^3 + 2x^2 - 3x + 7$
64. $2x^2 - 2xy + 3y^2$ **66.** 1 **68.** $13x^2 - 26x + 10$ **70.** $P = 2x + 2(2x + 3) = 6x + 6$

Practice Exercise 2.3

ODD   **1.** y^5 **3.** $10y^5$ **5.** $-24x^{20}$ **7.** $6u^{16}$ **9.** c^3d^4 **11.** $15x^2y^3z^5$ **13.** $y^2 + 7y$
15. $10y^2 - 35y$ **17.** $3a^5 + 6a^4$ **19.** $2y^3 + 4y^2 - 6y$ **21.** $7m^6 - 14m^5 - 7m^4 + 28m^3$ **23.** $10u^4v^3 - 15u^2v^4$
25. $2c^3d^4 - 4c^2d^4 + 8c^4d^5$ **27.** $6y^3 + 19y^2 + y - 6$ **29.** $m^3 - 2m^2n - 9mn^2 - 2n^3$
31. $6m^4 + 2m^3 - 5m^2 + 4m - 1$ **33.** $a^3 + b^3$ **35.** $2x^4 + x^3y - 7x^2y^2 + 5xy^3 - y^4$ **37.** $x^2 + 5x + 6$
39. $a^2 + 4a - 32$ **41.** $t^2 - 16$ **43.** $m^2 - n^2$ **45.** $4t^2 - 11t + 6$ **47.** $3x^2 - 7xy - 6y^2$ **49.** $4m^2 - 49$
51. $30x^2 - 2xy - 12y^2$ **53.** $6s^2 - 11st + 3t^2$ **55.** $9x^2 + 12x + 4$ **57.** $4x^2 - 20xy + 25y^2$
59. $36u^2 + 60uv + 25v^2$ **61.** $4m^2 - 20mn + 25n^2$ **63.** $-x^2 + 17x - 11$ **65.** $3x^2 - 10x - 8$
67. $x^3 + 6x^2y + 12xy^2 + 8y^3$ **69.** Area $= y(y - 8) = y^2 - 8y$

EVEN   **2.** x^5 **4.** $6x^5$ **6.** $-35u^{16}$ **8.** $24x^9$ **10.** a^3b^3 **12.** $-6x^4y^4z^2$ **14.** $x + x^2$
16. $6x^2 - 15x$ **18.** $2m^4 + 6m^3$ **20.** $4x^3 - 6x^2 + 2x$ **22.** $6x^5 + 9x^4 - 3x^3 - 6x^2$ **24.** $8m^5n^4 - 4m^3n^5$
26. $6x^3y^4 + 12x^3y - 3x^2y^3$ **28.** $2x^3 - 7x^2 + 13x - 5$ **30.** $x^3 - 6x^2y + 10xy^2 - 3y^3$ **32.** $2x^4 - 5x^3 + 5x^2 + 11x - 10$
34. $a^3 - b^3$ **36.** $a^4 - 2a^2b^2 + b^4$ **38.** $m^2 - 5m + 6$ **40.** $m^2 - 7m - 60$ **42.** $u^2 - 9$ **44.** $a^2 - b^2$
46. $6x^2 - 7x - 5$ **48.** $2x^2 + xy - 6y^2$ **50.** $9y^2 - 4$ **52.** $6m^2 - mn - 35n^2$ **54.** $6x^2 - 13xy + 6y^2$
56. $16x^2 + 24xy + 9y^2$ **58.** $4x^2 - 28x + 49$ **60.** $49p^2 + 28pq + 4q^2$ **62.** $16x^2 - 8x + 1$
64. $-7x^2 - x - 16$ **66.** $-4xy$ **68.** $8m^3 - 12m^2n + 6mn^2 - n^3$ **70.** Area $= y(2y - 3) = 2y^2 - 3y$

Practice Exercise 2.4

ODD   **1.** $3x + 1$ **3.** $2y^2 + y - 3$ **5.** $3x + 1$, R $= 3$ **7.** $4x - 1$ **9.** $3x - 4$, R $= -1$ **11.** $x + 2$
13. $4x + 1$, R $= -4$ **15.** $4x + 6$, R $= 25$ **17.** $x - 4$, R $= 3$ **19.** $x^2 + x + 1$ **21.** $x^3 + 3x^2 + 9x + 27$
23. $4a + 5$, R $= -7$ **25.** $x^2 + 3x - 5$ **27.** $x^2 + 3x + 8$, R $= 27$ **29.** $3x^3 + x^2 - 2$, R $= -4$
31. $4x^2 - 2x - 1$, R $= -2$ **33.** $2x^3 + 6x^2 + 32x + 84$, R $= 186x - 170$

EVEN   **2.** $2x - 3$ **4.** $x^2 - 3x - 5$ **6.** $2x + 3$, R $= 5$ **8.** $2x + 3$ **10.** $2x + 5$, R $= -2$ **12.** $y - 3$
14. $4x - 1$, R $= 5$ **16.** $3x + 2$, R $= -4$ **18.** $x + 5$, R $= -2$ **20.** $a^2 - 3a + 9$ **22.** $x^3 - 2x^2 + 4x - 8$
24. $5c - 2$, R $= 8$ **26.** $2y^2 + y - 3$ **28.** $2y^2 - 5y + 13$, R $= -27$ **30.** $2x^3 - 3x^2 - 5$, R $= 5$
32. $2x^2 - 3x + 2$, R $= 0$ **34.** $3x^3 - 4x + 3$, R $= -8x$

Practice Exercise 2.5

ODD   **1.** $A(2x + 3)$ **3.** $5x(2x + 3)$ **5.** $2u(7u - 3)$ **7.** $2u(3u - 5v)$ **9.** $5mn(2m - 3n)$
11. $2x^2y(x - 3y)$ **13.** $(x + 2)(3x + 5)$ **15.** $(m - 4)(3m - 2)$ **17.** $(x + y)(x - y)$ **19.** $3x^2(2x^2 - 3x + 1)$
21. $2xy(4x^2 - 3xy + 2y^2)$ **23.** $4x^2(2x^2 - 3xy + y^2)$ **25.** $(2x + 3)(3x - 5)$ **27.** $(x + 1)(x - 1)$
29. $(2x - 3)(4x - 1)$ **31.** $2x - 2$ **33.** $2x - 8$ **35.** $2u + 1$ **37.** $3x(x - 1) + 2(x - 1) = (x - 1)(3x + 2)$
39. $3x(x - 4) - 2(x - 4) = (x - 4)(3x - 2)$ **41.** $4u(2u + 1) - (2u + 1) = (2u + 1)(4u - 1)$ **43.** $(x - 1)(3x + 2)$
45. $(x - 4)(3x - 2)$ **47.** $(2u + 1)(4u - 1)$ **49.** $2m(m - 4) + 5(m - 4) = (m - 4)(2m + 5)$
51. $3x(2x - 3) - 2(2x - 3) = (2x - 3)(3x - 2)$ **53.** $3u(u - 4) - (u - 4) = (u - 4)(3u - 1)$
55. $3u(2u + v) - 2v(2u + v) = (2u + v)(3u - 2v)$ **57.** $3x(2x + y) - 5y(2x + y) = (2x + y)(3x - 5y)$

EVEN   **2.** $M(x - 4)$ **4.** $3y(3y - 2)$ **6.** $4m(5m + 3)$ **8.** $7x(2x - 3y)$ **10.** $3uv(3u + 2v)$
12. $6xy^2(x - y)$ **14.** $(y + 3)(4y + 7)$ **16.** $(x - 1)(x - 4)$ **18.** $(m - n)(m + n)$ **20.** $2m^2(3m^2 - 4m - 1)$
22. $5uv(2u^2 + 4uv - 3v^2)$ **24.** $3m^2(3m^2 - 2mn - 2n^2)$ **26.** $(3u - 8)(2u - 3)$ **28.** $(u - 1)(3u - 1)$
30. $(4y - 5)(3y - 1)$ **32.** $3x + 6$ **34.** $3y - 15$ **36.** $3x + 5$ **38.** $2x(x + 2) + 3(x + 2) = (x + 2)(2x + 3)$
40. $2y(y - 5) - 3(y - 5) = (y - 5)(2y - 3)$ **42.** $2x(3x + 5) - (3x + 5) = (3x + 5)(2x - 1)$ **44.** $(x + 2)(2x + 3)$
46. $(y - 5)(2y - 3)$ **48.** $(3x + 5)(2x - 1)$ **50.** $5x(x - 2) + 2(x - 2) = (x - 2)(5x + 2)$
52. $4x(3x + 2) - 3(3x + 2) = (3x + 2)(4x - 3)$ **54.** $2m(3m + 2) - (3m + 2) = (3m + 2)(2m - 1)$
56. $2x(x - 2v) - y(x - 2v) = (x - 2v)(2x - y)$ **58.** $4u(u - 4v) - 3v(u - 4v) = (u - 4v)(4u - 3v)$

Practice Exercise 2.6

ODD   **1.** $(x + 1)(x + 4)$ **3.** $(x + 2)(x + 3)$ **5.** $(x - 1)(x - 3)$ **7.** $(x - 2)(x - 5)$ **9.** Not factorable
11. Not factorable   **13.** $(x + 3y)(x + 5y)$ **15.** $(x - 3y)(x - 7y)$ **17.** Not factorable   **19.** $(3x + 1)(x + 2)$
21. $(3x - 4)(x - 1)$ **23.** $(x - 4)(3x - 2)$ **25.** $(3x - 2)(x - 3y)$ **27.** $(n - 4)(n + 2)$ **29.** Not factorable
31. $(x - 1)(3x + 2)$ **33.** $(x + 6y)(x - 2y)$ **35.** $(u - 4)(3u + 1)$ **37.** $(3x + 5)(2x - 1)$ **39.** $(3s + 1)(s - 2)$
41. Not factorable   **43.** $(x - 2)(5x + 2)$ **45.** $(2u + v)(3u - 2v)$ **47.** $(4x - 3)(2x + 3)$
49. $(3u - 2v)(u + 3v)$ **51.** $(u - 4v)(4u - 3v)$ **53.** $(6x + y)(2x - 7y)$ **55.** $(12x - 5y)(x + 2y)$

EVEN   **2.** $(x + 1)(x + 3)$ **4.** $(x + 2)(x + 5)$ **6.** $(x - 1)(x - 4)$ **8.** $(x - 2)(x - 3)$ **10.** Not factorable
12. Not factorable   **14.** $(x + 4y)(x + 5y)$ **16.** $(x - 2y)(x - 8y)$ **18.** Not factorable   **20.** $(2x + 1)(x + 3)$
22. $(2x - 3)(x - 2)$ **24.** $(y - 5)(2y - 3)$ **26.** $(2x - 3y)(x - 2y)$ **28.** $(n + 4)(n - 2)$ **30.** Not factorable
32. $(3m + 2)(2m - 1)$ **34.** $(x - 2y)(2x + y)$ **36.** $(2u + 1)(4u - 1)$ **38.** $(m - 4)(2m + 5)$
40. $(2s - 1)(s + 3)$ **42.** Not factorable   **44.** $(6x - 1)(2x + 3)$ **46.** $(2x + y)(3x - 5y)$ **48.** $(2x - 3)(3x - 2)$
50. $(4m - n)(m + 3n)$ **52.** $(3x + 2y)(4x - 3y)$ **54.** $(5x - y)(3x + 4y)$ **56.** $(8x + 3y)(3x - 5y)$

Practice Exercise 2.7

ODD   **1.** $(3x - 4)(x - 1)$ **3.** Not factorable   **5.** $(2x - 1)(x + 3)$ **7.** Not factorable   **9.** $(x - 4)(3x - 2)$
11. $(3x + 5)(2x - 1)$ **13.** Not factorable   **15.** $(m - 4)(2m + 5)$ **17.** $(u - 4)(3u + 1)$ **19.** $(2u + v)(3u - 2v)$
21. Not factorable   **23.** $(4x - 3)(2x + 3)$ **25.** $(2m - n)(m + 6n)$ **27.** Not factorable
29. $(u - 4v)(4u - 3v)$ **31.** $(6x + y)(2x - 7y)$

EVEN   **2.** $(2x - 3)(x - 2)$ **4.** Not factorable   **6.** $(3x + 1)(x - 2)$ **8.** Not factorable   **10.** $(y - 5)(2y - 3)$
12. $(x - 2)(5x + 2)$ **14.** Not factorable   **16.** $(6x - 1)(2x + 3)$ **18.** $(2u + 1)(4u - 1)$ **20.** $(2x + y)(3x - 5y)$
22. Not factorable   **24.** $(2x - 3)(3x + 2)$ **26.** $(3u - 2v)(u + 3v)$ **28.** Not factorable   **30.** $(3x + 2y)(4x - 3y)$
32. $(5x - y)(3x + 4y)$

Practice Exercise 2.8

ODD   **1.** $(v - 5)(v + 5)$ **3.** $(3x - 2)(3x + 2)$ **5.** Not factorable   **7.** $(3x - 4y)(3x + 4y)$
9. $(x + 1)(x^2 - x + 1)$ **11.** $(m - n)(m^2 + mn + n^2)$ **13.** $(2x + 3)(4x^2 - 6x + 9)$ **15.** $3uv^2(2u - v)$
17. $2(x - 2)(x + 2)$ **19.** $2x(x^2 + 4)$ **21.** $3x(2x - y)(2x + y)$ **23.** $2x(x + 1)(x^2 - x + 1)$ **25.** $6(x + 2)(x + 4)$
27. $3x(x^2 - 2x + 5)$ **29.** $(xy - 4)(xy + 4)$ **31.** $(ab + 2)(a^2b^2 - 2ab + 4)$ **33.** $2xy(2x + y)(x + 3y)$
35. $4(u + 2v)(u^2 - 2uv + 4v^2)$ **37.** $5y^2(6x + y)(2x - 7y)$ **39.** $(y + 2)(x + y)$ **41.** $(x - 5)(x + y)$
43. $(a - 2b)(x - y)$ **45.** $(3c - 4d)(5a + b)$ **47.** $(x - 2)(x + 1)(x - 1)$
49. $(y - x)[(y - x) - 1] = (y - x)(y - x - 1)$ **51.** $(r^2 + s^2)(r - s)(r + s)$ **53.** $(x^2 - 4)(x^2 + 1) = (x - 2)(x + 2)(x^2 + 1)$

55. $[(x - 3) - 4y][(x - 3) + 4y] = (x - 3 - 4y)(x - 3 + 4y)$ **57.** $[(a - b) - 2(c - d)][(a - b) + 2(c - d)]$

59. $[5(2x - 3y) - 3ab][5(2x - 3y) + 3ab]$ **61.** $(x - 1)(x^2 + x + 1)(x + 1)(x^2 - x + 1)$ **63.** $(2x - 1)(x - 2)(x + 2)$

65. $[5 - (a + b)][5 + (a + b)]$ **67.** $[4x^2 - (x - 3y)][4x^2 + (x - 3y)]$

EVEN **2.** $(x - 9)(x + 9)$ **4.** $(2m - 1)(2m + 1)$ **6.** Not factorable   **8.** $(5u - 2v)(5u + 2v)$

10. $(y - 1)(y^2 + y + 1)$ **12.** $(p + q)(p^2 - pq + q^2)$ **14.** $(u - 2v)(u^2 + 2uv + 4v^2)$ **16.** $2x^2y(x - 3y^2)$

18. $3(y - 3)(y + 3)$ **20.** $3x^2(x^2 + 9)$ **22.** $2uv(u - v)(u + v)$ **24.** $x(y + x)(y^2 - xy + x^2)$

26. $4(x - 3)(x + 2)$ **28.** $2x(x^2 - x + 4)$ **30.** $(mn - 6)(mn + 6)$ **32.** $(3 - xy)(9 + 3xy + x^2y^2)$

34. $3xy(x - 2y)(x - 3y)$ **36.** $2(3x - y)(9x^2 + 3xy + y^2)$ **38.** $4x^2(5x - y)(3x + 4y)$ **40.** $(x + 3)(x + y)$

42. $(x - 3)(x - y)$ **44.** $(x + y)(m - 2n)$ **46.** $(2m - 3n)(a + b)$ **48.** $(x - 2)(x^2 + 1)$ **50.** $(x - 1)^2(x + 1)$

52. $(4a^2 + b^2)(2a - b)(2a + b)$ **54.** $(x^2 + 2)(x - 3)(x + 3)$ **56.** $[(x + 2) - 3y][(x + 2) + 3y] = (x + 2 - 3y)(x + 2 + 3y)$

58. $[(x^2 - x) - 3(y^2 - y)][(x^2 - x) + 3(y^2 - y)]$ **60.** $2a[3a - 2(x + 4)][3a + 2(x + 4)]$

62. $(a + 2b)(a^2 - 2ab + 4b^2)(a - 2b)(a^2 + 2ab + 4b^2)$ **64.** $(y - 3)(2y - 3)(2y + 3)$ **66.** $(x - y + 3)(x - y - 3)$

68. $(x^2 - x + 2)(x + 2)(x - 1)$

Diagnostic (Review) Exercise 2.9

1. $5x^2 + x - 4$ (2.2)   **2.** $-2x^2 + x - 7$ (2.2)   **3.** $6x^5y - 9x^4y^2 + 3x^2y^3$ (2.3)   **4.** $6x^2 + 11x - 10$ (2.3)

5. $2x^3 - 7x^2 + 13x - 5$ (2.3)   **6.** $3x + 4$, R = 2 (2.4)   **7.** $2xy(2x - 3y)$ (2.5)   **8.** $(x - 2)(x - 7)$ (2.6)

9. $(3x - 2)^2$ (2.6)   **10.** Not factorable (2.6)   **11.** $(u + 8)(u - 8)$ (2.6)   **12.** $(3x - 4)(x - 2)$ (2.6)

13. $x(x - 2)(x - 3)$ (2.8)   **14.** $(x + y)^2$ (2.5)   **15.** $27x^4 + 63x^3 - 66x^2 - 28x + 24$ (2.3)   **16.** $2x^2 + 5x + 5$ (2.3)

17. $3x^2 + 2x - 2$, R = -2 (2.4)   **18.** (A) 5 (B) 3 (2.1)   **19.** $(m - 4n)(m + n)$ (2.6)   **20.** $2(m - 2n)(m + 2n)$ (2.8)

21. $3xy(4x^2 + 9y^2)$ (2.6)   **22.** $(2x - 3y)(x + y)$ (2.6)   **23.** $3n(2n - 5)(n + 1)$ (2.8)   **24.** Not factorable (2.6)

25. $(p + q)(x + y)$ (2.8)   **26.** $(x - y)(x - 4)$ (2.8)   **27.** $(y - b)(y - b - 1)$ (2.8)   **28.** $3(x - 2y)(x^2 + 2xy + 4y^2)$ (2.3)

29. $12x^5 - 19x^3 + 12x^2 + 4x - 3$ (2.3)   **30.** $4x^2 - 2x + 3$, R = -2 (2.4)   **31.** $-2x + 20$ (2.3)

32. $2x^3 - 4x^2 + 12x$ (2.3)   **33.** $3xy(6x - 5y)(2x + 3y)$ (2.8)   **34.** $4u^2(3u^2 - 3uv - 5v^2)$ (2.8)   **35.** $y(y - 2x)$ (2.8)

36. $(a^2 - b^2)^2 = (a - b)^2(a + b)^2$ (2.8)   **37.** $(m - n)(m^2 + mn + n^2)(m + n)(m^2 - mn + n^2)$ (2.8)

38. $(2x - 3m + 1)(2x + 3m - 1)$ (2.8)

Practice Test Chapter 2

1. C (2.1)   **2.** $6x^2 - 5x - 7$ (2.2)   **3.** $-x^2 + 3x - 5$ (2.2)   **4.** $2x^2 - 5x + 2$ (2.3)   **5.** $4x^3 - 12x^2 + 9x - 2$ (2.3)

6. $x - 2$ (2.4)   **7.** $x^2 - 3x + 5$ (2.2, 2.3)   **8.** $3x^2 + 4x + 8$, R = 21 (2.4)   **9.** $3x^3 - x^2 - 10x$ (2.2, 2.3)

10. $2x^2 + x - 2$, R = 3 (2.4)   **11.** $(3x - 2)(2x + 3)$ (2.6)   **12.** Not factorable (2.6)   **13.** $2xy(3x - 5)(2x + 1)$ (2.8)

14. $2(2u - 3)(2u + 3)$ (2.8)   **15.** $4(x^2 + 4)$ (2.8)   **16.** $(m + 2)(m^2 - 2m + 4)$ (2.8)   **17.** $2x(x - 2)(x^2 + 2x + 4)$ (2.8)

18. $(3x - y)(x - 2)$ (2.8)   **19.** $[(x - 3) - 2y][(x - 3) + 2y]$ (2.8)   **20.** $(x - 2)(x + 2)(x^2 + 4)$ (2.8)

CHAPTER 3

Practice Exercise 3.1

ODD **1.** $\dfrac{1}{2x^2}$ **3.** $\dfrac{2x^2}{3y}$ **5.** $\dfrac{3(x - 9)}{y}$ **7.** $\dfrac{2x - 1}{3x}$ **9.** $\dfrac{x}{2}$ **11.** $\dfrac{1}{n}$ **13.** $12xy$ **15.** $14x^3y$

17. $\dfrac{x + 2}{3x}$ **19.** $\dfrac{x - 3}{x + 3}$ **21.** $\dfrac{2x - 3y}{2xy}$ **23.** $\dfrac{x + 2}{x + y}$ **25.** $\dfrac{x + 5}{2x}$ **27.** $\dfrac{x^2 + 2x + 4}{x + 2}$ **29.** $3x^2 + 3xy$

31. $x^2 - y^2$ **33.** $\dfrac{x - y}{3x}$ **35.** $\dfrac{x - y}{2x + y}$ **37.** $\dfrac{x^2 + y^2}{(x + y)^2}$

EVEN **2.** $\dfrac{3}{u}$ **4.** $\dfrac{4n^4}{3m}$ **6.** $\dfrac{x}{3}$ **8.** $\dfrac{x + 3}{2x^2}$ **10.** $\dfrac{x}{2}$ **12.** a **14.** $6x^2y^2$ **16.** $16u^2v^3$ **18.** $\dfrac{x + 2}{2x}$

20. $\dfrac{x - 2}{x + 2}$ **22.** $\dfrac{a + 4b}{4b}$ **24.** $\dfrac{u - 2}{u + v}$ **26.** $\dfrac{3(x - 7)}{4x^2}$ **28.** $\dfrac{y + 3}{2y}$ **30.** $5mn - 5n^2$ **32.** $6x^2 + 17x + 5$

34. $\dfrac{2uv}{u + v}$ **36.** $\dfrac{m + n}{m - n}$ **38.** $\dfrac{(x - y)(x + y)}{x^2 + y^2}$

Practice Exercise 3.2

ODD
1. $\dfrac{8}{9}$ **3.** $\dfrac{6}{b}$ **5.** $\dfrac{y}{x}$ **7.** $\dfrac{3}{2}$ **9.** $\dfrac{3c}{a}$ **11.** $\dfrac{x}{9y^2}$ **13.** $\dfrac{16xy}{3}$ **15.** $\dfrac{9xy}{8c}$ **17.** $\dfrac{-45u^2}{16v^2}$

19. $\dfrac{c^3d^2}{a^6b^6}$ **21.** $\dfrac{x}{2}$ **23.** $\dfrac{x}{x-3}$ **25.** $\dfrac{3y}{x+3}$ **27.** $\dfrac{1}{2y}$ **29.** $t(t-4)$ **31.** $\dfrac{1}{m}$

33. $-x(x-2)$ or $2x-x^2$ **35.** $\dfrac{a^2}{2}$ **37.** 2 **39.** -1 **41.** $\dfrac{(x-y)^2}{y^2(x+y)}$

EVEN
2. $\dfrac{2}{3}$ **4.** $\dfrac{1}{z}$ **6.** $\dfrac{3x}{2y}$ **8.** 4 **10.** y **12.** $2y^2$ **14.** $\dfrac{3ad}{2c}$ **16.** $\dfrac{3v}{2u}$ **18.** $-\dfrac{2x^2}{3y}$

20. $\dfrac{u^2w^2}{25y^2}$ **22.** $\dfrac{2}{x}$ **24.** $a+1$ **26.** $\dfrac{x-2}{2x}$ **28.** $\dfrac{1}{y(x+4)}$ **30.** $\dfrac{1}{2y-1}$ **32.** $\dfrac{x}{x+5}$

34. $-3(x-2)$ or $6-3x$ **36.** $8d^6$ **38.** -2 **40.** $-\dfrac{1}{m}$ **42.** $\dfrac{x^2(x+y)}{(x-y)^2}$

Practice Exercise 3.3

ODD
1. $3x$ **3.** x **5.** v^3 **7.** $12x^2$ **9.** $(x+1)(x-2)$ **11.** $3y(y+3)$ **13.** $\dfrac{7x-2}{5x^2}$ **15.** 2

17. $\dfrac{1}{y+3}$ **19.** $\dfrac{3-2x}{k}$ **21.** $\dfrac{12x+y}{4y}$ **23.** $\dfrac{2+y}{y}$ **25.** $\dfrac{u^3+uv-v^2}{v^3}$ **27.** $\dfrac{9x^2+8x-2}{12x^2}$

29. $\dfrac{5x-1}{(x+1)(x-2)}$ **31.** $\dfrac{7y-6}{3y(y+3)}$ **33.** $24x^3y^2$ **35.** $75x^2y^2$ **37.** $18(x-1)^2$ **39.** $24(x-7)(x+7)^2$

41. $(x-2)(x+2)^2$ **43.** $12x^2(x+1)^2$ **45.** $\dfrac{8v-6u^2v^2+3u^3}{36u^3v^3}$ **47.** $\dfrac{15t^2+14t-6}{36t^3}$ **49.** $\dfrac{2}{t-1}$

51. $\dfrac{5a^2-2a-5}{(a+1)(a-1)}$ **53.** $\dfrac{5x+55}{12(x-5)^2(x+5)}$ **55.** $\dfrac{15x-11}{18(x-1)^2}$ **57.** $\dfrac{-4}{(x-1)(x+3)}$ **59.** $\dfrac{2s^2+s-2}{2s(s-2)(s+2)}$

61. $\dfrac{2(x+4)}{(x-2)(x+2)^2}$ **63.** $\dfrac{3}{x+3}$ **65.** $\dfrac{x+3}{(x-2)(x+7)}$ **67.** $\dfrac{(3x+1)(x+3)}{12x^2(x+1)^2}$ **69.** $\dfrac{xy^2-xy+y^2}{x^3-y^3}$

71. $\dfrac{7}{y-3}$ **73.** -1

EVEN
2. $4y$ **4.** y **6.** x^2 **8.** $24u^3$ **10.** $(x-2)(x+3)$ **12.** $2x(x-2)$ **14.** $\dfrac{3m-1}{2m^2}$ **16.** 5

18. $\dfrac{1}{2x-3}$ **20.** $\dfrac{1-b}{a^2}$ **22.** $\dfrac{6-x}{3x}$ **24.** $\dfrac{x^2+1}{x}$ **26.** $\dfrac{x^2-xy+y^2}{x^3}$ **28.** $\dfrac{20u^2-18u+3}{24u^3}$

30. $\dfrac{2x+1}{(x-2)(x+3)}$ **32.** $\dfrac{x+6}{2x(x-2)}$ **34.** $36u^3v^3$ **36.** $36m^4n^4$ **38.** $24(y-3)^2$ **40.** $12(x-5)^2(x+5)$

42. $(x-3)^2(x+3)$ **44.** $15m^2(m-1)^2$ **46.** $\dfrac{2y^2+9x-16x^2}{24x^3y^2}$ **48.** $\dfrac{y^2+8}{8y^3}$ **50.** $\dfrac{3x-5}{x-3}$

52. $\dfrac{3y^2-y-18}{(y+2)(y-2)}$ **54.** $\dfrac{13x-35}{24(x-7)(x+7)^2}$ **56.** $\dfrac{21-4y}{24(y-3)^2}$ **58.** $\dfrac{2x+7}{(2x-3)(x+2)}$ **60.** $\dfrac{5t-12}{3(t-4)(t+4)}$

62. $\dfrac{x+9}{(x-3)^2(x+3)}$ **64.** $\dfrac{2}{x+y}$ **66.** $\dfrac{5m^2+1}{6(m+1)^2}$ **68.** $\dfrac{17m^2+m-3}{15m^2(m-1)^2}$ **70.** $\dfrac{x^2}{x^3+y^3}$ **72.** $\dfrac{1}{x-1}$

74. -1

Practice Exercise 3.4

ODD
1. $\frac{3}{4}$ **3.** $\frac{9}{10}$ **5.** $\frac{8}{13}$ **7.** $\frac{22}{51}$ **9.** xy **11.** $\dfrac{3xy}{2}$ **13.** $\dfrac{1}{x-3}$ **15.** $\dfrac{x+y}{x}$ **17.** $\dfrac{1}{y-x}$

19. $\dfrac{x-y}{x+y}$ **21.** 1 **23.** $-\frac{1}{2}$ **25.** $\dfrac{1}{1-x}$ **27.** $-x$

EVEN **2.** $\frac{3}{8}$ **4.** $\frac{8}{25}$ **6.** $\frac{31}{22}$ **8.** $\frac{4}{3}$ **10.** $\frac{1}{ab}$ **12.** $\frac{6x^2}{5y}$ **14.** $\frac{1}{x+2}$ **16.** $\frac{a+b}{b}$ **18.** $a(a+b)$

20. $\frac{x-3}{x-1}$ **22.** $\frac{-m(m+n)}{n}$ **24.** -1 **26.** $\frac{x-2}{x}$ **28.** $\frac{t-1}{t}$

Diagnostic (Review) Exercise 3.5

1. $\frac{3x+2}{3x}$ (3.3) **2.** $\frac{2x+11}{6x}$ (3.3) **3.** $\frac{2-9x^2-8x^3}{12x^3}$ (3.3) **4.** $\frac{2xy}{ab}$ (3.2) **5.** $2(x+1)$ (3.3)

6. $\frac{2}{m+1}$ (3.3) **7.** $\frac{x+7}{(x-2)(x+1)}$ (3.3) **8.** $\frac{(d-2)^2}{d+2}$ (3.1) **9.** $\frac{-1}{(x+2)(x+3)}$ (3.3) **10.** $\frac{8}{9}$ (3.4)

11. $\frac{11}{6}$ (3.4) **12.** $\frac{y-2}{y+1}$ (3.4) **13.** $\frac{12a^2b^2-40a^2-5b}{30a^2b^3}$ (3.3) **14.** $\frac{5-2x}{2x-3}$ (3.3) **15.** $\frac{2y^4}{9a^4}$ (3.2)

16. $\frac{5x-12}{3(x-4)(x+4)}$ (3.2) **17.** $\frac{x}{x+1}$ (3.2) **18.** $\frac{x-y}{x}$ (3.4) **19.** $\frac{y}{x^3-y^3}$ (3.3) **20.** $\frac{x}{y(x+y)}$ (3.3)

21. $x+1$ (3.3) **22.** $\frac{x^2-24x-9}{12x(x-3)(x+3)^2}$ (3.3) **23.** $\frac{-1}{s+2}$ (3.3) **24.** -1 (3.2) **25.** $\frac{y^2}{x}$ (3.2)

Practice Test Chapter 3

1. $\frac{x^2-3x+4}{6x^2}$ (3.3) **2.** $\frac{2ab^2}{3x^2y}$ (3.2) **3.** $\frac{5}{x-3}$ (3.3) **4.** $\frac{m-3}{m+3}$ (3.4) **5.** $\frac{2x+14}{(x-5)(x+3)}$ (3.3) **6.** $\frac{1}{x}$ (3.2)

7. $\frac{11m-3n}{12m^2(m-n)^2}$ (3.3) **8.** $\frac{1}{x+y}$ (3.2) **9.** -1 (3.3) **10.** $\frac{u+v}{u}$ (3.4)

CHAPTER 4

Practice Exercise 4.1

ODD **1.** -2 **3.** $\frac{-13}{5}$ **5.** 0 **7.** $\frac{7}{2}$ **9.** 5 **11.** $\frac{-13}{2}$ **13.** No solution   **15.** 18
17. All real numbers   **19.** 9 **21.** 10 **23.** 1 **25.** 4 **27.** No solution   **29.** -3
31. $2x=x-8, x=-8$

EVEN **2.** -10 **4.** $\frac{-3}{4}$ **6.** 0 **8.** $\frac{-10}{3}$ **10.** 5 **12.** $\frac{11}{2}$ **14.** No solution   **16.** 4
18. All real numbers   **20.** $\frac{1}{3}$ **22.** 16 **24.** 4 **26.** 6 **28.** No solution   **30.** -2
32. $3x=12, x=4$

Practice Exercise 4.2

ODD **1.** 13 **3.** 8 **5.** -6 **7.** $\frac{-1}{12}$ **9.** 30 **11.** 20 **13.** 10 **15.** 3 **17.** $\frac{-7}{4}$ **19.** 3
21. 10 **23.** -9 **25.** 4 **27.** No solution   **29.** 5 **31.** $\frac{53}{11}$ **33.** No solution   **35.** 1
37. $\frac{31}{24}$ **39.** -4 **41.** No solution

EVEN **2.** 8 **4.** 12 **6.** 36 **8.** $\frac{-4}{3}$ **10.** 600 **12.** 30 **14.** -9 **16.** 4 **18.** 15
20. -3 **22.** 150 **24.** 8 **26.** -4 **28.** No solution   **30.** $\frac{-6}{5}$ **32.** 2 **34.** No solution
36. 8 **38.** $\frac{8}{5}$ **40.** $\frac{-3}{8}$ **42.** No solution

Practice Exercise 4.3

ODD **1.** $r=d/t$ **3.** $r=C/2\pi$ **5.** $\pi=C/D$ **7.** $x=-b/a$ $a\neq 0$ **9.** $x=\frac{y+5}{2}$

11. $y=\frac{3}{4}x-3$ **13.** $R=E/I$ **15.** $B=\frac{CL}{100}$ **17.** $G=\frac{Fd^2}{m_1 m_2}$ **19.** $C=\frac{5}{9}(F-32)$ **21.** $d=\frac{M-P}{Mt}$

23. $M=\frac{P}{1-dt}$ **25.** $f=\frac{ab}{a+b}$ **27.** $n=\frac{a_n-a_1+d}{d}$ **29.** $T_2=\frac{T_1 P_2 V_2}{P_1 V_1}$ **31.** $x=\frac{5y-3}{3y-2}$

EVEN   **2.** $t = \dfrac{d}{1,100}$ **4.** $t = \dfrac{I}{Pr}$ **6.** $m = \dfrac{e}{c^2}$ **8.** $a = \dfrac{p - 2b}{2}$ **10.** $m = \dfrac{y - b}{x}, x \neq 0$

12. $y = -\dfrac{A}{B}x - \dfrac{C}{B}, B \neq 0$ **14.** $a = \dfrac{b}{m}$ **16.** $(CA) = \dfrac{100(MA)}{IQ}$ **18.** $m_1 = \dfrac{Fd^2}{Gm_2}$ **20.** $F = \tfrac{9}{5}C + 32$

22. $t = \dfrac{M - P}{MD}$ **24.** $h = \dfrac{A}{a/2 + b/2} = \dfrac{2A}{a + b}$ **26.** $R = \dfrac{R_1 R_2}{R_1 + R_2}$ **28.** $d = \dfrac{a_n - a_1}{n - 1}$ **30.** $V_1 = \dfrac{T_1 P_2 V_2}{P_1 T_2}$

32. $x = \dfrac{4y + 2}{2y - 3}$

Practice Exercise 4.4

ODD   **1.** $6 > 3$ **3.** $-3 > -6$ **5.** $-6 < -3$ **7.** $5 > 0$ **9.** $-5 < 0$ **11.** $-8 < -4$ **13.** $e > a$
15. $c > b$ **17.** $0 < d$ **19.**

21.    **23.**

25. $x < 3$ **27.** $x > -7$ **29.** $x < 2$ **31.** $x > -2$ **33.** $x < -21$ **35.** $x < 10$ **37.** $x > 4$
39. $x < 1$ **41.** $x \geq -1$ **43.** $2 \leq x < 13$ **45.** $-4 \leq x \leq 1$

47. $8 > x > 3$ or $3 < x < 8$ **49.** $m > 3$

51. $x \geq 3$ **53.** $x \leq -11$

55. $-2 < t \leq 3$ **57.** $-2 < x \leq 3$

59. $-1 \leq x < 4$ **61.** $x > 4$

63. $x > 10$ **65.** $B \geq -4$

67. $p \geq 12$ **69.** $-20 \leq C \leq 20$

71. $23 \leq F \leq 50$ **73.** **(A)** Greater   **(B)** Right

75. $x \geq 4.5$ **77.** $-10 < x < 11$

EVEN   **2.** $5 < 7$ **4.** $-7 < -5$ **6.** $-5 > -7$ **8.** $0 < 8$ **10.** $0 > -8$ **12.** $-7 < 5$ **14.** $a < d$
16. $e < f$ **18.** $0 > a$ **20.**

22.   **24.**

26. $x > 7$ **28.** $x < -7$ **30.** $x > 4$ **32.** $x < -4$ **34.** $x > -10$ **36.** $x > 21$ **38.** $x < 2$
40. $x > 2$ **42.** $x < 1$ **44.** $-1 < x < 2$ **46.** $-2 < x < 3$ **48.** $7 > x \geq -1$ or $-1 \leq x < 7$
50. $u \leq \tfrac{2}{7}$ **52.** $x < 10$

54. $u > -9$ **56.** $3 \leq m \leq 7$

58. $-2 \leq x < 3$ **60.** $-3 < x \leq 4$

62. $x < 3$ **64.** $x > -4$

66. $y < -7$ **68.** $q < -14$

70. $-9 \le m \le 9$

72. $14 \le F \le 77$

74. (**A**) Less (**B**) Left

76. $x > \dfrac{-38}{9}$

78. $-13 \le x \le 15$

Practice Exercise 4.5

ODD **1.** $x = \pm 5$

3. $t = -1$ or 7

5. $x = -6$ or 4

7. $-5 \le t \le 5$

9. $-1 < t < 7$

11. $-6 \le x \le 4$

13. $x = -1, 4$

15. $-1 \le x \le 4$

17. $m = -\frac{11}{3}, \frac{2}{3}$

19. $-\frac{8}{9} < M < \frac{22}{9}$

21. $x \le -7$ or $x \ge 7$

23. $t < -1$ or $t > 7$

25. $x \le -6$ or $x \ge 4$

27. $u \le -\frac{7}{3}$ or $u \ge -\frac{1}{3}$

29. $-1 < x < 4$

EVEN **2.** $x = +7$

4. $y = 2$ or 8

6. $u = -11$ or -5

8. $-7 \le x \le 7$

10. $2 < y < 8$

12. $-11 \le u \le -5$

14. $x = -4, \frac{4}{3}$

16. $-\frac{9}{5} \le x \le 3$

18. $t = -\frac{4}{5}, \frac{18}{5}$

20. $-\frac{23}{7} < u < \frac{5}{7}$

22. $x \le -5$ or $x \ge 5$

24. $y < 2$ or $y > 8$

26. $u \le -11$ or $u \ge -5$

28. $y < 3$ or $y > 5$

30. $2 \le x \le 3$

Diagnostic (Review) Exercise 4.6

1. $x = -2$ (*4.1*) **2.** $x = 2$ (*4.1*) **3.** $x < -2$ (*4.4*) **4.** $1 < x < 6$ (*4.4*) **5.** $x = \pm 6$ (*4.5*)

6. $-6 < x < 6$ (*4.5*) **7.** $x < -6$ or $x > 6$ (*4.5*) **8.** $-14, -4$ (*4.5*) **9.** $-14 < y < -4$ (*4.5*)

10. $y < -14$ or $y > -4$ (*4.5*) **11.** 9 (*4.2*) **12.** $\frac{-10}{9}$ (*4.2*) **13.** 60 (*4.2*) **14.** $x \le -12$ (*4.4*)

15. $R = \dfrac{W}{I^2}$ (*4.3*) **16.** $b = \dfrac{2A}{h}$ (*4.3*) **17.** $-4 \le x < 3$ (*4.4*)

18. $1 < x \le 4$ (*4.4*) **19.** $x \ge 1$ (*4.4*)

20. 41 (*4.2*) **21.** -12 (*4.2*) **22.** 5 (*4.2*) **23.** No solution (*4.2*)

24. $\frac{1}{2}$, 3   (4.5)

25. $\frac{1}{2} \le x \le 3$ (4.5)

26. $x < \frac{1}{2}$ or $x > 3$ (4.5)

27. 11 (4.2)

28. $x \ge -19$ (4.4)

29. $-6 < x \le -1$ (4.4)

30. -2 (4.2)   **31.** -5 (4.2)   **32.** $\frac{-3}{5}$ (4.2)   **33.** No solution (4.4)   **34.** $\frac{3}{4}$ (4.2)

35. $L = \frac{2S}{n} - a$ or $L = \frac{2S - an}{n}$ (4.3)

36. $M = \frac{Q}{1 + T}$ (4.3)   **37.** (A) T   (B) T (4.4)   **38.** 1 (4.1)   **39.** $\frac{-13}{5}$ (4.2)

40. $x \ge \frac{25}{7}$ (4.4)

41. $-3 \le x \le 6$ (4.4)

42. -15 (4.2)   **43.** No solution (4.5)   **44.** $x = \frac{5y + 3}{2y - 4}$ (4.3)   **45.** $f_1 = \frac{ff_2}{f_2 - f}$ (4.3)

Practice Test Chapter 4

1. $x = -\frac{14}{3}$ (4.1)   **2.** $x \ge 2$ (4.4)   **3.** $-1 \le x \le 2$ (4.4)   **4.** $x = \pm 2$ (4.5)   **5.** $-2 < x < 2$ (4.5)

6. $x < -2$ or $x > 2$ (4.5)   **7.** No solution (4.5)   **8.** $x = \frac{3}{2}$ (4.2)   **9.** $x = 4$ (4.2)   **10.** No solution (4.2)

11. $x = 12$ (4.2)   **12.** $x = \frac{7}{4}$ (4.2)   **13.** $b = \frac{P - 2a}{2}$ (4.3)   **14.** $n = \frac{2S}{A + L}$ (4.3)

15. $x < -15$ (4.4)

16. $-4 \le x < 5$ (4.4)

17. $-6 \le x \le 10$ (4.4)

18. $-5 \le x \le 2$ (4.5)

19. $x = -3, 6$ (4.5)

20. $x < -1$ or $x > 4$ (4.5)

CHAPTER 5

Practice Exercise 5.2

ODD   **1.** $\frac{x}{3}$ or $\frac{1}{3}x$ **3.** $\frac{3x}{4}$ or $\frac{3}{4}x$ **5.** $\frac{2x}{3}$ or $\frac{2}{3}x$ **7.** $\frac{x}{3} - 9$ $\left(\text{not } 9 - \frac{x}{3}\right)$ **9.** $\frac{2(3x - 5)}{3}$ or $\frac{2}{3}(3x - 5)$

11. $\frac{x}{6} + 3 = \frac{2}{3}$; -14 **13.** $x + (x + 1) + (x + 2) = 96$; 31, 32, 33   **15.** $x + (x + 2) + (x + 4) = 42$; 12, 14, 16

17. $\frac{x}{3} - 3 = \frac{x}{4}$; 36   **19.** $\frac{x}{6} - 2 = \frac{x}{4} + 1$; -36 **21.** $\frac{3x}{5} - 4 = \frac{x}{3} + 8$; 45   **23.** $x + (x + 2) = (x + 4) + 5$; 7, 9, 11

25. $2(2x + 3) + 2x = 66$; 23 by 10 cm   **27.** $2x + 2 \cdot \frac{x}{6} = 84$; 36 by 6 meters   **29.** $2x + 2\left(\frac{3x}{8} - 11\right) = 264$; 104 by 28 cm

31. $\frac{2P}{5} + 70 + \frac{P}{4} = P$; 200 cm   **33.** $\frac{x}{3} + 6 + \frac{x}{2} = x$; 36 km

EVEN   **2.** $\frac{x}{5}$ or $\frac{1}{5}x$ **4.** $\frac{3x}{8} + 5$ or $5 + \frac{3x}{8}$ **6.** $\frac{2x}{5} + 7$ or $7 + \frac{2x}{5}$ **8.** $\frac{x}{4} - 11$ $\left(\text{not } 11 - \frac{x}{4}\right)$

10. $\frac{3(2x - 7)}{4}$ or $\frac{3}{4}(2x - 7)$ **12.** $\frac{x}{4} + 2 = \frac{1}{2}$; -6 **14.** $x + (x + 1) + (x + 2) = 78$; 25, 26, 27

16. $x + (x + 2) + (x + 4) = 54$; 16, 18, 20   **18.** $\frac{x}{2} - 2 = \frac{x}{3}$; 12   **20.** $\frac{x}{2} - 5 = \frac{x}{3} + 3$; 48   **22.** $\frac{2x}{3} + 5 = \frac{x}{4} - 10$; -36

24. $(x + 2) + (x + 4) = 3x + 1$; 5, 7, 9   **26.** $2(4x - 6) + 2x = 128$; 50 by 14 meters   **28.** $2x + 2 \cdot \frac{x}{3} = 72$; 27 by 9 cm

30. $2\left(\frac{2x}{5} - 7\right) + 2x = 112$; 45 by 11 cm   **32.** $\frac{P}{4} + 3 + \frac{P}{3} = P$; 7.2 meters   **34.** $\frac{x}{5} + 6 + \frac{x}{2} = x$; 20 meters

Practice Exercise 5.3

ODD **1.** 3/2 **3.** 5/2 **5.** 25/2 **7.** 20 **9.** 7 **11.** 20 **13.** 60 quarters **15.** 40 meters
17. 162 km **19.** $37.80 **21.** 10.17 ml **23.** 6.57 in **25.** 16.04 kg **27.** 3.77 liters **29.** 3.94 in
31. 1.61 km **33.** 19.23 liters **35.** 3,000 trout **37.** 3 kg

EVEN **2.** 1/3 **4.** 5/3 **6.** 80/3 **8.** 8 **10.** 18 **12.** 4 **14.** 169 pennies **16.** 32 cm
18. 60 mi **20.** $197.40 **22.** 5.4 cups **24.** 15.69 in **26.** 6.67 kg **28.** 4.55 kg **30.** 35 oz
32. 1.1 yards **34.** 10.64 qt **36.** 4,400 trout **38.** 24,000 mi

Practice Exercise 5.4

ODD **1.** 66 km/hr **3.** 6.5 hr **5.** 5 hr **7.** 6 hr **9.** 65 min **11.** 5 hr **13.** 71.25 min
15. 3.43 hr **17.** 7.5 days **19.** 1.5 hr **21.** 670 meters

EVEN **2.** 5,600 papers per hour **4.** 2.25 min **6.** 2.6 hr **8.** 4.5 hr **10.** 32 min **12.** 3 hr
14. 40 min **16.** 2.92 hr **18.** 10 A.M.; 24 km **20.** 10 A.M.; 90 km **22.** Air: 1,101 ft/sec; water: 4,954 ft/sec

Practice Exercise 5.5

ODD **1.** 30 quarters and 70 dimes **3.** 700 $2 tickets; 2,800 $4 tickets **5.** 20 dl **7.** 20 liters
9. 60 dl of 20% solution, 30 dl of 50% solution **11.** 25 kg of $5-per-kilogram tea and 50 kg of $6.50-per-kilogram tea
13. $5,000 at 8 percent and $15,000 at 12 percent **15.** 3.6 liters **17.** 80 12-gram packages, 60 16-gram packages

EVEN **2.** 30 nickels, 20 dimes **4.** 3,000 $10 tickets, 5,000 $6 tickets **6.** 2 ml **8.** 100 cl
10. 75 liters of 30% solution; 25 liters of 70% solution **12.** 60 kg of $7.00-per-kilogram coffee; 40 kg of $9.50-per-kilogram coffee
14. $7,500 at 8 percent and $2,500 at 12 percent **16.** 6 liters **18.** Not possible. By mixing a 20% solution with a 50%
solution we can only obtain a solution with a strength between 20 and 50%. A common error is to think that a 20% solution added to
a 50% solution yields a 70% solution.

Practice Exercise 5.6

ODD **1.** $2x - 3 \geq -6; x \geq \frac{-3}{2}$ **3.** $15 - 3x < 6; x > 3$ **5.** $10w > 65, w > 6.5$ cm **7.** $-40 \leq 70 - 0.0055h \leq 26$;
$8,000 \leq h \leq 20,000$ **9.** Solve $2x > 300 + 1.5x; x > 600$ **11.** $220 \leq 110I \leq 2,750; 2 \leq I \leq 25$

13. Solve $30 \leq \frac{5}{9}(F - 32) \leq 35; 86° \leq F \leq 95°$ **15.** Solve $1.7 \leq \frac{V}{740} \leq 2.4; 1,258 \leq V \leq 1,776$

17. Solve $70 \leq \frac{MA \cdot 100}{12} \leq 120; 8.4 \leq MA \leq 14.4$ **19.** Solve $75 \leq \frac{100B}{20} \leq 80; 15 \leq B \leq 16$

21. If r is the worker's maximum running rate and R is the train's rate, then he will escape running toward the train if $r > \frac{R}{3} = \frac{7R}{21}$,

and he will escape running away from the train if $r > \frac{3R}{7} = \frac{9R}{21}$. Thus, his chances are better if he runs toward the train!

EVEN **2.** $2x + 5 \leq 7; x \leq 1$ **4.** $3x - 5 \leq 4x; x \geq -5$ **6.** $2(10) + 2w < 30; w < 5$
8. Solve $-95 \leq 70 - 0.0055h \leq 37; 6,000 \leq h \leq 30,000$ **10.** Solve $2.5x > 490 + 1.8x; x > 700$ **12.** $\frac{W}{110} \leq 30; W \leq 3,300$

14. Solve $68 \leq \frac{9}{5}C + 32 \leq 77; 20° \leq C \leq 25°$ **16.** Solve $1.2 \leq \frac{V}{740} \leq 1.8; 888 \leq V \leq 1,332$

18. Solve $80 \leq \frac{MA \cdot 100}{12} \leq 160; 9.6 \leq MA \leq 19.2$ **20.** Solve $\frac{100B}{20} > 80; B > 16$

Diagnostic (Review) Exercise 5.7

1. 30 (5.2) **2.** 32 nickels, 18 dimes (5.5) **3.** 60 by 40 meters (5.2) **4.** 384 km (5.3) **5.** $x < -8$ (5.6)
6. 40 lb of $4 coffee, 20 lb of $7 coffee (5.5) **7.** 32, 34, 36 (5.2) **8.** $x > 8$ cm (5.6) **9.** 2 km/hr (5.4)

10. 24 ml *(5.3)*   **11.** 75 ml of 30% solution, 25 ml of 70% solution *(5.5)*   **12.** 50 min *(5.4)*   **13.** 55.9 min *(5.4)*
14. 39.37 in *(5.3)*   **15.** $25 \leq C \leq 30$ *(5.6)*

Practice Test Chapter 5

1. 27, 29, 31 *(5.2)*   **2.** 25 by 13 cm *(5.2)*   **3.** 25.2 ml of alcohol *(5.3)*   **4.** 35.5 min *(5.4)*   **5.** 50 in *(5.3)*
6. 30 lb of $3 candy, 60 lb of $6 candy *(5.5)*   **7.** $x > 160$ *(5.6)*   **8.** 4.2 hr *(5.4)*
9. 24 dl of 50% solution, 12 dl of 80% solution *(5.5)*   **10.** $50 \leq F \leq 59$ *(5.6)*

CHAPTER 6

Practice Exercise 6.1

ODD **1.** 1   **3.** 1   **5.** $\frac{1}{3^3}$ **7.** $\frac{1}{m^7}$ **9.** 4^3 **11.** y^5 **13.** 10^2 **15.** y **17.** 1   **19.** 10^{10}

21. x^{11} **23.** $\frac{1}{z^5}$ **25.** $\frac{1}{10^7}$ **27.** 10^{12} **29.** y^8 **31.** $u^{10}v^6$ **33.** $\frac{x^4}{y^6}$ **33.** $\frac{x^2}{y^3}$ **37.** 1

39. 10^2 **41.** y **43.** 10 **45.** 3×10^{16} **47.** y^6 **49.** $36m^3n^2$ **51.** $\frac{2^3m^3}{n^6}$ **53.** $\frac{n^{15}}{m^{12}}$ **55.** $\frac{3^3}{2}$

57. 1   **59.** $\frac{4y^3}{3x^5}$ **61.** $\frac{a^9}{8b^4}$ **63.** $\frac{1}{x^7}$ **65.** $\frac{n^5}{m^{12}}$ **67.** $\frac{m^5n^5}{8}$ **69.** $\frac{t^2}{x^2y^{10}}$ **71.** 4   **73.** $\frac{1}{a^2 - b^2}$

75. $\frac{1}{xy}$ **77.** $-cd$ **79.** $\frac{xy}{x + y}$ **81.** $\frac{(y - x)^2}{x^2y^2}$

EVEN **2.** 1   **4.** 1   **6.** $\frac{1}{2^2}$ **8.** $\frac{1}{x^4}$ **10.** 3^2 **12.** x^3 **14.** 10^2 **16.** x^4 **18.** 1

20. 10^{11} **22.** a^{12} **24.** $\frac{1}{b^8}$ **26.** $\frac{1}{10^6}$ **28.** 2^6 **30.** x^{10} **32.** x^3y^2 **34.** $\frac{y^6}{x^4}$ **36.** $\frac{y^3}{x^2}$

38. 1   **40.** 10^2 **42.** $\frac{1}{x}$ **44.** 10^{17} **46.** 4×10^2 **48.** x^6 **50.** $\frac{1}{2^3c^3d^6}$ **52.** $\frac{3^2x^6}{y^4}$ **54.** $\frac{x^6}{y^4}$

56. $\frac{2^6}{3^4}$ **58.** $\frac{1}{10^4}$ **60.** $\frac{3n^4}{4m^3}$ **62.** $\frac{2x}{y^2}$ **64.** n^2 **66.** $\frac{x^{12}}{y^8}$ **68.** $\frac{4x^8}{y^6}$ **70.** $\frac{w^{12}}{u^{20}v^4}$ **72.** $\frac{27y^3}{2x^3}$

74. $\frac{1}{(x + 2)^2}$ **76.** $\frac{1}{30}$ **78.** $\frac{144}{7}$ **80.** $\frac{36}{13}$ **82.** $\frac{1,000}{11}$

Practice Exercise 6.2

ODD **1.** 7×10 **3.** 8×10^2 **5.** 8×10^4 **7.** 8×10^{-3} **9.** 8×10^{-8} **11.** 5.2×10
13. 6.3×10^{-1} **15.** 3.4×10^2 **17.** 8.5×10^{-2} **19.** 6.3×10^3 **21.** 6.8×10^{-6} **23.** 800   **25.** 0.04
27. 300,000   **29.** 0.0009   **31.** 56,000   **33.** 0.0097   **35.** 430,000   **37.** 0.000 000 38   **39.** 5.46×10^9
41. 7.29×10^{-8} **43.** 10^{13} **45.** 10^{-5} **47.** 83,500,000,000   **49.** 0.000 000 000 006 14   **51.** 865,000
53. 0.000 000 000 000 000 000 000 001 7   **55.** 9×10^4 **57.** 6×10^{-4} **59.** 3×10^5 **61.** 5×10^4
63. 3×10 or 30   **65.** 3×10^{-4} or 0.0003   **67.** 6.6×10^{21} tons   **69.** 10^7; 6×10^8

EVEN **2.** 5×10 **4.** 6×10^2 **6.** 6×10^5 **8.** 6×10^{-2} **10.** 6×10^{-5} **12.** 3.5×10
14. 7.2×10^{-1} **16.** 2.7×10^2 **18.** 3.2×10^{-2} **20.** 5.2×10^3 **22.** 7.2×10^{-4} **24.** 500   **26.** 0.08
28. 6,000,000   **30.** 0.00002   **32.** 7,100   **34.** 0.00086   **36.** 8,800,000   **38.** 0.000 0061   **40.** 4.27×10^7
42. 7.23×10^{-5} **44.** 5.87×10^{12} **46.** 3×10^{-23} **48.** 3,460,000,000   **50.** 0.000 000 623   **52.** 93,000,000
54. 0.000 075   **56.** 8×10^2 **58.** 8×10^{-3} **60.** 3×10^3 **62.** 3×10^7 **64.** 2×10^4 or 20,000
66. 2×10^{-4} or 0.0002   **68.** 3.3×10^{18} lb   **70.** 0.0186 mi or 98.2 ft

Practice Exercise 6.3

ODD **1.** 5   **3.** Not a real number   **5.** 2   **7.** -2 **9.** -2 **11.** 64   **13.** 4   **15.** x
17. $\frac{1}{x^{1/5}}$ **19.** x^2 **21.** ab^3 **23.** $\frac{x^3}{y^4}$ **25.** x^2y^3 **27.** $\frac{2}{5}$ **29.** $\frac{8}{125}$ **31.** $\frac{1}{4}$ **33.** $\frac{1}{6}$ **35.** $\frac{1}{125}$

37. Not a real number **39.** $\frac{1}{9}$ **41.** $\dfrac{1}{x^{1/2}}$ **43.** $n^{1/12}$ **45.** x^4 **47.** $\dfrac{2v^2}{u}$ **49.** $\dfrac{1}{x^2 y^3}$ **51.** $\dfrac{x^4}{y^3}$

53. $\frac{5}{4}x^4 y^2$ **55.** $64y^{1/3}$ **57.** $12m - 6m^{35/4}$ **59.** $2x + 3x^{1/2}y^{1/2} + y$ **61.** $x + 2x^{1/2}y^{1/2} + y$

63. Not defined **65.** $\dfrac{2}{a} + \dfrac{5}{a^{1/2}b^{1/2}} - \dfrac{3}{b}$ **67.** $a^{1/2}b^{1/3}$ **69.** x **71.** Any negative number

EVEN **2.** 6 **4.** Not a real number **6.** 3 **8.** -3 **10.** -3 **12.** 125 **14.** 9 **16.** $y^{3/5}$

18. $a^{1/3}$ **20.** y^2 **22.** $x^2 y$ **24.** $\dfrac{m^3}{n^4}$ **26.** $\dfrac{u^6}{v^4}$ **28.** $\frac{3}{2}$ **30.** $\frac{27}{8}$ **32.** $\frac{1}{9}$ **34.** $\frac{1}{5}$ **36.** $\frac{1}{64}$

38. Not a real number **40.** $\frac{1}{8}$ **42.** d **44.** $m^{1/6}$ **46.** $\dfrac{1}{y^{1/2}}$ **48.** $\dfrac{2x}{y^2}$ **50.** $16xy^3$

52. $m^{1/2}n^{1/3}$ **54.** $\dfrac{2b^2}{3a^2}$ **56.** $3x^{1/2}$ **58.** $6x - 2x^{19/3}$ **60.** $x - y$ **62.** $x - 2x^{1/2}y^{1/2} + y$ **64.** $-\frac{1}{64}$

66. $\dfrac{1}{x} - \dfrac{2}{x^{1/2}y^{1/2}} + \dfrac{1}{y}$ **68.** $a^{1/n}b^{1/m}$ **70.** a **72.** All real numbers

Practice Exercise 6.4

ODD **1.** $\sqrt{11}$ **3.** $\sqrt[3]{5}$ **5.** $\sqrt[5]{u^3}$ **7.** $4\sqrt[5]{y^3}$ **9.** $\sqrt[7]{(4y)^3}$ **11.** $\sqrt[9]{(4ab^3)^2}$ **13.** $\sqrt{a + b}$

15. $6^{1/2}$ **17.** $m^{1/4}$ **19.** $y^{3/5}$ **21.** $(xy)^{3/4}$ **23.** $(x^2 - y^2)^{1/2}$ **25.** $-5\sqrt[3]{y^2}$ **27.** $\sqrt[7]{(1 + m^2 n^2)^3}$

29. $\dfrac{1}{\sqrt[3]{w^2}}$ **31.** $\dfrac{1}{\sqrt[5]{(3m^2 n^3)^3}}$ **33.** $\sqrt{a} + \sqrt{b}$ **35.** $\sqrt[3]{(a^3 + b^3)^2}$ **37.** $(a + b)^{2/3}$ **39.** $-3x(a^3 b)^{1/4}$

41. $(-2x^3 y^7)^{1/9}$ **43.** $\dfrac{3}{y^{1/3}}$ or $3y^{-1/3}$ **45.** $\dfrac{-2x}{(x^2 + y^2)^{1/2}}$ or $-2x(x^2 + y^2)^{-1/2}$ **47.** $m^{2/3} - n^{1/2}$ **49.** $5 \neq 7$

51. Any negative number

EVEN **2.** $\sqrt{7}$ **4.** $\sqrt[4]{6}$ **6.** $\sqrt[4]{x^3}$ **8.** $5\sqrt[3]{m^2}$ **10.** $\sqrt[3]{(5m)^2}$ **12.** $\sqrt[3]{(7x^2 y)^2}$ **14.** $\sqrt{a^2 + b^2}$

16. $3^{1/2}$ **18.** $m^{1/7}$ **20.** $a^{2/3}$ **22.** $(7m^3 n^3)^{4/5}$ **24.** $(1 + y^2)^{1/2}$ **26.** $-3\sqrt{x}$ **28.** $\sqrt[5]{(x^2 y^2 - w^3)^4}$

30. $\dfrac{1}{\sqrt[5]{y^3}}$ **32.** $\dfrac{1}{\sqrt[3]{(2xy)^2}}$ **34.** $\dfrac{1}{\sqrt{x}} + \dfrac{1}{\sqrt{y}}$ **36.** $\sqrt[3]{\sqrt{x} + \dfrac{1}{\sqrt{y}}}$ **38.** $(x - y)^{2/5}$ **40.** $-5(2x^2 y^2)^{1/3}$

42. $(-4m^2 n^3)^{1/5}$ **44.** $\dfrac{2x}{y^{1/2}}$ or $2xy^{-1/2}$ **46.** $\dfrac{2}{x^{1/2}} + \dfrac{3}{y^{1/2}}$ or $2x^{-1/2} + 3y^{-1/2}$ **48.** $\dfrac{-5u^2}{u^{1/2} + v^{3/5}}$ **50.** $5 \neq 7$

52. All real numbers

Practice Exercise 6.5

ODD **1.** y **3.** $2u$ **5.** $7x^2 y$ **7.** $3\sqrt{2}$ **9.** $m\sqrt{m}$ **11.** $2x\sqrt{2x}$ **13.** $\frac{1}{3}$ **15.** $\dfrac{1}{y}$ **17.** $\dfrac{\sqrt{5}}{5}$

19. $\dfrac{\sqrt{5}}{5}$ **21.** $\dfrac{\sqrt{y}}{y}$ **23.** $\dfrac{\sqrt{y}}{y}$ **25.** $3xy^2 \sqrt{xy}$ **27.** $3x^4 y^2 \sqrt{2y}$ **29.** $\dfrac{\sqrt{2x}}{2x}$ **31.** $2x\sqrt{3x}$ **33.** $\dfrac{3\sqrt{2ab}}{2b}$

35. $\dfrac{\sqrt{42xy}}{7y}$ **37.** $\dfrac{3m^2 \sqrt{2mn}}{2n}$ **39.** $2x^2 y$ **41.** $2xy^2 \sqrt[3]{2xy}$ **43.** $\sqrt{x}$ **45.** 4 **47.** $6m^3 n^3$ **49.** $2\sqrt[3]{9}$

51. $\dfrac{2a\sqrt{3ab}}{3b}$ **53.** Is in the simplest radical form **55.** $\dfrac{2x}{3y^2}$ **57.** $-3m^2 n^2 \sqrt[5]{3m^2 n}$ **59.** $\sqrt[3]{x^2(x - y)}$

61. $x^2 y \sqrt[3]{6xy}$ **63.** $2x^2 y \sqrt[3]{4x^2 y}$ **65.** $-\sqrt[5]{6x^2 y^2}$ **67.** $\sqrt[3]{(x - y)^2}$ **69.** $\dfrac{\sqrt[4]{12x^3 y^3}}{2x}$ **71.** $-x\sqrt{x^2 + 2}$

73. $4x^9 y \sqrt[3]{2y}$ **75.** **(A)** $7x$ **(B)** $3x$ **77.** **(A)** $-x$ **(B)** $13x$ **79.** **(A)** $2x$ **(B)** 0 **81.** **(A)** x **(B)** $-5x$

EVEN **2.** x **4.** $3m$ **6.** $5xy^2$ **8.** $2\sqrt{2}$ **10.** $x\sqrt{x}$ **12.** $3y\sqrt{2y}$ **14.** $\frac{1}{2}$ **16.** $\dfrac{1}{x}$ **18.** $\dfrac{\sqrt{3}}{3}$

20. $\dfrac{\sqrt{3}}{3}$ **22.** $\dfrac{\sqrt{x}}{x}$ **24.** $\dfrac{\sqrt{x}}{x}$ **26.** $2x^2 y \sqrt{xy}$ **28.** $2x^3 y^3 \sqrt{2x}$ **30.** $\dfrac{\sqrt{3y}}{3y}$

32. $2x\sqrt{2y}$ **34.** $\frac{2}{3}x\sqrt{3xy}$ **36.** $\dfrac{\sqrt{6mn}}{2n}$ **38.** $\dfrac{2a\sqrt{3ab}}{3b}$ **40.** $2mn^3$ **42.** $2ab^2 \sqrt[4]{a}$ **44.** $\sqrt[5]{x^3}$

46. 3 **48.** $3xy$ **50.** $\sqrt[3]{4}$ **52.** $\dfrac{3m^2 \sqrt{2mn}}{2n}$ **54.** Is in the simplest radical form **56.** $\dfrac{a^2 b}{2c^3}$

58. $-4x^3y^4\sqrt[3]{x^2y}$ **60.** $\sqrt[4]{2^3(x+y)^3}$ **62.** $u^2v\sqrt[4]{24v}$ **64.** $4u^2v^4\sqrt[3]{2uv}$ **66.** $2\sqrt[3]{6abc}$ **68.** $\dfrac{\sqrt[3]{x-y}}{x-y}$

70. $\dfrac{\sqrt[5]{8m^2n^2}}{2m}$ **72.** $m\sqrt[4]{1+4m^2}$ **74.** $2x\sqrt[3]{4y^2}$ **76.** (A) $6x$ (B) $-2x$ **78.** (A) $-2x$ (B) $10x$

80. (A) $2x$ (B) 0 **82.** (A) $2x$ (B) $8x$

Practice Exercise 6.6

ODD **1.** $9\sqrt{3}$ **3.** $-5\sqrt{a}$ **5.** $-5\sqrt{n}$ **7.** $4\sqrt{5}-2\sqrt{3}$ **9.** $\sqrt{m}-3\sqrt{n}$ **11.** $4\sqrt{2}$ **13.** $-6\sqrt{2}$
15. $7-2\sqrt{7}$ **17.** $3\sqrt{2}-2$ **19.** $y-8\sqrt{y}$ **21.** $4\sqrt{n}-n$ **23.** $3+3\sqrt{2}$ **25.** $3-\sqrt{3}$
27. $9+4\sqrt{5}$ **29.** $m-7\sqrt{m}+12$ **31.** $\sqrt{5}-2$ **33.** $\dfrac{\sqrt{5}-1}{2}$ **35.** $\dfrac{\sqrt{5}+\sqrt{2}}{3}$ **37.** $\dfrac{y-3\sqrt{y}}{y-9}$
39. $8\sqrt{2mn}$ **41.** $6\sqrt{2}-2\sqrt{5}$ **43.** $-\sqrt[5]{a}$ **45.** $5\sqrt[3]{x}-\sqrt{x}$ **47.** $\dfrac{9\sqrt{2}}{4}$ **49.** $\dfrac{-3\sqrt{6uv}}{2}$
51. $38-11\sqrt{3}$ **53.** $x-y$ **55.** $10m-11\sqrt{m}-6$ **57.** $5+\sqrt[3]{18}+\sqrt[3]{12}$
59. $(2-\sqrt{3})^2-4(2-\sqrt{3})+1=4-4\sqrt{3}+3-8+4\sqrt{3}+1=0$ **61.** $-7-4\sqrt{3}$ **63.** $5+2\sqrt{6}$
65. $\dfrac{x+5\sqrt{x}+6}{x-9}$ **67.** $\dfrac{6x+9\sqrt{x}}{4x-9}$ **69.** $3\sqrt{3}$ **71.** $\tfrac{10}{9}\sqrt[3]{9}$ **73.** $x+2\sqrt[3]{xy}-\sqrt[3]{x^2y^2}-2v$
75. $\dfrac{8x-22\sqrt{xy}+15y}{16x-25y}$

EVEN **2.** $8\sqrt{2}$ **4.** $-3\sqrt{y}$ **6.** $4\sqrt{x}$ **8.** $2\sqrt{2}-2\sqrt{3}$ **10.** $2\sqrt{x}+2\sqrt{y}$ **12.** $\sqrt{2}$
14. $-3\sqrt{3}$ **16.** $5-2\sqrt{5}$ **18.** $2\sqrt{3}-3$ **20.** $x-3\sqrt{x}$ **22.** $3\sqrt{m}-m$ **24.** $5\sqrt{2}+5$
26. $2\sqrt{2}-1$ **28.** $12-6\sqrt{3}$ **30.** $x-\sqrt{x}-6$ **32.** $\dfrac{\sqrt{11}+3}{2}$ **34.** $2\sqrt{6}+4$ **36.** $\sqrt{3}-\sqrt{2}$
38. $\dfrac{x+2\sqrt{x}}{x-4}$ **40.** $-\sqrt{x}$ **42.** $2\sqrt{6}+\sqrt{3}$ **44.** $-\sqrt[3]{u}$ **46.** $3\sqrt[5]{y}+3\sqrt[4]{y}$ **48.** $\dfrac{-\sqrt{6}}{6}$
50. $\dfrac{5\sqrt{2xy}}{2}$ **52.** 25 **54.** $4x-9$ **56.** $6u+8\sqrt{u}-8$ **58.** -2
60. $(2+\sqrt{3})^2-4(2+\sqrt{3})+1=4+4\sqrt{3}+3-8-4\sqrt{3}+1=0$ **62.** $3-2\sqrt{2}$ **64.** $\dfrac{6+\sqrt{a}-a}{a-4}$
66. $\dfrac{7-2\sqrt{10}}{3}$ **68.** $\dfrac{15\sqrt{a}+10a}{9-4a}$ **70.** $3\sqrt{2}$ **72.** $\tfrac{3}{2}\sqrt[4]{2}$ **74.** $u+\sqrt[5]{u^2v^2}-\sqrt[5]{u^3v^3}-v$
76. $\dfrac{6x+19\sqrt{xy}+10y}{4x-25y}$

Practice Exercise 6.7

ODD **1.** $8+3i$ **3.** $-5+3i$ **5.** $5+3i$ **7.** $6+13i$ **9.** $3-2i$ **11.** -15 or $-15+0i$
13. $-6-10i$ **15.** $15-3i$ **17.** $-4-33i$ **19.** 65 or $65+0i$ **21.** $\tfrac{2}{5}-\tfrac{1}{5}i$ **23.** $\tfrac{3}{13}+\tfrac{11}{13}i$
25. $5+3i$ **27.** $7-5i$ **29.** $-3+2i$ **31.** $8+25i$ **33.** $\tfrac{5}{3}-\tfrac{2}{3}i$ **35.** $\tfrac{2}{13}+\tfrac{3}{13}i$ **37.** $-\tfrac{2}{5}i$ or $0-\tfrac{2}{5}i$
39. $\tfrac{3}{2}-\tfrac{1}{2}i$ **41.** $4-7i$ **43.** 0 or $0+0i$ **45.** 0 **47.** For all $x\geq 10$

EVEN **2.** $8+4i$ **4.** $7-5i$ **6.** $7+2i$ **8.** $-3+2i$ **10.** $17-2i$ **12.** -8 or $-8+0i$
14. $-12-6i$ **16.** $-21+i$ **18.** $8+i$ **20.** 34 or $34+0i$ **22.** $\tfrac{3}{10}+\tfrac{1}{10}i$ **24.** $\tfrac{4}{13}-\tfrac{7}{13}i$ **26.** $3+3i$
28. $-5+3i$ **30.** $6+13i$ **32.** $13+i$ **34.** $3-4i$ **36.** $\tfrac{3}{25}+\tfrac{4}{25}i$ **38.** $-\tfrac{1}{3}i$ or $0-\tfrac{1}{3}i$ **40.** $-\tfrac{1}{3}-\tfrac{2}{3}i$
42. $-6i$ or $0-6i$ **44.** 0 **46.** 0 **48.** When b^2-4ac is negative

Diagnostic (Review) Exercise 6.8

1. 1 (6.1)   **2.** $\tfrac{1}{3}$ (6.1)   **3.** 8 (6.1)   **4.** $\tfrac{1}{2}$ (6.1)   **5.** Not a real number (6.3)   **6.** 4 (6.3)

7. (A) 4.28×10^9 (B) 3.18×10^{-5} (6.2)   **8.** (A) $729{,}000$ (B) $0.000\,603$ (6.2)   **9.** $6x^4y^7$ (6.1)   **10.** $\dfrac{3u^4}{v^2}$ (6.1)

11. $6x^5y^{15}$ (6.1)   **12.** $\dfrac{c^6}{d^{15}}$ (6.1)   **13.** $\dfrac{4x^4}{9y^6}$ (6.1)   **14.** x^{12} (6.1)   **15.** y^2 (6.1)   **16.** $\dfrac{y^3}{x^2}$ (6.1)   **17.** x^3 (6.3)

18. $\dfrac{1}{x^2}$ *(6.3)*   **19.** $\dfrac{1}{x^{1/3}}$ *(6.3)*   **20.** u *(6.3)*   **21.** **(A)** $\sqrt{3m}$ **(B)** $3\sqrt{m}$ *(6.4)*   **22.** **(A)** $(2x)^{1/2}$

(B) $(a+b)^{1/2}$ *(6.4)*   **23.** $2xy^2$ *(6.5)*   **24.** $\dfrac{5}{y}$ *(6.5)*   **25.** $6x^2y^3\sqrt{y}$ *(6.5)*   **26.** $\dfrac{\sqrt{2y}}{2y}$ *(6.5)*   **27.** $2b\sqrt{3a}$ *(6.5)*

28. $6x^2y^3\sqrt{xy}$ *(6.5)*   **29.** $\dfrac{\sqrt{2xy}}{2x}$ *(6.5)*   **30.** $-3\sqrt{x}$ *(6.6)*   **31.** $\sqrt{7}-2\sqrt{3}$ *(6.6)*   **32.** $5+2\sqrt{5}$ *(6.6)*

33. $1+\sqrt{3}$ *(6.6)*   **34.** $\dfrac{5+3\sqrt{5}}{4}$ *(6.6)*   **35.** $3-6i$ *(6.7)*   **36.** $15+3i$ *(6.7)*   **37.** $2+i$ *(6.7)*

38. $-\frac{1}{2}-i$ *(6.7)*   **39.** 2×10^{-3} or 0.002 *(6.2)*   **40.** $\dfrac{m^2}{2n^5}$ *(6.1)*   **41.** $\dfrac{x^6}{y^4}$ *(6.1)*   **42.** $\dfrac{4x^4}{y^6}$ *(6.1)*   **43.** $\dfrac{c}{a^2b^4}$ *(6.1)*

44. $\frac{1}{4}$ *(6.1)*   **45.** $\dfrac{n^{10}}{9m^{10}}$ *(6.1)*   **46.** $\dfrac{1}{(x-y)^2}$ *(6.1)*   **47.** $\dfrac{3a^2}{b}$ *(6.3)*   **48.** $\dfrac{3x^2}{2y^2}$ *(6.3)*   **49.** $\dfrac{1}{m}$ *(6.3)*

50. $6x^{1/6}$ *(6.3)*   **51.** $\dfrac{x^{1/12}}{2}$ *(6.3)*   **52.** $\frac{5}{8}$ *(6.1)*   **53.** $x+2x^{1/2}y^{1/2}+y$ *(6.3)*   **54.** $a^2=b$ *(6.3)*

55. **(A)** $\sqrt[3]{4m^2n^2}$ **(B)** $3\sqrt[5]{x^2}$ *(6.4)*   **56.** **(A)** $x^{5/7}$ **(B)** $-4(xy)^{2/3}$ *(6.4)*   **57.** $2x^2y$ *(6.5)*   **58.** $3x^2y\sqrt[3]{x^2y}$ *(6.5)*

59. $\dfrac{n^2\sqrt{6m}}{3}$ *(6.5)*   **60.** $\sqrt[4]{y^3}$ *(6.5)*   **61.** $-6x^2y^2\sqrt[5]{3x^2y}$ *(6.5)*   **62.** $x\sqrt[3]{2x^2}$ *(6.5)*   **63.** $\dfrac{\sqrt[5]{12x^3y^2}}{2x}$ *(6.6)*

64. $2x-3\sqrt{xy}-5y$ *(6.6)*   **65.** $\dfrac{x-4\sqrt{x}+4}{x-4}$ *(6.6)*   **66.** $\dfrac{6x+3\sqrt{xy}}{4x-y}$ *(6.6)*   **67.** $\dfrac{5\sqrt{6}}{6}$ *(6.6)*   **68.** $-1-i$ *(6.7)*

69. $\frac{4}{13}-\frac{7}{13}i$ *(6.7)*   **70.** $5+4i$ *(6.7)*   **71.** $\dfrac{xy}{x+y}$ *(6.1)*   **72.** $\dfrac{a^2b^2}{a^3+b^3}$ *(6.1)*   **73.** $y\sqrt[3]{2x^2y}$ *(6.5)*   **74.** 0 *(6.6)*

75. **(A)** x **(B)** $5x$ *(6.5)*

Practice Test Chapter 6

1. $\dfrac{v^3}{3u}$ *(6.1)*   **2.** $\dfrac{4x^6}{9y^4}$ *(6.1)*   **3.** $\frac{1}{4}$ *(6.3)*   **4.** $\dfrac{y^5}{2x^4}$ *(6.3)*   **5.** $\dfrac{x}{2y^4}$ *(6.3)*   **6.** $\dfrac{2m^5}{n}$ *(6.3)*   **7.** xy *(6.3)*

8. $\dfrac{2}{m^{1/3}}$ *(6.3)*   **9.** $\dfrac{1}{x}+\dfrac{2}{x^{1/2}y^{1/2}}+\dfrac{1}{y}$ or $\dfrac{x+2x^{1/2}y^{1/2}+y}{xy}$ *(6.3)*   **10.** $\dfrac{(7.5\times10^{-5})(4\times10^3)}{(2\times10^6)(5\times10^{-2})}=3\times10^{-6}$ *(6.2)*

11. $4(x-y)^{2/3}$ *(6.4)*   **12.** $6x^2y^2\sqrt[5]{2x^3y^2}$ *(6.5)*   **13.** $\dfrac{y\sqrt{6x}}{2x}$ *(6.5)*   **14.** $-2\sqrt[3]{2}$ *(6.5)*   **15.** $\sqrt[3]{(2xy)^2}$ *(6.5)*

16. 0 *(6.6)*   **17.** $\dfrac{x+2\sqrt{x}\sqrt{y}+y}{x-y}$ *(6.6)*   **18.** $-x$ *(6.5)*   **19.** $2-8i$ *(6.7)*   **20.** $\frac{4}{5}-\frac{7}{5}i$ *(6.7)*

CHAPTER 7

Practice Exercise 7.2

ODD   **1.** ±4 **3.** $\pm4i$ **5.** $\pm3\sqrt{5}$ **7.** $\pm\frac{3}{2}$ **9.** $\pm\frac{3}{4}$ **11.** $0,-5$ **13.** $0,-4$ **15.** $12,-1$

17. $1,-5$ **19.** $-\frac{2}{3},4$ **21.** $\pm\sqrt{2}$ **23.** $\pm\frac{3}{4}i$ **25.** $\pm\sqrt{\frac{7}{9}}$ or $\pm\dfrac{\sqrt{7}}{3}$ **27.** $8,-2$ **29.** $-1\pm3i$

31. $-\frac{1}{3},1$ **33.** $-1,3$ **35.** $-\frac{2}{3},1$ **37.** Not factorable in the integers   **39.** $-\frac{1}{2},3$ **41.** $-2,2$

43. $3,-4$ **45.** $-\frac{1}{2},2$ **47.** $\frac{1}{2},2$ **51.** $\dfrac{-5\pm\sqrt{10}}{2}$ **53.** $a=\sqrt{c^2-b^2}$ **49.** 11 by 3 in   **55.** 90 cents

EVEN   **2.** ±5 **4.** $\pm5i$ **6.** $\pm2\sqrt{3}$ **8.** $\pm\frac{4}{3}$ **10.** $\pm\frac{2}{3}$ **12.** $0,3$ **14.** $0,2$ **16.** $1,5$

18. $-2,6$ **20.** $\frac{1}{2},-8$ **22.** $\pm\sqrt{3}$ **24.** $\pm\frac{5}{2}i$ **26.** $\pm\sqrt{\dfrac{3}{4}}$ or $\dfrac{\pm\sqrt{3}}{2}$ **28.** $-2,-8$ **30.** $3\pm2i$

32. $-1,2$ **34.** $3,-5$ **36.** $\frac{1}{2},-3$ **38.** Not factorable in the integers   **40.** $\frac{2}{3},2$ **42.** $-3,3$

44. $-2,4$ **46.** $5,-3$ **48.** $-2,5$ **52.** $\dfrac{3\pm\sqrt{6}}{2}$ **54.** $t=\sqrt{\dfrac{2s}{g}}$ **50.** $h=1$ ft, $b=4$ ft

56. 2.83 ft/sec

Practice Exercise 7.3

ODD   **1.** $x^2+4x+4=(x+2)^2$ **3.** $x^2-6x+9=(x-3)^2$ **5.** $x^2+12x+36=(x+6)^2$

7. $-2 \pm \sqrt{2}$ **9.** $3 \pm 2\sqrt{3}$ **11.** $x^2 + 3x + \frac{9}{4} = (x + \frac{3}{2})^2$ **13.** $u^2 - 5u + \frac{25}{4} = (u - \frac{5}{2})^2$ **15.** $\dfrac{-1 \pm \sqrt{5}}{2}$

17. $\dfrac{5 \pm \sqrt{17}}{2}$ **19.** $2 \pm 2i$ **21.** $\dfrac{2 \pm \sqrt{2}}{2}$ **23.** $\dfrac{-3 \pm \sqrt{17}}{4}$ **25.** $\dfrac{3 \pm i\sqrt{7}}{4}$ **27.** $x = \dfrac{-m \pm \sqrt{m^2 - 4n}}{2}$

EVEN **2.** $x^2 + 8x + 16 = (x + 4)^2$ **4.** $x^2 - 10x + 25 = (x - 5)^2$ **6.** $x^2 + 2x + 1 = (x + 1)^2$ **8.** $-4 \pm \sqrt{13}$

10. $5 \pm 2\sqrt{7}$ **12.** $x^2 + x + \frac{1}{4} = (x + \frac{1}{2})^2$ **14.** $m^2 - 7m + \frac{49}{4} = (m - \frac{7}{2})^2$ **16.** $\dfrac{-3 \pm \sqrt{13}}{2}$ **18.** $\dfrac{3 \pm \sqrt{13}}{2}$

20. $1 \pm i\sqrt{2}$ **22.** $\dfrac{3 \pm \sqrt{3}}{2}$ **24.** $\dfrac{-1 \pm \sqrt{13}}{6}$ **26.** $\dfrac{5 \pm i\sqrt{11}}{6}$

Practice Exercise 7.4

ODD **1.** $a = 2, b = -5, c = 3$ **3.** $a = 3, b = 1, c = -1$ **5.** $a = 3, b = 0, c = -5$ **7.** $-4 \pm \sqrt{13}$

9. $5 \pm 2\sqrt{7}$ **11.** $\dfrac{-3 \pm \sqrt{13}}{2}$ **13.** $1 \pm i\sqrt{2}$ **15.** $\dfrac{3 \pm \sqrt{3}}{2}$ **17.** $\dfrac{-1 \pm \sqrt{13}}{6}$ **19.** $5 \pm \sqrt{7}$

21. $-1 \pm \sqrt{3}$ **23.** $0, -\frac{3}{2}$ **25.** $2 \pm 3i$ **27.** $5 \pm 2\sqrt{7}$ **29.** $\dfrac{2 \pm \sqrt{2}}{2}$ **31.** $1 \pm i\sqrt{2}$ **33.** $\frac{2}{5}, 3$

35. $t = \sqrt{2d/g}$ **37.** $r = -1 + \sqrt{A/P}$

EVEN **2.** $a = 3, b = -2, c = 1$ **4.** $a = 2, b = 3, c = -1$ **6.** $a = 2, b = -5, c = 0$ **8.** $-2 \pm \sqrt{2}$

10. $3 \pm 2\sqrt{3}$ **12.** $\dfrac{-1 \pm \sqrt{5}}{2}$ **14.** $2 \pm 2i$ **16.** $\dfrac{2 \pm \sqrt{2}}{2}$ **18.** $\dfrac{-3 \pm \sqrt{17}}{4}$ **20.** $-4 \pm \sqrt{11}$

22. $\dfrac{3 \pm \sqrt{13}}{2}$ **24.** $0, 2$ **26.** $-3 \pm 2i$ **28.** $3 \pm 2\sqrt{3}$ **30.** $\dfrac{3 \pm \sqrt{3}}{2}$ **32.** $2 \pm 2i$ **34.** $-50, 2$

36. $a = \sqrt{c^2 - b^2}$ **38.** $I = \dfrac{E \pm \sqrt{E^2 - 4RP}}{2R}$

Practice Exercise 7.5

ODD **1.** 12, 14 **3.** 0, 2 **5.** 127 mi **7.** 5.12 by 3.12 cm **9.** 1 ft **11.** 5.66 ft/sec **13.** 50 mi/hr

15. 1.41 min **17.** 2 hr; 3 hr **19.** 2 km/hr **21.** \$6

EVEN **2.** 8, 13 **4.** 3 or $\frac{1}{3}$ **6.** $h = 1$ meter, $b = 4$ meters (*Note*: $h = -4$ must be discarded.)

8. 4.61 by 2.61 meters **10.** 70 mi/hr **12.** 60 mi/hr **14.** (**A**) $t = 0, 11$; y is zero at the beginning and end of flight

(**B**) 10.91 sec, 0.09 sec **16.** 13.09 hr; 8.09 hr **18.** 5 km/hr; 12 km/hr **20.** 20 percent

Practice Exercise 7.6

ODD **1.** 4 **3.** 18 **5.** $\pm 1, \pm 3$ **7.** $\pm 3, \pm i\sqrt{2}$ **9.** $-1, 4$ **11.** 9, 16 **13.** 4

15. No solution **17.** No solution **19.** No solution **21.** 5, 13 **23.** $-1, 2$ **25.** $\pm 1, \pm 2, \pm 2i, \pm i$

27. $-8, 125$ **29.** 1, 16 **31.** $\frac{2}{3}, -\frac{3}{2}$ **33.** $\pm 2, \pm \frac{1}{2}$ **35.** $-2, 3, \frac{1}{2} \pm \dfrac{\sqrt{7}}{2}i$ **37.** $3 \pm 2i, 4, 2$ **39.** 4

EVEN **2.** 9 **4.** 8 **6.** $\pm 2, \pm 3$ **8.** $\pm 2, \pm i\sqrt{2}$ **10.** $-9, 1$ **12.** 4, 81 **14.** 6 **16.** 0, 4

18. 9 **20.** 1 **22.** $\frac{9}{4}, 3$ **24.** $-\sqrt[3]{5}, \sqrt[3]{2}$ **26.** $\pm\sqrt{\dfrac{7}{3}}$ or $\dfrac{\pm\sqrt{21}}{3}, \pm i$ **28.** $\frac{1}{8}, -8$ **30.** 16, 81

32. $-\frac{3}{4}, \frac{1}{5}$ **34.** $\pm 1, \pm 3$ **36.** $1, -3, -1 \pm i$ **38.** 2, 3, 7, 8

Practice Exercise 7.7

ODD **1.** $-4 < x < 3$ **3.** $x \leq -4$ or $x \geq 3$

5. $-4 < x < 3$ **7.** $x < 3$ or $x > 7$

9. $x \le -6$ or $x \ge 0$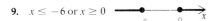

11. $x \le -3$ or $x \ge 3$

13. $-2 < x \le 5$

15. $x < -2$ or $x > 5$

17. $x < -2$ or $0 < x \le 4$

19. $x < 0$ or $x > \frac{1}{4}$

21. All real numbers; graph: whole real line   **23.** No solution   **25.** $x \le -\sqrt{3}$ or $x \ge \sqrt{3}$

27. $2 \le x < 3$

29. $-2 < x < 3$

31. $x \ge 2$ or $x \le 1$

33. $2 \le t \le 6$

EVEN **2.** $-2 < x < 4$

4. $x < -2$ or $x > 4$

6. $-5 < x < 2$

8. $x < -5$ or $x > -2$

10. $0 \le x \le 8$

12. $x < -2$ or $x > 2$

14. $-2 < x < 3$

16. $x \le -2$ or $x > 3$

18. $-5 \le x \le 0$ or $x > 3$

20. $0 < x < \frac{5}{3}$

22. All real numbers; graph: whole real line   **24.** No solution

26. $-\sqrt{2} < x < \sqrt{2}$

28. $x < -3$ or $x \ge 3$

30. $-1 < x < 2$ or $x \ge 5$

32. $x < -5$ or $x \ge 3$ **34.** $0 \le t \le 10$

Diagnostic (Review) Exercise 7.8

1. $0, 3$ (7.2)   **2.** ± 5 (7.2)   **3.** $2, 3$ (7.2)   **4.** $-3, 5$ (7.2)   **5.** $\pm \sqrt{7}$ (7.2)   **6.** $a = 3, b = 4, c = -2$ (7.4)

7. $x = \dfrac{-b \pm \sqrt{b^2 - 4ac}}{2a}$ (7.4)   **8.** $\dfrac{-3 \pm \sqrt{5}}{2}$ (7.4)   **9.** $3, 9$ (7.5)   **10.** $-5 < x < 4$  (7.7)

11. $x \le -5$ or $x \ge 4$ (7.7)   **12.** $0, 2$ (7.2)   **13.** $\pm 2\sqrt{3}$ (7.2)   **14.** $\pm 3i$ (7.2)   **15.** $-2, 6$ (7.2)

16. $-\frac{1}{3}, 3$ (7.2)   **17.** $\frac{1}{2}, -3$ (7.2)   **18.** $\dfrac{1 \pm \sqrt{7}}{3}$ (7.4)   **19.** $\dfrac{1 \pm \sqrt{7}}{2}$ (7.4)   **20.** $-4, 5$ (7.2)   **21.** $\frac{3}{4}, \frac{5}{2}$ (7.2)

22. $\dfrac{-1 \pm \sqrt{5}}{2}$ (7.4)   **23.** $1 \pm i\sqrt{2}$ (7.4)   **24.** $2, 3$ (7.6)   **25.** $9, 25$ (7.6)   **26.** $\pm 2, \pm 3i$ (7.6)   **27.** $64, -\frac{27}{8}$ (7.6)

28. $x \le -3$ or $x \ge 7$ (7.7)   **29.** $x < 0$ or $x > \frac{1}{2}$ (7.7)

30. No solution (7.7)   **31.** All real numbers; graph: real line (7.7)   **32.** 6 by 5 in (7.5)   **33.** $3 \pm 2\sqrt{3}$ (7.3)

34. $\dfrac{3 \pm i\sqrt{6}}{2}$ or $\dfrac{3}{2} \pm \dfrac{\sqrt{6}}{2}i$ (7.4)   **35.** $\dfrac{-3 \pm \sqrt{57}}{6}$ (7.4)   **36.** $\pm 1, \pm 2, \pm 2i, \pm i$ (7.6)   **37.** $3, \frac{9}{4}$ (7.6)

38. $x \le 1$ or $3 < x < 4$ (7.7)   **39.** 9 cm and 12 cm (7.5)   **40.** **(A)** 2,000 and 8,000   **(B)** 5,000 (7.5)

Practice Test Chapter 7

1. $\pm 2i$ (7.2)   **2.** $\dfrac{3 \pm \sqrt{17}}{4}$ (7.4)   **3.** $2 \pm 3i$ (7.4)   **4.** $\pm 1, \pm 2i$ (7.6)   **5.** $-1, 8$ (7.6)

6. 3 (-4 is extraneous) (7.6)   **7.** No solution (7.6)   **8.** $-2 \le x \le 3$ (7.7)

9. $-12 < x < -2$ or $x > 3$ (7.7)   **10.** 2.5 by 4 meters (7.5)

CHAPTER 8

Practice Exercise 8.1

ODD

1.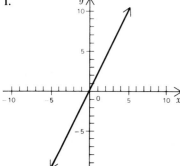

3.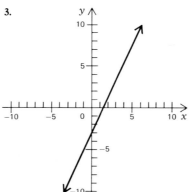

5.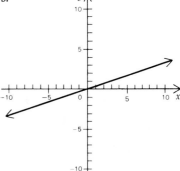

7.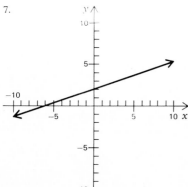

9.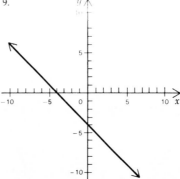

11.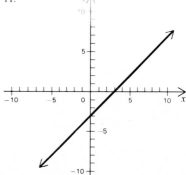

13.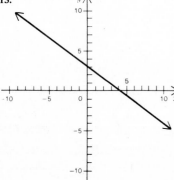

15.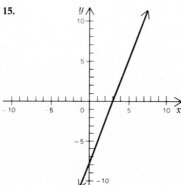

17.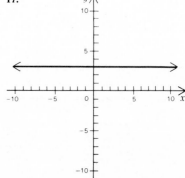

19.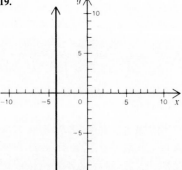

21.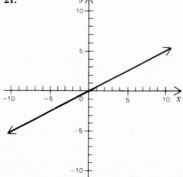

23.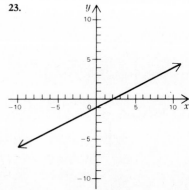

25.

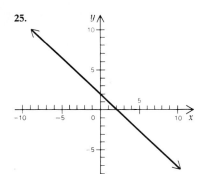

27.

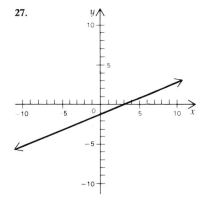

29.

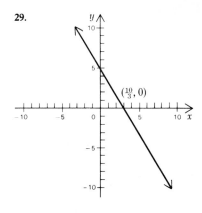

$\left(\frac{10}{3}, 0\right)$

31.

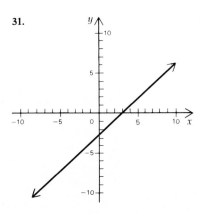

33.

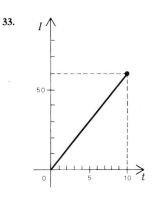

35.

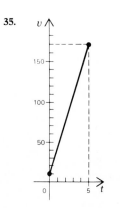

37.

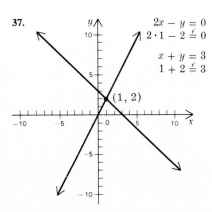

$2x - y = 0$
$2 \cdot 1 - 2 \stackrel{\checkmark}{=} 0$

$x + y = 3$
$1 + 2 \stackrel{\checkmark}{=} 3$

$(1, 2)$

39.

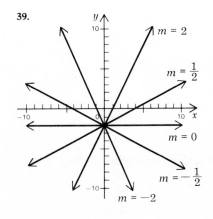

$m = 2$

$m = \frac{1}{2}$

$m = 0$

$m = -\frac{1}{2}$

$m = -2$

41.

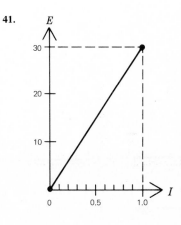

EVEN

2.

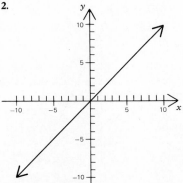

4.

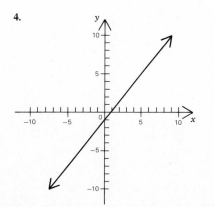

6.

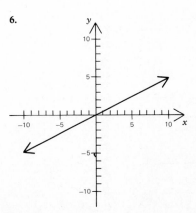

8.

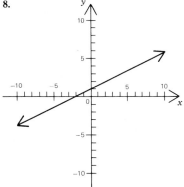

10.

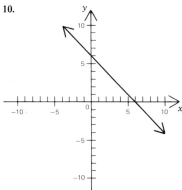

12.

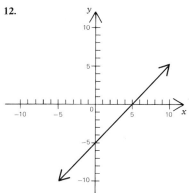

14.

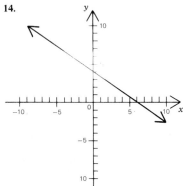

16.

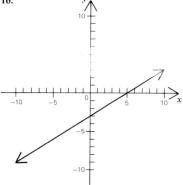

18.

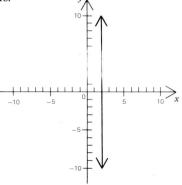

20.

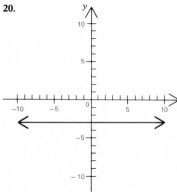

22.

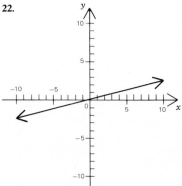

24.

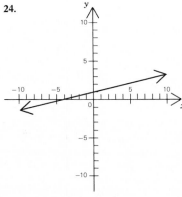

26.

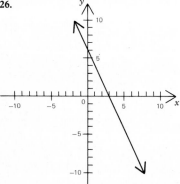

28.

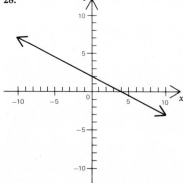

30.

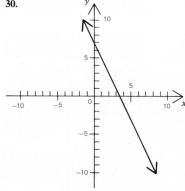

32.

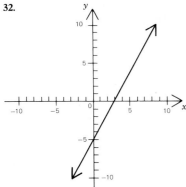

34.

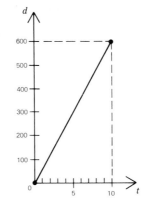

36.

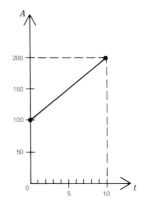

38.

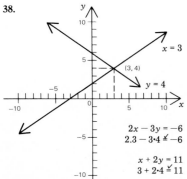

$2x - 3y = -6$
$2 \cdot 3 - 3 \cdot 4 \overset{?}{=} -6$

$x + 2y = 11$
$3 + 2 \cdot 4 \overset{?}{=} 11$

40.

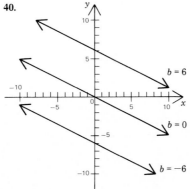

Practice Exercise 8.2

ODD

1.

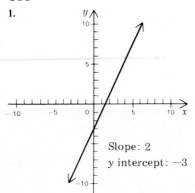

Slope: 2
y intercept: −3

3.

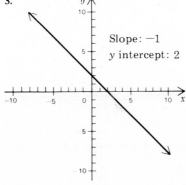

Slope: −1
y intercept: 2

5. $y = 5x - 2$ **7.** $y = -2x + 4$ **9.** $y - 4 = 2(x - 5)$ **11.** $y - 1 = -2(x - 2)$ **13.** 2   **15.** $\frac{1}{2}$

17. $y - 6 = 2(x - 5)$ or $y - 2 = 2(x - 3)$ **19.** $y - 5 = \frac{1}{2}(x - 10)$ or $y - 1 = \frac{1}{2}(x - 2)$

21. 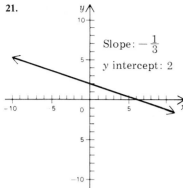 Slope: $-\frac{1}{3}$ y intercept: 2

23. 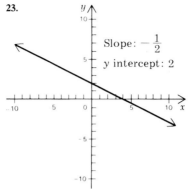 Slope: $-\frac{1}{2}$ y intercept: 2

25. 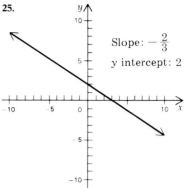 Slope: $-\frac{2}{3}$ y intercept: 2

27. $y = -\dfrac{x}{2} - 2$ **29.** $y = \frac{2}{3}x + \frac{3}{2}$ **31.** $y - 2 = -2(x + 3),\ y = -2x - 4$ **33.** $y - 3 = \frac{1}{2}(x + 4),\ y = \dfrac{x}{2} + 5$

35. $\frac{1}{1}$ **37.** $-\frac{1}{1}$ **39.** $y - 4 = \frac{1}{3}(x + 6)$ or $y - 7 = \frac{1}{3}(x - 3),\ y = \dfrac{x}{3} + 6$ **41.** $y = -\frac{1}{4}(x + 4)$ or $y + 2 = -\frac{1}{4}(x - 4),$

$y = -\dfrac{x}{4} - 1$ **43.** $x = -3,\ y = 5$ **45.** $x = -1,\ y = 22$ **47.** (A) $y = -4x + 1$ (B) $y = 4x + 18$

49. (A) $y = \frac{1}{3}x - \frac{13}{3}$ (B) $y = -3x + 9$ **51.** (A) $R = \frac{3}{2}C + 6$ (B) \$366

EVEN

2. 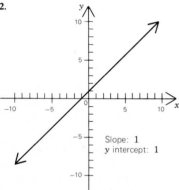 Slope: 1 y intercept: 1

4. 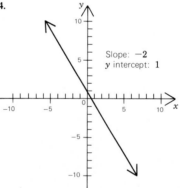 Slope: -2 y intercept: 1

6. $y = 3x - 5$ **8.** $y = -x + 2$ **10.** $y - 5 = 3(x - 2)$ **12.** $y - 3 = -3(x - 1)$ **14.** 1 **16.** $\frac{1}{3}$
18. $y - 4 = x - 2$ or $y - 3 = x - 1$ **20.** $y - 5 = \frac{1}{3}(x - 7)$ or $y - 3 = \frac{1}{3}(x - 1)$

22. 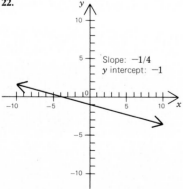 Slope: $-1/4$ y intercept: -1

24. 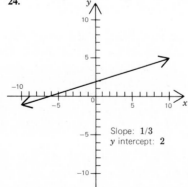 Slope: $1/3$ y intercept: 2

26. 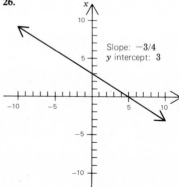 Slope: $-3/4$ y intercept: 3

28. $y = -\dfrac{x}{3} - 5$ **30.** $y = -\frac{3}{2}x + \frac{5}{2}$ **32.** $y + 1 = -3(x - 4), y = -3x + 11$ **34.** $y + 5 = \frac{2}{3}(x + 6), y = \frac{2}{3}x - 1$

36. Undefined   **38.** $-\frac{1}{3}$ **40.** $y + 4 = -\frac{1}{5}(x - 5)$ or $y + 2 = -\frac{1}{5}(x + 5); y = -\dfrac{x}{5} - 3$

42. $y + 4 = -\frac{2}{3}(x - 3)$ or $y = -\frac{2}{3}(x + 3); y = -\frac{2}{3}x - 2$ **44.** $x = 6, y = -2$ **46.** $x = 5, y = 0$

48. **(A)** $y = -2x + 1$ **(B)** $y = \frac{1}{2}x - 4$ **50.** **(A)** $y = -\frac{2}{3}x + \frac{2}{3}$ **(B)** $y = \frac{3}{2}x + 5$

Practice Exercise 8.3

ODD

1.

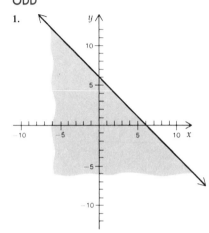

3.

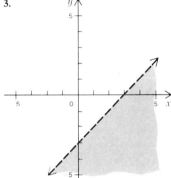

5.

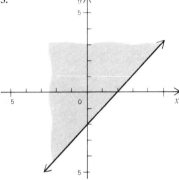

7.

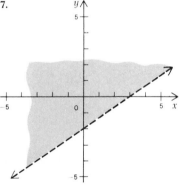

9.

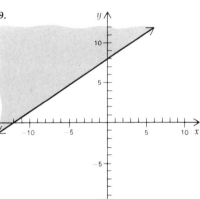

11.

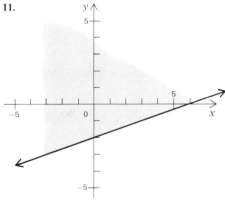

13.

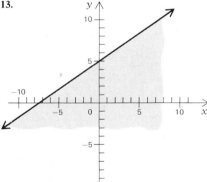

15.

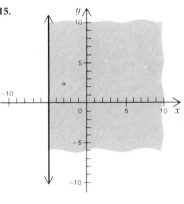

17.

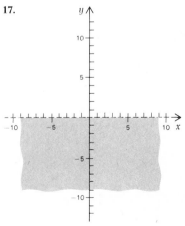

19.

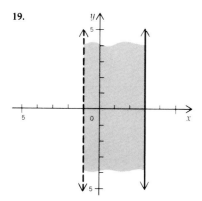

21.

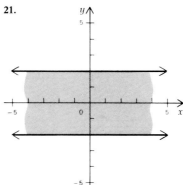

23.

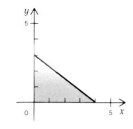

EVEN

2.

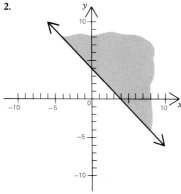

4.

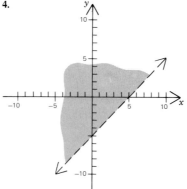

6.

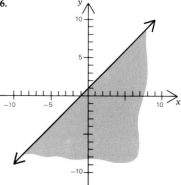

8.

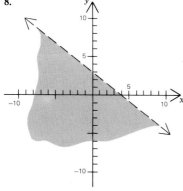

10.

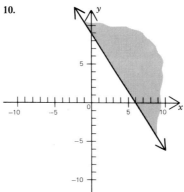

12.

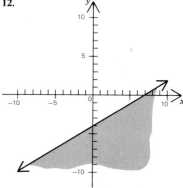

14.
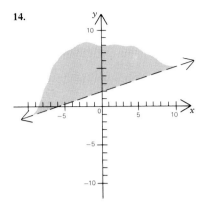

16.
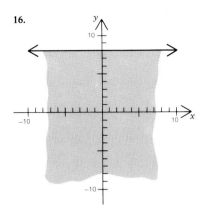

18.
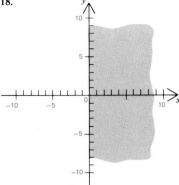

20.
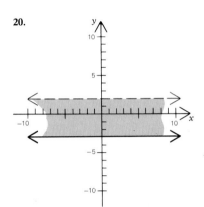

22.
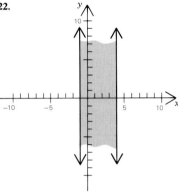

24.
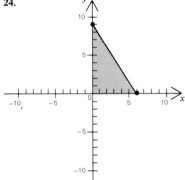

Practice Exercise 8.4

ODD   **1.** $\sqrt{5}$ **3.** $\sqrt{89}$ **5.** $\sqrt{18}$ or $3\sqrt{2}$

7.
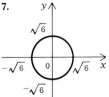

9.
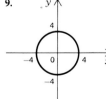

11. $x^2 + y^2 = 49$ **13.** $x^2 + y^2 = 5$ **15.** Yes, since two sides have length $\sqrt{17}$.

17.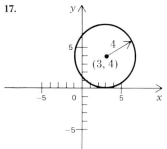

19.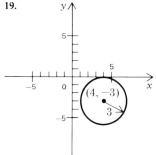

21.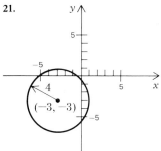

23. $(x - 3)^2 + (y - 5)^2 = 49$ **25.** $(x + 3)^2 + (y - 3)^2 = 64$ **27.** $(x + 4)^2 + (y + 1)^2 = 3$

29. $(x - 2)^2 + (y - 3)^2 = 9$ **31.** $(x - 3)^2 + (y + 3)^2 = 16$ **33.** $(x + 3)^2 + (y + 2)^2 = 9$

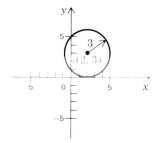

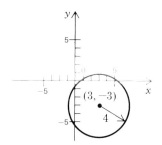

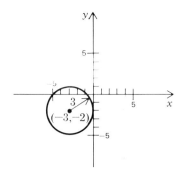

35. $x^2 + y^2 = 50^2$

EVEN   **2.** $\sqrt{10}$ **4.** $\sqrt{41}$ **6.** $\sqrt{52}$ or $2\sqrt{13}$

8.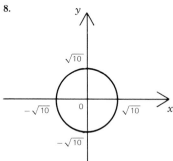

10.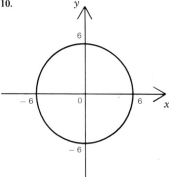

12. $x^2 + y^2 = 64$ **14.** $x^2 + y^2 = 10$ **16.** No; the length of the third side is $\sqrt{18}$, while the other two sides have length $\sqrt{17}$.

18.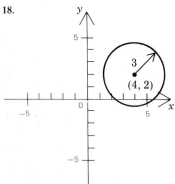

20.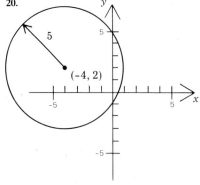

22.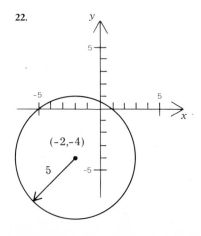

24. $(x-4)^2 + (y-1)^2 = 4$ **26.** $(x-5)^2 + (y+2)^2 = 36$ **28.** $(x+7)^2 + (y+5)^2 = 14$
30. $(x-3)^2 + (y-2)^2 = 9$ **32.** $(x+3)^2 + (y-2)^2 = 16$ **34.** $(x+2)^2 + (y+2)^2 = 16$

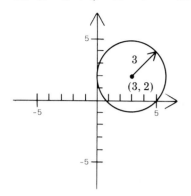
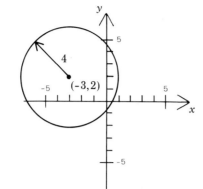
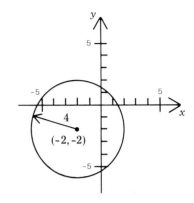

Practice Exercise 8.5

ODD

1.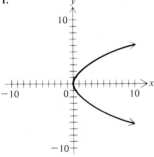

3.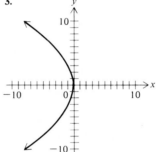

5.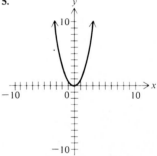

7.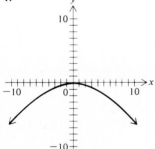

9.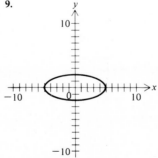

11.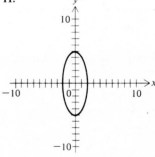

13.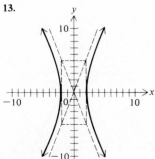

15.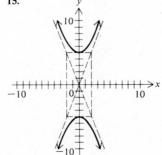

17.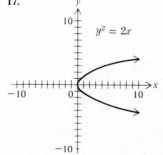
$y^2 = 2x$

19.

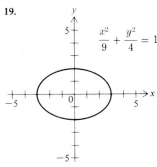

$$\frac{x^2}{9} + \frac{y^2}{4} = 1$$

21.

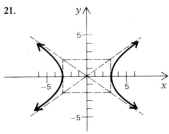

EVEN

2.

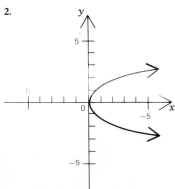

4.

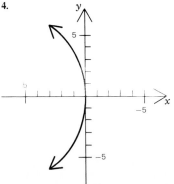

6.

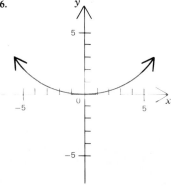

8.

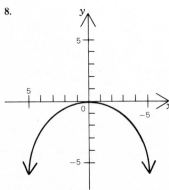

10.

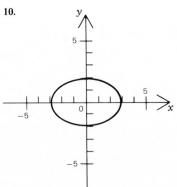

12.

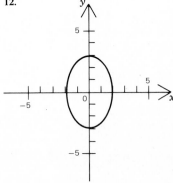

14.

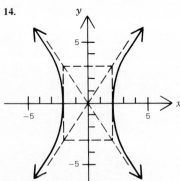

16.

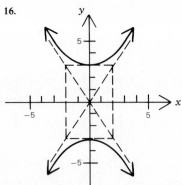

18.

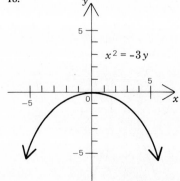

$$x^2 = -3y$$

20.

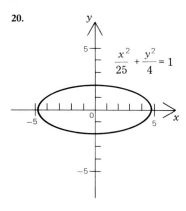

$$\frac{x^2}{25} + \frac{y^2}{4} = 1$$

22.

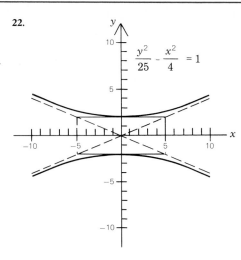

$$\frac{y^2}{25} - \frac{x^2}{4} = 1$$

Diagnostic (Review) Exercise 8.6

1. *(8.1)*

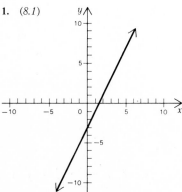

2. *(8.1)*

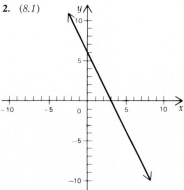

3. *(8.3)*

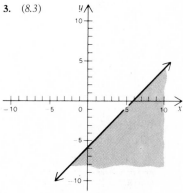

4. *(8.3)*

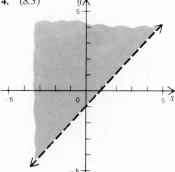

5. *(8.4)*

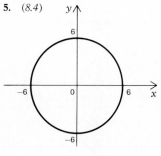

6. Slope $= -2$, y intercept $= -3$ *(8.2)* **7.** $y = -2x + 8$ *(8.2)* **8.** 2 *(8.2)* **9.** $y = 2x + 1$ *(8.2)*

10. $\sqrt{20}$ or $2\sqrt{5}$ *(8.4)* **11.** $x^2 + y^2 = 25$ *(8.4)*

12. *(8.1)*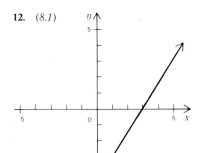

13. *(8.1)*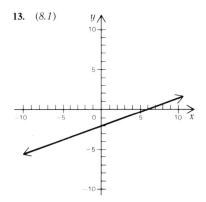

14. *(8.1)*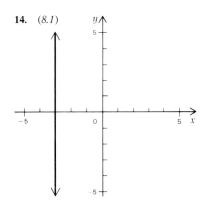

15. *(8.3)*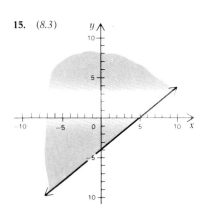

16. *(8.3)*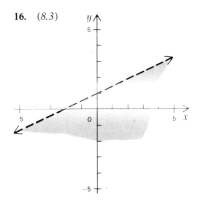

17. *(8.3)*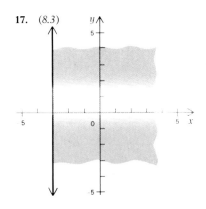

18. *(8.3)*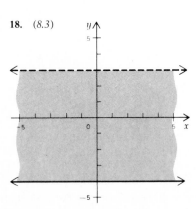

19. *(8.4)*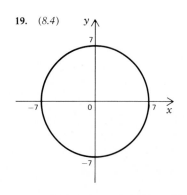

20. *(8.4)*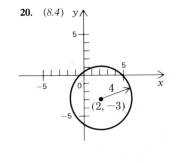

21. *(8.5)*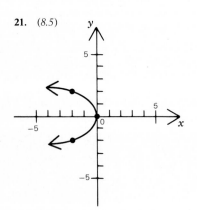

22. *(8.5)*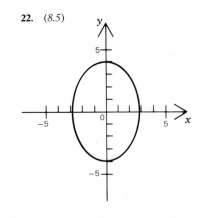

23. *(8.5)*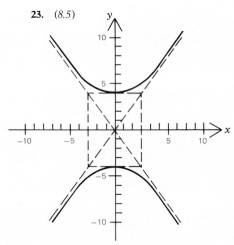

24. Slope $= -\frac{1}{2}$, y intercept $= -3$ (8.2) **25.** $y = -\frac{1}{3}x + 1$ (8.2) **26.** $2x + 3y = 0$ (8.2) **27.** $y = 2x - 10$ (8.2)
28. Vertical: $x = 5$; horizontal: $y = -2$ (8.2) **29.** $(x + 3)^2 + (y - 4)^2 = 49$ (8.4) **30.** $x^2 + y^2 = 169$ (8.4)
31. $y = \frac{3}{2}x + 11$ (8.2) **32.** $(x + 3)^2 + (y - 4)^2 = 25$; radius $= 5$, center $= (-3, 4)$ (8.4)
33. (8.3)

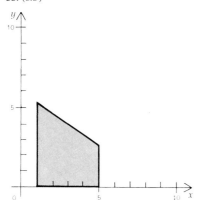

Practice Test Chapter 8

1. (8.1)
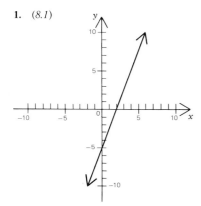

2. (8.3)
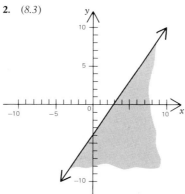

3. (8.4)
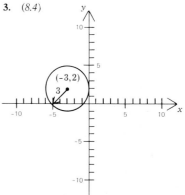

4. (8.5)
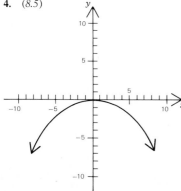

5. (8.5)
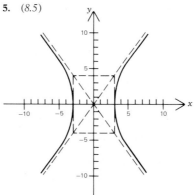

6. $y = -2x + 5$ (8.2) **7.** $y = \frac{1}{2}x - 2$ (8.2) **8.** $y = -2$ (8.2) **9.** $y = -\frac{2}{3}x + 2$ (8.2)
10. $(x - 4)^2 + (y + 3)^2 = 16$; radius $= 4$, center $= (4, -3)$ (8.4)

CHAPTER 9

Practice Exercise 9.1

ODD   **1.** $(2, -3)$ **3.** No solution   **5.** $(1, 4)$ **7.** $(-2, -3)$ **9.** $(-\frac{4}{3}, 1)$ **11.** Infinite number of solutions, $(x, 3x - 9)$ for any real number x **13.** $(-2, 3)$ **15.** $(1, -5)$ **17.** $(\frac{1}{3}, -2)$ **19.** Infinite number of solutions, $(x, \frac{1}{2}x + 3)$ for any real number x **21.** $(-2, 2)$ **23.** $(1.1, 0.3)$ **25.** $(8, 6)$

EVEN   **2.** $(-5, -1)$ **4.** Infinite number of solutions (lines coincide)   **6.** No solution   **8.** $(-3, -5)$ **10.** $(-2, -3)$ **12.** $(\frac{3}{2}, -\frac{2}{3})$ **14.** $(1, -1)$ **16.** $(2, -4)$ **18.** $(-2, -3)$ **20.** No solution   **22.** $(2, 3)$ **24.** $(1, 0.2)$ **26.** $(-6, 12)$

Practice Exercise 9.2

ODD   **1.** 14 nickels and 8 dimes   **3.** $52°, 38°$ **5.** 84 $\frac{1}{4}$-lb packages, 60$\frac{1}{2}$-lb packages   **7.** 60 oz of mix A and 80 oz of mix B   **9.** 60 ml of 80% solution and 40 ml of 50% solution   **11.** (A) $1\frac{9}{13}$ sec, $7\frac{9}{13}$ sec (B) Approximately 8,462 ft

EVEN   **2.** 18 4-cent stamps and 12 5-cent stamps   **4.** 16 by 20 in   **6.** 48 $\frac{1}{4}$-lb packages and 32 $\frac{1}{2}$-lb packages   **8.** 80 oz of mix A and 40 oz of mix B   **10.** 60 ml of 50% solution and 40 ml of 80% solution   **12.** 40 sec, 24 sec, 120 mi

Practice Exercise 9.3

ODD   **1.** $x = -1, y = -1, z = -2$ **3.** $x = 0, y = -2, z = 5$ **5.** $x = 2, y = 0, z = -1$ **7.** $a = -1, b = 2, c = 0$ **9.** $x = 0, y = 2, z = -3$ **11.** $x = 1, y = -2, z = 1$ **13.** No solution   **15.** $w = 1, x = -1, y = 0, z = 2$ **17.** $D = -4, E = -4, F = -17$ **19.** Oldest press: 12 hr; middle press: 6 hr; newest press: 4 hr

EVEN   **2.** $x = -4, y = -3, z = 2$ **4.** $x = -3, y = 2, z = 1$ **6.** $x = 3, y = -1, z = -2$ **8.** $u = 1, v = -1, w = 2$ **10.** $x = -2, y = 1, z = 0$ **12.** $x = -1, y = 1, z = 2$ **14.** Infinitely many solutions   **16.** $r = 2, s = 0, t = 0, u = -1$ **18.** A: 60 grams; B: 50 grams; C: 40 grams

Practice Exercise 9.4

ODD   **1.** $(-3, -4), (3, -4)$ **3.** $(\frac{1}{2}, 1)$ **5.** $(4, -2), (-4, 2)$ **7.** $(3 - i, 4 - 3i), (3 + i, 4 + 3i)$ **9.** $(2, 1), (2, -1), (-2, 1), (-2, -1)$ **11.** $(-3, -2), (-3, 2), (3, -2), (3, 2)$ **13.** $(2 + \sqrt{10}, -2 + \sqrt{10}), (2 - \sqrt{10}, -2 - \sqrt{10})$ **15.** $(-1, 2), (4, 7)$ **17.** $(i, 2i), (i, -2i), (-i, 2i), (-i, -2i)$ **19.** $(2, 4), (-2, 4), (i\sqrt{5}, -5), (-i\sqrt{5}, -5)$ **21.** $(4, 0), (-3, \sqrt{7}), (-3, -\sqrt{7})$ **23.** 2 by 16 ft   **25.** $(2\sqrt{2}, \sqrt{2}), (-2\sqrt{2}, -\sqrt{2}), (1, 4), (-1, -4)$ **27.** $(3, 3), (-3, -3), (0, 6), (0, -6)$ **29.** $(4, 4), (-4, -4), (\frac{4}{5}\sqrt{5}, -\frac{4}{5}\sqrt{5}), (-\frac{4}{5}\sqrt{5}, \frac{4}{5}\sqrt{5})$

EVEN   **2.** $(-12, 5), (-12, -5)$ **4.** $(2, 4), (-2, -4)$ **6.** $(5i, -5i), (-5i, 5i)$ **8.** $(3 + 4i, -1 + 2i), (3 - 4i, -1 - 2i)$ **10.** $(2, 4), (2, -4), (-2, 4), (-2, -4)$ **12.** $(1, 3), (1, -3), (-1, 3), (-1, -3)$ **14.** $(-1 + i\sqrt{3}, 1 + i\sqrt{3}), (-1 - i\sqrt{3}, 1 - i\sqrt{3})$ **16.** $(0, -1), (-4, -3)$ **18.** $(2, 2i), (2, -2i), (-2, 2i), (-2, -2i)$ **20.** $(2, \sqrt{2}), (2, -\sqrt{2}), (-1, i), (-1, -i)$ **22.** $(2, 1), (-2, 1), (2i, 3), (-2i, 3)$ **24.** $\frac{1}{2} + \frac{\sqrt{3}}{2}i$ and $\frac{1}{2} - \frac{\sqrt{3}}{2}i$ **26.** $(2, 1), (-2, -1), (i, -2i), (-i, 2i)$ **28.** $(2, 2), (-2, -2), (\sqrt{2}, -\sqrt{2}), (-\sqrt{2}, \sqrt{2})$ **30.** $(-3, 1), (3, -1), (-i, i), (i, -i)$

Diagnostic (Review) Exercise 9.5

1. $x = 6, y = 1$ (9.1)   **2.** $x = 2, y = 1$ (9.1)   **3.** $x = 2, y = 1$ (9.1)   **4.** $x = -1, y = 2, z = 1$ (9.3)   **5.** $(1, -1), (\frac{7}{5}, -\frac{1}{5})$ (9.4)   **6.** $(4, 3), (-4, 3), (4, -3), (-4, -3)$ (9.4)   **7.** $x = 4, y = 3$ (9.1)   **8.** $x = 4, y = 3$ (9.1)   **9.** $m = -1, n = -3$ (9.1)   **10.** No solution (9.1)   **11.** $(2, -1, 2)$ (9.3)   **12.** $(1, 3), (1, -3), (-1, 3), (-1, -3)$ (9.4)   **13.** $(1 + i, 2i), (1 - i, -2i)$ (9.4)   **14.** Lines coincide—infinitely many solutions   **15.** $(0, 2), (0, -2), (i\sqrt{2}, i\sqrt{2}), (-i\sqrt{2}, -i\sqrt{2})$ (9.4)   **16.** 16 dimes and 14 nickels (9.3)   **17.** \$4,000 at 6 percent and \$2,000 at 10 percent (9.3)   **18.** 6 by 5 cm (9.3)   **19.** 30 grams of 70% solution and 70 grams of 40% solution   **20.** 48 $\frac{1}{4}$-lb packages and 72 $\frac{1}{3}$-lb packages

Practice Test Chapter 9

1. (9.1)

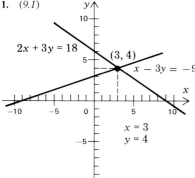

$2x + 3y = 18$

$(3, 4)$

$x - 3y = -9$

$x = 3$
$y = 4$

2. $x = 3, y = 4$ (9.1)   **3.** $x = -2, y = 1$ (9.1)   **4.** No solution (9.1)

5. $(1, -1, 2)$ (9.3)   **6.** $(0, i\sqrt{7}), (0, -i\sqrt{7})$ (9.4)   **7.** $(1, -1),$

$(\frac{5}{7}, -\frac{1}{7})$ (9.4)   **8.** $(i\sqrt{5}, 0), (-i\sqrt{5}, 0), (1, -2), (-1, 2)$ (9.4)

9. $2,500 at 9 percent and $5,500 at 5 percent (9.2)

10. 20 ml of 30% solution and 30 ml of 80% solution (9.2)

CHAPTER 10

Practice Exercise 10.1

ODD 1. Function   **3.** Not a function   **5.** Function   **7.** Function   **9.** Not a function   **11.** Function
13. Function   **15.** Function   **17.** Not a function   **19.** Not a function   **21.** Function   **23.** Function

25.

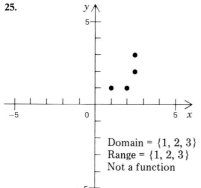

Domain = {1, 2, 3}
Range = {1, 2, 3}
Not a function

27.

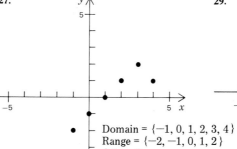

Domain = {−1, 0, 1, 2, 3, 4}
Range = {−2, −1, 0, 1, 2}
A function

29.

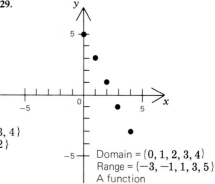

Domain = (0, 1, 2, 3, 4)
Range = (−3, −1, 1, 3, 5)
A function

31.

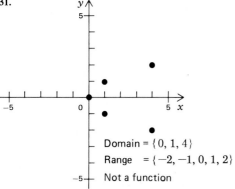

Domain = {0, 1, 4}
Range = {−2, −1, 0, 1, 2}
Not a function

33.

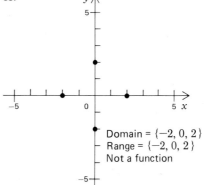

Domain = {−2, 0, 2}
Range = {−2, 0, 2}
Not a function

EVEN **2.** Function   **4.** Not a function   **6.** Function   **8.** Function   **10.** Not a function
12. Not a function   **14.** Function   **16.** Function   **18.** Not a function   **20.** Not a function
22. Function   **24.** Function

26.
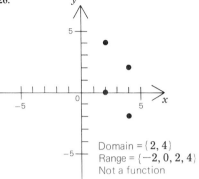
Domain = $(2, 4)$
Range = $(-2, 0, 2, 4)$
Not a function

28.
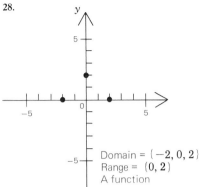
Domain = $(-2, 0, 2)$
Range = $(0, 2)$
A function

30.
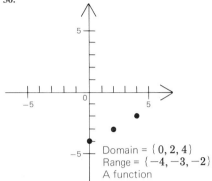
Domain = $(0, 2, 4)$
Range = $(-4, -3, -2)$
A function

32.
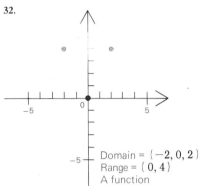
Domain = $(-2, 0, 2)$
Range = $(0, 4)$
A function

34.
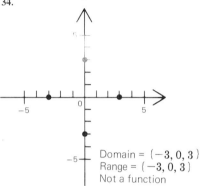
Domain = $(-3, 0, 3)$
Range = $(-3, 0, 3)$
Not a function

Practice Exercise 10.2

ODD **1.** 4   **3.** -8 **5.** -2 **7.** -2 **9.** -12 **11.** -6 **13.** -27 **15.** 2
17. 6   **19.** 25   **21.** 22   **23.** -91 **25.** $2 - 2h$ **27.** -2 **29.** 10   **31.** 48   **33.** 0
35. -7 **37.** $\frac{1}{5}, -\frac{3}{5}$, not defined   **39.** 10   **41.** $1 - 2u - h$ **43.** $C(x) = 5x$ **45.** $C(F) = \frac{5}{9}(F - 32)$
47. **(A)** 30 mi, 300 mi   **(B)** 30

EVEN **2.** 1   **4.** -5 **6.** 10   **8.** 0   **10.** -20 **12.** -2 **14.** 3   **16.** -12 **18.** 0
20. 6   **22.** -36 **24.** $-\frac{3}{4}$ **26.** $3(2 + h)^2$ **28.** $12 + 3h$ **30.** $-3 - h$ **32.** -6 **34.** 6
36. -2 **38.** 3, 0, not defined   **40.** -2 **42.** $2t + h - 1$ **44.** $C(x) = 800 + 60x$ **46.** $P(d) = 15\left(1 + \dfrac{d}{33}\right)$
48. **(A)** 0, 16, 64, 144   **(B)** $64 + 16h$; tends to 64 as h tends to zero. The average speed from $t = 2$ to $t = 2 + h$ tends to 64 as h tends to zero.

Practice Exercise 10.3

ODD

1.
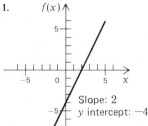
Slope: 2
y intercept: -4

3.
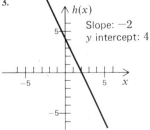
Slope: -2
y intercept: 4

5.
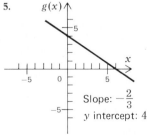
Slope: $-\dfrac{2}{3}$
y intercept: 4

7.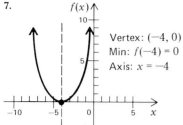

Vertex: $(-4, 0)$
Min: $f(-4) = 0$
Axis: $x = -4$

9.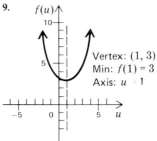

Vertex: $(1, 3)$
Min: $f(1) = 3$
Axis: $u = 1$

11.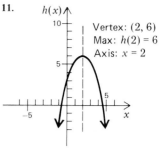

Vertex: $(2, 6)$
Max: $h(2) = 6$
Axis: $x = 2$

13.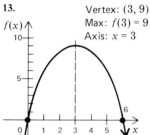

Vertex: $(3, 9)$
Max: $f(3) = 9$
Axis: $x = 3$

15.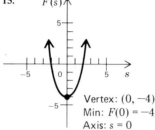

Vertex: $(0, -4)$
Min: $F(0) = -4$
Axis: $s = 0$

17.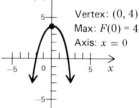

Vertex: $(0, 4)$
Max: $F(0) = 4$
Axis: $x = 0$

19. $f(2) = -7$, a minimum   **21.** $f(12) = -288$, a minimum   **23.** $A(25) = 625$, a maximum

25.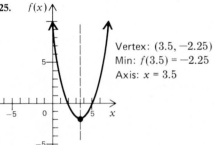

Vertex: $(3.5, -2.25)$
Min: $f(3.5) = -2.25$
Axis: $x = 3.5$

27.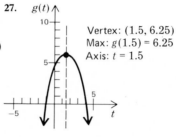

Vertex: $(1.5, 6.25)$
Max: $g(1.5) = 6.25$
Axis: $t = 1.5$

29.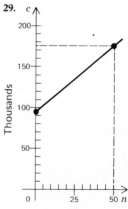

EVEN

2.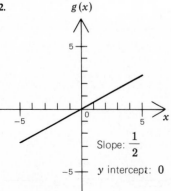

Slope: $\dfrac{1}{2}$
y intercept: **0**

4.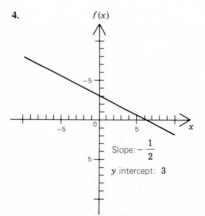

Slope: $-\dfrac{1}{2}$
y intercept: **3**

6.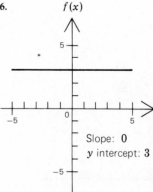

Slope: **0**
y intercept: **3**

8.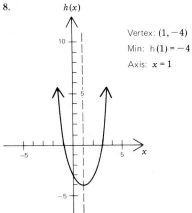

$h(x)$

Vertex: $(1, -4)$
Min: $h(1) = -4$
Axis: $x = 1$

10.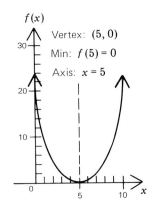

$f(x)$

Vertex: $(5, 0)$
Min: $f(5) = 0$
Axis: $x = 5$

12.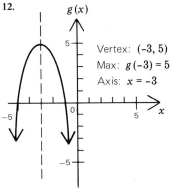

$g(x)$

Vertex: $(-3, 5)$
Max: $g(-3) = 5$
Axis: $x = -3$

14.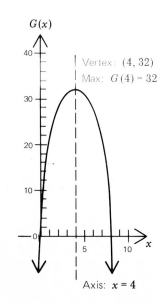

$G(x)$

Vertex: $(4, 32)$
Max: $G(4) = 32$

Axis: $x = 4$

16.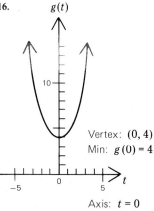

$g(t)$

Vertex: $(0, 4)$
Min: $g(0) = 4$

Axis: $t = 0$

18.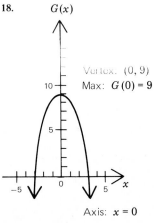

$G(x)$

Vertex: $(0, 9)$
Max: $G(0) = 9$

Axis: $x = 0$

20. $h(3) = 30$, a maximum   **22.** $g(-6) = -108$, a minimum   **24.** $A(25) = 1,250$, a maximum

26.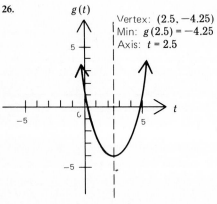

$g(t)$

Vertex: $(2.5, -4.25)$
Min: $g(2.5) = -4.25$
Axis: $t = 2.5$

28.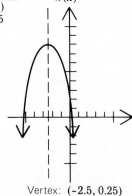

$h(x)$

Vertex: $(-2.5, 0.25)$
Max: $h(-2.5) = 8.25$
Axis: $x = -2.5$

30.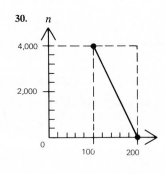

n

Practice Exercise 10.4

ODD   **1.** $R^{-1} = \{(1, -2), (3, 0), (2, 2)\}$ **3.** $G^{-1} = \{(4, -2), (1, -1), (0, 0), (1, 1), (4, 2)\}$

5.

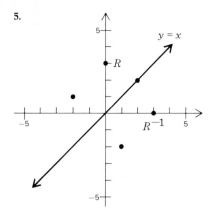

7.

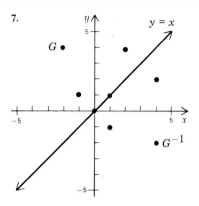

9. Both are functions   **11.** G is a function; G^{-1} is not.   **13.** f^{-1}: $x = 3y - 2$ or $y = \dfrac{x + 2}{3}$

15. F^{-1}: $x = \dfrac{y}{3} - 2$ or $y = 3(x + 2)$ **17.** h^{-1}: $x = \dfrac{y^2}{2}$ or $y = \pm\sqrt{2x}$

19.

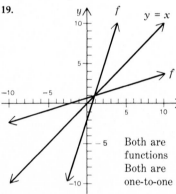

Both are functions Both are one-to-one

21.

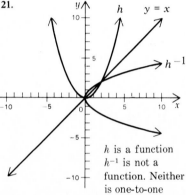

h is a function h^{-1} is not a function. Neither is one-to-one

23. **(A)** $f^{-1}(x) = \dfrac{x + 2}{3}$ **(B)** $\frac{4}{3}$ **(C)** 3   **25.** **(A)** $F^{-1}(x) = 3(x + 2)$ **(B)** 3   **(C)** 4   **27.** **(A)** $f^{-1}(x) = 3(x - 2)$

(B) a

EVEN   **2.** $F^{-1} = \{(-1, -3), (1, 0), (2, 3)\}$ **4.** $H^{-1} = \{(0, -5), (1, -2), (0, 0), (1, 2), (0, 5)\}$

6.

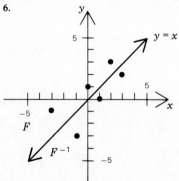

8.

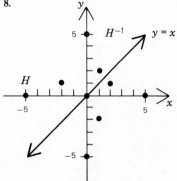

10. Both are functions.
12. H is a function; H^{-1} is not.

14. g^{-1}: $x = 2y + 3$ or $y = \dfrac{x - 3}{2}$ **16.** G^{-1}: $x = \dfrac{y}{2} + 5$ or $y = 2(x - 5)$ **18.** H^{-1}: $x = |2y|$

20.

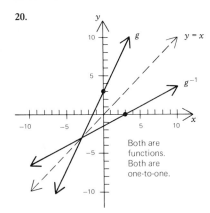

Both are functions. Both are one-to-one.

22.

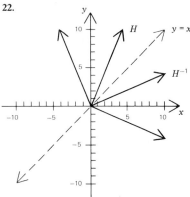

H is a function; H^{-1} not a function
Neither is one-to-one.

24. (A) $g^{-1}(x) = \dfrac{x-3}{2}$ (B) 1 (C) 4 **26.** (A) $G^{-1}(x) = 2(x-5)$ (B) 6 (C) -4 **28.** (A) $g^{-1}(x) = \dfrac{x-2}{4}$

(B) a

Practice Exercise 10.5

ODD **1.** $F = kv^2$ **3.** $f = k\sqrt{T}$ **5.** $y = \dfrac{k}{\sqrt{x}}$ **7.** $t = \dfrac{k}{T}$ **9.** $R = kSTV$ **11.** $v = khr^2$ **13.** 4

15. $9\sqrt{3}$ **17.** $U = k\dfrac{ab}{c^3}$ **19.** $L = k\dfrac{wh^2}{l}$ **21.** -12 **23.** 83 lb **25.** 20 amperes

EVEN **2.** $u = kv$ **4.** $P = kv^6$ **6.** $I = \dfrac{k}{t}$ **8.** $p = \dfrac{k}{n}$ **10.** $g = kxy^2$ **12.** $Q = ktRI^2$ **14.** 125

16. $\tfrac{1}{2}$ **18.** $w = k\dfrac{x^2}{\sqrt{y}}$ **20.** $n = k\dfrac{P_1P_2}{d}$ **22.** 3 **24.** 60 mi/hr or greater **26.** 90

Diagnostic (Review) Exercise 10.6

1. Not a function (10.1) **2.** A function (10.1) **3.** A function (10.1) **4.** A function (10.1)
5. Not a function (10.1) **6.** A function (10.1) **7.** A function (10.1) **8.** Not a function (10.1)
9. Not a function (10.1) **10.** Not a function (10.1) **11.** A function (10.1) **12.** A function (10.1)
13. Domain $= \{0, 1\}$, range $= \{-2, 2, 3\}$ (10.1) **14.** Domain $= \{-1, 1, 2\}$, range $= \{2, 3, 4\}$ (10.1)
15. Domain $= \{-2, 0, 2\}$, range $= \{3\}$ (10.1) **16.** (A) 0 (B) 6 (C) 9 (D) $6 - m$ (10.2) **17.** (A) -6 (B) 0
(C) -3 (D) $c - 2c^2$ (10.2)
18. (10.3) **19.** (10.3) **20.** (10.4)

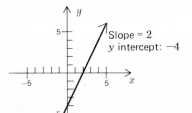

Slope = 2
y intercept: -4

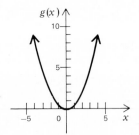

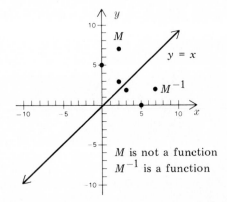

M is not a function
M^{-1} is a function

21. Domain = $\{3, 5, 7\}$, range = $\{0, 2\}$, *(10.4)* **22.** $m = kn^2$ *(10.5)* **23.** $P = \dfrac{k}{Q^3}$ *(10.5)* **24.** $A = kab$ *(10.*

25. $y = k\dfrac{x^3}{\sqrt{z}}$ *(10.5)* **26.** No. A function is a relation with the added restriction that each domain element correspon

exactly one range element. *(10.1)* **27.** **(A)** 7 **(B)** 1 **(C)** $x^2 + x - 7$ **(D)** 3 *(10.2)* **28.** **(A)** $3 + 2h$ **(B**

29. *(10.3)* **30.** *(10.3)*

g(t) — Slope: $-\dfrac{3}{2}$, y intercept: 6

f(x) — Vertex (2, 1), Min: $f(2) = 1$, Axis: $x = 2$

31. Maximum occurs at $x = -\dfrac{b}{2a} = -2$; that is, $g(-2) = 11$ is the maximum *(10.3)* **32.** Problem 3 *(10.4)*

33. Problem 6 *(10.4)* **34.** None *(10.4)* **35.** Problem 11 *(10.4)* **36.** Problem 6 *(10.4)* **37.** Problem 3

38. $f^{-1}: x = 1 - \dfrac{y}{2}$ or $y = 2 - 2x$ *(10.4)* **39.** **(A)** $M^{-1}(x) = 2x - 3$ **(B)** Yes **(C)** 1 **(D)** 3 *(10.4)* **40.** **(A)**

(B) $y = \tfrac{4}{3}$ *(10.5)* **41.** **(A)** $-3 - 4h - h^2$ **(B)** $-4 - h$ *(10.2)*

42. *(10.3)* **43.** *(10.3)* **44.**

g(t) — Vertex: (3, 144), Max: $g(3) = 144$, Axis: $t = 3$

C(x) — $C(x) = 200 + 0.05x$

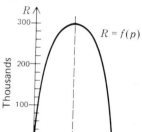

(A) **(B)** $p = \$100$, maximum $R = f(100) = \$300{,}000$ (

45. $t = k\dfrac{wd}{P}$, $t = 24$ sec *(10.5)*

Practice Test Chapter 10

1. **(A)** A function **(B)** Not a function **(C)** A function **(D)** Not a function *(10.1)* **2.** $f(3) = 4, f(0) = 1$ *(10.2)*

3. $f(-2) = 9, f(a) = a^2 - 2a + 1$ *(10.2)* **4.** $f(4) + g(2) = 4, 3g(0) = 18$ *(10.2)* **5.** $f(m) - g(m) = m^2 + m - 8$,

$f[g(2)] = 0$ *(10.2)* **6.** *(10.1)*

Domain   Range; no

$-1 \longrightarrow 2$

$\longrightarrow 3$

$0 \longrightarrow 0$

7. $f^{-1}(x) = 4x + 2, f^{-1}(1) = 6$ *(10.4)* **8.** $f^{-1}[f(2)] = 2, f[f^{-1}(a)] = a$ *(10.4)* **9.** *(10.4)*

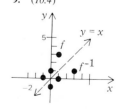

5)

ds to

2 (10.2)

12. (*10.3*)

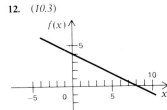

13. Slope $= -\frac{1}{2}$, y intercept $= 4$ (*10.3*)
14. (*10.3*)

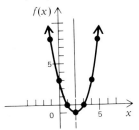

0.3)
)

16. Min $f(x) = f(2) = -1$ (*10.3*)

$n = \dfrac{k}{n^{2/3}}$ (*10.5*)   **20.** 54 (*10.5*)

(*10.4*)

0.3)

3.

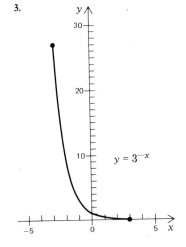

$y = 3^{-x}$

5.

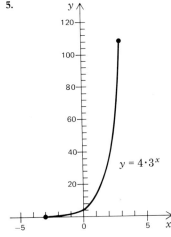

$y = 4 \cdot 3^{x}$

x + 3

9.

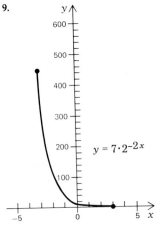

$y = 7 \cdot 2^{-2x}$

11.

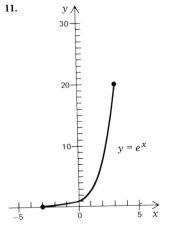

$y = e^{x}$

ION

13.

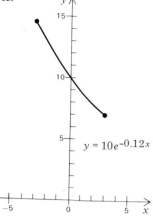

$y = 10e^{-0.12x}$

28. $-3 \log_b M$ **30.** $5 \log_b m + 3 \log_b n - \frac{1}{2} \log_b p$

$\log_b z)$ **34.** $\log_b \dfrac{m}{\sqrt{n}}$ **36.** $\log_b \dfrac{\sqrt[9]{w}}{x^3 y^5}$ **38.** $\log_b \sqrt[3]{\dfrac{x}{y}}$ **40.** 0.41   **42.** 0.55

.5145   **5.** $0.3385 - 2 = -1.6615$ **7.** 2.35   **9.** 3.18×10^4 **11.** 2.08×10^{-4}

$- 2 = -1.2083$ **17.** 1.30×10^9 **19.** 3.87×10^{-4} **21.** 4.9177   **23.** -1.2080

,000   **27.** 0.009 687   **29.** (A) 0.8544   (B) 0.8544   **31.** (A) 8.5622   (B) 8.5627

6404   **35.** 22.26   **37.** 0.080 38

.3263   **6.** $0.5705 - 3 = -2.4295$ **8.** 6.57   **10.** 1.27×10^7 **12.** 6.05×10^{-5}

$- 3 = -2.3215$ **18.** 2.74×10^7 **20.** 1.58×10^{-5} **22.** 5.4842   **24.** -2.3216

9   **30.** (A) 1.9488   (B) 1.9488   **32.** (A) 13.3114   (B) 13.3105   **34.** (A) -5.7541

38. 0.04920

8   **5.** 2.37   **7.** 80   **9.** 18.9   **11.** -2.48 **13.** 20.2   **15.** 10

21. Approx. 3.8 hr   **23.** Approx. 35 years   **25.** Approx. 28 years

1   **6.** 2.07   **8.** 20   **10.** 14.2   **12.** 3.74   **14.** -1.83 **16.** 5

22. (A) Approx. 12.8   (B) Approx. 206 in   **24.** Approx. 533 years

EVEN

2.

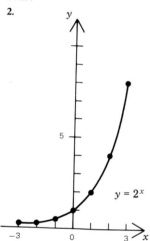

$y = 2^x$

xercise 11.6

$g_{10} m = n$ (11.2)   **3.** 4 (11.2)   **4.** 5 (11.2)   **5.** 3 (11.2)   **6.** 1.24 (11.5)

11.2)   **9.** $x = \ln y$ (11.2)   **10.** -2 (11.2)   **11.** $\frac{1}{3}$ (11.2)   **12.** 64 (11.2)

?)   **15.** 1 (11.2)   **16.** 2.32 (11.5)   **17.** 92.1 (11.5)   **18.** 30 (11.2)

8.

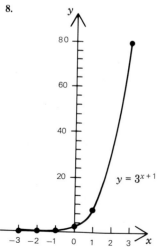

$y = 3^{x+1}$

10.

$\Rightarrow x$

all pos. real no.
all real no.
$^{-1}$: all real no.
$^{-1}$: pos. real no.

Approx. 37,100 years (11.5)

11

$= 10^{0.2x}$ (11.2)   **3.** $\ln A = 0.08t$ (11.2)   **4.** $x = 16$ (11.2)   **5.** $x = \frac{1}{2}$ (11.2)

(11.5)   **8.** $x = 2$ (11.3)   **9.** 5.6355 (11.4)   **10.** -4.6925 (11.4)

2.   0.539 (11.4)   **13.** 6.78×10^{-5} or 0.000 067 8 (11.4)   **14.** 5.3613 (11.4)

15. -7.1271 (11.4) **16.** $x = 0.701$ (11.5) **17.** $x = 1.63$ (11.5) **18.** $x = 5.75$ (11.5) **19.** 18 years (11.5)
20. $(11.1, 11.2)$

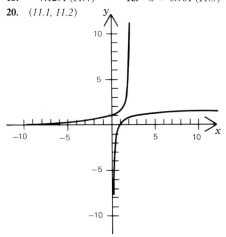

APPENDIX A

Practice Exercise

ODD **1.** 0.3649   **3.** 5.8472   **5.** 1.8131   **7.** $0.6027 - 3$ **9.** $0.9546 - 1$ **11.** 5.8403   **13.** 5.204
15. $3.514 \times 10^5 = 351{,}400$ **17.** $2.693 \times 10^3 = 2{,}693$ **19.** $4.016 \times 10^{-3} = 0.004\,016$ **21.** $6.906 \times 10^{-1} = 0.6906$
23. $1.474 \times 10^{-1} = 0.1474$

EVEN **2.** 0.7112   **4.** 4.4538   **6.** 1.3086   **8.** $0.5697 - 2$ **10.** $0.8071 - 1$ **12.** 4.9277   **14.** 2.562
16. $2.975 \times 10^2 = 297.5$ **18.** $8.206 \times 10 = 82.06$ **20.** $1.651 \times 10^{-1} = 0.1651$ **22.** $7.094 \times 10^{-4} = 0.000\,7094$
24. $2.162 \times 10^{-3} = 0.002\,162$

APPENDIX B

Practice Exercise

ODD **1.** -14 **3.** 2   **5.** -26 **7.** $\begin{vmatrix} a_{22} & a_{23} \\ a_{32} & a_{33} \end{vmatrix}$ **9.** $\begin{vmatrix} a_{11} & a_{12} \\ a_{31} & a_{32} \end{vmatrix}$ **11.** $(-1)^{1+1}\begin{vmatrix} a_{22} & a_{23} \\ a_{32} & a_{33} \end{vmatrix}$

13. $(-1)^{2+3}\begin{vmatrix} a_{11} & a_{12} \\ a_{31} & a_{32} \end{vmatrix}$ **15.** $\begin{vmatrix} 1 & -2 \\ -4 & 8 \end{vmatrix}$ **17.** $\begin{vmatrix} -2 & 0 \\ 5 & -2 \end{vmatrix}$ **19.** $(-1)^{1+1}\begin{vmatrix} 1 & -2 \\ -4 & 8 \end{vmatrix} = 0$

21. $(-1)^{3+2}\begin{vmatrix} -2 & 0 \\ 5 & -2 \end{vmatrix} = -4$ **23.** 10   **25.** -21 **27.** -120 **29.** -40 **31.** -12 **33.** 0

35. **(A)** $\begin{vmatrix} x & y & 1 \\ 2 & 3 & 1 \\ -1 & 2 & 1 \end{vmatrix} = 0$ **(B)** $x - 3y = -7$

EVEN **2.** 8   **4.** -20 **6.** 4.3   **8.** $\begin{vmatrix} a_{11} & a_{12} \\ a_{21} & a_{22} \end{vmatrix}$ **10.** $\begin{vmatrix} a_{11} & a_{13} \\ a_{31} & a_{33} \end{vmatrix}$ **12.** $(-1)^{3+3}\begin{vmatrix} a_{11} & a_{12} \\ a_{21} & a_{22} \end{vmatrix}$

14. $(-1)^{2+2}\begin{vmatrix} a_{11} & a_{13} \\ a_{31} & a_{33} \end{vmatrix}$ **16.** $\begin{vmatrix} -2 & 0 \\ 7 & 8 \end{vmatrix}$ **18.** $\begin{vmatrix} 3 & 0 \\ -4 & 8 \end{vmatrix}$ **20.** $(-1)^{2+2}\begin{vmatrix} -2 & 0 \\ 7 & 8 \end{vmatrix} = -16$ **22.** $(-1)^{2+1}\begin{vmatrix} 3 & 0 \\ -4 & 8 \end{vmatrix} = -24$
24. -24 **26.** -40 **28.** 22   **30.** -43 **32.** -1 **34.** -18 **36.** $\frac{23}{2}$

APPENDIX C

Practice Exercise

ODD **1.** $x = 5, y = -2$ **3.** $x = 1, y = -1$ **5.** $x = -1, y = 1$ **7.** $x = 2, y = -2, z = -1$
9. $x = 2, y = -1, z = 2$ **11.** $x = 2, y = -3, z = -1$ **13.** $x = 1, y = -1, z = 2$ **15.** $a = -1, b = 2, c = 0$

EVEN   **2.** $x = -1, y = 2$ **4.** $x = 4, y = -1$ **6.** $x = 1, y = -1$ **8.** $x = -3, y = -1, z = 2$
10. $x = -4, y = 0, z = 3$ **12.** $x = 3, y = -2, z = 1$ **14.** $x = 1, y = 0, z = -2$ **16.** $u = 1, v = -1, w = 2$

INDEX